Introduction to Quantum Mechanics

Quantum mechanics is a subfield of physics that studies how the universe works at atomic and subatomic levels. It is an essential part of undergraduate and graduate courses in physics and undergraduate engineering courses in India.

This book comes from the desk of authors who have decades of graduate-level teaching experience in quantum mechanics. It covers the syllabus requirements set forth by the University Grants Commission (UGC), India, and has additional several unique features that will set it apart. It introduces the students to vector space in the beginning to give a comprehensive idea of the mathematics involved in the concepts, has separate appendices emphasizing the techniques of differential equations, and covers applications of quantum mechanics related to atomic physics, condensed matter physics, particle physics, and so on. The text is carefully designed and is not completely mathematical; it gives equal emphasis to the discussions around physics and creates a nice balance.

The book will be suitable for both undergraduate and graduate students taking courses in physics and chemistry. Several advanced topics have been covered in this book. They are expected to be beneficial particularly to graduate students and researchers.

Krishnendu Sengupta is Senior Professor at the Indian Association for the Cultivation of Science, Kolkata. His research interest includes theoretical condensed matter physics with emphasis on non-equilibrium dynamics of interacting quantum systems. He is a member of Indian National Science Academy, National Academy of Sciences, and Indian Academy of Sciences. He was awarded the Shanti Swarup Bhatnagar Award in 2012.

Palash B. Pal retired from the Saha Institute of Nuclear Physics, Kolkata, and is now Emeritus Professor at the University of Calcutta. His main area of research is elementary particle physics, with emphasis on neutrino properties, grand unification and thermal field theory. He has also published several pedagogical books, including *A Physicist's Introduction to Algebraic Structures* with the Press in 2019.

T0295256

Introduction to Quantum Mechanics

Krishnendu Sengupta

Palash B. Pal

CAMBRIDGE
UNIVERSITY PRESS

Shaftesbury Road, Cambridge CB2 8EA, United Kingdom

One Liberty Plaza, 20th Floor, New York, NY 10006, USA

477 Williamstown Road, Port Melbourne, VIC 3207, Australia

314–321, 3rd Floor, Plot 3, Splendor Forum, Jasola District Centre, New Delhi 110025, India

103 Penang Road, #05–06/07, Visioncrest Commercial, Singapore 238467

Cambridge University Press is part of Cambridge University Press & Assessment, a department of the University of Cambridge.

We share the University's mission to contribute to society through the pursuit of education, learning and research at the highest international levels of excellence.

www.cambridge.org

Information on this title: www.cambridge.org/9781009338424

First published 2023

Printed in India by Nutech Print Services, New Delhi 110020

A catalogue record for this publication is available from the British Library

ISBN 978-1-009-33842-4 Paperback

To the memories of

Pushan Majumdar and Manoranjan Saha

Contents

Part Four: Advanced topics   349

Figures

Tables

Notation

p 3-vector in the coordinate space.

p   Magnitude of the 3-vector p, i.e., $|p|$. We have used p^2 and p^2 interchangeably.

$\mathring{p}$ Unit 3-vector in the direction of p, i.e., p/p.

$|u\rangle$ Ket vector.

$\langle u|$ Bra vector.

$|\Omega\rangle$ The null vector.

$\langle u|v\rangle$ Inner product of two vectors.

$\widehat{A}$ Operator A on a vector space. In many places where confusion is not likely to arise, we have omitted the hat sign.

$\widehat{L}$ An operator on a vector space that is also a vector in the coordinate space.

$A^\dagger$ Adjoint of the operator A, defined in §2.8.

$[A,B]_P$ Poisson bracket of A and B.

$[A,B]$ Commutator $AB - BA$.

$\{A,B\}$ Anticommutator $AB + BA$.

σ_i Pauli matrices, given in Eq. (4.26, p. 87).

τ_i Same as σ_i, but used when there is no connection with spin.

$\mathcal{S}$ Action of a system $(= \int dt\, L)$.

e Electric charge of proton. Electron carries charge $-e$.

α Fine structure constant, first appearing in Eq. (9.63, p. 231).

$\lfloor x \rfloor$ The largest integer less than or equal to x.

Preface

Quantum mechanics is an essential part of any physics undergraduate or graduate curriculum for its wide range of applicability to different branches of physics. It is taught at different levels, starting from second or third year undergraduate to final year graduate programmes. Through these courses students learn to appreciate the details of the subject at various levels. It is therefore difficult to have a single book which caters to students at all levels. This gap is precisely what we aim to fill here.

The book bore out of several courses on quantum mechanics taught by both the authors at several institutions such as the Saha Institute of Nuclear Physics, the University of Calcutta, and the Indian Association for the Cultivation of Science. These included both undergraduate and graduate level courses. These courses made us realize the necessity of a book which not only provides a clear introduction to the basic tenets of the subject but also presents a thorough discussion of several advanced topics. The latter seemed particularly difficult to find in a single book, which served as a motivation for writing this book.

We have not included a discussion of the old quantum theory in our book. We assumed that the reader would be familiar with that development, which was initiated by Planck and Einstein. That theory deals with particle-like properties of energy. The subject matter of our book is the other side of the wave–particle duality that is the hallmark of modern physics, i.e., the theory which treats matter as waves. The title of this book clearly shows that this theory, 'quantum mechanics', will be the sole concern of this book.

Any book on quantum mechanics requires introduction to certain mathematical techniques such as group theory, linear algebra, and differential equations. We provide a somewhat detailed introduction to the first two topics since it is our understanding that these topics may not be taught in detail in other courses that physics students usually encounter before taking their first course on quantum mechanics. However, the last topic is usually well discussed in a generic physics curriculum; we have therefore relegated its discussion to the appendices, making every effort to keep it self-contained.

The book is divided into several parts. The first part introduces the formalism, while the second discusses exactly solvable problems. This is followed by a discussion of various approximation techniques in the third part and a discussion of several advanced topics in the fourth. Each part has several chapters with varying degrees of sophistication; however, we have made every attempt to make each of them self-contained so that it is possible for an instructor to choose the chapters according to the course requirement. The fifth part of the book contains appendices.

We have provided a large number of exercises for the reader. These will definitely help the beginner gain confidence in the subject. Instead of compiling all exercises at the end of each chapter, we have strewn them across the text. Solving any exercise at the place that it appears in the text, in our opinion, would be advantageous to the learner. If for some reason the reader does not want to solve the exercises, we recommend that at least the statement of the exercise should be read. In some cases, they contain important notes.

Over the years, many people have influenced our understanding of the subject, including our teachers, collaborators, and students. We would like to take this opportunity to express sincere thanks to all of them. This preface is too short to put forth all the names. During the time that we wrote this book, we have also received a lot of support from others. We are particularly thankful to Kumar Gupta, who has kindly read through all chapters of an earlier version and made important comments which resulted in a lot of rewriting. We also acknowledge useful discussions with, and helpful comments from, Jayanta Kumar Bhattacharjee, Arnab Das, Sumit Das, Asit De, Indrajit Mitra, Abhijit Mukherjee, Koushik Ray, Arnab Sen, and Parongama Sen. In addition, we thank the referees for their important comments and the publication team of Cambridge University Press for their support. Last but not least, we acknowledge the support of our spouses, Dora Saha and Shukla Sanyal, who tolerated our long working hours during the project.

The book was prepared camera-ready by us. This means that we are responsible for all mistakes that might have crept in, including typographical ones. If any reader finds a mistake, we will appreciate a message. We will maintain a webpage for corrections where such messages will be acknowledged. The address of the webpage is `https://sites.google.com/view/qmerrata` and is accessible to everyone.

Krishnendu Sengupta
Palash B. Pal

April 2023

Part One

Formalism

Chapter 1

Uncertainty

Classical mechanics is governed by Newton's laws of motion. It has been very successful, for over three centuries, in explaining motions of objects that we see around us. However, around the beginning of the twentieth century, when it came to understanding the properties of small systems like an atom, classical mechanics seemed inadequate. In this chapter, we will review the basic formulas of classical mechanics and indicate why it could not describe small systems.

1.1 Classical mechanics

Classical mechanics seeks a description of the motion of a particle, by specifying the path of motion of the particle, i.e., the position and velocity of the particle at any given instant. In the Newtonian formulation, this is done by invoking the idea of forces, and using Newton's second law of motion, which says that the rate of change of momentum of a particle is equal to the force that acts on the particle:

$$\frac{\mathrm{d}}{\mathrm{d}t}\left(m\frac{\mathrm{d}\boldsymbol{r}}{\mathrm{d}t}\right) = \boldsymbol{F}. \tag{1.1}$$

If we know the forces as a function of time, we can in principle solve this equation. It is a second-order differential equation in time, so we will need two initial conditions to solve the position vector $\boldsymbol{r}$ as a function of time. In particular, if we know the position $\boldsymbol{r}$ and velocity $\boldsymbol{v} = \mathrm{d}\boldsymbol{r}/\mathrm{d}t$ at an initial instant, we can solve for the position and velocity of the particle at any instant, given the knowledge of the force $\boldsymbol{F}$.

There are alternative ways of formulating classical mechanics. One of these is the Lagrangian formulation. In this formulation, one defines a function L of coordinates and velocities of all particles in the system, called the Lagrangian. The equation of motion is then given by

$$\frac{\mathrm{d}}{\mathrm{d}t}\left(\frac{\partial L}{\partial \dot{x}_a}\right) = \frac{\partial L}{\partial x_a}, \tag{1.2}$$

where the x_a's denote different independent coordinates and $\dot{x}_a$'s their time derivatives, i.e., the corresponding velocities. For example, consider a particle constrained to move in one dimension (1D), so that there is only one relevant coordinate, and define

$$L = \frac{1}{2}m\dot{x}^2 - V(x), \tag{1.3}$$

i.e., just the difference of the kinetic and potential energies of the particle. The equation of motion is then

$$\frac{\mathrm{d}}{\mathrm{d}t}\left(m\dot{x}\right) = -\frac{\partial V}{\partial x}, \tag{1.4}$$

which is the same as Newton's equations if we identify the force through the equation

$$F = -\frac{\partial V}{\partial x}. \tag{1.5}$$

Once again, the equations of motion are second-order differential equations in time. Thus, if one knows the position and velocity at one given instant, one can find out the positions and velocities at all other times.

Another way of approaching the problem is through the definition of the Hamiltonian, which is a function of the coordinates and momenta. The motion of the particles is then encoded in the Hamilton's equations:

$$\dot{x}_a = \frac{\partial H}{\partial p_a}, \qquad \dot{p}_a = -\frac{\partial H}{\partial x_a}. \tag{1.6}$$

These are first-order differential equations in time. However, once the momenta are eliminated to obtain equations involving the coordinates only (or vice versa), one encounters second-order differential equations. The point can be exemplified with the help of a simple system with one particle moving in a single dimension, for which the Hamiltonian is given by the sum of the kinetic and potential energies:

$$H = \frac{p^2}{2m} + V(x). \tag{1.7}$$

The Hamilton's equations then read

$$\dot{x} = p/m \,, \qquad \dot{p} = -\frac{\partial V}{\partial x}. \tag{1.8}$$

If we eliminate p from these equations, we obtain Eq. (1.4). The comments made earlier regarding the correspondence with Newton's equation of motion remain the same.

We thus see that the different formulations are equivalent. However, we will see that none of these formulations is suitable for quantum mechanics. The main problem is not in defining the Lagrangian or the Hamiltonian. We can borrow the definition of the Lagrangian as a function of position and velocity, or of the Hamiltonian as a function of position and momentum, from classical mechanics. The problem lies in how to use the Lagrangian or the Hamiltonian to define the equation of motion. In deriving the classical equations of motion, we need to take derivatives with respect to the position coordinate as well as the velocity or the momentum. The problem is that such derivatives are not allowed in quantum mechanics in the same sense as in classical mechanics, because position and velocity/momentum cannot be defined as simple functions, having some specific numerical value at a given time. Not being able to define position as a numerical function of time amounts to saying that the notion of path of motion cannot be defined for very small systems. We discuss this point in detail in §1.3.

1.2 Wave–particle duality

At the end of the nineteenth century, it was generally believed that matter was composed of elementary particles, whereas energy manifested itself in the form of waves. In the beginning of the twentieth century, Max Planck showed that the energy distribution of blackbody radiation can be explained by conjecturing that heat is radiated in the form of energy packets, or quanta. For a monochromatic wave for which the electric field and the magnetic field have the generic form

$$(\text{constant}) \times e^{i(\boldsymbol{k}\cdot\boldsymbol{r}-\omega t)}, \tag{1.9}$$

each packet, or quantum, has the same energy, given by

$$E = \hbar\omega, \tag{1.10}$$

where $\hbar$ is a universal constant, now known as the *Planck constant*.

Historically, Planck used not ω but the frequency of the wave, ν, to write his formula in the form

$$E = h\nu, \tag{1.11}$$

and therefore the constant h that appears in this formula should be truly called the Planck constant. Since

$$\omega = 2\pi\nu, \tag{1.12}$$

we see that it is related to $\hbar$, introduced in Eq. (1.10), through the equation

$$\hbar \equiv \frac{h}{2\pi}. \tag{1.13}$$

Since many of the basic formulas look neater if $\hbar$ is used instead of Planck's original h, we will loosely use the phrase "Planck constant" to denote $\hbar$, ignoring the historical nomenclature.

In 1905, Albert Einstein used the idea of quanta to explain the photoelectric effect, treating light as a collection of particles whose energy is related to the frequency of light through Eq. (1.10). It was then agreed upon that light, as well as other forms of energy, shows dual characteristics: sometimes that of a wave and sometimes that of a particle. This is called the *wave–particle duality*.

Wavelength of light, λ, is related to its frequency by the relation

$$\nu\lambda = c, \tag{1.14}$$

where c is the speed of light. Also, the energy–momentum relation that Einstein obtained for light particles, or *photons*, in his special theory of relativity, is

$$E = c\mathrm{p}, \tag{1.15}$$

where p denotes the magnitude of the momentum vector $\boldsymbol{p}$. Combining Eqs. (1.14) and (1.15) and using the Planck relation, Eq. (1.11), we can write down the magnitude of the momentum of a photon in terms of the wavelength of light in the form

$$\mathrm{p} = \frac{h}{\lambda} = \frac{2\pi\hbar}{\lambda} = \hbar\mathrm{k}, \tag{1.16}$$

where

$$\mathrm{k} = \frac{2\pi}{\lambda} \tag{1.17}$$

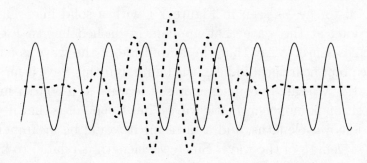

Figure 1.1 Two waves: one localized and one spread out in space. We show the amplitudes at a certain instant of time at different positions along one coordinate axis.

is the magnitude of the *wave vector* $\boldsymbol{k}$ that appears in Eq. (1.9). Since the direction of the photon momentum is the direction of the wave vector, we can write the momentum vector for the photon as

$$\boldsymbol{p} = \hbar\boldsymbol{k}. \tag{1.18}$$

Eqs. (1.10) and (1.18) constitute the basic relations of wave–particle duality, expressing the energy and momentum of a photon in terms of the wave characteristics.

In 1924, Louis de Broglie felt uncomfortable with the state of affairs that existed at that time. Light, or other forms of energy, showed dual characteristics, whereas matter was still known to be composed of particles. He suggested that a more symmetric situation can be obtained if matter showed wave properties in addition to particle properties. He suggested that Eqs. (1.10) and (1.18) would bridge the particle and wave properties of matter as well, although, of course, energy and momentum of matter should not be connected through Eq. (1.15). Direct experimental evidence of the existence of matter waves came only a few years later, and we will discuss it in §1.5.

☐ **Exercise 1.1** *Estimate the de-Broglie wavelength of an electron, a proton, and a ball of mass 1 g, each at a speed of 0.1c, where c is the speed of light in the vacuum.*

1.3 The uncertainty principle

The problem alluded to at the end of §1.1 appears as soon as we talk of waves. For a monochromatic wave, the wavelength is fixed, but the wave is spread all over, and we cannot talk about the *position* of the wave. In fact, the wave is everywhere, so to say, as seen in Figure 1.1 with a solid line. On the other hand, if we look at the wave that appears in dashed line in Figure 1.1, it certainly makes sense to say that the amplitude of the wave is concentrated around some point near the middle of the figure, but this wave is far from being monochromatic. We can, of course, express the wave as a Fourier integral, i.e., as a superposition of monochromatic waves, but the point is that it will involve waves of various wavelengths, and as a result it would be meaningless to talk about *the wavelength* of the wave. Since wavelength is related to momentum, we can also say that the more it makes sense to talk about the position of a wave, the less we can specify the momentum of the wave quanta. In the extreme case where the momentum is perfectly known because we have a monochromatic wave, it makes absolutely no sense to talk about the position of the wave. In the opposite extreme in which case the position is perfectly known, the 'wave' would look like a spike, and it would take infinite number of Fourier components to describe such a spike, implying that the momentum will be completely obscure. This inverse relation between the uncertainties in the knowledge of position and the momentum of a system is called the *uncertainty principle*, which was discovered by Werner Heisenberg.

We can arrive at the same conclusion by considering the observable object to be a particle. Consider, for example, that there is an electron somewhere within a box. If we want to know its position, we have to illuminate it with some probe. The word 'illuminate' here has to be understood in a general sense, because it is not necessarily visible light that can be used for the purpose. Whatever it is, if it is a wave with wavelength λ, the position cannot be known to an accuracy much smaller than λ. In other words, the position uncertainty along a given direction, Δx, should obey the relation

$$\Delta x \lesssim \lambda. \tag{1.19}$$

Certainly, we can make the position uncertainty as small as we want by choosing waves with sufficiently short wavelength. However, if we make the wavelength small, the momentum of the quanta would be large, as seen from Eq. (1.16). When the electron collides with a photon carrying high momentum, its momentum can change by a large amount, thereby making the momentum measurement quite uncertain. Thus, using probes that can systematically access increasingly shorter wavelengths, one can improve on the position uncertainty but at the same time sacrifice the momentum uncertainty. Clearly, with increasingly larger wavelengths, momentum uncertainty can be improved at the cost of losing position uncertainty. Both uncertainties cannot be reduced arbitrarily, and this is precisely what Heisenberg's uncertainty principle tells us.

The principle can be put in a mathematical form by using our discussion above. If we denote the uncertainty in the position along a certain direction by Δx and the uncertainty in the component of the wave vector along the same direction by Δk_x, we seem to obtain a relation of the form

$$\Delta x \, \Delta k_x \geqslant c_0, \tag{1.20}$$

where c_0 is some number. It will have to be dimensionless because the wave vector has dimensions of inverse length. We will shortly comment on the magnitude of this number. Before that, we note that we can use Eq. (1.18) to write Eq. (1.20) in the equivalent form

$$\Delta x \, \Delta p_x \geqslant c_0 \hbar, \tag{1.21}$$

so that this can be called the position–momentum uncertainty relation. Similar inequalities should hold for position and momentum components in any other direction.

The value of the number c_0 that appears on the right side of Eq. (1.21) cannot be specified until we set up a definition of the word *uncertainty*. Later, in §3.2, we will use a precise definition and derive a value of the number. However, even without a precise value for the numerical factor, we can try to appreciate the importance of the uncertainty principle by assuming that this factor is a number of order unity (which is a loose expression to mean that the number is neither very large nor very small). For this, we rewrite Eq. (1.21) in terms of velocities instead of momenta:

$$\Delta x \Delta v_x \geqslant c_0 \hbar/m, \tag{1.22}$$

where m is the mass of the particle. The value of the Planck constant, known experimentally, is:

$$\hbar = 1.054 \times 10^{-27} \, \mathrm{g\, cm^2\, s^{-1}} \equiv 1.054 \times 10^{-34} \, \mathrm{kg\, m^2\, s^{-1}}. \tag{1.23}$$

Consider now a small object of mass 1 g, whose position is known to an incredible accuracy of 10^{-8} cm, which is about the size of a typical atom. Eq. (1.22) then tells us that, near the limit of the inequality, the velocity of the object will be uncertain by an amount

$$\Delta v_x \sim 10^{-19} \, \mathrm{cm\, s^{-1}}. \tag{1.24}$$

This uncertainty in the velocity would imply an uncertainty of position of a few nuclear dimensions after the object moves for a full year. Certainly, no known measuring instrument would be able to detect such differences. In fact, the uncertainty in the velocity due to experimental inaccuracies would be much bigger, and therefore the effects of uncertainty relation can be neglected.

Consider now an electron, whose mass is roughly 9×10^{-28} g. In this case, the same value of Δx gives

$$\Delta v_x \sim 10^8 \, \text{cm s}^{-1}, \tag{1.25}$$

through Eq. (1.22). This is stunning! If we know that an electron is in an atom and thereby say that its position uncertainty is about the atomic size, our knowledge about its velocity is fuzzy by an amount which is a good fraction of the speed of light. Certainly such uncertainties cannot be neglected.

When the effect of uncertainty principle cannot be neglected, we cannot meaningfully talk about the path of a particle. Notions such as position and velocity of a particle make little sense, since any knowledge about them is bound to be very fuzzy. In such situations, classical mechanics cannot be used: the basic questions asked in classical mechanics make no sense at all. For systems where the uncertainty limits are not important, we can use classical mechanics.

If there is uncertainty in specifying the position and momentum of a particle, can we at least talk about a probability of finding the particle at a specified position with a given momentum? It turns out that specifying probabilities is not enough for describing and explaining various phenomenological consequences of a system. In particular, probabilities cannot explain interference phenomena, which will be described in §1.5.

□ **Exercise 1.2** *An atomic nucleus has a size of order 10^{-13} cm. This means that the position of the particles in the nucleus is known to that accuracy. Find the uncertainty in the momentum of a proton in the nucleus.*

□ **Exercise 1.3** *An electron is emitted from a nucleus of radius 1 femtometer (10^{-15} m). Assuming that its momentum must at least be equal to uncertainty in its momentum p and taking $c_0 = 1/2$, find the minimum energy of the emitted electron using $E = \sqrt{m^2c^4 + p^2c^2}$, where $m = 9.1 \times 10^{-31}$kg is the electron rest mass and $c = 3 \times 10^8$m/s is the speed of light.*

□ **Exercise 1.4** *An electron resides in a box of length L. If the minimum velocity of this electron is estimated to be $1 \, \text{cm/s}$, what is the largest possible value of L? You may assume $c_0 = 1/2$.*

□ **Exercise 1.5** *A boy throws a ball of mass $1 \, \text{g}$ on the wall such that it creates a visible spot at the point of contact. The uncertainty in the velocity of the* *ball parallel to the wall (in any direction) is 1 m/ s. What is the uncertainty in the position of the spot? How would it change if the boy had an electron instead of a ball?*

1.4   Time–energy uncertainty

The uncertainty is not restricted to position and momentum only. Later on, we will see that there can be uncertainties in angular momentum, and in many other quantities. Ultimately, all such uncertainties can be traced back to the position–momentum uncertainty, so in that sense the latter can be called somewhat more fundamental.

There is another kind of uncertainty that has nothing to do with the position–momentum uncertainty. The way we introduced position–momentum uncertainty has to do with the fact that the coordinate x and the wavenumber $k = p/\hbar$ appear in a product in the expression for a plane wave, as shown in Eq. (1.9). We notice that ω and time t are present as a product in the same expression, and therefore the arguments which lead to Eq. (1.20) can be used to conclude that

$$\Delta t \, \Delta \omega \geqslant c_0. \tag{1.26}$$

Using Eq. (1.10) now, we can express this relation in the form

$$\Delta t \, \Delta E \geqslant c_0 \hbar, \tag{1.27}$$

which can be called the *time–energy uncertainty relation*.

What would it mean? Refer to Figure 1.1 (p. 7) to seek an analogy. In that figure, the dashed line shows that the wave is localized in space, which results in the uncertainty in wavelength and consequently to the uncertainty in momentum. Similarly, uncertainty in energy would have its root in the fact that certain waves might be spread only over a finite amount of time.

But why should waves be spread only over a finite time window? Let us say that we are thinking about de Broglie's suggestion of matter waves. Is not matter supposed to exist forever, and therefore the wave associated with it should be infinitely extended?

Not really. To be sure, the statement to the contrary is true. Although one can contemplate exactly solvable and absolutely closed systems which have infinite lifetime, most things in the real world do not live happily ever after. Even large things decay, or break up in collisions, or bind with other things to make a composite system. If we look at the atomic level, we know that there are some atomic levels which are unstable. There are elementary particles which live a short life after they are created, decaying into other particles. For an unstable system with a finite lifetime, certainly we can say that the wave associated with it is spread only over the extent of its lifetime. Then Δt, in Eq. (1.27), can at best be equal to the lifetime τ. The time–energy uncertainty relation then tells us that the energy of that system must have a minimum uncertainty. This is the essence of the time–energy uncertainty relation.

This makes a lot of sense. The point is that if some system is unstable, i.e., it decays after an average time τ, we do not have all eternity to measure its energy. We must perform the measurement within a time of order τ. Within this time, energy cannot be measured more accurately compared to what Eq. (1.27) dictates. As for the position–momentum uncertainty, the effect of time–energy uncertainty is a negligible fraction of the total energy for a macroscopic system. But, if the system is small, and the lifetimes are small, the uncertainty in energy can be appreciable.

□ **Exercise 1.6** *When heated, the sodium atom emits a radiation of wavelength 589 nm ("nm" means nanometer, i.e., 10^{-9} m). It results from transition from an excited state to the ground state.*

  (a) *What is the energy of the photon emitted?*

  (b) *The excited state has a lifetime of 1.6×10^{-8} s. Using $c_0 = \frac{1}{2}$ in Eq. (1.27), estimate the uncertainty in energy of the excited state.*

□ **Exercise 1.7** *Particles known as Δ-particles have a lifetime of about 5×10^{-24} s. What is the best accuracy at which its energy can be measured?*

1.5   Interference experiments

The most distinctive feature of a wave is that a wave can undergo interference with another wave. If matter shows wave properties, matter beams must demonstrate interference. Indeed, only a few years after de Broglie's bold hypothesis, interference of electron beams was demonstrated in an experiment by Clinton Davisson and Lester Germer, and simultaneously in another experiment by George Thomson.

We will start by describing a typical interference experiment using light, not electrons. In Figure 1.2, we show the arrangement for a double-slit experiment. Light comes from the source marked S and passes through two slits in an opaque wall W and falls on a screen D, which is shown with a vertical line to the right of the diagram. In the figure, we show schematically the intensity of light falling on different positions of the screen.

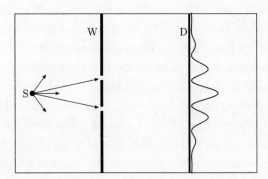

Figure 1.2 A two-slit interference experiment. Light from source S goes through the slits on the wall W and falls on the screen D. The curved line to the right of D shows the intensity of light at different points on the screen.

The important point is that the pattern has maxima and minima, and that the intensity even vanishes at some points. If we consider the probabilities of light coming through each slit and add them up, we can never obtain zero intensity at any point on the screen.

For the case of light, we know how to escape this impasse. We should not add the probabilities. The intensity of light reaching a particular point on the screen is proportional to the probability of light reaching there. Intensity is proportional to $|\boldsymbol{E}|^2$, where $\boldsymbol{E}$ is the electric field that shows wave behavior. For light coming from the two slits, we should not add the intensities. Rather, we should add the electric fields and then square the sum to find the intensity at any given point. Thus, the intensity of light falling on any point on the screen is given by

$$I \propto |\boldsymbol{E}_1 + \boldsymbol{E}_2|^2, \tag{1.28}$$

where $\boldsymbol{E}_1$ and $\boldsymbol{E}_2$ are the electric fields corresponding to the waves that come from the two slits. The position of a point on the screen determines the path lengths of the point from the two slits and consequently the phase of the waves coming from the two holes. The two waves will add constructively or destructively, producing either a bright or a dark spot on the screen, depending on the path difference from the two slits.

If we want to describe an electron interference experiment, the idea would be just the same, only the terminology would be a little different. Here again, probabilities will not suffice. We need something more fundamental than probability which can add constructively as well as destructively. The relationship between this *something* to the probability of finding an electron at a certain point would be the same as the relationship between the electric field vector and the intensity of light. The more familiar name of this *something* is the *wavefunction*, but it is only a representation, valid in special circumstances, of a more general concept called the *state vector*. The relation between the two concepts will be clarified later in the book. For now, it is enough to note that the general name implies that, like the electric field, the state vector is to be treated as a vector in some sense. We therefore discuss the general mathematical properties of vectors in Chapter 2, before coming back to the physics of the state vectors in Chapter 3.

Chapter 2

State vectors

As stated in the concluding sections of Chapter 1, in the light of the interference experiments, we would like to treat the states of a system as vectors. In this chapter, we discuss the general theory of vectors, and how the state vectors fit into that theoretical structure.

2.1 Introduction

During the first encounter with vectors, in high school, a student learns that a vector is something that has both magnitude and direction. This definition is unsatisfactory for reasons that we will discuss now. In order to apply this definition, one needs to first define the magnitude of a vector, which is not a vector. The direction is represented by the angles subtended by the vector with some fixed vectors. So we need to develop the notion of the angle between two vectors, and the angle is again not a vector. Alternatively, a vector is described in terms of its components, which are projections of the vector on some fixed axes, or basis vectors. This just means that the definition is incomplete, because now we need to define the basis vectors, i.e., the fixed axes. Rather, a vector should be defined by the fundamental properties of the set of vectors itself.

Let us then look for operations on vectors that produce vectors. We can take some hint from the set of numbers. Given two numbers, we can add them and get a number. We can also multiply them and get a number. There are certain properties that addition and multiplication of two numbers possess, for example, both should be commutative and both should be associative.

In elementary texts on vectors, we certainly learn how to add them, so addition is a property defined on vectors as well. However, multiplication cannot be defined. Although in high school or early college we learn about dot products and cross products of vectors, they do not qualify for our purpose. Dot product does not qualify because the result of dot product of two vectors is not a vector: it is a scalar. The reason cross product does not qualify is that it can be defined only in three dimensional (3D) space, where there can be a unique direction that is perpendicular to two given vectors. In no other dimensions can one define a cross product, i.e., a product of two vectors that is a vector. In summary, no product of vectors can be defined for vectors in general.

There is another thing that can be done on vectors to produce vectors. One can multiply any vector by a number. Thus we find two operations that can be done on vectors which produce vectors: addition and scaling. The formal definitions and properties of these two operations are summarized in §2.2.

2.2   Vector spaces

2.2.1   Definition

The natural mathematical structure that is used for dealing with vectors is called a *vector space*. The idea is to summarize the essential properties of vectors like those encountered in particle dynamics – e.g., velocity, force – into a set of axioms. Once this is done, the idea of vectors can be extended and applied to many other mathematical constructs.

To define a vector space formally, let us start with the ingredients that are necessary for the definition. These are:

- a set $\mathbb{V}$ of elements called *vectors*;

- a set $\mathbb{S}$ of elements called *scalars*;

- two binary operations involving only the scalars, to be called ordinary addition and multiplication and to be denoted by the symbols '+' and '·';

- a binary operation called vector addition (to be denoted also by the simple '+' sign);

- an operation called scalar multiplication, which is a multiplication of a vector by a scalar.

It is to be noted that the first two operations mentioned here do not involve the vectors at all. They are defined only between the scalars. In a very mathematical language, these two operations can be called mappings of the type $\mathbb{S} \times \mathbb{S} \to \mathbb{S}$, meaning that each of these operations needs two scalars, and the result of the operation is one scalar. In this language, the vector addition is a $\mathbb{V} \times \mathbb{V} \to \mathbb{V}$ mapping, i.e., it produces a vector from two given vectors. And finally, the operation termed scalar multiplication is an $\mathbb{S} \times \mathbb{V} \to \mathbb{V}$ mapping, i.e., it can define a vector from a given scalar and a given vector.

All these operations should be closed, i.e., the result of the operation should be defined no matter which two elements we choose from the relevant sets (e.g., for vector addition, both should be elements of the set $\mathbb{V}$). In addition, the operations should possess some other properties, which we describe now.

The scalars mentioned above are ordinary numbers, with the operations of usual addition and multiplication defined on them. We can restrict ourselves to only real numbers, in which case the resulting vector space will be called a *real vector space* and the set $\mathbb{S}$ will be denoted by $\mathbb{R}$. Or we can consider complex numbers, in which case the vector space is called a *complex vector space*. In this case, we will denote the set $\mathbb{S}$ by $\mathbb{C}$. We will argue in §3.4 that for quantum mechanics, we will need a complex vector space.

□ **Exercise 2.1** *Any complex number $r + is$ can be written as an ordered pair (r, s). Find the ordered pair which equals the product of two complex numbers $r_1 + is_1$ and $r_2 + is_2$.*

□ **Exercise 2.2** *A complex number $r + is$ can also be represented by a 2×2 matrix given by*

$$\begin{pmatrix} r & s \\ -s & r \end{pmatrix}. \tag{2.1}$$

Find the matrix corresponding to the product of two complex numbers $r_1 + is_1$ and $r_2 + is_2$.

Before we give the properties of the two operations involving vectors, we want to set up a notation for denoting vectors. There are many conventional notations. In handwriting, one often uses an arrow on the top of a lettered symbol to indicate that the letter stands for a vector. In printed text, a boldface letter is often used to denote vectors. For our purpose in this book, it will be most convenient if, from the beginning, we use the symbol $|u\rangle$ to indicate that u is a vector. We will also use a matrix notation, which will be introduced in §2.4.

We now define the properties of the operation on vectors.

1. Vector addition is commutative, i.e., for any two vectors $|u\rangle$ and $|v\rangle$, we have

$$|u\rangle + |v\rangle = |v\rangle + |u\rangle. \tag{2.2}$$

2. Vector addition is associative, i.e.,

$$|u\rangle + \Big(|v\rangle + |w\rangle\Big) = \Big(|u\rangle + |v\rangle\Big) + |w\rangle \qquad (2.3)$$

for any three vectors.

3. There is a *null vector*, to be denoted by $|\Omega\rangle$, which has the property

$$|v\rangle + |\Omega\rangle = |v\rangle \qquad (2.4)$$

for any $|v\rangle$.

4. For any vector $|v\rangle$, there exists a vector $|\bar{v}\rangle$ such that

$$|v\rangle + |\bar{v}\rangle = |\Omega\rangle. \qquad (2.5)$$

We will denote the vector $|\bar{v}\rangle$ by $-|v\rangle$, and $|u\rangle + |\bar{v}\rangle$ by $|u\rangle - |v\rangle$.

5. Scalar multiplication is associative in the sense that

$$\alpha(\beta\,|v\rangle) = (\alpha\beta)\,|v\rangle \qquad (2.6)$$

for arbitrary two scalars and a vector.

6. Scalar multiplication is distributive over vector addition, i.e.,

$$(\alpha + \beta)\,|u\rangle = \alpha\,|u\rangle + \beta\,|u\rangle. \qquad (2.7)$$

The properties stated in Eqs. 2.2 to 2.7 formalize our intuitive notions of vector addition and scaling of vectors. Note that nothing in the definitions implies that the "vectors", i.e., elements of the set $\mathbb{V}$, should have a *magnitude* and a *direction*, as are demanded in school-level definitions of vectors. In fact, we have not even defined the notions of *magnitude* and *direction* of a vector yet. The definition given here includes many more general objects that can be called vectors but for which magnitudes and directions may not be necessarily defined. Examples of such objects appear in some of the exercises that follow.

☐ **Exercise 2.3** *(a) Show that all $n \times n$ matrices with real elements form a real vector space.*

(b) Show that all $n \times n$ matrices with complex elements form a complex vector space.

☐ **Exercise 2.4** *Show that*

$$0\,|v\rangle = |\Omega\rangle \qquad (2.8)$$

for any vector $|v\rangle$. [**Hint:** *Take $\beta = -\alpha$ in Eq. (2.7).*]

☐ **Exercise 2.5** *Show that the set of all n-dimensional integer-valued column matrices does not form a real vector space while the set of all n dimensional real-valued column matrices does.*

☐ **Exercise 2.6** *Show that the set of polynomials of degree $d \leqslant n$ forms a vector space.*

☐ **Exercise 2.7** *Show that the set of functions $f(x)$ satisfying a linear differential equation $d^n f(x)/dx^n + f = 0$ forms a vector space.*

2.2.2 Linear independence and basis

A number of vectors, denoted by $\left|v_{(n)}\right\rangle$ for a range of values of n, form a linearly independent set if, for the equation

$$\sum_n \alpha_n \left|v_{(n)}\right\rangle = |\Omega\rangle, \tag{2.9}$$

the only solution is given by

$$\alpha_n = 0 \qquad \forall n. \tag{2.10}$$

If any other solution is obtained for Eq. (2.9) for a given set of vectors, the vectors are called linearly dependent.

The maximum number of linearly independent vectors that can be defined in a vector space is called the *dimension* of the vector space. If the dimension is a natural number, the vector space is called *finite dimensional*. Otherwise, if one can define infinite number of linearly independent vectors, the vector space is *infinite dimensional*. In this chapter we make statements which are appropriate for finite dimensional vector spaces. Some statements may need modification for infinite dimensional vector spaces. We will occasionally give warnings about such cases.

A set of vectors $\left|b_{(i)}\right\rangle$ constitute a *basis* in a vector space if an arbitrary vector can be written as a linear superposition of vectors in the set, i.e., if any vector $|v\rangle$ can be written in the form

$$|v\rangle = \sum_i v_i \left|b_{(i)}\right\rangle, \tag{2.11}$$

and the set contains linearly independent vectors. The objects v_i are scalars which are called the *components* of the vector $|v\rangle$ in the given basis.

☐ **Exercise 2.8** *Suppose, in a 3D space, we have chosen a basis. The components of two vectors in this basis are as follows: $A : (1, 2, 1)$, $B : (2, 1, 3)$. What condition must be satisfied by the components of a third vector if it has to be linearly independent of these two?*

□ **Exercise 2.9** *Show that for two vector spaces V_1 and V_2 to be isomorphic (in the sense there exists a bijective linear map between each of their elements), they have to be of the same dimension.*

□ **Exercise 2.10** *Show that all real functions of a single variable form a real vector space.* [**Note:** *This space is infinite dimensional, unlike the vector spaces indicated in earlier examples.*]

2.3 Vector spaces with extra structure

Ultimately, we need our theory to compare with experiments. Measurement of any quantity in an experiment produces a number. Thus, we need some numbers, i.e., scalars, to be defined from the vectors in order to compare our theory with experiments. There are two operations on vectors in the definition of a vector space, and the results of both these operations are vectors. Thus, those operations will not suffice: we need some extra structure to define scalars out of vectors. In this section, we outline two such structures. The first one assigns a real number corresponding to a vector, and the second one assigns a complex number, in general, corresponding to two vectors.

2.3.1 Normed vector spaces

A vector space is called *normed* when there is a norm defined on its elements. A norm for any vector $|v\rangle$ will be denoted by $\||v\rangle\|$ or simply as $\|v\|$. There will be no confusion if we omit the vector sign, because only norms of vectors will be denoted by the double-line delimiters. If we want to write the absolute value of a real or complex number, we will use single-line delimiters, as in $|\alpha|$.

The *norm* is a real number corresponding to any vector (i.e., a mapping of the sort $\mathbb{V} \to \mathbb{R}$) that has the following properties:

- $\|v\| \geqslant 0$ for all $|v\rangle$;

- $\|v\| = 0$ if and only if $|v\rangle$ is the null vector $|\Omega\rangle$;

- $\|\alpha v\| = |\alpha|\,\|v\|$, where $|\alpha v\rangle \equiv \alpha\,|v\rangle$;

- $\|u + v\| \leqslant \|u\| + \|v\|$ for any two vectors $|u\rangle$ and $|v\rangle$.

The notion of the norm is a formalization of the concept of the magnitude of a vector. Note that the definition above does not tell us how $\|v\|$ can be found from a given vector $|v\rangle$. Any prescription that satisfies the conditions listed above is acceptable.

To give examples of such prescriptions, we will denote the vector $|v\rangle$ by writing $|v\rangle = (v_1, v_2, \cdots, v_n)$. One can think of these n numbers as the 'components' of a vector in some basis, or simply by the points in an n-dimensional space. Vector addition is defined by

$$|u\rangle + |v\rangle = (u_1 + v_1, u_2 + v_2, \cdots, u_n + v_n), \tag{2.12}$$

and scalar multiplication by

$$\alpha |v\rangle = (\alpha v_1, \alpha v_2, \cdots, \alpha v_n). \tag{2.13}$$

Then, a definition known as the p-norm:

$$\|v\|_p = \left(\sum_i |v_i|^p \right)^{1/p}, \tag{2.14}$$

satisfies the conditions for any $p \geqslant 1$. For example, for $p \to \infty$, the norm is the element with the highest absolute value. The $p = 1$ norm is called the *Taxicab norm*. In this book, we will exclusively use the *Euclidean norm*, obtained by putting $p = 2$ in Eq. (2.14),

$$\|v\|_E = \sqrt{\sum_i |v_i|^2}. \tag{2.15}$$

The modulus sign on v_i might seem unnecessary if we think only of real components, but it is essential for a complex vector space.

2.3.2 Inner product spaces

Another kind of extra structure on a vector space is a map of the form $\mathbb{V} \times \mathbb{V} \to \mathbb{S}$. In other words, given two vectors (which may be same or different), one can find a scalar. For a complex vector space, the scalar should be a complex number in general. Because the result is not a vector, it is not called simply a *product*: the name given is *inner product*.

We will write the inner product of two vectors $|u\rangle$ and $|v\rangle$ as $\langle u|v\rangle$. With other notations for vectors, one writes $\boldsymbol{u} \cdot \boldsymbol{v}$, for which the inner product is also referred to as the *dot product*. The result must satisfy the following properties.

1. It should be linear, i.e., $\langle u|v + w\rangle = \langle u|v\rangle + \langle u|w\rangle$.

2. If the vector $\alpha |v\rangle$ is denoted by $|\alpha v\rangle$, then $\langle u|\alpha v\rangle = \alpha \langle u|v\rangle$.

3. Any two vectors $|u\rangle$ and $|v\rangle$ must satisfy the rule

$$\langle u\,|\,v\rangle = (\langle v\,|\,u\rangle)^*, \qquad (2.16)$$

where the star means complex conjugation, an operation defined in $\mathbb{C}$. In a real vector space, the star has no effect.

4. For any vector $|u\rangle$, the quantity $\sqrt{\langle u|u\rangle}$ should have all properties for being a norm, i.e., one should be able to assign

$$\|u\| = \sqrt{\langle u\,|\,u\rangle}. \qquad (2.17)$$

☐ **Exercise 2.11** *Show that* $\langle \alpha u|v\rangle = \alpha^* \langle u|v\rangle$.

☐ **Exercise 2.12** *Using the axioms above, show that*

$$\langle u\,|\,\beta v + \gamma w\rangle = \beta\,\langle u\,|\,v\rangle + \gamma\,\langle u\,|\,w\rangle. \qquad (2.18)$$

☐ **Exercise 2.13** *Show that the set of vectors* $|v_{(i)}\rangle$, $i = 1, 2, \ldots, n$, *are linearly independent if* $\langle v_{(i)}|v_{(j)}\rangle = 0$ *for* $i \neq j$.

It seems that the notation for inner product is a bit uncomfortable, since the first factor in the product of the form $\langle u|v\rangle$ does not appear in the usual vector notation, but rather as $\langle u|$. This, however, is not a bad thing. Interchanging the two vectors participating in the inner product produces in general a different result, as can be seen from property 3 in the definition. It is therefore good that the different ways that the two vectors participate in the product are indicated in the notation for the inner product. We will define an object called $\langle u|$ later, which will be related to, but different from, $|u\rangle$.

Property 4 in the definition of inner products shows that, once we define an inner product, a norm is automatically defined on the vector space through Eq. (2.17). In other words, an inner product space is also a normed space.

It is interesting to go back at this point to the more intuitive high-school definition of vectors, viz., that a vector is something which has a magnitude and a direction. The intuitive notion of magnitude corresponds to the formal definition of the norm of a vector, and therefore can be defined only in vector spaces in which a norm can be defined.

In fact, the intuitive definition is even more restricted, because in general a normed vector space need not have the concept of a 'direction'. In an inner product space, however, one can at least define the concept of *orthogonality* of two vectors, viz., two vectors are orthogonal if their inner product is zero:

$$\langle u\,|\,v\rangle = 0. \qquad (2.19)$$

Using this, one can easily prove the Pythagoras theorem, which is given in Ex. 2.14.

□ **Exercise 2.14** *Show that, if two vectors* $|u\rangle$ *and* $|v\rangle$ *are orthogonal, then*

$$\|u + v\|^2 = \|u\|^2 + \|v\|^2. \tag{2.20}$$

□ **Exercise 2.15** *Consider three vectors given by*

$$|1\rangle = \begin{pmatrix} 2 \\ 3 \\ 1 \end{pmatrix}, \quad |2\rangle = \begin{pmatrix} 1 \\ 2 \\ 3 \end{pmatrix}, \quad |3\rangle = \begin{pmatrix} 3 \\ 1 \\ 2 \end{pmatrix}. \tag{2.21}$$

(a) *Are the three vectors linearly independent?*

(b) *If a vector* $|4\rangle$ *is orthogonal to* $|1\rangle$ *and* $|2\rangle$ *and satisfies* $\langle 4|3\rangle = 1$, *find the norm of* $|4\rangle$.

(c) *Is it possible to write* $|4\rangle$ *as a linear combination of the vectors given in Eq. (2.21)?*

Now, for arbitrary $|u\rangle$ and $|v\rangle$, it is easy to see that the vector

$$|w\rangle = |v\rangle - \frac{\langle u|v\rangle}{\langle u|u\rangle} |u\rangle \tag{2.22}$$

is orthogonal to $|u\rangle$, and therefore to any $\alpha |u\rangle$ where α is a scalar. Using the Pythagoras theorem on the orthogonal vectors $|w\rangle$ and $\frac{\langle u|v\rangle}{\langle u|u\rangle} |u\rangle$, we obtain

$$\|w\|^2 + \frac{\left|\langle u|v\rangle\right|^2}{\langle u|u\rangle} = \|v\|^2. \tag{2.23}$$

Noting that $\langle u|u\rangle = \|u\|^2$ and using the fact that $\|w\|^2 \geqslant 0$, we obtain an inequality

$$\left|\langle u|v\rangle\right| \leqslant \|u\| \, \|v\| \tag{2.24}$$

for any two vectors. This is called the *Cauchy–Schwarz inequality*. If we deal with a real vector space in which the inner product of two vectors is real, we can define the angle θ between two vectors $|u\rangle$ and $|v\rangle$ through the relation

$$\cos\theta = \frac{\langle u|v\rangle}{\|u\| \, \|v\|}, \tag{2.25}$$

since the expression on the right side is guaranteed to be a real number in the range from -1 to $+1$. The direction of a vector can be defined by the angle it subtends with some fixed vectors in the vector space, for example, the basis vectors. Thus, the high school definition of a vector works *only* in a real vector space where an inner product is defined. In quantum mechanics, we will see that we must deal with complex vector spaces, so the concept of angle between two vectors will make no sense.

☐ **Exercise 2.16** *The general definition of vector spaces requires the existence of a vector $|\bar{v}\rangle$ corresponding to each vector $|v\rangle$ such that $|v\rangle + |\bar{v}\rangle = |\Omega\rangle$. Argue that $\| \, |\bar{v}\rangle \, \| = \| \, |v\rangle \, \|$.*

2.4   Orthogonality, normalization, and completeness

As we have said in §2.3.2, two vectors $|u\rangle$ and $|v\rangle$ are called *orthogonal* when their inner product is zero, i.e., if Eq. (2.19) holds. If two non-null vectors are orthogonal, they must be linearly independent. This can be seen by assuming the contrary, i.e., there exists two non-zero scalars α and β such that $\alpha\,|u\rangle + \beta\,|v\rangle = |\Omega\rangle$. Taking the inner product with $\langle u|$ and using the orthogonality condition of Eq. (2.19), we obtain $\alpha\langle u|u\rangle = 0$. This is impossible, since $|u\rangle$ is non-null and we have assumed that α is non-zero.

Earlier, we said that the maximum number of linearly independent vectors that one can choose in a vector space is equal to the dimension of the space. Therefore, the same can be said about the maximum number of mutually orthogonal vectors in a vector space. Of course, the first definition is more general, since it does not even depend on the existence of an inner product. But, for the purpose of quantum mechanics, there is no harm using the second definition since we will be using vector spaces where an inner product is defined.

We call a vector *normalized* if its norm is unity, i.e., if

$$\langle v\,|\,v\rangle = 1. \tag{2.26}$$

Given any vector $|u\rangle$ for which $\langle u|u\rangle$ is finite, we can find a scalar α such that $\alpha\,|u\rangle$ is normalized. The process of multiplying with a proper factor to produce a unit norm is called *normalizing the vector*.

> Note that in a finite dimensional vector space, all vectors with finite components have a finite norm, and therefore any vector can be normalized by multiplying it with a number. The same is not obviously true for an infinite dimensional vector space, because the norm of a vector $|v\rangle$, being a sum over infinite number of *components* of $|v\rangle$, may diverge. We will address this question later.

A linearly independent set of vectors is called *complete* in a vector space if any vector of that space can be written as a linear superposition of the members of the set. Any complete set of vectors can be used as a *basis* of the vector space. Further, in an inner product space, we can always make superpositions of linearly independent vectors such that each one is orthogonal to all the rest. An example of such orthogonalization was provided in Eq. (2.22), where, starting from two arbitrary non-null vectors $|u\rangle$ and $|v\rangle$, we constructed an orthogonal set comprising $|u\rangle$ and $|w\rangle$, the latter being a linear superposition of the two initial vectors. The process can be suitably adopted if one starts with a larger number of linearly independent vectors.

☐ **Exercise 2.17** *Start with three linearly independent vectors $|u\rangle$, $|v\rangle$ and $|w\rangle$. Now define*

$$\left|v'\right\rangle = |v\rangle + a\,|u\rangle, \qquad \left|w'\right\rangle = |w\rangle + b\,|u\rangle + c\,|v\rangle, \qquad (2.27)$$

such that $|u\rangle$, $|v'\rangle$ and $|w'\rangle$ are mutually orthogonal. [**Note:** *This procedure is called the* Gram–Schmidt *orthogonalization procedure.*]

Thus, in an inner product space, we can take a number of mutually orthogonal vectors, the number being equal to the dimension of the vector space, and use them as a basis. We can normalize all the basis vectors to have unit norm:

$$\langle i\,|\,j\rangle = \delta_{ij}. \qquad (2.28)$$

Here, the notation δ_{ij} stands for the *Kronecker delta*, defined as

$$\delta_{ij} = \begin{cases} 1 & \text{if } i = j, \\ 0 & \text{otherwise.} \end{cases} \qquad (2.29)$$

Examples of manipulations with this symbol have been summarized in Appendix A. If the basis vectors satisfy Eq. (2.28), the basis is called an *orthonormal basis*.

In an orthonormal basis, any vector $|v\rangle$ can be written in the form

$$|v\rangle = \sum_i v_i\,|i\rangle, \qquad (2.30)$$

where the numbers v_i are the called the components of the vector $|v\rangle$. Taking the inner product of both sides of Eq. (2.30) with $|j\rangle$, we obtain

$$\langle j\,|\,v\rangle = \sum_i v_i\,\langle j\,|\,i\rangle = \sum_i v_i \delta_{ij} = v_j. \qquad (2.31)$$

Substituting this expression back into Eq. (2.30), we obtain

$$|v\rangle = \sum_i |i\rangle \langle i | v\rangle. \qquad (2.32)$$

Because this has to be true for any vector $|v\rangle$, the basis vectors must satisfy the relation

$$\sum_i |i\rangle \langle i| = \mathbb{1}. \qquad (2.33)$$

This can be taken as the definition of completeness for a set of orthonormal vectors.

If we start from Eq. (2.33) and take an inner product with $\langle u|$, we obtain

$$\sum_i \langle u | i\rangle \langle i| = \langle u|. \qquad (2.34)$$

Comparing this equation with Eq. (2.32), we conclude that if the components of $\langle u|$ are denoted by $\tilde{u}_i$, then

$$\tilde{u}_i = \langle u | i\rangle = \langle i | u\rangle^* = u_i^*, \qquad (2.35)$$

using Eq. (2.31). Therefore we will not use a different notation for the components of $\langle u|$; just denote them by u_i^*.

This statement shows quite clearly the distinction between $|u\rangle$ and $\langle u|$: their components are complex conjugates of one another. The notation $|u\rangle$ is usually said to denote a *ket vector*, or simply *ket*. Complex conjugate of the form $\langle u|$ is called a *bra vector*, or just *bra*. Together, when a bra and a ket form an inner product as seen in Eq. (2.31) or Eq. (2.35) for example, the product is called a *bracket*.

We can now also deduce the definition of the inner product that we are going to adopt:

$$\langle u | v\rangle = \sum_i \langle u | i\rangle \langle i | v\rangle = \sum_i u_i^* v_i. \qquad (2.36)$$

Eq. (2.36) suggests that we can use an attractive matrix notation to denote vectors. We can think of a vector $|u\rangle$ as a column matrix. And suppose $\langle u|$ is a row matrix, each of whose elements is the complex conjugate of the corresponding element of $|u\rangle$. Then the inner product $\langle u|v\rangle$ will correspond to ordinary matrix multiplication,

$$\langle u | v\rangle = \mathrm{u}^\dagger \mathrm{v}, \qquad (2.37)$$

where v is a column matrix whose elements are v_i, whereas $\mathrm{u}^\dagger$ is a row matrix defined by the elements

$$\left(\mathrm{u}^\dagger\right)_i \equiv \left(u_i\right)^*. \tag{2.38}$$

This analogy further bolsters the point that the two objects taking part in the inner product are not really the same kind of objects, and therefore different notations for them are justified. However, in the bra–ket notation, no dagger sign has to be used: just the type of delimiters, the bra in one case and the ket in the other, is enough to tell us that the objects are of different kinds.

Anyone who is familiar with matrix algebra might feel comfortable in thinking of ket vectors as column matrices and row vectors as row matrices and the inner product as simple matrix multiplication, but there is a catch. In quantum mechanics, we will also have to deal with infinite-dimensional vector spaces. The matrix notation does not apply there. However, for any system which involves a finite-dimensional vector space, there is no harm thinking in terms of matrices.

☐ **Exercise 2.18** *Show that the inner product defined in Eq. (2.36) is independent of the basis vectors used. In other words, if we had used some other orthonormal set of basis vectors $|i'\rangle$ to write the vector $|u\rangle$ as $|u\rangle = \sum_i u_i' |i'\rangle$ and similarly for $|v\rangle$, then $\sum_i u_i^* v_i = \sum_i u_i'^* v_i'$.*

☐ **Exercise 2.19** *Of course, there is no compulsion that one must use an orthonormal basis. If, instead, one uses a basis in which the basis vectors are mutually orthogonal but not necessarily normalized, we can write, instead of Eq. (2.28), a relation of the sort*

$$\langle i\,|\,j\rangle = a_i \delta_{ij} \tag{2.39}$$

with some constants a_i. Show how Eq. (2.33) will look in this basis.

☐ **Exercise 2.20** *Consider a vector space of two-dimensional column vectors with real elements. Any transformation T which preserves the inner product is defined as an* orthogonal transformation: *$\langle Tv|Tu\rangle = \langle v|u\rangle$ for all column vectors $|u\rangle$ and $|v\rangle$. Find an explicit 2×2 matrix representation of T.*

2.5 States as vectors

The crucial property of vectors, as given in §2.2, is that vectors can be superposed linearly. It means that two vectors can be added, and a vector can be multiplied by a scalar. These are exactly the properties that we need for a state vector. If we get two beams to interfere, the resulting state vector should be the sum of the two individual state vectors.

But an act of measurement does not measure a state vector. It measures observables for which individual measurements give real numbers. In §2.9.3, we will see that the definition of the average value of an observable in a state uses the norm of the state vector. Hence, we would at least need a normed vector space. Moreover, in order to discuss transition from one state to another, we will also need the inner product of two different state vectors. Hence, an inner product space is the appropriate mathematical structure for the discussions that will be carried out in the rest of the book. In addition, the inner product should have the property of Cauchy completeness, something that we will discuss in §3.4. A vector space equipped with an inner product that is Cauchy complete is called a *Hilbert space*.

Do we need a real vector space or a complex one? We will argue later, in §3.4, that we will need a complex vector space. In any case, there is no harm if we consider the vector space to be complex: we do not lose anything by employing a more general framework.

Physics deals with the change of the state of a system. For a classical system, the state is defined by the positions and momenta of the particles in the system. For a quantum mechanical system, the state is defined by a state vector, as indicated at the end of Chapter 1. So it is imperative that we consider objects which can change one vector into another within a vector space. The resulting vector can be in some special case the same as the original vector, but in general the two will be different. Objects that can inflict such changes on vectors are called *operators*.

2.6   Operators on vector spaces

An operator on a vector space is a mapping of the vector space onto itself, i.e., a mapping of the type $\mathbb{V} \to \mathbb{V}$. In other words, an operator acting on a vector produces a vector. We will be concerned, almost exclusively, with linear operators, a term that can be explained as follows. If $|u\rangle$ and $|v\rangle$ are two arbitrary vectors and α and β two arbitrary scalars, $\alpha |u\rangle + \beta |v\rangle$ must also be a vector, according to the axioms that define a vector space. A linear operator obeys the rule

$$\mathscr{O}\Big(\alpha |u\rangle + \beta |v\rangle \Big) = \alpha \mathscr{O} |u\rangle + \beta \mathscr{O} |v\rangle \qquad (2.40)$$

for arbitrary vectors and scalars. We will mostly deal with linear operators. The only operator considered in this book that does not follow this rule is the time-reversal operator, which will be discussed in detail in §5.4.3.

Calligraphic letters have been used here and elsewhere to denote operators. This convention has not been maintained throughout the book. In some cases, just an uppercase letter has been used. In yet other cases where even that cannot be used because of prevalent convention or otherwise, we use a caret ($\widehat{}$). Even this indicator will sometimes be omitted when there will be no ambiguity.

We can use the linearity property of Eq. (2.40) to show that operators themselves form a vector space. For example, we can define the addition of two operators by the relation

$$(\mathscr{O}_1 + \mathscr{O}_2)\,|u\rangle = \mathscr{O}_1\,|u\rangle + \mathscr{O}_2\,|u\rangle \tag{2.41}$$

for any vector $|u\rangle$. Multiplying an operator by a scalar can be defined equally easily:

$$(\alpha\mathscr{O})\,|u\rangle = \alpha(\mathscr{O}\,|u\rangle). \tag{2.42}$$

The null operator $\widehat{0}$ can be defined by the condition

$$\mathscr{O} + \widehat{0} = \mathscr{O} \tag{2.43}$$

for any operator $\mathscr{O}$, and $\mathscr{O}_2$ will be the additive inverse of the operator $\mathscr{O}_1$ if in Eq. (2.41) the right side yields the null vector.

□ **Exercise 2.21** *Show that the null operator acting on any vector gives the null vector, i.e.,*

$$\widehat{0}\,|v\rangle = |\Omega\rangle \tag{2.44}$$

for any vector $|v\rangle$.

However, in general there is no concept of multiplication of two vectors. The operators on a vector space form a special kind of vector space on which a multiplication can be defined, as we demonstrate now. Consider an operator $\mathscr{O}_1$. Acting on a vector $|v\rangle$, it produces a vector. Instead of denoting the resulting vector by a new letter, we devise a suggestive notation for it:

$$\mathscr{O}_1\,|v\rangle \equiv |\mathscr{O}_1 v\rangle. \tag{2.45}$$

Now, another operator $\mathscr{O}_2$, acting on this vector which we just called $|\mathscr{O}_1 v\rangle$, produces a vector again. We can use the same method for naming this vector:

$$\mathscr{O}_2\,|\mathscr{O}_1 v\rangle = |\mathscr{O}_2\mathscr{O}_1 v\rangle. \tag{2.46}$$

This vector, however, can be seen as the result of the product of two operators, $\mathscr{O}_2\mathscr{O}_1$, acting on $|v\rangle$. This means that we define the product of two vectors by the rule

$$(\mathscr{O}_2\mathscr{O}_1)\,|v\rangle = |\mathscr{O}_2\mathscr{O}_1 v\rangle = \mathscr{O}_2(\mathscr{O}_1\,|v\rangle) \qquad (2.47)$$

for any vector $|v\rangle$.

This multiplication has some features which are similar to ordinary multiplication of numbers. It is easy to see that the product of operators is associative. There is also an *identity operator*, which does nothing, i.e., which maps any vector onto itself. This operator will be denoted by $\mathbb{1}$. Thus,

$$\mathbb{1}\,|u\rangle = |u\rangle \qquad (2.48)$$

for any $|u\rangle$. Obviously, this implies that for any operator $\mathscr{O}$,

$$\mathscr{O}\mathbb{1} = \mathbb{1}\mathscr{O} = \mathscr{O}. \qquad (2.49)$$

Also, if there exists an operator $\mathscr{O}^{-1}$ that satisfies the relation

$$\mathscr{O}^{-1}\mathscr{O}\,|v\rangle = |v\rangle\,, \qquad (2.50)$$

for all vectors $|v\rangle$, this operator is called the inverse of the operator $\mathscr{O}$. Of course, not all operators will have inverses. Given an operator A, if there exists a vector $|v\rangle$ such that $A\,|v\rangle$ is the null vector $|\Omega\rangle$, then any other operator operating further on it cannot recover $|v\rangle$: the result of all such operations will be $|\Omega\rangle$, as shown in Ex. 2.23. Thus, an operator A will have an inverse only if

$$A\,|v\rangle \neq |\Omega\rangle \qquad (2.51)$$

for all non-null vectors $|v\rangle$. But this is not an impediment in the way of defining multiplication of operators: even for the case of numbers, the number zero does not have a multiplicative inverse.

□ **Exercise 2.22** *Show that, for any operator A,*

$$A\,|\Omega\rangle = |\Omega\rangle\,, \qquad (2.52)$$

where $|\Omega\rangle$ is the null vector.

□ **Exercise 2.23** *If $A\,|v\rangle = |\Omega\rangle$, show that for any other operator B, $BA\,|v\rangle = |\Omega\rangle$.*

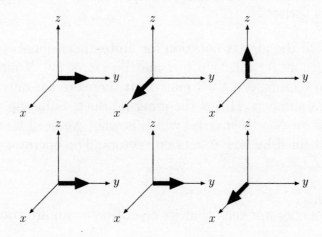

Figure 2.1 Two rows of diagrams show two different outcomes of rotating a vector. The vector is denoted by a thick arrow, which initially lies along the y-axis. In the top row, we first rotate the vector clockwise by $90°$ about the z-axis and then the result by a clockwise $90°$ rotation about the y-axis. In the bottom row, the order of the two rotations has been reversed. The final outcome is different in the two cases.

However, note that there is nothing in the properties that indicates that the operators should commute, i.e., $AB\,|u\rangle$ should be equal to $BA\,|u\rangle$. In fact, in general operators on a vector space do not commute, as can be seen by applying two rotations in orthogonal planes to a vector in ordinary coordinate space. The results are seen in Figure 2.1.

Clearly, on a finite dimensional vector space, an operator can be represented by a matrix. We have already said that we want to represent vectors as column matrices. In an N-dimensional vector space, such columns will have N elements, i.e., will be $N \times 1$ matrices. An operator will then be represented by an $N \times N$ square matrix, so that an operator acting on a vector gives an $N \times 1$ matrix, which is a column matrix representing a vector. Note that matrix multiplication automatically satisfies the linearity property given in Eq. (2.40) and also the associative property. Moreover, matrix multiplication is not automatically commutative, so we are also not introducing any extra structure by thinking of operators as matrices.

Once again we pause to say that with infinite-dimensional spaces, the concept of matrices as arrays of numbers do not hold in any obvious manner. We will discuss the proper generalization in Chapter 3.

Continuing with the matrix notation for finite-dimensional spaces, we can say that a bra such as $\langle v|$ will be a row matrix, i.e., a $1 \times N$ matrix, so that $\langle u|v\rangle$, for any two vectors, is a 1×1 matrix. It therefore has only one element and behaves like a number. This is the inner product. Similarly, a product of the form $|u\rangle \langle v|$ is an $N \times N$ matrix, which is what we need for an operator. Any operator is defined by how it acts on vectors. The operator

$$A = |u\rangle \langle v| \tag{2.53}$$

is defined to be an operator such that its operation on an arbitrary vector $|w\rangle$ yields the result

$$A|w\rangle = |u\rangle \langle v|w\rangle, \tag{2.54}$$

which is a number multiplied by $|u\rangle$. In fact, this is the technique that was used in Eq. (2.33), where the object on the right side is really not a number, but the identity operator defined in Eq. (2.48).

If we think of an operator as a matrix, we might also wonder about the elements of this matrix. The *matrix elements* of an operator can be defined between any two vectors $|u\rangle$ and $|v\rangle$, and is written in the form $\langle u|\mathcal{O}|v\rangle$. The meaning of this definition is quite obvious from the notation: $\mathcal{O}|v\rangle$ is a vector, and the matrix element is the inner product of this vector with $|u\rangle$. If we take the vectors appearing in this definition from the members of an orthonormal basis, then an operator $\mathcal{O}$ can be written as a matrix in this vector space whose ij^{th} element, i.e., element in the ith row and j^{th} column, would indeed be $\langle i|\mathcal{O}|j\rangle$:

$$\mathcal{O}_{ij} = \langle i|\mathcal{O}|j\rangle. \tag{2.55}$$

In fact, using the completeness of the basis vectors, Eq. (2.33), we can write, for any operator $\mathcal{O}$,

$$\mathcal{O} = \sum_i |i\rangle \langle i| \, \mathcal{O} \sum_j |j\rangle \langle j|$$

$$= \sum_i \sum_j |i\rangle \underbrace{\langle i|\mathcal{O}|j\rangle}_{=\mathcal{O}_{ij}} \langle j|. \tag{2.56}$$

Since $\mathscr{O}_{ij}$ is a number, we can put it anywhere: it does not matter. So the final result is:

$$\mathscr{O} = \sum_i \sum_j \mathscr{O}_{ij} |i\rangle \langle j| . \tag{2.57}$$

☐ **Exercise 2.24** *Do it the opposite way now. Show that if an operator is defined by Eq. (2.57), then its matrix elements in an orthonormal basis would be given by Eq. (2.55). [***Note:*** In doing such problems, do not use the same symbol to denote the basis vectors in both equations, as that would confuse what exactly is summed over. In other words, since i and j are summed over on the right side, do not find $\langle i|\mathscr{O}|j\rangle$ from Eq. (2.57); rather, find $\langle k|\mathscr{O}|l\rangle$.]*

☐ **Exercise 2.25** *Show that Eq. (2.55), the definition of a the matrix corresponding to an operator, is consistent with the usual matrix multiplication rule:*

$$\left(\mathscr{O}_1\mathscr{O}_2\right)_{ij} = \sum_k \left(\mathscr{O}_1\right)_{ik}\left(\mathscr{O}_2\right)_{kj}. \tag{2.58}$$

[**Hint:** *Use Eq. (2.33).*]

2.7 Eigenvalues and eigenvectors of operators

2.7.1 Definition and algorithm

Given an operator A, if we can find a non-null vector $|\Lambda\rangle$ such that

$$A|\Lambda\rangle = \lambda|\Lambda\rangle, \tag{2.59}$$

where λ is a number, then $|\Lambda\rangle$ is called an *eigenvector* of the operator A. The number λ, which can in general be complex, is called the *eigenvalue* corresponding to the eigenvector $|\Lambda\rangle$. A relation of the sort given in Eq. (2.59) is called an eigenvalue equation.

Clearly, eigenvectors satisfy the homogeneous equations

$$(A - \lambda\mathbb{1})|\Lambda\rangle = 0. \tag{2.60}$$

Non-null vectors will exist as solutions of these equations if

$$\det(A - \lambda\mathbb{1}) = 0. \tag{2.61}$$

If an operator can be expressed as an $N \times N$ square matrix, Eq. (2.61) gives an Nth degree polynomial equation in λ, which has N solutions. It does not mean that there are N distinct solutions. If a solution appears multiple times, the eigenvalue is called *degenerate*.

Once the eigenvalues are obtained, one can plug any one of them into Eq. (2.60) and try to find the corresponding eigenvector. The eigenvectors are not unique. First of all, any eigenvector can be defined only up to an overall multiplicative constant. The constant can be chosen by a suitable normalization condition. Even if we disregard an overall factor, there is no guarantee that there will be N eigenvectors. Certainly if A is the identity operator, Eq. (2.59) is satisfied by any vector, so there are an infinite number of eigenvetors. On the other hand, it is also possible that an $N \times N$ matrix would have less than N eigenvetors, an example of which will be given through Ex. 2.29 later. For matrices relevant for quantum mechanics, this problem is irrelevant, so we will assume there are indeed N eigenvectors.

Let us give an example with a 2×2 matrix

$$A = \begin{pmatrix} a & c \\ d & b \end{pmatrix}, \tag{2.62}$$

where all elements are real. The eigenvalue equation would be

$$(a - \lambda)(b - \lambda) - cd = 0, \tag{2.63}$$

whose solutions are

$$\lambda_\pm = \frac{1}{2} \left[a + b \pm \sqrt{(a - b)^2 + 4cd} \right]. \tag{2.64}$$

These are the eigenvalues. The eigenvector corresponding to any of these eigenvalues λ_k will have two components. Let us call the upper component x_k and the lower one y_k. Then, from Eq. (2.60), we obtain the equation

$$(a - \lambda_k)x_k + cy_k = 0, \tag{2.65}$$

so

$$x_k = \alpha c, \qquad y_k = \alpha(\lambda_k - a) \tag{2.66}$$

for arbitrary α. If we normalize such that $\langle \Lambda | \Lambda \rangle = 1$, the eigenvectors are obtained to be

$$|\Lambda\rangle_\pm = \begin{pmatrix} \cos\theta_\pm \\ \sin\theta_\pm \end{pmatrix}, \tag{2.67}$$

where

$$\tan\theta_\pm = \frac{b - a \pm \sqrt{(a - b)^2 + 4cd}}{2c}. \tag{2.68}$$

☐ **Exercise 2.26** *In writing Eq. (2.65), we equated the upper component of the vector on the left side of Eq. (2.60) to zero. We could have equated the lower component to zero as well. Show that this line of action gives the same eigenvectors.*

☐ **Exercise 2.27** *Show that, if the eigenvalues of an operator A are λ_i, the eigenvalues of A^n are λ_i^n for any positive integer n.*

☐ **Exercise 2.28** *For the matrix of Eq. (2.62) worked out above, suppose $d = 0$. The two eigenvalues now will be a and b, as seen from Eq. (2.64).*

   (a)   *Find the eigenvectors corresponding to the two eigenvalues.*

   (b)   *Are these eigenvectors orthogonal to each other?*

☐ **Exercise 2.29** *Consider the matrix obtained by putting $a = b$ and $d = 0$ in Eq. (2.62). Show that it has only one eigenvector.*

☐ **Exercise 2.30** *Consider a set of eigenvalues λ_i and corresponding eigenvectors*

$$|c_i\rangle = \begin{pmatrix} c_{1i} \\ c_{2i} \\ \vdots \\ c_{Ni} \end{pmatrix} \tag{2.69}$$

for an operator A represented by a $N \times N$ matrix. Assume that N eigenvectors exist, and they have been normalized so that each has unit norm. Construct the $n \times n$ matrix Q in terms of elements of $|c_i\rangle$ by writing the eigenvectors one after another in columns. Define the matrix $A_D = Q^{-1}AQ$. Find the elements of A_D.

2.7.2   Linear independence of eigenvectors

One important property of the eigenvectors of an operator is that eigenvectors corresponding to distinct eigenvalues are linearly independent. In other words, for a set of distinct eigenvalues $\lambda_1, \lambda_2, \cdots, \lambda_k$ of an operator A, the corresponding eigenvectors $|\Lambda_{(1)}\rangle, |\Lambda_{(2)}\rangle, \cdots, |\Lambda_{(k)}\rangle$ must be linearly independent.

To prove this statement, we assume the contrary. Suppose we can find a positive integer j such that the eigenvectors $|\Lambda_{(1)}\rangle, |\Lambda_{(2)}\rangle, \cdots, |\Lambda_{(j-1)}\rangle$ are linearly independent, but as soon as we add another one, $|\Lambda_{(j)}\rangle$, to the set, the linear independence is lost. This means that there exist numbers $c_1, c_2, \cdots, c_j$, not all of which are zero, that satisfy the equation

$$c_1 |\Lambda_{(1)}\rangle + c_2 |\Lambda_{(2)}\rangle + \cdots + c_j |\Lambda_{(j)}\rangle = 0. \tag{2.70}$$

Operating this equation by A and using the eigenvalue equation, we obtain

$$c_1 \lambda_1 \left| \Lambda_{(1)} \right\rangle + c_2 \lambda_2 \left| \Lambda_{(2)} \right\rangle + \cdots + c_j \lambda_j \left| \Lambda_{(j)} \right\rangle = 0. \qquad (2.71)$$

Subtracting λ_j times of Eq. (2.70) from this equation, we obtain

$$\begin{aligned} c_1(\lambda_1 - \lambda_j) \left| \Lambda_{(1)} \right\rangle + c_2(\lambda_2 - \lambda_j) \left| \Lambda_{(2)} \right\rangle + \cdots \\ + c_{j-1}(\lambda_{j-1} - \lambda_j) \left| \Lambda_{(j-1)} \right\rangle = 0. \end{aligned} \qquad (2.72)$$

Since the eigenvalues are distinct, none of the factors of the form $(\lambda_i - \lambda_j)$ is zero. Since the c_i's are also not all zero, it shows that the set of eigenvectors $\left| \Lambda_{(1)} \right\rangle, \left| \Lambda_{(2)} \right\rangle, \cdots, \left| \Lambda_{(j-1)} \right\rangle$ are linearly dependent, contrary to their definition. Therefore, our initial assumption must be wrong, i.e., no subset of the eigenvectors can be linearly dependent.

The importance of this result is the following. Suppose an $N \times N$ matrix indeed possesses N eigenvectors, and all eigenvalues are different. Then this result guarantees that all N eigenvectors are linearly independent. They can therefore be used as the basis in the vector space.

For restricted classes of matrices, stricter results may apply. It might be possible to use the eigenvectors as basis even if there is degeneracy in the eigenvalues. We will soon define such operators.

☐ **Exercise 2.31** *If there are N linearly independent vectors in an N-dimensional vector space, show that they can be used as a basis, i.e., any vector can be written as a linear superposition of those N vectors.* [**Hint:** *Assume that the opposite is true.*]

☐ **Exercise 2.32** *Show that the eigenvectors of an operator corresponding to a particular eigenvalue, augmented by the null vector, form a vector space.* [**Note:** *If the eigenvalue is non-degenerate, the proof is trivial.*]

2.7.3 Eigenvectors of commuting operators

There is an important property of eigenvectors that we will use often in this book. Consider two operators A and B which commute, i.e.,

$$AB = BA. \qquad (2.73)$$

Now suppose we have found the eigenvalues and eigenvectors of the operator A:

$$A \left| \Psi_{(n)} \right\rangle = a_n \left| \Psi_{(n)} \right\rangle. \qquad (2.74)$$

Then,

$$AB \left| \Psi_{(n)} \right\rangle = BA \left| \Psi_{(n)} \right\rangle = a_n B \left| \Psi_{(n)} \right\rangle. \tag{2.75}$$

This shows that $B \left| \Psi_{(n)} \right\rangle$ is also an eigenvector of the operator A, with the same eigenvalue that $\left| \Psi_{(n)} \right\rangle$ has.

To explore the meaning of this statement, suppose that all eigenvalues are distinct (i.e., there is no degeneracy) and non-zero . Then, apart from a possible multiplicative factor, the eigenvector corresponding to a particular eigenvalue is unique. Therefore, $B \left| \Psi_{(n)} \right\rangle$ must be proportional to $\left| \Psi_{(n)} \right\rangle$:

$$B \left| \Psi_{(n)} \right\rangle = b_n \left| \Psi_{(n)} \right\rangle, \tag{2.76}$$

where b_n is a number. This would mean that $\left| \Psi_{(n)} \right\rangle$ is an eigenvector of B as well, i.e., A and B have simultaneous eigenvectors.

If we now relax the condition of distinct eigenvalues, definition of eigenvectors corresponding to a degenerate eigenvalue contains some arbitrariness. Thus, Eq. (2.75) still tells us that $B \left| \Psi_{(n)} \right\rangle$ is an eigenvector of A, but if the corresponding eigenvalue is degenerate then Eq. (2.76) would not necessarily hold. Instead, all we can say is that $B \left| \Psi_{(n)} \right\rangle$ must be a linear combination of all eigenvectors of A that have the same eigenvalue a_n. However, whatever this combination might be, it is also an eigenvector of A with eigenvalue a_n, so the theorem holds in the general form that, if A and B commute, any eigenvector of A is also an eigenvector of B.

Note that we never said that non-commuting operators cannot have simultaneous eigenstates. We show here in passing that two non-commuting operators A and B may have simultaneous zero eigenvalues. This may happen in two distinct ways. For the first case, let $C = AB - BA$. Now consider a state vector $|\psi\rangle$ such that $C|\psi\rangle = B|\psi\rangle = 0$. If such a state vector exists, then it necessarily follows that $A|\psi\rangle = 0$ or $(BA)|\psi\rangle = 0$. This means that B and either A or BA have a simultaneous zero eigenstate $|\psi\rangle$; it is to be noted that we have nowhere assumed that B commutes with either A or BA.

The second case constitutes a slightly more non-trivial possibility. Let us consider an operator S which has the property that, for two operators A and B,

$$AS = SA, \quad BS = -SB. \tag{2.77}$$

Then if a state vector $|\psi\rangle$ is a zero eigenstate of S and if S does not annihilate the states $A|\psi\rangle$ or $B|\psi\rangle$, then $|\psi\rangle$ is also a zero eigenstate of both A and B.

□ **Exercise 2.33** *Two operators A and B anticommute (i.e., satisfy the relation $AB + BA = 0$), and have simultaneous eigenvalues a and b corresponding to an eigenstate $|\psi\rangle$. Show that either a or b must be zero.*

2.8   Adjoint of an operator

2.8.1   Definition

Corresponding to any operator $\mathcal{O}$, one defines another operator $\mathcal{O}^\dagger$, called the *adjoint* or the *Hermitian conjugate* of $\mathcal{O}$, which has the property that

$$\langle u\,|\,\mathcal{O}\,|\,v\rangle^* \equiv \langle v\,|\,\mathcal{O}^\dagger\,|\,u\rangle \qquad (2.78)$$

for any two vectors $|u\rangle$ and $|v\rangle$. Using the notation introduced in Eq. (2.45), we can rewrite this definition in the form

$$\langle v\,|\,\mathcal{O}^\dagger\,|\,u\rangle = \langle u\,|\,\mathcal{O}v\rangle^* = \langle \mathcal{O}v\,|\,u\rangle\,, \qquad (2.79)$$

using Eq. (2.16), which forms part of the definition of the inner product. Since this relation is valid for any vector $|u\rangle$, we can omit it and write

$$\langle v|\,\mathcal{O}^\dagger = \langle \mathcal{O}v|\,, \qquad (2.80)$$

which can also be used as the definition of the adjoint operator.

The matrix elements of the adjoint operator are related simply with the matrix elements of the original operator. If we take the two vectors in Eq. (2.78) as the basis vectors, we obtain

$$\left(\mathcal{O}^\dagger\right)_{ij} = \left(\mathcal{O}_{ji}\right)^*. \qquad (2.81)$$

We can also see what is the adjoint of a product of two operators. Using Eq. (2.33) for an orthonormal basis, we can write

$$\begin{aligned}
\langle u\,|\,(AB)^\dagger\,|\,v\rangle &= \langle v\,|\,AB|\,u\rangle^* \\
&= \sum_k \left(\langle v\,|\,A|\,k\rangle\langle k\,|\,B|\,u\rangle\right)^* \\
&= \sum_k \langle k\,|\,A^\dagger\,|\,v\rangle\langle u\,|\,B^\dagger\,|\,k\rangle. \qquad (2.82)
\end{aligned}$$

The ordering of the two factors is not important here since both are numbers. Therefore,

$$\begin{aligned}
\langle u\,|\,(AB)^\dagger\,|\,v\rangle &= \sum_k \langle u\,|\,B^\dagger\,|\,k\rangle\,\langle k\,|\,A^\dagger\,|\,v\rangle \\
&= \langle u\,|\,B^\dagger A^\dagger\,|\,v\rangle\,, \qquad (2.83)
\end{aligned}$$

using Eq. (2.33) once again. Since this relation is true for any pair of vectors, we can write the operator relation

$$\left(AB\right)^{\dagger} = B^{\dagger}A^{\dagger}. \tag{2.84}$$

The same word *adjoint*, or alternatively *adjugate* or *adjunct*, is used in the context of finding the determinant of a square matrix. That is a different use of the word, and has nothing to do with what we have discussed here.

□ **Exercise 2.34** *Show that, for any operator* $\mathcal{O}$,

$$\left(\mathcal{O}^{\dagger}\right)^{\dagger} = \mathcal{O}. \tag{2.85}$$

□ **Exercise 2.35** *If* A, B *are operators and* α, β *are numbers (complex, to be general), then show that*

$$\left(\alpha A + \beta B\right)^{\dagger} = \alpha^* A^{\dagger} + \beta^* B^{\dagger}. \tag{2.86}$$

□ **Exercise 2.36** *Show that*

$$\left(\left|u\right\rangle\left\langle v\right|\right)^{\dagger} = \left|v\right\rangle\left\langle u\right|. \tag{2.87}$$

[**Hint:** *Call* $\left|u\right\rangle\left\langle v\right| = A$ *and use the definition of Eq. (2.78).*]

2.8.2 Normal operators

One of the possible uses of eigenvectors is for defining a basis in a vector space. This cannot be done with the eigenvectors of any operator. In an N-dimensional space, one would need N eigenvectors to define a basis. Not all operators have N eigenvectors, as emphasized through an example given in Ex. 2.29 (p. 35). So we need to restrict ourselves to the class of operators which have N eigenvectors. If all of these eigenvectors correspond to non-degenerate eigenvalues, they are linearly independent and can form a basis. If some eigenvalues are degenerate, the corresponding eigenvectors form a vector space, as indicated in Ex. 2.32 (p. 36), so one can find a set of orthogonal vectors in this subspace through the Gram-Schmidt procedure. Thus, if N eigenvectors exist, they can be used as a basis. But this basis need not be orthogonal. A *normal operator* in an N-dimensional vector space is defined to be an operator which has N mutually orthogonal eigenvectors. We will take them to be normalized and denote them by $\left|\Lambda_{(i)}\right\rangle$, for $i = 1, 2, \cdots, N$. Therefore, we can write

$$\sum_i \left|\Lambda_{(i)}\right\rangle \left\langle\Lambda_{(i)}\right| = \mathbb{1}. \tag{2.88}$$

This is the same as the definition of an orthonormal basis given in Eq. (2.33), except that now we have in mind a specific orthonormal basis consisting of the eigenvectors of a normal operator A. Operating both sides by A, we obtain

$$\begin{aligned} A &= A \sum_i \left|\Lambda_{(i)}\right\rangle \left\langle\Lambda_{(i)}\right| = \sum_i A \left|\Lambda_{(i)}\right\rangle \left\langle\Lambda_{(i)}\right| \\ &= \sum_i \lambda_i \left|\Lambda_{(i)}\right\rangle \left\langle\Lambda_{(i)}\right|. \end{aligned} \tag{2.89}$$

This shows how the operator can be specified completely in terms of its eigenvalues and eigenvectors. If an operator is not normal, the corresponding relation is somewhat more complicated, but we need not get into that because we will be exclusively using normal operators.

Later in the book, when we discuss infinite-dimensional vector spaces, we will have to make some obvious adjustments in the notations. For example, at least the sum over n has to be turned into an integral over a variable that takes different values for different eigenvectors.

An important property of normal operators is worth mentioning here. Taking the adjoint of Eq. (2.89) and using Eqs. (2.86) and (2.87), we obtain

$$A^\dagger = \sum_i \lambda_i^* \left|\Lambda_{(i)}\right\rangle \left\langle\Lambda_{(i)}\right|. \tag{2.90}$$

In that case,

$$\begin{aligned} AA^\dagger &= \left(\sum_i \lambda_i \left|\Lambda_{(i)}\right\rangle \left\langle\Lambda_{(i)}\right|\right) \left(\sum_j \lambda_j^* \left|\Lambda_{(j)}\right\rangle \left\langle\Lambda_{(j)}\right|\right) \\ &= \sum_i \sum_j \lambda_i \lambda_j^* \left|\Lambda_{(i)}\right\rangle \left\langle\Lambda_{(i)}\,\middle|\,\Lambda_{(j)}\right\rangle \left\langle\Lambda_{(j)}\right|. \end{aligned} \tag{2.91}$$

Using the orthonormality condition of the eigenvectors, we can sum over j and obtain

$$AA^\dagger = \sum_i \left|\lambda_i\right|^2 \left|\Lambda_{(i)}\right\rangle \left\langle\Lambda_{(i)}\right|. \tag{2.92}$$

However, by following more or less similar steps, we can get to the conclusion that $A^\dagger A$ also equals to the same expression. Therefore, for normal operators,

$$AA^\dagger = A^\dagger A, \tag{2.93}$$

or, in other words,

$$\left[A, A^\dagger\right] = 0, \tag{2.94}$$

i.e., a normal operator commutes with its adjoint.

Commuting operators share eigenvectors. Therefore, the eigenvectors of A can also be eigenvectors of $A^\dagger$. Eq. (2.90) then tells us that the eigenvalues of $A^\dagger$ are complex conjugates of the eigenvalues of A.

> There is a little bit of sloppiness in the way that we have written Eq. (2.94). The product of two operators is an operator. The sum or difference of two such products is therefore also an operator. So, the right side of Eq. (2.94) should have been the null operator defined in Eq. (2.43), not the number zero.
>
> Indeed, that is what it should have been. We have taken a little liberty in the notation because the null operator acting on any vector gives the null vector, and the number zero multiplied by any vector gives the same. So, even if one forgets the distinction between the null operator and the number zero, no harm will be done in any deduction that involves the matrix elements of both sides.

Two obvious classes of normal operators come to mind immediately. First, if the adjoint of an operator is equal to the operator itself, i.e., if $\mathscr{O} = \mathscr{O}^\dagger$, then $\mathscr{O}$ must be a normal operator. Such an operator is called a *Hermitian operator*, and will be discussed in §2.9 in some detail. The other kind contains operators whose adjoints are equal to their inverses, i.e., $\mathscr{O}^\dagger = \mathscr{O}^{-1}$. These are called *unitary operators*, and will be discussed in §2.10. Both these kinds of operators are very important for quantum mechanics, as we will see in subsequent chapters of this book.

We have defined normal operators as operators whose eigenvectors form an orthonormal complete set, and showed that such operators must commute with their adjoints. In view of the paramount importance of such operators in the study of quantum mechanics, we now prove the converse, i.e., the statement that if an operator satisfies Eq. (2.94), then its eigenvectors form an orthonormal set.

For this, we use the fact that commuting operators share eigenstates, as shown in §2.7.3. Therefore, if

$$A\left|\Lambda_{(i)}\right\rangle = \lambda_i \left|\Lambda_{(i)}\right\rangle, \tag{2.95}$$

we can write the eigenvalue equation for $A^\dagger$ as

$$A^\dagger \left|\Lambda_{(i)}\right\rangle = \lambda_i^* \left|\Lambda_{(i)}\right\rangle. \tag{2.96}$$

Therefore,

$$\left\langle\Lambda_{(j)}\left|A\right|\Lambda_{(i)}\right\rangle = \lambda_i \left\langle\Lambda_{(j)}\left|\Lambda_{(i)}\right.\right\rangle \tag{2.97}$$

and

$$\left\langle\Lambda_{(i)}\left|A^\dagger\right|\Lambda_{(j)}\right\rangle = \lambda_j^* \left\langle\Lambda_{(i)}\left|\Lambda_{(j)}\right.\right\rangle. \tag{2.98}$$

Taking the complex conjugate of this equation and using Eq. (2.78), we obtain

$$\left\langle\Lambda_{(j)}\left|A\right|\Lambda_{(i)}\right\rangle = \lambda_j \left\langle\Lambda_{(j)}\left|\Lambda_{(i)}\right.\right\rangle. \tag{2.99}$$

Subtracting Eq. (2.99) from Eq. (2.97), we obtain

$$0 = (\lambda_i - \lambda_j) \left\langle\Lambda_{(j)}\left|\Lambda_{(i)}\right.\right\rangle. \tag{2.100}$$

Thus, eigenvectors corresponding to different eigenvalues are orthogonal. This, in fact, is all we wanted to prove. The vectors can always be normalized, and eigenvectors corresponding to degenerate eigenvalues can be chosen to be orthogonal to each other, as outlined earlier. In any defenerate subspace, one can choose an orthonormal basis through the Gram–Schmidt method.

☐ **Exercise 2.37** *Show that the most general 2×2 normal matrix has the form*

$$\begin{pmatrix} a & re^{i\theta} \\ re^{i\phi} & b \end{pmatrix}, \tag{2.101}$$

where a, b can be complex, whereas the other parameters are real, and either $a = b$ or $2\arg(a - b) = \theta + \phi$.

2.9   Hermitian operators

A special subclass of normal operators, called *Hermitian operators*, will play a very important role in the study of quantum mechanics. Here we discuss such operators in some detail. This discussion will not only equip us in dealing with mathematical manipulation involving such operators but also indicate why they are important for physics.

2.9.1 Definition and basic properties

Consider an operator A. When it acts on a vector $|u\rangle$, the result of the operation is a vector. According to the notation introduced in Eq. (2.45), this vector should be called $|Au\rangle$. Similarly, starting from a vector $|v\rangle$, we can define a vector $|Av\rangle$. The operator A is called *Hermitian* if the relation

$$\langle v \,|\, Au\rangle = \langle Av \,|\, u\rangle \qquad (2.102)$$

is satisfied for any two vectors $|u\rangle$ and $|v\rangle$. Recalling the notation that we introduced in Eq. (2.45), we can write the left side of Eq. (2.102) in an alternative form:

$$\langle v \,|\, Au\rangle = \langle v \,|A|\, u\rangle . \qquad (2.103)$$

As for the expression on the right side of Eq. (2.102), we use property 3 in the definition of inner products in §2.3.2 and write

$$\langle Av \,|\, u\rangle = \langle u \,|\, Av\rangle^{*} = \langle u \,|A|\, v\rangle^{*}. \qquad (2.104)$$

Thus, a Hermitian operator is one whose matrix elements satisfy the property

$$\langle v \,|A|\, u\rangle = \langle u \,|A|\, v\rangle^{*} \qquad (2.105)$$

for any vectors $|u\rangle$ and $|v\rangle$. Using the definition of the adjoint of an operator given in Eq. (2.78), we can write this equation as

$$\langle v \,|A|\, u\rangle \equiv \langle v \,\big|A^{\dagger}\big|\, u\rangle . \qquad (2.106)$$

Since this should be obeyed for any two vectors $|u\rangle$ and $|v\rangle$, we can say that the two operators must be the same:

$$A = A^{\dagger}. \qquad (2.107)$$

Such operators are therefore normal operators, since any operator commutes with itself.

2.9.2 Eigenvalues and eigenvectors

Hermitian operators are interesting because of their eigenvalues and eigenvectors. Consider a Hermitian operator A on a finite-dimensional vector space. The number of eigenvectors will be finite, and let us denote them by

$|n\rangle$, with $n = 0, 1, 2, \ldots$ enumerating the states. Let the eigenvalue of the state $|n\rangle$ be a_n, i.e.,

$$A|n\rangle = a_n |n\rangle. \qquad (2.108)$$

Taking the inner product of both sides with $|n\rangle$, we obtain

$$\langle n|A|n\rangle = a_n \langle n|n\rangle. \qquad (2.109)$$

By Eq. (2.105), the left side of this equation must be real. On the right side, the inner product is a norm, which is real. Thus, a_n must be real, implying that any eigenvalue of a Hermitian operator is real. In addition, the eigenvectors of Hermitian matrices corresponding to unequal eigenvalues are orthogonal, a result that is obeyed by any normal operator, as shown in §2.8.

The converse of the statement on eigenvalues is not true, meaning that it is possible to find non-Hermitian operators all of whose eigenvalues are real. While some such operators will be discussed in §17.6, here we can mention that if an operator has only real eigenvalues *and* is a normal operator, then it must be Hermitian. This follows from Eq. (2.89), when we take the matrix element of both sides between two arbitrary vectors $|u\rangle$ and $|v\rangle$. Using Eq. (2.88) that applies for a normal matrix, we can write

$$\langle u|\mathcal{O}|v\rangle = \sum_n \langle u|\mathcal{O}|n\rangle \langle n|v\rangle = \sum_n \lambda_n \langle u|n\rangle \langle n|v\rangle. \qquad (2.110)$$

Similarly,

$$\langle v|\mathcal{O}|u\rangle = \sum_n \lambda_n \langle v|n\rangle \langle n|u\rangle$$

$$= \sum_n \lambda_n \langle u|n\rangle^* \langle n|v\rangle^*, \qquad (2.111)$$

using the complex conjugation property of inner products. Complex conjugate of this equation gives

$$\langle v|\mathcal{O}|u\rangle^* = \sum_n \lambda_n^* \langle u|n\rangle \langle n|v\rangle. \qquad (2.112)$$

If all eigenvalues are real, clearly one sees that the operator $\mathcal{O}$ would satisfy Eq. (2.105), i.e., it will be Hermitian.

Another property of the eigenvectors of Hermitian operators should be noted. Since the Hermitian opertators are a subclass of normal operators, any Hermitian operator in an N-dimensional vector space admits of N mutually orthogonal eigenvectors. We will use this property to set up the basis in many vector spaces that will be discussed.

2.9.3   Probabilistic interpretation of state vectors

In any state Ψ, the expectation value of an operator $\mathscr{O}$ is defined by

$$\langle \mathscr{O} \rangle \equiv \frac{\langle \Psi | \mathscr{O} | \Psi \rangle}{\langle \Psi | \Psi \rangle}. \tag{2.113}$$

Note that Ψ may or may not be an eigenstate of the operator $\mathscr{O}$. If it happens to be an eigenstate, the expectation value is the same as the eigenvalue. If $\mathscr{O}$ is a Hermitian operator, Ψ can be written as superpositions of eigenvectors of $\mathscr{O}$, since we have proved that the eigenvectors of a Hermitian operator form a complete set of vectors. So, we can write

$$|\Psi\rangle = \sum_k a_k |k\rangle, \tag{2.114}$$

where the vectors $|k\rangle$ are the eigenstates of $\mathscr{O}$, meaning that

$$\mathscr{O} |k\rangle = \lambda_k |k\rangle, \tag{2.115}$$

where the λ_k's are the eigenvalues. Therefore,

$$\mathscr{O} |\Psi\rangle = \mathscr{O} \sum_k a_k |k\rangle = \sum_k a_k \mathscr{O} |k\rangle = \sum_k a_k \lambda_k |k\rangle. \tag{2.116}$$

On the other hand,

$$\langle \Psi | = \sum_l a_l^* \langle l |, \tag{2.117}$$

allowing for the possibility that the coefficient a_k can be complex. Therefore,

$$\langle \Psi | \mathscr{O} | \Psi \rangle = \sum_l \sum_k a_l^* a_k \lambda_k \langle l | k \rangle. \tag{2.118}$$

Earlier, we have shown that if $|k\rangle$ and $|l\rangle$ are eigenstates of a Hermitian operator, $\langle l|k \rangle = 0$ unless the two states are the same one. If we normalize the states such that $\langle k|k \rangle = 1$, we can write, in general,

$$\langle l | k \rangle = \delta_{kl}, \tag{2.119}$$

the Kronecker delta. Putting this into Eq. (2.118) and performing the sum over l, we obtain

$$\langle \Psi | \mathscr{O} | \Psi \rangle = \sum_k \lambda_k |a_k|^2. \tag{2.120}$$

Proceeding exactly the same way, we would obtain

$$\langle \Psi \,|\, \Psi \rangle = \sum_k |a_k|^2. \tag{2.121}$$

Therefore, the expectation value of the operator $\mathscr{O}$ in the state Ψ can be rewritten in the form

$$\langle \mathscr{O} \rangle = \frac{\sum_k \lambda_k |a_k|^2}{\sum_k |a_k|^2}. \tag{2.122}$$

This expression shows that the expectation value of a Hermitian operator is real in any state. In addition, there is another very important significance. To understand it, note first that the quantity

$$P_k = \frac{|a_k|^2}{\sum_k |a_k|^2} \tag{2.123}$$

appearing in Eq. (2.122) looks very much like a probability, with

$$\sum_k P_k = 1. \tag{2.124}$$

If the result of an individual measurement of the physical quantity associated with the operator $\mathscr{O}$ always yields only an eigenvalue of the operator, then clearly the expectation value should be given by

$$\langle \mathscr{O} \rangle = \sum_k \lambda_k P_k, \tag{2.125}$$

where P_k would be the probability that the system is found in the state $|k\rangle$. Eq. (2.122) tells us that the probability of finding the system in the eigenstate $|k\rangle$ is given by the expression P_k shown in Eq. (2.123). This is the probabilistic interpretation of the state vectors.

> It should be commented that the vector space structure appears in classical systems also, for example classical electrodynamics, where the wave operator can be Hermitian. Mathematically, the structures will be identical. However, no such probabilistic interpretation is applied there. The probabilistic interpretation comes with the extra assumption in quantum mechanics, viz., that the result of each measurement can only be the eigenvalue of the associated operator.

We can write this probability by using the state vector $|\Psi\rangle$ directly. For this, we first note that, for any eigenstate $|l\rangle$ of the operator $\mathscr{O}$, we can write

$$\langle l \,|\, \Psi \rangle = \sum_k a_k \, \langle l \,|\, k \rangle. \tag{2.126}$$

Using Eq. (2.119), we see that only the term with $k = l$ contributes to the sum, and we obtain

$$a_l = \langle l \,|\, \Psi \rangle \,.$$

(2.127)

Thus, the numerator of Eq. (2.123) is $|\langle k | \Psi \rangle|^2$. For the denominator, we can use Eq. (2.121) and write

$$P_k = \frac{|\langle k | \Psi \rangle|^2}{\langle \Psi | \Psi \rangle} \,.$$

(2.128)

Note that the denominator in Eq. (2.128) is essential when the norm of the state has not been set equal to unity. If we use states with unit norm, we can just say that the probability of being in the eigenstate $|k\rangle$ of an operator $\mathcal{O}$ is given by $|\langle k | \Psi \rangle|^2$. Since the square is involved in the probability, the quantity $\langle k | \Psi \rangle$ is often called the *probability amplitude*, or simply the *amplitude*, of being in the state $|k\rangle$. This use is a bit different from the use of the same word in classical wave theory. In classical theory, one characterizes a wave with amplitude and phase. The word *amplitude* is differently used in quantum mechanics: it is in general complex and contains phase information.

This completes the probabilistic interpretation of the state vector. Whether or not one normalizes the states to unity, there is one important point to remember. The probabilistic interpretation requires that $\langle \Psi | \Psi \rangle$ is finite, a statement that is succinctly written as

$$\langle \Psi \,|\, \Psi \rangle < \infty.$$

(2.129)

For any finite-dimensional vector space, this condition is automatically satisfied. But if we are talking about an infinite-dimensional vector space, this condition will put important restrictions on the state vectors, which will be discussed later.

□ **Exercise 2.38** *Consider the following operator in a 2D vector space:*

$$\sigma \equiv \begin{pmatrix} 1 & 0 \\ 0 & -1 \end{pmatrix} .$$

(2.130)

(a) *Find the eigenvalues and eigenvectors of this operator.*

(b) *Find the expectation value of this operator in the state $\frac{1}{\sqrt{2}} \binom{1}{1}$.*

(c) *Interpret your result in terms of eigenvalues and probabilities.*

2.9.4 A relation between standard deviations

Consider two Hermitian operators A and B. We said before that operators on a vector space do not necessarily commute. We can define their *commutator* as follows:

$$\left[A, B \right] \equiv AB - BA = iC \tag{2.131}$$

for some operator C. The factor of i on the right side is notational: it merely ensures that C is also a Hermitian operator.

□ **Exercise 2.39** *In Eq. (2.131), show that C is Hermitian if A and B are.*

In an arbitrary state $|\Psi\rangle$, the *standard deviation* of A is defined as

$$\Delta A \equiv \sqrt{\left\langle \Psi \left| (A - \langle A \rangle)^2 \right| \Psi \right\rangle \Big/ \langle \Psi | \Psi \rangle}, \tag{2.132}$$

where

$$\langle A \rangle = \langle \Psi |A| \Psi \rangle \Big/ \langle \Psi | \Psi \rangle, \tag{2.133}$$

the expectation value of the operator A in that state. Then one can prove that

$$\Delta A \, \Delta B \geqslant \frac{1}{2} \left| \langle C \rangle \right|. \tag{2.134}$$

Proof: Consider two vectors $|u\rangle$ and $|v\rangle$ which are both normalized. By the general postulates of a vector space, $|ru + v\rangle$ is also a vector for an arbitrary real parameter r. Then

$$\langle ru + v \, | \, ru + v \rangle = r^2 \langle u | u \rangle + r(\langle u | v \rangle + \langle v | u \rangle) + \langle v | v \rangle. \tag{2.135}$$

If $|u\rangle$ and $|v\rangle$ are not proportional, i.e., there does not exist any scalar r for which $|ru + v\rangle$ is the null vector, then expression on the right side of Eq. (2.135) is the norm of a non-zero vector and hence must be positive. This means that the expression on the right side of Eq. (2.135) cannot be equal to zero for any real value of r. This expression, as we see, is a quadratic polynomial in r, and the condition for its not having any real root is

$$\langle u | u \rangle \langle v | v \rangle > \frac{1}{4}(\langle u | v \rangle + \langle v | u \rangle)^2. \tag{2.136}$$

If, on the other hand, $|u\rangle$ is proportional to $|v\rangle$, the two sides of Eq. (2.136) are easily seen to be equal. So, for unrestricted $|u\rangle$ and $|v\rangle$, we should write

$$\langle u | u \rangle \langle v | v \rangle \geqslant \frac{1}{4}(\langle u | v \rangle + \langle v | u \rangle)^2, \tag{2.137}$$

or

$$\sqrt{\langle u | u \rangle \langle v | v \rangle} \geqslant \frac{1}{2} \left| \langle u | v \rangle + \langle v | u \rangle \right|. \tag{2.138}$$

Now, with the operators A and B, define

$$|u\rangle = (A - \langle A \rangle)|\Psi\rangle, \qquad |v\rangle = -i(B - \langle B \rangle)|\Psi\rangle. \tag{2.139}$$

For an arbitrary operator, this would mean that

$$\langle u| = \langle\Psi|\left(A^\dagger - \langle A \rangle^*\right), \qquad \langle v| = i\langle\Psi|\left(B^\dagger - \langle B \rangle^*\right). \tag{2.140}$$

Since we are interested in the case where A is Hermitian, we can put $A^\dagger = A$ and use the fact that $\langle A \rangle$ is real. Using also similar relations regarding B, we can write

$$\langle u \, | \, u \rangle = (\Delta A)^2, \qquad \langle v \, | \, v \rangle = (\Delta B)^2. \tag{2.141}$$

In writing these equations and whatever follows of this proof, we assume $|\Psi\rangle$ is normalized to unity, so that we can omit the denominators of expressions like in Eqs. (2.132) and (2.133). If $|\Psi\rangle$ is not normalized, the modifications would be trivial. Also,

$$\langle u \, | \, v \rangle = -i\langle\Psi|(A - \langle A \rangle)(B - \langle B \rangle)|\Psi\rangle = -i\langle\Psi|AB|\Psi\rangle + i\langle A \rangle\langle B \rangle,$$
$$\langle v \, | \, u \rangle = i\langle\Psi|(B - \langle B \rangle)(A - \langle A \rangle)|\Psi\rangle = i\langle\Psi|BA|\Psi\rangle - i\langle A \rangle\langle B \rangle, \tag{2.142}$$

so

$$\langle u \, | \, v \rangle + \langle v \, | \, u \rangle = -i\langle\Psi|AB - BA|\Psi\rangle. \tag{2.143}$$

Putting this back into Eq. (2.138) and using the notation of the commutator from Eq. (2.131), we obtain the inequality of Eq. (2.134).

2.9.5 Importance of Hermitian operators

From the discussion so far in this section, we have found many indications of the importance of Hermitian operators for dealing with physics problems. Let us try to collect all the indications and guess which way they point.

We start from the indication that we have obtained most recently. We found that the product of standard deviations of two Hermitian operators cannot be reduced to arbitrarily small values if the operators do not commute. In Chapter 1, we discussed that the uncertainties in measurements of position and momentum cannot be arbitrarily reduced. The similarity between the two statements is striking, and it suggests that dynamical variables like position and momentum can probably be thought of as Hermitian operators on the vector space of states of a system.

This suggestion is reinforced by the result that the expectation value of a Hermitian operator is a real number in any state. It would be tempting, or almost imperative, to suppose that the expectation value of an operator would correspond to the average of measured values if one tries to measure the dynamical variable represented by that operator. Of course, the result of any measurement is a real number, so the expectation value must be real in any state.

Once we accept the physical interpretation of the expectation value, the analysis of §2.9.3 leads us to conclude that the result of a singular measurement is always an eigenvalue of an operator. Fortunately, all eigenvalues of a Hermitian operator are real, which is a necessity for the suggested physical interpretation to make sense. A measurement process will yield a particular eigenvalue with a certain probability that was given in §2.9.3.

☐ **Exercise 2.40** *Consider the following two statements about an operator A.*

- *All eigenvalues of A are real.*
- *The expectation value of A is real in any state.*

Show that these two statements are equivalent, i.e., each can be derived from the other.

☐ **Exercise 2.41** *Show that the determinant of any Hermitian matrix is real.*

☐ **Exercise 2.42** *Show that if a Hermitian matrix has an inverse, the inverse is also Hermitian.*

☐ **Exercise 2.43** *Find the condition under which the product of two Hermitian operators A and B is also Hermitian.*

2.10 Unitary operators

For reasons to become clear presently, another kind of operators are very important in quantum mechanics. In the notation introduced in Eq. (2.45), an operator U is called a *unitary operator* if it satisfies the property

$$\langle Ua \,|\, Ub \rangle = \langle a \,|\, b \rangle \tag{2.144}$$

for any two vectors $|a\rangle$ and $|b\rangle$. In other words, a unitary operation preserves the inner product of any two vectors: the inner product of the original vectors is equal to the inner product of the vectors obtained after applying the operation.

We can write the definition in an alternative manner without using any vector. For this, note that

$$\langle Ua \,|\, Ub \rangle = \langle Ub \,|\, Ua \rangle^* = \langle b \,|\, U^\dagger \,|\, Ua \rangle^* = \langle b \,|\, U^\dagger U \,|\, a \rangle^*, \tag{2.145}$$

using the definition of the Hermitian conjugate of an operator as given in Eq. (2.80). The right side of Eq. (2.144), on the other hand, can be written as $\langle b|a\rangle^*$, using the basic property of inner products. Comparing both sides of

Eq. (2.144) and remembering that the equation must hold for arbitrary vectors $|a\rangle$ and $|b\rangle$, we can write

$$U^\dagger U = \mathbb{1}, \tag{2.146}$$

where the symbol $\mathbb{1}$, as before, denotes the *identity operator*.

Of course, Eq. (2.146) implies that $U^\dagger = U^{-1}$, so the equation can also be written as

$$UU^\dagger = \mathbb{1}, \tag{2.147}$$

which is an equivalent definition of a unitary operator. Obviously, a unitary operator is a normal operator, so it has an orthonormal set of eigenvectors.

Possible eigenvalues of a unitary operator can be deduced easily. An eigenstate of a unitary operator satisfies the property

$$U|n\rangle = \lambda_n |n\rangle, \tag{2.148}$$

where λ_n is the eigenvalue. Taking the Hermitian conjugate of this equation in the manner shown in Eq. (2.80), we obtain

$$\langle n| U^\dagger = \langle Un| = \langle \lambda_n n| = \lambda_n^* \langle n|. \tag{2.149}$$

Multiplying these two equations, we obtain

$$\langle n |U^\dagger U| n\rangle = |\lambda_n|^2 \langle n|n\rangle. \tag{2.150}$$

Using Eq. (2.146) now, we see that

$$|\lambda_n|^2 = 1, \tag{2.151}$$

which means that any eigenvalue of a unitary operator has modulus equal to unity, i.e., is of the form $e^{i\theta}$ for some real θ.

Let us now discuss why unitary operators are important for the formulation of physical problems. As Eq. (2.144) indicates, the inner product of two vectors remains unchanged under a unitary transformation applied on the vectors. So, unitary operators are important for studying symmetry properties of matrix elements. Moreover, it is easily seen that if we consider a Hermitian operator A, the operator denoted by e^{iA} would be unitary. In Chapter 3, we will see that the equations for time evolution of a system are associated with the operator

$$U = \exp(-iHt/\hbar), \tag{2.152}$$

where H is the Hamiltonian of the system and t is the time. Since H must be a Hermitian operator, this operator must be unitary, and this constitutes another reason for the importance of unitary operators.

Lastly, unitary operators are important for finding a basis in which a normal operator is diagonal. To see how this is possible, let us start with two different sets of orthonormal vectors in the same vector space. One of them will be denoted by $E_{(i)}$ for $i = 1, 2, \cdots, N$, where N is the dimension of the vector space, and the other will be denoted by $e_{(i)}$ for the same range of the index i. Now let us define the operator

$$U = \sum_i |E_{(i)}\rangle \langle e_{(i)}| . \tag{2.153}$$

Using Eq. (2.87), we then obtain

$$U^\dagger = \sum_i |e_{(i)}\rangle \langle E_{(i)}| . \tag{2.154}$$

Then,

$$UU^\dagger = \sum_i \sum_j |E_{(i)}\rangle \underbrace{\langle e_{(i)} | e_{(j)}\rangle}_{=\,\delta_{ij}} \langle E_{(j)}|$$

$$= \sum_i |E_{(i)}\rangle \langle E_{(i)}| = \mathbb{1}. \tag{2.155}$$

In an exactly similar manner, we can show that $U^\dagger U$ is also equal to the identity operator. Together, the two results mean that U is a unitary operator.

Now suppose the $|E\rangle$'s are the eigenvectors of a normal matrix A. By the definition of normal matrices, they form an orthonormal set, and we can write

$$A = \sum_i \lambda_i |E_{(i)}\rangle \langle E_{(i)}| , \tag{2.156}$$

where the λ's are the eigenvalues. Consider now the operator $U^\dagger A U$.

$$U^\dagger A U = \sum_i \sum_j \sum_k \lambda_j |e_{(i)}\rangle \langle E_{(i)} | E_{(j)}\rangle \langle E_{(j)} | E_{(k)}\rangle \langle e_{(k)}|$$

$$= \sum_i \sum_j \sum_k \lambda_j |e_{(i)}\rangle \delta_{ij} \delta_{jk} \langle e_{(k)}|$$

$$= \sum_i \lambda_i |e_{(i)}\rangle \langle e_{(i)}| . \tag{2.157}$$

This means that if we consider the matrix element of this operator $U^\dagger A U$ between two basis vectors in the $|e\rangle$ basis, we would obtain

$$\langle e_{(j)} | U^\dagger A U | e_{(k)} \rangle = \sum_i \lambda_i \langle e_{(j)} | e_{(i)} \rangle \langle e_{(i)} | e_{(k)} \rangle$$

$$= \sum_i \lambda_i \delta_{ij} \delta_{ik} = \lambda_j \delta_{jk}. \qquad (2.158)$$

The off-diagonal elements are all zero. Hence, this is a diagonal operator. The operator U is called the *diagonalizing operator* and can be obtained from the eigenvectors through the definition of Eq. (2.153).

☐ **Exercise 2.44** *Show that any unitary matrix U satisfies $|\mathrm{Tr}(U)| \leq n$, where n is the dimensionality of U.*

☐ **Exercise 2.45** *Show that the product of two unitary operators is also a unitary operator.*

☐ **Exercise 2.46** *Show that for any diagonalizable operator A represented by a square matrix,*

$$\det(e^A) = \exp[\mathrm{tr}(A)], \qquad (2.159)$$

where 'det' indicates the determinant and 'tr' indicates the trace of a matrix. [**Hint:** *First prove it in the basis in which A is diagonal.*]

☐ **Exercise 2.47** *Prove the statement made after Eq. (2.152), that U will be unitary if H is Hermitian.*

2.11 Basis and basis-independence

In any vector space, the basis vectors cannot be chosen uniquely. Thus, physical implications of vectors must be independent of any basis used for writing them.

Let us therefore check how things change with the choice of basis. Suppose we have two sets of basis vectors, one denoted by $e_{(i)}$ and the other by $e'_{(i)}$. Let us assume that the members of each set are orthonormal. Let us suppose that the vectors in the two sets are related by

$$|e'_{(i)}\rangle = \sum_j U_{ij} |e_{(j)}\rangle. \qquad (2.160)$$

It can then be easily shown that the matrix U, formed by these coefficients U_{ij}, is a unitary matrix.

A vector $|v\rangle$ can be written in terms of components in either basis:

$$|v\rangle = \sum_j v_j \left|e_{(j)}\right\rangle = \sum_i v_i' \left|e_{(i)}'\right\rangle. \tag{2.161}$$

Using Eq. (2.160) now, we can write

$$\sum_j v_j \left|e_{(j)}\right\rangle = \sum_{i,j} v_i' U_{ij} \left|e_{(j)}\right\rangle. \tag{2.162}$$

Since the basis vectors are independent, this means the components in the two systems are related by

$$v_j = \sum_j v_i' U_{ij}. \tag{2.163}$$

The components are different in a different basis, as expected.

Let us now see the same thing for the matrix elements of an operator. In analogy with Eq. (2.162), for an operator $\mathscr{O}$ we can write

$$\mathscr{O} = \sum_{k,l} O_{kl} \left|e_{(k)}\right\rangle \left\langle e_{(l)}\right| = \sum_{i,j} O_{ij}' \left|e_{(i)}'\right\rangle \left\langle e_{(j)}'\right|. \tag{2.164}$$

In order to proceed, we note that Eq. (2.160) implies

$$\left\langle e_{(i)}'\right| = \sum_j (U_{ij})^* \left\langle e_{(j)}\right|. \tag{2.165}$$

Putting this into Eq. (2.164), we obtain

$$\sum_{k,l} O_{kl} \left|e_{(k)}\right\rangle \left\langle e_{(l)}\right| = \sum_{i,j} \sum_{k,l} O_{ij}' U_{ik}(U_{jl})^* \left|e_{(k)}\right\rangle \left\langle e_{(l)}\right|, \tag{2.166}$$

so that we obtain the following relation for the components in the two systems:

$$O_{kl} = \sum_{i,j} O_{ij}' U_{ik}(U_{jl})^*. \tag{2.167}$$

So, if the array containing the matrix elements of $\mathscr{O}$ is called O in the unprimed basis and O$'$ in the primed basis, the relation between them can be written as

$$\mathrm{O} = U^\top \mathrm{O}' U^*, \tag{2.168}$$

where $U^\top$ denotes the transpose of the matrix U, defined by the relation

$$\left\langle u \left|U^\top\right| v\right\rangle = \left\langle v \left|U\right| u\right\rangle \tag{2.169}$$

for any two vectors $|u\rangle$ and $|v\rangle$. Obviously, the elements of the array O are basis-dependent.

However, it should be noted that combinations of the form $\sum_{i,j} u_i^* O_{ij} v_j$ are independent of the choice of basis for any operator $\mathscr{O}$ and any vectors $|u\rangle$ and $|v\rangle$. This is easily seen by combining Eqs. (2.163) and (2.167), and using the property that the matrix U is unitary. More directly, we can say that the combination is basis-independent because

$$\sum_{i,j} u_i^* O_{ij} v_j = \langle u | \mathscr{O} | v \rangle, \tag{2.170}$$

and the right side of this equation makes no reference to the basis vectors at all.

To summarize, all physically measurable consequences of the state vectors must involve combinations like $\langle u | \mathscr{O} | v \rangle$. In its most general form, it represents the matrix element of an operator between two different state vectors. As special cases, it includes the following kinds of combinations as well:

1. Inner product of two state vectors, obtained from the general form when $\mathscr{O}$ is the identity operator.

2. Expectation value of an operator in a state, obtained when $|u\rangle = |v\rangle$ in the general form.

3. Eigenvalues of operators, obtained when the expectation value is taken in an eigenstate of the operator $\mathscr{O}$.

□ **Exercise 2.48** *Prove the statement made in the text that Eq. (2.160) defines a unitary matrix U.*

Chapter 3

Quantum dynamics

In this chapter, we will formulate the basic equations of quantum mechanics that govern the time evolution of a quantum system.

3.1 Quantum bracket

In Chapter 1, we discussed several formulations for classical mechanics. We now see that none of them is suitable for quantum mechanics because they involve position, velocity or momentum, none of which can be defined for a particle because of the uncertainty principle.

What is worse, the discussion of Chapter 2 seems to indicate that dynamical variables should be represented by operators on a vector space. The Lagrangian and the Hamiltonian formalisms require taking derivatives with respect to dynamical variables. One cannot take derivatives with respect to an operator.

Faced with these problems, we may wonder if there is a formulation of classical mechanics where one does not have to take derivatives with respect to position and other dynamical variables. Indeed, such a formalism exists, and it involves constructs called *Poisson brackets*.

In classical mechanics, any dynamical variable F, i.e., any function F of positions x and momenta p, evolves according to the rule

$$\frac{dF}{dt} = \frac{\partial F}{\partial t} + \left[F, H\right]_P, \tag{3.1}$$

where H is the Hamiltonian of the system (which, by the way, is a dynamical variable itself), and the Poisson bracket for any two dynamical variables is defined to be

$$\left[A, B\right]_P = \sum_a \left(\frac{\partial A}{\partial x_a} \frac{\partial B}{\partial p_a} - \frac{\partial A}{\partial p_a} \frac{\partial B}{\partial x_a} \right), \tag{3.2}$$

where x_a indicate generalized coordinates, p_a the corresponding momenta, and the sum is over all such generalized coordinates.

☐ **Exercise 3.1** *Verify that the Hamilton's equations of motion,*

$$\dot{x}_a = \frac{\partial H}{\partial p_a}, \qquad \dot{p}_a = -\frac{\partial H}{\partial x_a}, \tag{3.3}$$

follow trivially from Eqs. (3.1) and (3.2).

☐ **Exercise 3.2** *For an arbitrary dynamical variable $F(x, p)$ whose arguments can be all coordinates and all momenta, use the Hamilton's equations and the chain rule of differentiation to arrive at Eq. (3.1).*

Obviously, this definition is meaningless in the realm of quantum mechanics, where x_a and p_a are operators in a Hilbert space. To obtain a suitable modification of the definition that can be used in quantum mechanics, let us first note that the Poisson bracket, defined above, has the following properties:

$$\left[A, B\right]_P = -\left[B, A\right]_P, \tag{3.4a}$$

$$\left[A, B_1 + B_2\right]_P = \left[A, B_1\right]_P + \left[A, B_2\right]_P, \tag{3.4b}$$

$$\left[A_1 + A_2, B\right]_P = \left[A_1, B\right]_P + \left[A_2, B\right]_P, \tag{3.4c}$$

$$\left[A, B_1 B_2\right]_P = \left[A, B_1\right]_P B_2 + B_1 \left[A, B_2\right]_P, \tag{3.4d}$$

$$\left[A_1 A_2, B\right]_P = A_1 \left[A_2, B\right]_P + \left[A_1, B\right]_P A_2. \tag{3.4e}$$

And, in addition, if there is a constant K, which can be seen as a special case of a dynamical variable, we should have

$$\left[A, K\right]_P = 0 \tag{3.5}$$

for any dynamical variable A. These are the properties that can be used to find all Poisson brackets, starting from the basic brackets

$$\left[x_a, x_b\right]_P = 0, \qquad \left[p_a, p_b\right]_P = 0, \qquad \left[x_a, p_b\right]_P = \delta_{ab}. \tag{3.6}$$

The important point is that as long as one needs to take Poisson brackets between dynamical variables which involve only positive integral powers of position and momenta, it is not necessary to invoke the definition of Eq. (3.2). One can simply start from the basic Poisson brackets given in Eq. (3.6) and use the algebraic properties listed in Eq. (3.4). The procedure has been indicated in Ex. 3.4.

☐ **Exercise 3.3** *Show that Eqs. (3.4c) and (3.4e) are not independent: they can be derived from the other relations in Eq. (3.4).*

☐ **Exercise 3.4** *By employing the basic brackets in Eq. (3.6) and the properties given in Eq. (3.4) and not invoking the definition of Eq. (3.2) at any stage, show that*

$$\left[x^n, p\right]_P = nx^{n-1}, \tag{3.7a}$$

$$\left[x, p^n\right]_P = np^{n-1}, \tag{3.7b}$$

where n is a positive integer. [**Hint:** *Use mathematical induction.*]

☐ **Exercise 3.5** *Now show that Eq. (3.7) can be extended to negative integers n as well, by using Eq. (3.4).* [**Hint:** *Start by evaluating the bracket of $1/x$ with p, starting from $[x \cdot \frac{1}{x}, p]$.*]

☐ **Exercise 3.6** *Show that the properties described in Eq. (3.4) imply that three variables A, B, and C must satisfy the condition*

$$\left[A, [B, C]_P\right]_P + \left[B, [C, A]_P\right]_P + \left[C, [A, B]_P\right]_P = 0. \tag{3.8}$$

Such cyclic conditions are called Jacobi identities.

☐ **Exercise 3.7** *Show that if the Hamiltonian H of a classical system is independent of a generalized coordinate x_0, the corresponding generalized momentum p_0 must be a constant of motion. These coordinates are termed as* cyclic coordinates.

☐ **Exercise 3.8** *If $\mathbf{L} = \mathbf{r} \times \mathbf{p}$ is the classical angular momentum of a particle, where $\mathbf{r} = (x, y, z)$ are the coordinates and $\mathbf{p} = (p_x, p_y, p_z)$ are the momenta, find $[L_x, L_y]_P$.*

A modification for the Poisson bracket can be obtained if the properties mentioned in Eqs. (3.4), (3.5), and (3.6) can be respected by the modified definition, even though the definition is not the same as in Eq. (3.2). Let us call this prospective definition *quantum bracket* and denote it by $[A, B]_Q$. Now notice that

$$\left[A_1 A_2, B_1 B_2\right]_Q = A_1\left[A_2, B_1 B_2\right]_Q + \left[A_1, B_1 B_2\right]_Q A_2$$

$$= A_1\left[A_2, B_1\right]_Q B_2 + A_1 B_1\left[A_2, B_2\right]_Q$$

$$+ \left[A_1, B_1\right]_Q B_2 A_2 + B_1\left[A_1, B_2\right]_Q A_2. \qquad (3.9)$$

However, we could have evaluated the bracket in another way as well, viz.,

$$\left[A_1 A_2, B_1 B_2\right]_Q = B_1\left[A_1 A_2, B_2\right]_Q + \left[A_1 A_2, B_1\right]_Q B_2$$

$$= B_1 A_1\left[A_2, B_2\right]_Q + B_1\left[A_1, B_2\right]_Q A_2$$

$$+ A_1\left[A_2, B_1\right]_Q B_2 + \left[A_1, B_1\right]_Q A_2 B_2. \qquad (3.10)$$

Since the two expressions must be the same, we can equate them and obtain

$$\left(A_1 B_1 - B_1 A_1\right)\left[A_2, B_2\right]_Q = \left[A_1, B_1\right]_Q\left(A_2 B_2 - B_2 A_2\right). \qquad (3.11)$$

To be sure, this condition should be valid for any definition of a binary operation $[A, B]$ that satisfies the conditions given in Eq. (3.4). In particular, the conditions should be satisfied even if we use Poisson brackets instead of the quantum brackets in Eq. (3.11). In classical physics, this is guaranteed by the fact that all dynamical variables are represented by numbers, and hence they commute. For quantum mechanics, we indicated earlier that operators on a vector space might be the appropriate objects. So we cannot take commutativity for granted. Therefore, the condition of Eq. (3.11) can hold for arbitrary operators $\widehat{A}_1$, $\widehat{A}_2$, $\widehat{B}_1$, $\widehat{B}_2$ only if we define

$$\left[\widehat{A}, \widehat{B}\right]_Q = (\text{constant}) \times \left(\widehat{A}\widehat{B} - \widehat{B}\widehat{A}\right). \qquad (3.12)$$

We now make some observations about the constant that can go into this definition. Real dynamical variables of classical mechanics correspond to Hermitian operators in quantum mechanics, as mentioned in §2.9. If we want the bracket of two real dynamical variables to be real as well, we must put in a factor of i in the constant, because the expression $(\widehat{A}\widehat{B} - \widehat{B}\widehat{A})$ is not Hermitian for Hermitian operators $\widehat{A}$ and $\widehat{B}$. Moreover, we should put in a factor with the inverse dimension of action, so that the quantum bracket can have same unit as the Poisson bracket. We can then write

$$\left[\widehat{A}, \widehat{B}\right]_Q = -\frac{i}{\hbar}\left(\widehat{A}\widehat{B} - \widehat{B}\widehat{A}\right). \qquad (3.13)$$

Of course, our argument cannot guarantee that there will not be an extra numerical factor on the right side of this equation. An overall numerical factor has been fixed by the requirement that the results derived from this formalism should agree with experiments.

Once the quantum bracket is defined, we will stick to the same basic brackets given in Eq. (3.6). Also, since the quantum bracket is so similar to the commutator bracket, defined by

$$\left[\widehat{A}, \widehat{B}\right] = \widehat{A}\widehat{B} - \widehat{B}\widehat{A}, \tag{3.14}$$

we might as well not overburden ourselves with notations and use the commutator bracket itself. Thus, instead of writing $[\widehat{x}_a, \widehat{p}_b]_Q = \delta_{ab}$ that follows directly from Eq. (3.6), we will write $(-i/\hbar)[\widehat{x}_a, \widehat{p}_b] = \delta_{ab}$, using the commutator brackets. Putting in the brackets between the operators corresponding to two coordinates and two momenta, we summarize the basic commutation relations as follows:

$$\left[\widehat{x}_a, \widehat{x}_b\right] = 0, \qquad \left[\widehat{p}_a, \widehat{p}_b\right] = 0, \qquad \left[\widehat{x}_a, \widehat{p}_b\right] = i\hbar\delta_{ab}. \tag{3.15}$$

Of course, the commutator of two operators is always an operator. In the first two commutators of Eq. (3.15), the '0' appearing on the right sides ought to be interpreted as the null operator. In the x-p commutator, one should assume that there is the identity operator sitting on the right side along with the factors explicitly written there.

□ **Exercise 3.9** *Show that the associative property of operators guarantee the identities*

$$\left[\widehat{A}, \widehat{B}\widehat{C}\right] = \left[\widehat{A}, \widehat{B}\right]\widehat{C} + \widehat{B}\left[\widehat{A}, \widehat{C}\right], \tag{3.16}$$

$$\left[\widehat{A}\widehat{B}, \widehat{C}\right] = \widehat{A}\left[\widehat{B}, \widehat{C}\right] + \left[\widehat{A}, \widehat{C}\right]\widehat{B}, \tag{3.17}$$

which are just Eqs. (3.4c) and (3.4d) written in terms of the commutator brackets instead of the Poisson brackets.

□ **Exercise 3.10** *Show that the Jacobi identity, Eq. (3.8), is also valid because of the associativity of the operators if we use commutator brackets.*

□ **Exercise 3.11** *The exponential of any operator can be defined through the power series:*

$$e^{\widehat{A}} = \mathbb{1} + \widehat{A} + \frac{1}{2!}\widehat{A}^2 + \cdots. \tag{3.18}$$

Show that for any two operators $\widehat{A}$ and $\widehat{B}$

$$e^{\widehat{A}}\widehat{B}e^{-\widehat{A}} = \widehat{B} + [\widehat{A}, \widehat{B}] + \frac{1}{2}[\widehat{A}, [\widehat{A}, \widehat{B}]] + \cdots. \tag{3.19}$$

Write down the nth term of the series. This identity is called the Hadamard *identity.* [**Hint:** *Take $\widehat{A} = \lambda\widehat{C}$ and make Taylor expansion in powers of λ.*]

3.2   Heisenberg equation

The equation of motion of any operator $\mathcal{O}$ should be given, in analogy with Eq. (3.1), by

$$\frac{d\mathcal{O}}{dt} = \frac{\partial\mathcal{O}}{\partial t} - \frac{i}{\hbar}\Big[\mathcal{O}, H\Big], \tag{3.20}$$

where H is the Hamiltonian of the system. This is called the *Heisenberg equation* of motion of a quantum system. Later in the book, we will see various examples of the application of this equation and solutions obtained thereof.

> The phrase 'equation of motion' should be taken with a grain of salt. Classically, a description of motion involves the specification of position and momenta of particles. In quantum mechanics, position and momentum cannot be specified simultaneously, as explained in Chapter 1. Certainly, the Heisenberg equation does not describe motion in the classical sense. We use the term loosely, to indicate the time evolution of a system. The Heisenberg equation tells us that the time evolution of any observable quantity can be viewed as the time evolution of the operator that describes that quantity.

It should be noticed that once we use the commutator brackets, the uncertainty relation is automatically ingrained in it. Because the commutator of position and momentum is non-zero, the theorem presented in the form of Eq. (2.134) implies that the product of the uncertainties in these operators must have a minimum. To be precise, comparing Eq. (2.131, p. 48) with the x-p commutation relation of Eq. (3.15), we see that the role of the operator C is played by $\hbar$ times the unit operator. Hence,

$$\Delta x \, \Delta p_x \geqslant \frac{1}{2}\hbar, \tag{3.21}$$

with similar relations for other components of coordinates and momenta. Note that if we take a certain component of coordinate and a different component of momentum, there is no minimum in the product of uncertainties, since the commutator vanishes in this case.

3.3   Schrödinger equation

We have given the Heisenberg equation of motion in Eq. (3.20), which describes
time evolution of quantum systems. The time evolution of physical observables
can be viewed in a different way as well. We do not observe operators:
we observe matrix elements of operators between states. In the Heisenberg
picture, the states are considered to be time-independent, so the time evolution
of matrix elements comes solely from the time evolution of the operators. In
other words,

$$\frac{d}{dt} \langle \Psi_1 | \mathscr{O} | \Psi_2 \rangle = \left[\left\langle \Psi_1 \left| \frac{d\mathscr{O}}{dt} \right| \Psi_2 \right\rangle \right]_H, \tag{3.22}$$

where $|\Psi_1\rangle$ and $|\Psi_2\rangle$ are two states, and the subscript 'H' is indicative of the
fact that we are using the Heisenberg picture.

Let us now use an operator that does not explicitly depend on time, i.e.,
$\partial \mathscr{O}/\partial t = 0$. Then, using Eq. (3.20), we get

$$\frac{d}{dt} \langle \Psi_1 | \mathscr{O} | \Psi_2 \rangle = -\frac{i}{\hbar} \langle \Psi_1 | \mathscr{O}H - H\mathscr{O} | \Psi_2 \rangle. \tag{3.23}$$

Note that we have removed the subscript 'H' from the right side. Since it is a
matrix element, it should be the same in any representation.

We now see that we could have taken the time derivative on the left side in
a different manner, viz., by assuming that the states evolve with time and the
operator stays fixed. This is what is done in the Schrödinger picture. Thus,

$$\frac{d}{dt} \langle \Psi_1 | \mathscr{O} | \Psi_2 \rangle = \left[\left\langle \dot{\Psi}_1 \left| \mathscr{O} \right| \Psi_2 \right\rangle + \left\langle \Psi_1 \left| \mathscr{O} \right| \dot{\Psi}_2 \right\rangle \right]_S, \tag{3.24}$$

where

$$\left| \dot{\Psi} \right\rangle \equiv \frac{d}{dt} |\Psi\rangle. \tag{3.25}$$

Comparing this with Eq. (3.23), we see that this can be achieved for arbitrary
state vectors $|\Psi_1\rangle$ and $|\Psi_2\rangle$ provided

$$\left| \dot{\Psi} \right\rangle = -\frac{i}{\hbar} H |\Psi\rangle \tag{3.26}$$

or equivalently

$$\left\langle \dot{\Psi} \right| = \frac{i}{\hbar} \langle \Psi | H \tag{3.27}$$

for any state vector. Of course, the two relations are equivalent, so we might as well use only the ket relation, which can be written as

$$i\hbar\frac{d}{dt}\left|\Psi(t)\right\rangle = H\left|\Psi(t)\right\rangle, \tag{3.28}$$

where we have now explicitly indicated that the state vector is a function of time. This is the *Schrödinger equation*. This equation will be solved for many systems with different Hamiltonians in the book.

3.4 Hilbert space

In Chapter 2, we said that we want to treat the states as vectors in a vector space. A very general kind of vector space, with no extra structure on it, will not do for this purpose. We have collected bits and pieces of information about the kind of vector space that we want. Some new information has also arrived in disguise in §3.3. Let us decode this information, and summarize all the information received earlier, to specify the kind of vector spaces needed for quantum mechanics.

3.4.1 What kind of a vector space do we need?

Already in Chapter 2, we mentioned that the vector space will in general be complex. We can see the reason from Eq. (3.26), for example. On the right side, we see $H\left|\Psi\right\rangle$, which is a vector, because the operator H, acting on a vector $\left|\Psi\right\rangle$, produces a vector. But then we need to have the structure to multiply the resulting vector by $-i/\hbar$, which is an imaginary factor. Therefore, a complex vector space is essential.

Then, remember that we do not *measure* states, or vectors for that matter. The result of a measurement is a number. Thus, we must need some kind of vector spaces where, given an initial state and a final state, we can arrive at a number. This means that we need an inner product defined on the vector space. Once we have an inner product, it automatically defines a norm of any vector, as was pointed out in §2.3.2. Let us now see what more is needed.

3.4.2 Cauchy sequences and convergent sequences

A sequence on any set $\mathbb{W}$ is an ordered list of elements of the set. An infinite sequence is an infinite list of this sort. More formally, we can say that it is a map of the form $\mathbb{N} \to \mathbb{W}$, i.e., corresponding to each natural number there is an element of the set.

A *metric* on any set $\mathbb{W}$ is a function of the form $\mathbb{W} \times \mathbb{W} \to \mathbb{R}$, i.e., given any two elements of the set a and b, one can define a real number $d(a, b)$, which has the following properties for arbitrary elements a, b, and c that appear in the statements:

1. $d(a, b) \geqslant 0$.

2. $d(a, b) = 0$ if and only if $a = b$.

3. $d(a, b) = d(b, a)$.

4. $d(a, b) \leqslant d(a, c) + d(c, b)$.

Intuitively, a metric defines a distance between two elements of the set. In conformity with the intuitive notion, the distance between two elements is non-negative unless the two elements are the same and does not depend on the order in which we consider the elements. The inequality in step 4 of the definition is the *triangle inequality*.

Whenever a metric can be defined, we can define *Cauchy sequences*. These are infinite sequences $t_1, t_2, t_3, \cdots$, such that, given an arbitrary positive number ϵ, we can find an integer ν such that

$$d(t_n, t_{n+1}) < \epsilon \qquad \forall n > \nu. \tag{3.29}$$

Said in words, a Cauchy sequence is a sequence in which adjacent elements come arbitrarily close to each other as the sequence progresses.

Note what the definition does *not* say. It does not say that as the sequence progresses, the elements tend to be arbitrarily close to a limit point. If a sequence has a limit point, i.e., if there is an element t of $\mathbb{W}$ such that

$$\lim_{n \to \infty} t_n = t, \tag{3.30}$$

then the sequence is called a *convergent sequence*.

To understand the difference in the two definitions, we give an example with the interval $(0, 1)$, i.e., the set of numbers satisfying the inequality $0 < x < 1$. In this set, we define a sequence

$$\frac{1}{2}, \frac{1}{3}, \cdots, \frac{1}{n}, \cdots, \tag{3.31}$$

whose n^{th} term is $t_n = \frac{1}{n+1}$. Then,

$$d(t_n, t_{n+1}) = \frac{1}{n+1} - \frac{1}{n+2} = \frac{1}{(n+1)(n+2)}, \tag{3.32}$$

which becomes smaller and smaller as n becomes larger and larger, implying that this sequence is a Cauchy sequence. However,

$$\lim_{n \to \infty} t_n = 0, \tag{3.33}$$

which is not a member of the set $(0, 1)$. Thus, the sequence shown in Eq. (3.31) is a Cauchy sequence but not a convergent sequence in the interval $(0, 1)$.

It can be easily argued that every convergent sequence is a Cauchy sequence. But every Cauchy sequence is not a convergent sequence, as our example shows.

It should be understood that the question of convergence of a Cauchy sequence does not depend on only the elements. It also depends on the set W whose elements they are. For example, if we take the same sequence as that in Eq. (3.31) in the set $0 \leqslant x \leqslant 1$, the limit point, 0, would belong to the set, and the sequence would therefore be convergent. A little thought shows that in this set $0 \leqslant x \leqslant 1$, any Cauchy sequence is a convergent sequence. Such sets are called *Cauchy-complete*. On the other hand, we have already seen that all Cauchy sequences are *not* convergent sequences in the set $0 < x < 1$, and this set is therefore not Cauchy-complete.

☐ **Exercise 3.12** *Show that every convergent sequence is a Cauchy sequence.* [**Hint:** *Use triangle inequality.*]

3.4.3 Sequences in a vector space

In a vector space equipped with a norm, it is easy to see that a metric can be defined by the rule

$$d(u, v) = \|u - v\|. \tag{3.34}$$

Everything that we have said about Cauchy sequences, convergent sequences, and Cauchy convergence is therefore valid for a vector space. A vector space will be Cauchy-complete if all Cauchy squences are convergent.

In order to understand what kind of a vector space we need for quantum mechanics, we look back at the Schrödinger equation, Eq. (3.28). As we said, Ψ is an element of a vector space, and H is an operator on that vector space. Thus, H, acting on Ψ, produces a vector in the same vector space that Ψ belongs to. So the right side of Eq. (3.28) is an element of the vector space of states. The left side must then be also an element of the same vector space.

The left side contains a time-derivative. Using the standard definition of a derivative, we can write

$$\frac{d}{dt} |\Psi(t)\rangle = \lim_{n \to \infty} \frac{|\Psi(t + \tau/n)\rangle - |\Psi(t)\rangle}{\tau/n}, \tag{3.35}$$

where τ has the dimension of time but is otherwise arbitrary. For each value of n in the expression on the right side, the numerator contains the difference between two vectors. According to the postulates that define a vector space, the difference is guaranteed to lie in the vector space. Then we divide by τ/n, which is like multiplying by a scalar, and the result is also guaranteed to belong to the vector space. So, for each n, the ratio appearing on the right side of Eq. (3.35) belongs to the vector space of states. If we list the ratios for all n, the list will represent an infinite sequence in the vector space. It can be seen easily that it will be a Cauchy sequence.

But then, finally, we need to take the limit of n going to infinity. What is the guarantee that the limit would exist? None really, unless the vector space is Cauchy-complete, where every Cauchy sequence converges to a limit point. Therefore, this is the kind of vector space that we need for the state vectors. Let us summarize the requirements.

- The vector space must have an inner product defined on it. This guarantees the existence of a norm for each vector.

- Cauchy sequences must converge in the vector space.

A vector space equipped with these extra properties is called a *Hilbert space*. This is the kind of space on which the state vectors must be defined in quantum mechanics.

3.5   Coordinate representation

"Coordinate representation" of states means that we use the eigenstates of position, or coordinate, as basis states in the Hilbert space of state vectors. To illustrate the point, we can use a 1D space where x is the only coordinate and p is the corresponding momentum. Generalization to three or any other dimensional space will be obvious. Note that in this paragraph, we are talking about the dimension of space. It has nothing to do with the dimension of the vector space of state vectors that we had discussed earlier. In fact, the spatial dimensions considered are small positive integers, whereas the state vectors appropriate for the present discussion belong to an infinite-dimensional vector space, as we will soon explain.

3.5.1   The wavefunction

Let us first define the eigenstates of the coordinate operator, $\widehat{x}$. Obviously, these are defined through equations of the type

$$\widehat{x}\,|x\rangle = x\,|x\rangle. \tag{3.36}$$

In other words, we define the vector $|x\rangle$ to be an eigenstate of the operator $\widehat{x}$ with eigenvalue x. These are the states that are used as basis states in the coordinate representation.

In analogy with Eq. (2.32, p. 26), we can write any state vector $|\Psi\rangle$ in the form

$$|\Psi\rangle = \int dx\;|x\rangle\,\langle x\,|\,\Psi\rangle. \tag{3.37}$$

Quite obviously, the only difference with Eq. (2.32, p. 26), apart from the letters used in the notation, is that the sum over the basis states has been changed here to an integration. This is necessary because the number of basis states is infinite, i.e., we are talking about an infinite-dimensional Hilbert space. Completeness relation of the basis states can be easily read from Eq. (3.37), viz.,

$$\int dx\;|x\rangle\,\langle x| = \mathbb{1}. \tag{3.38}$$

The quantity $\langle x|\Psi\rangle$ that appears in Eq. (3.37) is a complex number. It can be written, doing away with the bra and ket notation, simply as a function of x:

$$\langle x\,|\,\Psi\rangle \equiv \Psi(x). \tag{3.39}$$

More explicitly, since $|\Psi\rangle$ is a time-dependent state vector in the Schrödinger picture, we should write

$$\langle x\,|\,\Psi(t)\rangle \equiv \Psi(x,t). \tag{3.40}$$

This is the definition of the *wavefunction* in the coordinate representation.

Let us now consider $\widehat{x}\,|\Psi\rangle$. The operator $\widehat{x}$ acts on the vector $|\Psi\rangle$. According to the general definition of operators, this should be a vector as well. What should be the coordinate representation of this vector? Clearly, it should be $\langle x|\widehat{x}|\Psi\rangle$. Now, note that Eq. (3.36) implies

$$\langle x|\,\widehat{x} = x\,\langle x|, \tag{3.41}$$

so we can write

$$\langle x \,|\widehat{x}|\, \Psi(t)\rangle = x \,\langle x \,|\, \Psi(t)\rangle = x\Psi(x,t). \tag{3.42}$$

Thus, we find that if we act on the wave vector with the position operator, it acts multiplicatively on the wavefunction.

Eqs. (3.37) and (3.38) show, in an implicit fashion, a big difference between the finite-dimensional case and the infinite-dimensional case. The integral involves a measure dx, which has the dimension of length for Cartesian coordinates x. Therefore, while the basis vectors in the finite-dimensional cases are dimensionless, here we have

$$\dim\left[\,|x\rangle\,\right] = L^{-\frac{1}{2}}. \tag{3.43}$$

Clearly, this shows that the basis vectors cannot be orthogonal in the same sense as that expressed in Eq. (2.28, p. 25). Rather, from Eq. (3.37), we can get

$$\langle x' \,|\, \Psi\rangle = \int dx \, \langle x' \,|\, x\rangle \,\langle x \,|\, \Psi\rangle , \tag{3.44}$$

or

$$\Psi(x',t) = \int dx \, \langle x' \,|\, x\rangle \, \Psi(x,t), \tag{3.45}$$

using the wavefunction. We see that the integration picks up only the value of the wavefunction at $x = x'$. It means that $\langle x'|x\rangle$ vanishes at all points where $x \neq x'$, and is yet non-zero at $x = x'$ because otherwise the result of the integration would have been zero. This is somewhat similar to the definition of Kronecker delta given earlier in Chapter 2, except that the arguments of this inner product take continuous values. Such a function is called the Dirac delta function, and its definition and properties have been explained in detail in Appendix A. In particular, it has the property that for any function $f(x)$,

$$\int dx \, \delta(x - x')f(x) = f(x'). \tag{3.46}$$

Comparing this property with Eq. (3.45), we conclude that

$$\langle x' \,|\, x\rangle = \delta(x - x'), \tag{3.47}$$

which can serve as the orthogonality relation between the basis states. Note that this relation conforms with the dimension of the state vectors given in Eq. (3.43), since the Dirac delta has dimension of inverse length, as indicated in Ex. 3.13.

□ **Exercise 3.13** *From Eq. (3.46), argue that*

$$\dim\left[\delta(x)\right] = \frac{1}{\dim(x)}. \tag{3.48}$$

3.5.2 Probabilistic interpretation

The wavefunction in the coordinate space has a simple probabilistic interpretation. The idea was already presented in §2.9 for states belonging to finite-dimensional vector spaces. Here, we reopen the discussion with some obvious adjustments in notation.

Consider the position operator $\widehat{x}$. Its expectation value in a state Ψ is given by

$$\langle\widehat{x}\rangle = \langle\Psi\,|\widehat{x}|\,\Psi\rangle. \tag{3.49}$$

Using the completeness relation of the position eigenstates, Eq. (3.38), we can write

$$\langle\widehat{x}\rangle = \int dx\,\langle\Psi\,|\widehat{x}|\,x\rangle\,\langle x\,|\,\Psi\rangle$$
$$= \int dx\,x\,\langle\Psi\,|\,x\rangle\,\langle x\,|\,\Psi\rangle, \tag{3.50}$$

using Eq. (3.36) for the definition of the position eigenstates $|x\rangle$. Using the definition of the wavefunction for the state $|\Psi\rangle$ from Eq. (3.39), we can now write

$$\langle\widehat{x}\rangle = \int dx\,x\Psi^*(x)\Psi(x) = \int dx\,x\left|\Psi(x)\right|^2. \tag{3.51}$$

Compare this equation with Eq. (2.120, p. 45) now. The operator $\mathcal{O}$ is $\widehat{x}$ here. The summation of Eq. (2.120) has been replaced by the integration. The other factor in Eq. (3.51), therefore, is like the factor $|a_k|^2$ of Eq. (2.120). This means that if a system is in a state $|\Psi\rangle$, the quantity $|\Psi(x)|^2\,dx$ is the probability of finding the system between the positions x and $x + dx$. This probabilistic interpretation of the wavefunction was first advanced by Max Born.

Note that we did not write the time dependence explicitly while writing Eq. (3.51). In general, the expectation value will be time-dependent, unless the time dependence is in the form of a phase factor, as for the stationary states to be introduced in Eq. (4.23, p. 86).

□ **Exercise 3.14** *Show that the inner product between any two state vectors, $|\psi_1\rangle$ and $|\psi_2\rangle$, can be written, using coordinate representation in one-dimension, as*

$$\langle\psi_2|\psi_1\rangle = \int_{-\infty}^{\infty} dx\,\psi_2^*(x)\,\psi_1(x). \tag{3.52}$$

3.5.3   Square integrability of wavefunctions

When we identified the expression on the right side of Eq. (3.51) as the expectation value of the operator $\widehat{x}$, we assumed that the wavefunction $\Psi(x)$ is normalized to unity, i.e.,

$$\int dx \left| \Psi(x) \right|^2 = 1. \tag{3.53}$$

If not, the formula for the expectation value should read

$$\langle \widehat{x} \rangle = \frac{\int dx \, x \left| \Psi(x) \right|^2}{\int dx \left| \Psi(x) \right|^2}, \tag{3.54}$$

an expression that follows from Eq. (2.113, p. 45). If we start with a wavefunction for which the integral on the left side of Eq. (3.53) has some value different from unity, we can always scale the wavefunction, by dividing it by an appropriate number, so that the newly defined wavefunction satisfies Eq. (3.53). But if the integral appearing on the left side of Eq. (3.53) is infinite, the situation cannot be salvaged: Eq. (3.54) would have an infinity as the denominator and will therefore be meaningless. Hence, it is necessary that the wavefunction of a physical state produces a finite value for the integral shown on the left side of Eq. (3.53), a condition that is often written as

$$\int dx \left| \Psi(x) \right|^2 < \infty. \tag{3.55}$$

Functions satisfying this condition are called *square-integrable* functions, for obvious reasons.

If the limits on integration extend over the entire x-axis for a physical system, then this condition has the important consequence that

$$\lim_{x \to \pm \infty} \Psi(x) = 0. \tag{3.56}$$

The proof is simple. Suppose the limit for $x \to +\infty$ is non-zero, say some number a. This means that for an arbitrary ϵ, we can find an x_0 such that

$$\left| \Psi(x) - a \right| < \epsilon \qquad \forall x > x_0. \tag{3.57}$$

So now consider the integral

$$\int_{x_0}^{\infty} dx \left| \Psi(x) \right|^2. \tag{3.58}$$

For all values of x in the range,

$$|\Psi(x)| > |a - \epsilon|, \tag{3.59}$$

and therefore the integral of Eq. (3.58) gives an infinite contribution for the integral on the left side of Eq. (3.55). There is no way that contributions from other values of x will cancel the infinity, because the modulus of $\Psi(x)$ is always non-negative. Hence the contradiction with Eq. (3.55), which shows that the limit of $\Psi(x)$ must vanish for arbitrarily large values of x, as said in Eq. (3.56).

Note that if a physical system is confined to a finite domain, the wavefunction need not vanish at the boundaries. The integral on the left side of Eq. (3.55) will then extend over a finite region, and the result of the integration will always be finite.

> In some situations, such as the case of free particle, we need to use wavefunctions which are not square integrable. We comment on such wavefunctions after Eq. (3.72).

3.5.4   Momentum operator

What if we act on the state with the momentum operator? How does it affect the wavefunction? To see this, we start with the identities

$$\langle x\,|\widehat{x}\widehat{p}|\,x'\rangle = x\,\langle x\,|\widehat{p}|\,x'\rangle,$$
$$\langle x\,|\widehat{p}\widehat{x}|\,x'\rangle = x'\,\langle x\,|\widehat{p}|\,x'\rangle, \tag{3.60}$$

both of which follow from the definition of position eigenstates. Subtracting one of these from the other and using the commutation relation of Eq. (3.15), we obtain

$$i\hbar\,\langle x\,|\,x'\rangle = (x - x')\,\langle x\,|\widehat{p}|\,x'\rangle. \tag{3.61}$$

On the left side, $\langle x|x'\rangle$ is merely the delta function $\delta(x - x')$, as argued in Eq. (3.47). In Appendix A, we have shown that the derivative of the Dirac delta function is given by

$$\frac{d}{dx}\,\delta(x) = -\frac{\delta(x)}{x}. \tag{3.62}$$

Therefore, we can write

$$\langle x\,|\widehat{p}|\,x'\rangle = -i\hbar\frac{d}{d(x - x')}\delta(x - x') = -i\hbar\frac{d}{dx}\delta(x - x'). \tag{3.63}$$

Let us now see the action of the momentum operator on an arbitrary state vector $|\Psi\rangle$. The resulting state vector would be $\widehat{p}|\Psi\rangle$, so its coordinate space representation would be

$$\langle x \,|\widehat{p}| \,\Psi\rangle = \int dx' \,\langle x \,|\widehat{p}| \,x'\rangle \,\langle x' \,| \,\Psi\rangle$$
$$= -i\hbar \int dx' \,\frac{d}{dx}\delta(x - x') \,\langle x' \,| \,\Psi\rangle. \qquad (3.64)$$

The derivative can be taken outside the integration sign, and then the integral over x' can be easily performed to give

$$\langle x \,|\widehat{p}| \,\Psi\rangle = -i\hbar \frac{\partial}{\partial x} \,\langle x \,| \,\Psi\rangle. \qquad (3.65)$$

The content of this equation is worth emphasizing. It says that the coordinate representation of $\widehat{p}|\Psi\rangle$ is the same as $-i\hbar\frac{\partial}{\partial x}$ acting on the coordinate representation of $|\Psi\rangle$. This statement is often abbreviated by saying that the momentum operator has a coordinate representation $-i\hbar\frac{\partial}{\partial x}$.

> It should be noted that the x-derivative has changed from a total derivative to a partial derivative while passing on from Eq. (3.63) to Eq. (3.65). The reason is that in Eq. (3.63) we were concerned with functions of x alone, so total vs partial derivative makes no difference. However, the state vector is a function of time, and therefore $\langle x|\Psi\rangle$ depends both on x and time. Thus, the derivative on $\langle x|\Psi\rangle$ must be a partial derivative, otherwise it loses all meaning.

□ **Exercise 3.15** *Show that*

$$\langle x \,\big|\widehat{p}^2\big| \,\Psi\rangle = -\hbar^2 \frac{\partial^2}{\partial x^2} \,\langle x \,| \,\Psi\rangle. \qquad (3.66)$$

[Hint: Call $\widehat{p}|\Psi\rangle = |\phi\rangle$ and use Eq. (3.65).]

□ **Exercise 3.16** *The momentum operator $\widehat{p}$ and the position operator $\widehat{x}$ satisfy the relation*

$$e^{a\widehat{p}+b\widehat{x}} = e^{a\widehat{p}}e^{b\widehat{x}}e^{\alpha}. \qquad (3.67)$$

Find α. [**Note:** *This relation constitutes a special case of the* Baker–Campbell–Hausdorff *formula, to be discussed in Chapter 5.*]

While Eq. (3.65) is valid for any state $|\Psi\rangle$, we can extract more information if the state happens to be a momentum eigenstate, denoted by $|p\rangle$. In this case as well, we can write

$$\langle x \,|\widehat{p}| \,p\rangle = -i\hbar \frac{\partial}{\partial x} \,\langle x \,| \,p\rangle, \qquad (3.68)$$

which is a special case of Eq. (3.65). On the other hand, since $|p\rangle$ is a momentum eigenstate, it satisfies an equation

$$\widehat{p}\,|p\rangle = p\,|p\rangle,\qquad(3.69)$$

where p is the eigenvalue. Using this, we obtain

$$\langle x\,|\widehat{p}|\,p\rangle = p\,\langle x\,|p\rangle.\qquad(3.70)$$

Thus,

$$-i\hbar\frac{\partial}{\partial x}\langle x\,|p\rangle = p\,\langle x\,|p\rangle.\qquad(3.71)$$

Integration of this equation gives us

$$\langle x\,|p\rangle = \exp(ipx/\hbar).\qquad(3.72)$$

Note that in general there can be an overall constant on the right side. We fix the normalization of the momentum eigenstate by choosing this constant to be equal to unity. It should be noted that this choice of normalization is somewhat exceptional. The usual process of normalization was described in Eq. (2.26, p. 24), where the norm of the vector is set equal to unity. This procedure cannot be applied to the momentum eigenstates because the integration over the square of the modulus of $\langle x|p\rangle$ is infinite. One needs to exercise some caution if such wavefunctions have to be used for physical states of a system. In these cases, it is only possible to compare the ratios where the overall normalization would cancel. Examples of such ratios will be encountered in scattering problems, in §7.5 and also in Chapter 13.

3.5.5 Schrödinger equation

It will be useful to see how the Schrödinger equation looks in the coordinate representation, because we will be using it extensively in the rest of the book. We can start by taking the inner product of the Schrödinger equation, Eq. (3.28), with an eigenstate of the position vector to write

$$i\hbar\left\langle x\,\middle|\,\dot{\Psi}\right\rangle = \langle x\,|H|\,\Psi\rangle.\qquad(3.73)$$

On the left side of the equation, we have written $\left|\dot{\Psi}\right\rangle$, which is just a notation for the time derivative of $|\Psi\rangle$. As discussed in §3.4, this is also a vector in a Hilbert space. Note that the position eigenvectors, $|x\rangle$, have been used here as basis states. Since the operators do not depend on time in the Schrödinger picture, their eigenstates are also time-independent, a topic that will be discussed in some detail in §4.6. So we can write the left side of Eq. (3.73) as

$$i\hbar \frac{\partial}{\partial t} \langle x \,|\, \Psi \rangle = i\hbar \frac{\partial}{\partial t} \Psi(x,t). \qquad (3.74)$$

Note that here we have to use partial derivatives because, unlike the state vector $|\Psi\rangle$ which is a function of time only, the wavefunction in the coordinate representation is a function of time and the coordinates.

On the right side, the Hamiltonian is in general a function of position and momentum. The kinetic energy part depends on the momentum. If we consider a non-relativistic particle in continuum, the kinetic energy is given by $p^2/2m$ for a particle of mass m. If we assume that the potential energy depends only on the position, we can write

$$H = \frac{\widehat{p}^2}{2m} + V(\widehat{x}). \qquad (3.75)$$

This is an operator equation: since x and p are operators on the vector space of states, the Hamiltonian is an operator. Let us see now what is the matrix element of this operator that appears on the right side of Eq. (3.73). The position operator acts multiplicatively, as shown in Eq. (3.42). Thus,

$$\langle x \,|V(\widehat{x})|\, \Psi \rangle = V(x) \langle x \,|\, \Psi \rangle = V(x)\Psi(x,t). \qquad (3.76)$$

For the kinetic energy part, we can use Eq. (3.66) to write

$$\left\langle x \,\middle|\, \frac{\widehat{p}^2}{2m} \,\middle|\, \Psi \right\rangle = -\frac{\hbar^2}{2m} \frac{\partial^2}{\partial x^2} \langle x \,|\, \Psi \rangle = -\frac{\hbar^2}{2m} \frac{\partial^2}{\partial x^2} \Psi(x,t). \qquad (3.77)$$

Thus, Eq. (3.73) reduces to a differential equation for the wavefunction $\Psi(x,t)$:

$$i\hbar \frac{\partial}{\partial t} \Psi(x,t) = -\frac{\hbar^2}{2m} \frac{\partial^2}{\partial x^2} \Psi(x,t) + V(x)\Psi(x,t). \qquad (3.78)$$

If, instead of a 1D system, we consider a system in the 3D space, the modification is obvious. Assuming that the potential energy is a function of the position vector $\boldsymbol{r}$, the equation will be

$$i\hbar \frac{\partial}{\partial t} \Psi(\boldsymbol{r},t) = -\frac{\hbar^2}{2m} \nabla^2 \Psi(\boldsymbol{r},t) + V(\boldsymbol{r})\Psi(\boldsymbol{r},t), \qquad (3.79)$$

where ∇^2 denotes the Laplacian:

$$\nabla^2 = \frac{\partial^2}{\partial x^2} + \frac{\partial^2}{\partial y^2} + \frac{\partial^2}{\partial z^2}. \tag{3.80}$$

This is how the Schrödinger equation looks in the coordinate space, for a non-relativistic particle whose potential energy depends only on the position. Variations with velocity-dependent potentials will be discussed in Chapter 10.

Eq. (3.79) involves second derivatives of the wavefunction. Even if an infinite second derivative is possible if the potential is infinite at some point, this equation at least says that the wavefunction must be continuous. It should be pointed out that continuous functions, by themselves, do not form a Hilbert space. For example, consider the sequence of functions:

$$f_n(x) = \begin{cases} 1 & \text{for } |x| \leqslant 1, \\ 1 + n(1-|x|) & \text{for } 1 < |x| < 1 + \frac{1}{n}, \\ 0 & \text{otherwise.} \end{cases} \tag{3.81}$$

It is easily seen that this is a Cauchy sequence, but the limit, for $n \to \infty$, is not a continuous function. Such functions need to be included in order to obtain a Hilbert space.

3.5.6 Probability current

Earlier in §3.5.2, we established that the absolute square of the wavefunction at a point gives the probability density of finding the particle at that point. In other words, if we take a small volume dV around a point at $\boldsymbol{r}$, the probability of finding the particle within that volume would be $\int dV \, |\Psi(\boldsymbol{r})|^2$.

Let us now see how this probability changes with time. Obviously, for a volume fixed in time, we have

$$\frac{d}{dt} \int dV \, |\Psi(\boldsymbol{r})|^2 = \int dV \left[\Psi^*(\boldsymbol{r}) \frac{\partial \Psi(\boldsymbol{r})}{\partial t} + \frac{\partial \Psi^*(\boldsymbol{r})}{\partial t} \Psi(\boldsymbol{r}) \right]. \tag{3.82}$$

Using Eq. (3.78) and its complex conjugate for the time derivative, we obtain

$$\frac{d}{dt} \int dV \, |\Psi(\boldsymbol{r})|^2 = \frac{i\hbar}{2m} \int dV \left[\Psi^* \nabla^2 \Psi - \Psi \nabla^2 \Psi^* \right]$$
$$= \frac{i\hbar}{2m} \int dV \, \boldsymbol{\nabla} \cdot \left[\Psi^* \boldsymbol{\nabla} \Psi - \Psi \boldsymbol{\nabla} \Psi^* \right]. \tag{3.83}$$

The volume integral can be converted to the integral over the bounding surface by using a well-known theorem of vector calculus. This gives

$$\frac{d}{dt} \int dV \, |\Psi(\boldsymbol{r})|^2 = - \oint d\boldsymbol{S} \cdot \boldsymbol{J}, \tag{3.84}$$

where $d\boldsymbol{S}$ is an element on the closed surface, and

$$\boldsymbol{J} = -\frac{i\hbar}{2m}\left(\Psi^*\boldsymbol{\nabla}\Psi - \Psi\boldsymbol{\nabla}\Psi^*\right) \tag{3.85}$$

or equivalently

$$\boldsymbol{J} = \frac{\hbar}{m}\,\mathrm{Im}\left(\Psi^*\boldsymbol{\nabla}\Psi\right). \tag{3.86}$$

The interpretation of Eq. (3.84) is clear: probability changes in a volume because the probability is transported through its bounding surface. This kind of equation is called the *equation of continuity*, and it expresses some kind of conservation law. In this context, it signifies that the total probability is conserved: it changes in a given volume only if it slips out of the bounding surface. The vector $\boldsymbol{J}$ defined in Eq. (3.86) should therefore be interpreted as the probability current density.

It should be remembered that the particular expression for it, given in Eq. (3.86), is valid only for Hamiltonians of the form shown in Eq. (3.75). If the Hamiltonian of a system does not conform to this form, the expression for the probability current density will be different.

3.6   Hermiticity of the momentum operator

We defined Hermitian operators in §2.9. Applying Eq. (2.105, p. 43) for the momentum operator between two arbitrary state vectors, we can write

$$\langle\Psi_1\,|\widehat{p}|\,\Psi_2\rangle = \langle\Psi_2\,|\widehat{p}|\,\Psi_1\rangle^* \tag{3.87}$$

if the momentum operator $\widehat{p}$ has to be Hermitian. Let us see what this relation implies on the wavefunctions in the coordinate representation.

By introducing a complete set of position eigenstates, the left side of Eq. (3.87) can be written as

$$\langle\Psi_1\,|\widehat{p}|\,\Psi_2\rangle = \int dx\,\langle\Psi_1\,|\,x\rangle\,\langle x\,|\widehat{p}|\,\Psi_2\rangle. \tag{3.88}$$

Using the definition of the wavefunction from Eq. (3.39) as well as the coordinate representation of the momentum operator given in Eq. (3.65), we obtain

$$\langle\Psi_1\,|\widehat{p}|\,\Psi_2\rangle = \int dx\,\Psi_1^*(x)\left(-i\hbar\frac{\partial}{\partial x}\right)\Psi_2(x). \tag{3.89}$$

Similarly, the right side of Eq. (3.87) can be written as

$$\langle\Psi_2\,|\widehat{p}|\,\Psi_1\rangle^* = \left(\int dx\,\Psi_2^*(x)\left(-i\hbar\frac{\partial}{\partial x}\right)\Psi_1(x)\right)^*$$

$$= \int dx\,\Psi_2(x)\left(+i\hbar\frac{\partial}{\partial x}\right)\Psi_1^*(x). \tag{3.90}$$

We have been, somewhat lazily, omitting the limits of integration in all these formulas. Suppose the limits are from x_1 to x_2. Then, performing integration by parts on the expression obtained in Eq. (3.90), we obtain

$$\langle\Psi_2\,|\widehat{p}|\,\Psi_1\rangle^* = \Psi_2(x_2)\Psi_1^*(x_2) - \Psi_2(x_1)\Psi_1^*(x_1)$$

$$- \int dx\,\Psi_1^*(x)\left(+i\hbar\frac{\partial}{\partial x}\right)\Psi_2(x). \tag{3.91}$$

Comparing this expression with the one in Eq. (3.89) and remembering Eq. (3.87), we see that the momentum operator is Hermitian, provided

$$\Psi_2(x_2)\Psi_1^*(x_2) - \Psi_2(x_1)\Psi_1^*(x_1) = 0 \tag{3.92}$$

for any two state vectors $|\Psi_1\rangle$ and $|\Psi_2\rangle$.

If we are talking about a system that lives on the entire real line extending from $-\infty$ to $+\infty$, then this condition is automatically satisfied. The reason was discussed in §3.5.3: the wavefunctions must vanish at spatial infinities. The condition stated in Eq. (3.56) therefore ensures that the momentum operator is Hermitian.

If, however, we are talking of a coordinate that does not extend all the way up to infinity, then Eq. (3.92) is not automatically satisfied for any square-integrable function. Therefore, such constraints have to be imposed on the wavefunctions in order to obtain a Hermitian momentum operator. A general constraint of this sort is given by

$$\Psi(x_2) = e^{i\alpha}\Psi(x_1), \tag{3.93}$$

where α is some real constant. For $\alpha = 0$, this condition is called the *periodic boundary condition*. For $\alpha = \pi$, we get *antiperiodic boundary condition*, provided the wavefunction does not vanish at the endpoints.

3.7   Momentum representation

In Eq. (3.39), we defined the wavefunction in the coordinate representation by taking the inner product of the state vector with a position eigenstate.

Likewise, we can define a wavefunction in the momentum representation, $\tilde{\Psi}(p)$, through the relation

$$\tilde{\Psi}(p) \equiv \langle p \,|\, \Psi \rangle, \tag{3.94}$$

where $|p\rangle$ is a momentum eigenstate as defined in Eq. (3.69). The orthogonality relation for the momentum eigenstates is not exactly like that of position eigenstates given in Eq. (3.47). The reason is that we can use the completeness relation of position eigenstates, Eq. (3.38), to write

$$\langle p \,|\, p' \rangle = \int dx \; \langle p \,|\, x \rangle \, \langle x \,|\, p' \rangle. \tag{3.95}$$

Using Eq. (3.72), we obtain

$$\langle p \,|\, p' \rangle = \int dx \; \exp\left(i(p' - p)x/\hbar \right). \tag{3.96}$$

Using Eq. (A.16, p. 427) from Appendix A, we obtain

$$\langle p \,|\, p' \rangle = 2\pi\hbar \, \delta(p - p'). \tag{3.97}$$

The factor of $2\pi\hbar$ on the right side shows the basic difference of this normalization with that of the position eigenstates shown in Eq. (3.47).

The completeness relation of the momentum eigenstates should also have the same factor of $2\pi\hbar$ in it, i.e., we should have

$$\int \frac{dp}{2\pi\hbar} \; |p\rangle \, \langle p| = 1. \tag{3.98}$$

The necessity of this factor can be appreciated by considering the inner product of two position eigenstates:

$$\begin{aligned}
\langle x \,|\, x' \rangle &= \int \frac{dp}{2\pi\hbar} \; \langle x \,|\, p \rangle \, \langle p \,|\, x' \rangle \\
&= \int \frac{dp}{2\pi\hbar} \; \exp\left(ip(x - x')/\hbar \right).
\end{aligned} \tag{3.99}$$

Consulting Eq. (A.15, p. 427) from Appendix A, we see that Eq. (3.47) is reproduced. This justifies the factor of $2\pi\hbar$ in the completeness relation of Eq. (3.98).

The interpretation for the wavefunction in momentum space is quite similar to that of the wavefunction in coordinate space, viz., $|\tilde{\Psi}(p)|^2 \, dp/(2\pi\hbar)$ is the probability of finding the system having momentum between p and $p + dp$.

It seems that there is a lack of symmetry between the formulas of position eigenstates and those of momentum eigenstates. It should be commented that this is a matter of convention and is rooted in our choice of the normalization of the momentum eigenstate wavefunctions in Eq. (3.72). The point is that our choice of the completeness condition of Eq. (3.38) implies that the position eigenvectors $|x\rangle$ have dimensions of $L^{-1/2}$, as mentioned in §3.5. Then, in Eq. (3.72), we chose the inner product of a position eigenstate and a momentum eigenstate to be a pure number, which is dimensionless. As a result, the momentum eigenstates, $|p\rangle$, acquired dimensions of $L^{+1/2}$. The factors of $2\pi\hbar$ present in Eqs. (3.97) and (3.98) merely set the dimensions right and could be avoided if we had put a factor of $(2\pi\hbar)^{-1/2}$ on the right side of Eq. (3.72).

☐ **Exercise 3.17** *Show that in the momentum representation, the operator $\widehat{x}$ can be represented by $i\hbar\frac{\partial}{\partial p}$, i.e., the analog of Eq. (3.65) is*

$$\langle p|\widehat{x}|\Psi\rangle = i\hbar\frac{\partial}{\partial p}\langle p|\Psi\rangle. \tag{3.100}$$

☐ **Exercise 3.18** *Show that, using commutation relation between $\widehat{x}$ and $\widehat{p}$ and resorting to coordinate representation,*

$$[f(\widehat{x}),\widehat{p}] = i\hbar\frac{df(x)}{dx}, \tag{3.101}$$

where $f(x)$ is an arbitrary function of x. [**Note:** *There is an implied unit operator on the right side.*]

☐ **Exercise 3.19** *The coordinate space representation of a wavefunction $\psi(x)$ is given by*

$$\psi(x) = \langle x|\psi\rangle = Ne^{-x^2/a^2}, \quad -\infty \leq x \leq \infty, \tag{3.102}$$

where N is the normalization and a is a constant having dimension of length. Find N. Also find the momentum-space representation of $|\psi\rangle$.

3.8 Another look at time–energy uncertainty

We have talked about time–energy uncertainty in §1.4. We reopen the discussion here and discuss some points associated with it.

3.8.1 Is time an operator?

In §2.9.4, we showed that the product of the standard deviations of two operators in any state is related to the expectation value of the commutator of these operators in the same state. The basic relation, presented in

Eq. (2.134, p. 48), was crucial in establishing the mathematical form of the position–momentum uncertainty relation in Eq. (3.21). Since the time–energy uncertainty relation looks the same as the position–momentum uncertainty relation, one may wonder whether the former arises the same way as the latter, i.e., as a consequence of some non-vanishing commutator of two operators. In analogy with the commutator relation between position and momentum given in Eq. (3.15), we are really contemplating whether time can be thought of as an operator that satisfies the commutation relation

$$\left[\hat{t}, H \right] = i\hbar. \tag{3.103}$$

In writing this equation, we use the notation $\hat{t}$ for the proposed time operator, and use the Hamiltonian operators whose eigenvalues are the energies of a system. On the right side, the sign can also be negative, since both signs can give the uncertainty relation.

We will show now that Eq. (3.103) is an impossible statement. For this, we go back to the arguments given in §3.6 about the hermiticity of the momentum operator. We showed that hermiticity can be ensured with square-integrable wavefunctions if we work on the entire real line from $-\infty$ to $+\infty$, but is not guaranteed if space extends up to a finite point in even one of the two ends. Now, energy values cannot range all the way to $-\infty$ because then we will have a completely unstable system. So, by the same argument, the conjugate time operator cannot be Hermitian.

It is nevertheless true that one can find operators satisfying the commutation relation of Eq. (3.103). For example, consider the operator

$$\hat{t} = \frac{m}{2} \left(p^{-1}x + xp^{-1} \right) \tag{3.104}$$

for a free particle, with $H = p^2/2m$. It can be easily seen to satisfy Eq. (3.103). However, such a definition would imply that each system can have its own time, a position that is hard to reconcile with our intuitive notion of time.

□ **Exercise 3.20** *Find the commutator of the operator given in Eq. (3.104) with the free Hamiltonian.*

□ **Exercise 3.21** *For a simple harmonic oscillator, the Hamiltonian is*

$$H = \frac{p^2}{2m} + \frac{1}{2}m\omega^2 x^2. \tag{3.105}$$

Consider this as a classical system and find its Poisson bracket with the combination

$$\hat{t} = -\frac{1}{\omega} \tan^{-1} \left(\frac{p}{m\omega x} \right). \tag{3.106}$$

[**Note:** *Looks like the commutator will obey Eq. (3.103), but it is difficult to treat the combination defined in Eq. (3.106) as a Hilbert space operator.*]

3.8.2   Uncertainty through any operator

Another way of understanding the time–energy uncertainty relation was proposed by Leonid Mandelstam and Igor Tamm. Consider any dynamical operator A. According to the Heisenberg equation, its time derivative is related to its commutator with the Hamiltonian, as shown in Eq. (3.20). From the basic operator uncertainty relation of Eq. (2.134, p. 48), we can write

$$\Delta A \, \Delta E \geqslant \frac{1}{2} \left| \left\langle \left[A, H \right] \right\rangle \right|. \tag{3.107}$$

And now, using the Heisenberg equation for the operator A, we obtain

$$\Delta A \, \Delta E \geqslant \frac{1}{2} \hbar \left| \langle dA/dt \rangle \right|. \tag{3.108}$$

This inequality reduces to the time–energy uncertainty relation if we define

$$\Delta t = \frac{\Delta A}{\left| \langle dA/dt \rangle \right|}. \tag{3.109}$$

Defined this way, Δt means the time it takes for a system to change the expectation value of an operator A by an amount that is equal to its standard deviation, assuming that the rate of change remains the same. In a more qualitative tone, we can say that Δt is the time taken for the expectation value of any operator A to change by a *large* amount.

This statement is consistent with the earlier explanation of the time–energy relation, given in §1.4. There, we said that Δt represents the lifetime of an unstable system. Certainly, when a system decays, there are large changes in the system. The lifetime is a measure of the timescale over which such large changes are taking place. This is also exactly what Eqs. (3.108) and (3.109) tell us.

Chapter 4

Time evolution of quantum systems

It is often said in science fiction as well as in books for science popularization that quantum mechanics makes everything probabilistic: nothing is certain. This is not quite true. The state vector of a quantum system evolves in time according to very precise rules, as shown in §3.3. Alternatively, we can think of operators to evolve in time according to well-determined rules, as described in §3.2. In this chapter, we show various ways of discussing the time evolution of quantum systems, often giving examples with a system whose state vectors belong to a 2D vector space.

4.1 Heisenberg equations and classical mechanics

We already commented that the Heisenberg equation looks very similar to the classical equation of motion written in terms of Poisson brackets. As we saw, it can be obtained from the classical equation by making the replacement

$$\left[A, H\right]_P \longrightarrow -\frac{i}{\hbar}\left[\widehat{A}, \widehat{H}\right], \tag{4.1}$$

for any dynamical variable A, which is represented by an operator $\widehat{A}$ in quantum mechanics. Once the brackets are evaluated, the Heisenberg equations are the same as the classical Hamilton's equations, although they do not mean exactly the same thing.

Take, for example, a free particle in 1D. The Hamiltonian is

$$H = \frac{p^2}{2m}. \tag{4.2}$$

Hamilton's equations of motion for a classical free particle are:

$$\dot{p} = 0, \qquad \dot{x} = p/m. \tag{4.3}$$

The Heisenberg equations for the quantum free particle are also the same, in terms of the operators $\widehat{x}$ and $\widehat{p}$:

$$\frac{d\widehat{p}}{dt} = 0, \qquad \frac{d\widehat{x}}{dt} = \frac{\widehat{p}}{m}. \tag{4.4}$$

The first of these equations implies that p is time-independent:

$$\widehat{p}(t) = \widehat{p}(0). \tag{4.5}$$

Putting it in the equation for $\widehat{x}$, we obtain

$$\widehat{x}(t) = \widehat{x}(0) + \frac{\widehat{p}t}{m}, \tag{4.6}$$

which looks exactly like the classical solution.

Of course, the interpretation is not the same as in classical mechanics. In Eq. (4.6), we have operators that are not directly measurable. But an interesting point should be noticed. If we take the expectation values of both sides of this equation in any state $|\Psi\rangle$, we would obtain

$$\langle \Psi | \widehat{x}(t) | \Psi \rangle = \langle \Psi | \widehat{x}(0) | \Psi \rangle + \frac{\langle \Psi | \widehat{p} | \Psi \rangle t}{m}. \tag{4.7}$$

The expectation values are numbers. This equation then shows that the expectation values of the quantum operators in any state obey the classical equation of motion for the corresponding dynamical variable.

Although we showed this result using a free particle, it is easy to see that it applies to all quantum systems. Suppose we have a potential $V(x)$ in addition to the kinetic energy term shown in Eq. (4.2). The classical Hamilton's equation for $\dot{x}$ would still be the same, whereas the other equation will now become

$$\dot{p} = -V'(x), \tag{4.8}$$

where $V'(x)$ is the derivative of the potential with respect to x. This statement becomes meaningless if we think of x as an operator, as should be done for quantum mechanics. However, suppose we write the potential as a power series in x:

$$V(x) = \sum_{n} v_n x^n. \tag{4.9}$$

Then we will obtain

$$\left[p, V(x)\right] = -i\hbar \sum_n n v_n x^{n-1} \tag{4.10}$$

by applying Eq. (3.7a, p. 58) even if we interpret the brackets as commutator brackets. This commutator gives the same expression in the equation of motion as is obtained by taking the derivative with respect to x. Hence, the Heisenberg equation would look the same as the classical Hamilton equations. When we take the expectation value of Eq. (4.8), we obtain

$$\frac{d}{dt} \langle \widehat{p} \rangle = - \langle V'(\widehat{x}) \rangle . \tag{4.11}$$

This should be obeyed in any state, and that is why the state vector has been omitted in writing this equation. This result is called the *Ehrenfest theorem*.

It has to be pointed out that, although the similarity is striking with the classical equation, Eq. (4.11) is *not* the classical equation. If we think of the expectation value as the corresponding classical variable, the classical equation would be the same as the equation

$$\frac{d}{dt} \langle \widehat{p} \rangle = -V'(\langle \widehat{x} \rangle). \tag{4.12}$$

The expectation value of a function is not the same as the function of an expectation value. However, if the function looks like a sharp spike concentrated at one value of its argument, then the two are very close. It is in this sense that one should think of the Ehrenfest theorem.

☐ **Exercise 4.1** *Consider a classical system for which x is distributed by a probability distribution*

$$P(x) = \frac{1}{\ell\sqrt{2\pi}} \exp(-x^2/2\ell^2). \tag{4.13}$$

Define

$$\langle f(x) \rangle = \int_{-\infty}^{+\infty} dx \, P(x) f(x) \tag{4.14}$$

for any function $f(x)$. Find $\langle x^2 \rangle$ and verify that it is not the same as $\langle x \rangle^2$.

4.2 Formal solutions of the basic equations

We have derived the basic equation of quantum mechanics in two forms: first, the Heisenberg equation of motion in Eq. (3.20), in which the operators are supposed to change with time, and then the Schrödinger equation, Eq. (3.28), in which the state vectors change with time.

If the Hamiltonian is time-dependent, the solution is complicated and will be discussed in Chapter 12. For time-independent Hamiltonians, however, it is easy to present a formal solution for both of these equations. For the Schrödinger equation, it is easily seen that the solution is of the form

$$|\Psi(t)\rangle = e^{-i\widehat{H}t/\hbar}|\Psi(0)\rangle. \tag{4.15}$$

If one encounters a differential equation of the type

$$\frac{df}{dt} = af \tag{4.16}$$

for some function $f(t)$, one usually rewrites it in the form

$$\frac{df}{f} = a\,dt \tag{4.17}$$

and then integrates both sides. This cannot be done for the differential equation involving the state vector, because one cannot divide by a vector. So, one makes a guess that the solution will be of the form

$$|\Psi(t)\rangle = e^{\beta t}|\Psi(0)\rangle \tag{4.18}$$

and then plugs this trial form into the Schrödinger equation, Eq. (3.28), to determine β. This procedure yields the answer given in Eq. (4.15).

For the Heisenberg equation of motion, the formal solution for $\mathscr{O}$ is

$$\mathscr{O}(t) = e^{+i\widehat{H}t/\hbar}\mathscr{O}(0)e^{-i\widehat{H}t/\hbar}, \tag{4.19}$$

provided $\mathscr{O}$ does not have any explicit time dependence, i.e., $\partial\mathscr{O}/\partial t = 0$.

☐ **Exercise 4.2** *Verify that the expression of Eq. (4.19) satisfies Eq. (3.20, p. 61) if the $\partial\mathscr{O}/\partial t$ term is zero.* [**Note:** *Be careful about the fact that $\mathscr{O}(0)$ as well as $\mathscr{O}(t)$ need not commute with the Hamiltonian, but a factor like $e^{-i\widehat{H}t/\hbar}$ certainly will, since the Hamiltonian must commute with itself.*]

A look at Eqs. (4.15) and (4.19) reconfirms the fact that the time evolution of matrix elements of operators does not depend on our choice of the Heisenberg or the Schrödinger picture. The point is that if we look at an expression of the form

$$\left\langle \Psi_2(0)\left|e^{+i\widehat{H}t/\hbar}\mathscr{O}(0)e^{-i\widehat{H}t/\hbar}\right|\Psi_1(0)\right\rangle, \tag{4.20}$$

we can either use Eq. (4.15) to interpret it as

$$\langle\Psi_2(t)|\mathscr{O}(0)|\Psi_1(t)\rangle \tag{4.21}$$

or use Eq. (4.19) to interpret it as

$$\langle \Psi_2(0) \, | \, \mathscr{O}(t) | \, \Psi_1(0) \rangle. \tag{4.22}$$

Either way, it describes the time evolution of the matrix element of the operator $\mathscr{O}$ between the states Ψ_1 and Ψ_2.

Although the formal solutions are aesthetically pleasing and reassuring because they show us that solutions can be obtained, they are of little practical importance. This is because in practice it is not easy to analytically exponentiate the Hamiltonian operator, which contains all integral powers of H, and in the end to add up the effects that all these powers inflict on the state vector or on the operator. So, in practice, one often solves the Schrödinger equation by turning it into a differential equation of the wavefunction, which is a complex function of space coordinates and time, introduced in §3.5.

4.3   Stationary states

It is worthwhile to note that one kind of solutions assumes a very important role in the discussion of quantum mechanical systems. These are solutions of the form

$$|\Psi(t)\rangle = e^{-iEt/\hbar} |\Psi(0)\rangle \tag{4.23}$$

for some real number E. In other words, the time-dependence of the state vector is purely a phase factor. Plugging the solution into Eq. (3.28, p. 63), we obtain

$$\hat{H} \, |\Psi(0)\rangle = E \, |\Psi(0)\rangle, \tag{4.24}$$

i.e., the state vector $|\Psi(0)\rangle$ is an eigenstate of the Hamiltonian operator with eigenvalue E. Such states are called *stationary states*, and accordingly Eq. (4.24) is called the stationary state Schrödinger equation. Since $|\Psi(0)\rangle$ does not depend on time (it involves the state vector only at time $t = 0$), the equation is also called the *time-independent Schrödinger equation*.

The word *stationary* means time-independent. It should be noted that the state vectors are not independent of time. Indeed, their time-dependence has been shown in Eq. (4.23). The use of the word in this context refers to the time-independence of some things related to the state. First, the probabilities of finding the state $|\Psi\rangle$ in a time-independent state $|a\rangle$ is given by $|\langle a|\Psi\rangle|^2$, as argued in §2.9.3. These probabilities do not depend on time. Second, the expectation value of any operator $\widehat{A}$ in the state $|\Psi\rangle$ is given by

$$\left\langle \widehat{A} \right\rangle = \left\langle \Psi \left| \widehat{A} \right| \Psi \right\rangle. \tag{4.25}$$

In general, this is a function of time because $|\Psi\rangle$ is. However, for a state of the form given in Eq. (4.23), the time dependence cancels from $\langle\Psi|$ and $|\Psi\rangle$, so the expectation value is independent of time as long as $\widehat{A}$ does not have explicit time dependence.

4.4 Equivalence of Schrödinger and Heisenberg pictures

Our presentation of the Heisenberg and Schrödinger equations of motion makes it clear that the two are equivalent. We have in fact motivated the Schrödinger equation from the Heisenberg equation. The reverse can also be done easily.

Here, we give a small example to show explicitly how the equivalence of the two formalisms work. Consider a system for which the states span a 2D vector space. Thus, the states can be represented by column matrices with two rows.

Operators are 2×2 matrices which operate on such states. The Hamiltonian must be a Hermitian matrix, as argued earlier. Here are some examples of 2×2 Hermitian matrices:

$$\sigma_x = \begin{pmatrix} 0 & 1 \\ 1 & 0 \end{pmatrix}, \quad \sigma_y = \begin{pmatrix} 0 & -i \\ i & 0 \end{pmatrix}, \quad \sigma_z = \begin{pmatrix} 1 & 0 \\ 0 & -1 \end{pmatrix}. \tag{4.26}$$

These three matrices are called *Pauli matrices*, after Wolfgang Pauli, who realized their importance in physics. Any Hermitian operator $\mathscr{O}$ in a 2×2 vector space can be written as

$$\mathscr{O} = a_0 \mathbb{1} + a_x \sigma_x + a_y \sigma_y + a_z \sigma_z, \tag{4.27}$$

where a_0, a_x, a_y, and a_z are real numbers, and $\mathbb{1}$ is the 2×2 unit matrix.

☐ **Exercise 4.3** *Show that Eq. (4.27) indeed gives the most general form for 2×2 Hermitian matrices.*

☐ **Exercise 4.4** *Show that the square of each Pauli matrix is the unit matrix:*

$$(\sigma_x)^2 = \mathbb{1}, \quad (\sigma_y)^2 = \mathbb{1}, \quad (\sigma_z)^2 = \mathbb{1}. \tag{4.28}$$

□ **Exercise 4.5** *Show that the Pauli matrices satisfy the commutation relation*

$$\left[\sigma_i, \sigma_j\right] = 2i\varepsilon_{ijk}\sigma_k, \tag{4.29}$$

where the indices take the value x, y, z, and ε_{ijk} is the Levi-Civita symbol which is completely antisymmetric in the indices and equal to +1 if the indices are in cyclic order.

□ **Exercise 4.6** *Show that the Pauli matrices anticommute among themselves, i.e., obey the relation*

$$\sigma_x\sigma_y + \sigma_y\sigma_x = 0, \tag{4.30}$$

and other similar equations.

□ **Exercise 4.7** *Show that the number of independent real parameters in an $n \times n$ Hermitian matrix is n^2.*

For our illustrative example, we do not take the Hamiltonian in the most general form given in Eq. (4.27). Rather, we consider the Hamiltonian in the form

$$\widehat{H} = \hbar\omega\sigma_y, \tag{4.31}$$

where ω is a constant. In the definition of the Hamiltonian, we have put an explicit factor of $\hbar$, since in Eqs. (4.15) and (4.19) the Hamiltonian appears in the combination $\widehat{H}/\hbar$. Thus, ω has the physical dimension of inverse time.

At $t = 0$, suppose the system is in the eigenstate of σ_z with eigenvalue +1. Using the representation of σ_z from Eq. (4.26), we can write this state as

$$\chi_+ \equiv \begin{pmatrix} 1 \\ 0 \end{pmatrix}. \tag{4.32}$$

The question that we ask is this: what is the expectation value of σ_z after the system evolves for a time t?

We first answer the question using the Schrödinger equation. Since H is independent of time, the solution of the Schrödinger equation is given by Eq. (4.15). For the Hamiltonian at hand, the exponential factor can be written as

$$\exp(-i\omega\sigma_y t) = \sum_{n=0}^{\infty} \frac{1}{n!}\left(-i\omega\sigma_y t\right)^n$$

$$= \sum_{n=0}^{\infty} \frac{(-1)^n(\omega t)^{2n}}{(2n)!}\mathbb{1} - i\sigma_y \sum_{n=0}^{\infty} \frac{(-1)^n(\omega t)^{2n+1}}{(2n+1)!}. \tag{4.33}$$

In the last step, we have separated the sum into odd powers and even powers, and used the fact, encoded in Eq. (4.28), that each even power of σ_y is equal to the unit matrix and therefore each odd power is equal to σ_y itself. We now notice that the two summations are nothing but the Taylor series of cosine and sine functions:

$$\exp(-i\omega\sigma_y t) = \cos(\omega t)\mathbb{1} - i\sin(\omega t)\sigma_y. \tag{4.34}$$

Now, using the explicit form for σ_y given in Eq. (4.26), we obtain

$$\exp(-i\omega\sigma_y t) = \begin{pmatrix} \cos\omega t & -\sin\omega t \\ \sin\omega t & \cos\omega t \end{pmatrix}. \tag{4.35}$$

Since $|\Psi(0)\rangle = \chi_+$, given in Eq. (4.32), the general solution of Eq. (4.15) takes the form

$$|\Psi(t)\rangle = \begin{pmatrix} \cos\omega t & -\sin\omega t \\ \sin\omega t & \cos\omega t \end{pmatrix} \begin{pmatrix} 1 \\ 0 \end{pmatrix} = \begin{pmatrix} \cos\omega t \\ \sin\omega t \end{pmatrix}. \tag{4.36}$$

The expectation value of σ_z in this state is given by

$$\langle\Psi(t)|\sigma_z|\Psi(t)\rangle = (\cos\omega t \quad \sin\omega t) \begin{pmatrix} 1 & 0 \\ 0 & -1 \end{pmatrix} \begin{pmatrix} \cos\omega t \\ \sin\omega t \end{pmatrix}$$
$$= \cos^2\omega t - \sin^2\omega t = \cos 2\omega t. \tag{4.37}$$

Let us now pursue the same problem in the Heisenberg picture. In this picture, the state does not change with time; it is always χ_+ as given in Eq. (4.32). The operator σ_z, however, evolves with time. At $t = 0$, it is given by the matrix in Eq. (4.26). Its matrix form at other times is governed by the Heisenberg equation, Eq. (3.20, p. 61), so

$$\frac{d\sigma_z}{dt} = -i\omega\left[\sigma_z, \sigma_y\right] = 2i\omega\sigma_y\sigma_z, \tag{4.38}$$

where we have made use of the fact that σ_y and σ_z anticommute. Note that the similar equation for the operator σ_y tells us that σ_y is in fact independent of time, since the Hamiltonian is proportional to this matrix and therefore the commutator appearing in Heisenberg equation vanishes. The equation for σ_z is therefore of the form $d\sigma_z/dt = (\text{constant}) \times \sigma_z$. As argued through Eq. (4.18), it can be integrated to obtain

$$\sigma_z(t) = \exp\left(2i\omega\sigma_y t\right)\sigma_z(0). \tag{4.39}$$

Using again the power series expansion of the exponential, we find

$$\sigma_z(t) = \begin{pmatrix} \cos 2\omega t & \sin 2\omega t \\ -\sin 2\omega t & \cos 2\omega t \end{pmatrix} \begin{pmatrix} 1 & 0 \\ 0 & -1 \end{pmatrix}$$

$$= \begin{pmatrix} \cos 2\omega t & -\sin 2\omega t \\ -\sin 2\omega t & -\cos 2\omega t \end{pmatrix}. \tag{4.40}$$

The expectation value in the state χ_+ is therefore given by

$$\langle \chi_+ | \sigma_z(t) | \chi_+ \rangle = \cos 2\omega t, \tag{4.41}$$

the same result that was obtained by solving the Schrödinger equation.

□ **Exercise 4.8** *Find the matrix element $\langle \chi_+ | \sigma_z | \chi_- \rangle$ using both Schrödinger and Heisenberg representations and verify that the results come out to be the same.*

□ **Exercise 4.9** *Find the equation of motion for $\sigma_x(t)$ for H given in Eq. (4.31). Find the time t_0 so that $\langle \chi_+ | \sigma_x(t_0) | \chi_+ \rangle = 0$.*

4.5   Interaction picture

So far, we have talked only about the Heisenberg and Schrödinger pictures of time evolution. In the first one, the operators evolve in time while the states are time-independent, whereas in the second one it is just the opposite. However, looking at the way the Schrödinger picture has been obtained, it is possible to consider a more general formulation. As we said, Eq. (3.23, p. 62) is fundamental, because the matrix elements are related to physically measurable quantities. The Heisenberg and the Schrödinger formulations are extreme ways of satisfying this condition by assuming that either the states are time-independent or the operators are. In a more general formalism, we can assume that the states and the operators both evolve with time. Such formulations are usually called *interaction formulation*, and the states and operators are denoted with a subscripted 'I'. In an interaction formalism,

$$i\hbar \frac{d}{dt} \mathcal{O}_I(t) = \left[\mathcal{O}_I, \widehat{H}_0 \right], \tag{4.42a}$$

$$i\hbar \frac{d}{dt} | \Psi_I(t) \rangle = \widehat{H}' | \Psi_I(t) \rangle, \tag{4.42b}$$

where the Hamiltonian of the system is given by

$$\widehat{H} = \widehat{H}_0 + \widehat{H}'. \tag{4.43}$$

Here and henceforth in this section, whenever an operator is mentioned without any time variable as its argument, we should understand that it is the operator at time $t = 0$. This also coincides with the operator in the Schrödinger formalism since operators do not change with time in that formalism. In particular, these comments apply to H_0 and H': both of them can be time dependent.

It is easy to see that Eq. (3.23, p. 62) is satisfied for any $\widehat{H}'$ and $\widehat{H}_0$ satisfying Eq. (4.43). The Schrödinger formalism is the extreme case obtained with $\widehat{H}_0 = 0$, whereas the Heisenberg formalism is the other extreme obtained with $\widehat{H}' = 0$. Intermediate forms for H' and H_0 are sometimes useful, and will be discussed later.

□ **Exercise 4.10** *Check that Eq. (4.42) gives Eq. (3.23, p. 62), irrespective of whether $\widehat{H}'$ and $\widehat{H}_0$ commute.*

Looking at the similarity of form between the interaction picture equation Eq. (4.42b) for the evolution of the state vector and the Schrödinger equation in Eq. (3.28, p. 63), one should not hastily conclude that the solution of Eq. (4.42b) is $|\Psi(t)\rangle = e^{-i\widehat{H}'t/\hbar}|\Psi_I(0)\rangle$. In fact, that would be wrong. The reason is that in the Schrödinger picture the solution of Eq. (4.15) is obtained for a time-independent Hamiltonian $\widehat{H}$. The object $\widehat{H}'$ appearing in Eq. (4.42b) is an operator, and therefore, like all other operators, it has a time dependence governed by Eq. (4.42a).

On the other hand, the similarity between Eq. (3.20, p. 61) and Eq. (4.42a) is perfect, and we can write the time dependence of an operator in the form

$$\mathscr{O}_I(t) = e^{+i\widehat{H}_0 t/\hbar}\mathscr{O}_I(0)e^{-i\widehat{H}_0 t/\hbar} \tag{4.44}$$

in the interaction picture. Using this solution for $H'(t)$ that appears in Eq. (4.42b), we can rewrite that equation in the form

$$i\hbar\frac{d}{dt}|\Psi_I(t)\rangle = e^{+i\widehat{H}_0 t/\hbar}H'(0)e^{-i\widehat{H}_0 t/\hbar}|\Psi_I(t)\rangle. \tag{4.45}$$

Defining now

$$|\Psi_I(t)\rangle = e^{+i\widehat{H}_0 t/\hbar}|\Psi_S(t)\rangle \tag{4.46}$$

and substituting into Eq. (4.45), it is easily seen that $|\Psi_S(t)\rangle$ satisfies the equation

$$i\hbar\frac{d}{dt}|\Psi_S(t)\rangle = \left(\widehat{H}_0 + \widehat{H}'(0)\right)|\Psi_S(t)\rangle = \widehat{H}|\Psi_S(t)\rangle. \tag{4.47}$$

This shows that $|\Psi_S(t)\rangle$ is nothing but the state vector in the Schrödinger picture. The relation given in Eq. (4.46) is exact, which is to say that it does not depend on how the division of the total Hamiltonian has been performed in Eq. (4.43). It can be taken as an alternative definition of the interaction picture.

The time evolution can be written in yet another useful way. At $t = 0$, $|\Psi_I\rangle$ and $|\Psi_S\rangle$ are equal. Starting from Eq. (4.46) and using the time evolution of the Schrödinger picture state vector from Eq. (4.15), we can write

$$|\Psi_I(t)\rangle = \widehat{U}_I(t,0)\,|\Psi_I(0)\rangle , \tag{4.48}$$

where

$$\widehat{U}_I(t,0) = e^{+i\widehat{H}_0 t/\hbar}\, e^{-i\widehat{H}t/\hbar}. \tag{4.49}$$

Then,

$$i\hbar \frac{d}{dt}\widehat{U}_I(t,0) = e^{+i\widehat{H}_0 t/\hbar}\left(-\widehat{H}_0 + \widehat{H}\right)e^{-i\widehat{H}t/\hbar}$$
$$= e^{+i\widehat{H}_0 t/\hbar}\widehat{H}'_I(0)e^{-i\widehat{H}t/\hbar}, \tag{4.50}$$

which is the same as

$$i\hbar \frac{d}{dt}\widehat{U}_I(t,0) = \widehat{H}'_I(t)\widehat{U}_I(t,0) \tag{4.51}$$

in view of the definition in Eq. (4.49) and the form of $\widehat{H}'_I(t)$ obtained through Eq. (4.44). Since $\widehat{U}(0,0) = \mathbb{1}$ by definition, we can also write this relation in the form of an integral equation:

$$\widehat{U}_I(t,0) = \mathbb{1} - \frac{i}{\hbar}\int_0^t dt'\,\widehat{H}'_I(t')\widehat{U}_I(t',0), \tag{4.52}$$

whose usefulness will be seen in Chapter 12.

4.6   Behavior of basis states

In the vector space of the states of a system, one can choose any complete set of vectors as the basis. In practice, one often chooses the eigenstates of some Hermitian operator for this purpose. As has been argued in §2.9.5, it is indeed possible to do so, since the eigenstates form a complete set. This kind of choice introduces some peculiarity of the basis states in the Schrödinger and Heisenberg pictures, which we now discuss.

Let us say that we choose, as basis states, the eigenvectors of an operator $\widehat{A}$. The eigenvectors are defined by the equation

$$\widehat{A}\,|a\rangle = a\,|a\rangle, \tag{4.53}$$

which means that the eigenvector corresponding to the eigenvalue a is denoted by $|a\rangle$. In the Schrödinger picture, the operator $\widehat{A}$ is independent of time. Therefore, its eigenvectors are also independent of time. Thus, unlike the state vectors which evolve in time in the Schrödinger picture, the basis vectors are time-independent.

Let us now consider the Heisenberg picture. As said in Eq. (4.19, p. 85), the operator $\widehat{A}$ in this picture has a time evolution given by

$$\widehat{A}(t) = \widehat{U}^{\dagger}(t)\widehat{A}(0)\widehat{U}(t), \tag{4.54}$$

where

$$\widehat{U}(t) = \exp(-i\widehat{H}t/\hbar). \tag{4.55}$$

Accordingly, their eigenvectors, taken as basis vectors, would be time-dependent as well. Let us denote the eigenvector at time t by $|a, t\rangle$. The eigenvalues would remain the same, so we can write the eigenvalue equation as

$$\widehat{A}(t)\,|a, t\rangle = a\,|a, t\rangle, \tag{4.56}$$

or

$$\widehat{U}^{\dagger}(t)\widehat{A}(0)\widehat{U}(t)\,|a, t\rangle = a\,|a, t\rangle. \tag{4.57}$$

Operating both sides by $U(t)$ from the left, we obtain

$$\widehat{A}(0)\widehat{U}(t)\,|a, t\rangle = a\widehat{U}(t)\,|a, t\rangle. \tag{4.58}$$

At $t = 0$, the Heisenberg and the Schrödinger pictures coincide, so the $\widehat{A}(0)$ appearing in this equation must be the same as what we had written simply as $\widehat{A}$ in Eq. (4.53). In that case, comparing Eq. (4.53) with Eq. (4.58), we obtain

$$|a\rangle = \widehat{U}(t)\,|a, t\rangle, \tag{4.59}$$

or

$$|a, t\rangle \equiv \widehat{U}^{\dagger}(t)\,|a\rangle. \tag{4.60}$$

Thus, although the state vectors do not evolve in time in the Heisenberg equation, the basis vectors do.

□ **Exercise 4.11** *We made the statement that the eigenvalues do not change when the operator changes by Eq. (4.54). Prove this statement.*

But it is also a weird time dependence. In Eq. (4.15), we saw that the time evolution of state vectors is given by

$$|\Psi(t)\rangle = \widehat{U}(t)\,|\Psi(0)\rangle. \tag{4.61}$$

On the other hand, Eq. (4.60) tells us that the basis vectors evolve by $\widehat{U}^\dagger(t)$. We can say that the basis vectors evolve in the opposite way compared to state vectors of the Schrödinger picture.

There is no great mystery in this opposite behavior. Rather, one can see that it is expected from some simple arguments. Consider, for example, that we have prepared a system in the state in a state $|\psi\rangle$. At time t, we ask what is the amplitude of finding the system in state $|a\rangle$ given in Eq. (4.53). In the Schrödinger picture, we will say that, at time t, the state has evolved into $\widehat{U}(t)\,|\psi\rangle$, according to Eq. (4.15). So the amplitude of finding the system in state $|a\rangle$ would be $\langle a|\widehat{U}(t)|\psi\rangle$. In the Heisenberg picture, the state vector would not evolve at all. Thus, the state of the system remains $|\psi\rangle$ even at time t. The amplitude of finding the system in the basis state corresponding to the eigenvalue a must then be given by $\langle a, t|\psi\rangle$. Equating the two expressions, we obtain

$$\left\langle a\left|\widehat{U}(t)\right|\psi\right\rangle = \langle a, t\,|\,\psi\rangle. \tag{4.62}$$

Since this should be true for any initial state $|\psi\rangle$, we can write

$$\langle a|\,\widehat{U}(t) = \langle a, t|\,, \tag{4.63}$$

which is equivalent to Eq. (4.60) through Eq. (2.80, p. 38).

An example with a real vector space might help make the distinction more intuitive. Suppose we have an arrow, drawn from the origin of a 2D space. We check the arrow at time $t = 0$, and then check back at another time t, and make the statement that the arrow has rotated in this period by 15°. For example, suppose we had measured the angle between the arrow and a coordinate axis (a basis vector) and found it to be 30° at the beginning, and measuring at time t we get the angle to be 45°. One way that this can happen is if we have the fixed coordinate axes, and the arrow rotates by 15°. This is the Schrödinger approach. The other possibility is that the arrow has stayed fixed, but the coordinate axes have moved in the opposite way by 15°. This will be the Heisenberg approach. Both of them have been illustrated in Figure 4.1. The

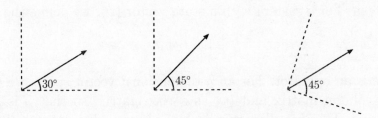

Figure 4.1 What is meant by the statement that "an arrow has rotated"? The left panel shows the original configuration. The central and the right panels show the changed configurations. See text for details.

leftmost panel in the figure shows the direction of a vector at time $t = 0$, which makes an angle $30°$ with the x-axis. At a later time t, suppose we hear that the vector makes an angle $45°$ with the same axis. That can happen if the axes have stayed fixed and the vector itself has rotated by $15°$ in the anticlockwise direction, which is the situation shown in the middle panel. But it can also happen that the direction of the vector has not changed and the axes have turned by $15°$ in the clockwise direction, as shown in the right panel. The two panels correspond to the Schrödinger and Heisenberg pictures. A mixture of the two effects is also possible, which would correspond to the characterization of states and operators given in Eqs. (4.42) and (4.43).

4.7 Density matrix

4.7.1 Definition and formalism

We said that the state vectors of a quantum system belong to a complex vector space. This means that the components of the state vector can be complex numbers. The magnitudes of these complex numbers give information about the probabilities that the system can be found in the basis states, as shown in §3.5.2. As regards the phases, the relative phases between components are important. However, an overall phase multiplying all components does not carry any crucial information about the system. The difference of the overall phases of two state vectors is important when there is an interference. But even there, the actual values of the phases are unimportant. Can we not therefore discuss the time evolution of a system without paying attention to the phases?

Yes, we can. For a system with a state vector $|\psi\rangle$, we define the object

$$\widehat{\rho} = |\psi\rangle\langle\psi|. \tag{4.64}$$

Clearly, this is an operator. For an n-dimensional vector space, we can think of $|\psi\rangle$ as a column matrix and $\langle\psi|$ as a row matrix, so that $\widehat{\rho}$ is an $n \times n$ square matrix. This is called the *density matrix*. For infinite-dimensional vector spaces, it is better to use the name *density operator*. The overall phase of $|\psi\rangle$ cancels with the phase of $\langle\psi|$, so $\widehat{\rho}$ does not contain the overall phase.

Let us now discuss some other properties of the density operator.

1. Using Eq. (2.87, p. 39), it is easy to see that

$$\widehat{\rho}^\dagger = \widehat{\rho}, \tag{4.65}$$

i.e., ρ is a Hermitian operator.

2. Since trace is a cyclic operation, we can say

$$\text{Tr}\,\widehat{\rho} = \text{Tr}(|\psi\rangle\langle\psi|) = \text{Tr}(\langle\psi|\psi\rangle). \tag{4.66}$$

But $\langle\psi|\psi\rangle$ is just a number, so it is the trace of itself. We assume that we are dealing with normalized states, so we obtain

$$\text{Tr}\,\widehat{\rho} = 1. \tag{4.67}$$

3. Also, because the state is normalized, we obtain

$$\widehat{\rho}^2 = |\psi\rangle\langle\psi|\psi\rangle\langle\psi| = |\psi\rangle\langle\psi|, \tag{4.68}$$

i.e.,

$$\widehat{\rho}^2 = \widehat{\rho}. \tag{4.69}$$

□ **Exercise 4.12** *A system with 2D state vectors is in a state*

$$|\psi\rangle = \frac{1}{\sqrt{2}}\begin{pmatrix} 1 \\ 1 \end{pmatrix}. \tag{4.70}$$

Find the density matrix and verify that all properties mentioned above are satisfied.

□ **Exercise 4.13** *Do the same problem, but this time with the general form of the state vector:*

$$|\psi\rangle = \begin{pmatrix} \cos\alpha \\ \sin\alpha \end{pmatrix}. \tag{4.71}$$

The important point is that insofar as the expectation value of any operator is needed, the density matrix has enough information for determining that. The expectation value of an operator A in the state ψ is given by $\langle\psi|\widehat{A}|\psi\rangle$, which is a number. It can therefore be written as the trace of a 1×1 matrix, and, using the cyclic property of traces, we can write

$$\left\langle \psi \left| \widehat{A} \right| \psi \right\rangle = \mathrm{Tr}\left(\left\langle \psi \left| \widehat{A} \right| \psi \right\rangle\right) = \mathrm{Tr}\left(|\psi\rangle\langle\psi|\widehat{A}\right), \tag{4.72}$$

or

$$\left\langle \psi \left| \widehat{A} \right| \psi \right\rangle = \mathrm{Tr}\left(\widehat{\rho}\widehat{A}\right), \tag{4.73}$$

where $\widehat{\rho}$ is the density matrix for the state $|\psi\rangle$.

Let us now discuss how the density operator evolves with time. Obviously,

$$\frac{d\widehat{\rho}}{dt} = \left|\dot{\psi}\right\rangle\langle\psi| + |\psi\rangle\left\langle\dot{\psi}\right|, \tag{4.74}$$

where

$$\left|\dot{\psi}\right\rangle \equiv \frac{d}{dt}|\psi\rangle = -\frac{i}{\hbar}\widehat{H}|\psi\rangle, \tag{4.75}$$

using the Schrödinger equation. Similarly,

$$\left\langle\dot{\psi}\right| \equiv \frac{d}{dt}\langle\psi| = +\frac{i}{\hbar}\langle\psi|\widehat{H}. \tag{4.76}$$

Putting these back, we obtain

$$\frac{d\widehat{\rho}}{dt} = \frac{i}{\hbar}\left(-\widehat{H}|\psi\rangle\langle\psi| + |\psi\rangle\langle\psi|\widehat{H}\right) = \frac{i}{\hbar}\left[\widehat{\rho}, \widehat{H}\right]. \tag{4.77}$$

This is the equation for time evolution. This equation looks like the Heisenberg equation, Eq. (3.20, p. 61), except for a different sign for the commutator term. Like the Heisenberg equation, it can be formally integrated to give

$$\widehat{\rho}(t) = e^{-i\widehat{H}t/\hbar}\widehat{\rho}(0)e^{+i\widehat{H}t/\hbar}. \tag{4.78}$$

In principle, the density matrix at any time t can be determined from this equation, which can be used to find the expectation value of any operator at any time. We will show an example in §4.7.2.

We have described how the density matrix can be constructed from the state vector. It is useful to know that the reverse is not fully possible, as we argue now. We first note that the hermiticity of the density matrix tells us that its eigenvalues must be real. Moreover, Eq. (4.69) tells us that the eigenvalues can be only 0 and 1, and Eq. (4.67) confirms that there can be only one eigenvalue equal to 1. The state vector is the eigenvector corresponding to this non-degenerate eigenvalue. However, the normalization of eigenvectors is arbitrary. In fact, even if we normalize the magnitude, the phase of the eiegnvectors remain arbitrary. Therefore, the phase information cannot be obtained from the density matrix. It should not be, since the overall phase disappears in the definition of the density matrix, as commented earlier.

□ **Exercise 4.14** *The density matrix of a system is given by*

$$\rho = \begin{pmatrix} \cos^2\theta & \cos\theta\sin\theta \\ \cos\theta\sin\theta & \sin^2\theta \end{pmatrix}. \tag{4.79}$$

What is the state vector that corresponds to this density matrix?

4.7.2   An example

We now give an example of time evolution by using the density matrix. For this example, we use the same system as was used in §4.4. The Hamiltonian of the system was given in Eq. (4.31, p. 88). Instead of specifying the state at $t = 0$, we now specify the density matrix at initial time:

$$\widehat{\rho}(0) = \begin{pmatrix} 1 & 0 \\ 0 & 0 \end{pmatrix}. \tag{4.80}$$

For a general time, let us write

$$\widehat{\rho}(t) = \begin{pmatrix} r & s \\ s^* & 1-r \end{pmatrix}, \tag{4.81}$$

in conformity with the properties of density matrices given in Eq. (4.67). From Eq. (4.65), we know that r must be real, but s can be complex.

We now use the evolution equation, Eq. (4.77). It gives

$$\frac{d\widehat{\rho}}{dt} = i\omega[\widehat{\rho}, \sigma_y] = \omega \begin{pmatrix} -s-s^* & -1+2r \\ -1+2r & s+s^* \end{pmatrix}. \tag{4.82}$$

The right side is real, and ρ is real at $t = 0$. Hence, ρ is real at all times, i.e., $s^* = s$. Equating the components of both sides of Eq. (4.82), we then obtain the differential equations

$$\frac{dr}{dt} = -2\omega s, \qquad \frac{ds}{dt} = -\omega(1 - 2r). \qquad (4.83)$$

We can turn them into second-order differential equations:

$$\frac{d^2r}{dt^2} = -2\omega\frac{ds}{dt} = 2\omega^2(1 - 2r), \qquad (4.84a)$$

$$\frac{d^2s}{dt^2} = 2\omega\frac{dr}{dt} = -4\omega^2 s. \qquad (4.84b)$$

Using now the initial conditions,

$$r(0) = 1, \qquad s(0) = 0, \qquad (4.85)$$

which imply

$$\frac{dr}{dt}(0) = -2\omega s(0) = 0,$$

$$\frac{ds}{dt}(0) = -\omega(1 - 2r(0)) = \omega, \qquad (4.86)$$

we find the solution for r and s, and write the density matrix:

$$\hat{\rho} = \frac{1}{2}\begin{pmatrix} 1 + \cos 2\omega t & \sin 2\omega t \\ \sin 2\omega t & 1 - \cos 2\omega t \end{pmatrix}. \qquad (4.87)$$

At time t, the expectation values of σ_z and σ_x are given by:

$$\langle \sigma_z(t) \rangle = \mathrm{Tr}(\rho\sigma_z) = \rho_{11} - \rho_{22} = \cos 2\omega t, \qquad (4.88a)$$

$$\langle \sigma_x(t) \rangle = \mathrm{Tr}(\rho\sigma_x) = 2\rho_{12} = \sin 2\omega t. \qquad (4.88b)$$

These are identical with the expectation values calculated in §4.4.

☐ **Exercise 4.15** *Consider the density matrix* $\hat{\rho} = |\psi\rangle\langle\psi|$ *constructed out of quantum state*

$$|\psi\rangle = \sum_n c_n|n\rangle$$

Find $\rho_{nm} = \langle n|\hat{\rho}|m\rangle$ *in terms of the coefficients* c_n *and* c_m.

4.7.3 Mixed states

We have said that the disadvantage of density matrices is that the phase
information of the state vector is lost. This feature turns out to be an
advantage if the state is a mixture of several states, where we know the
probabilities with which the different states are present but do not know the
phases with which they combine. Such a state is called a *mixed state*. In order
to distinguish from them, a state with a well-defined state vector is called a
pure state.

Suppose we have a mixed state in which different states $|\psi_a\rangle$ are mixed
with probabilities P_a. Then the density matrix will be given by

$$\widehat{\rho} = \sum_a P_a \,|\psi_a\rangle \langle\psi_a| . \qquad (4.89)$$

It is easy to see, since the P_a's are real, that the density matrix is still
Hermitian, i.e., Eq. (4.65) is satisfied. Also, since the probabilities P_a add
up to 1, the trace condition, Eq. (4.67), is satisfied. However, Eq. (4.69) is not
satisfied:

$$\widehat{\rho}^2 \neq \widehat{\rho}. \qquad (4.90)$$

The eigenvalues of the matrix ρ are the probabilities P_a, and the eigenvectors
are the $|\psi_a\rangle$'s that appear in Eq. (4.89).

The evolution equation, Eq. (4.77), is valid even for mixed states.
Therefore, the differential equations in Eq. (4.84) are still valid. However,
let us suppose that the initial state is a mixed state,

$$\widehat{\rho}(0) = \begin{pmatrix} p & 0 \\ 0 & 1-p \end{pmatrix}. \qquad (4.91)$$

This means that, with the parametrization of Eq. (4.81), we have

$$r(0) = p, \qquad s(0) = 0. \qquad (4.92)$$

The solutions for r and s can now easily be obtained:

$$r = \frac{1}{2} + \frac{1}{2}(2p-1)\cos 2\omega t,$$

$$s = \frac{1}{2}(2p-1)\sin 2\omega t. \qquad (4.93)$$

Therefore, for this case,

$$\langle \sigma_z(t)\rangle = \mathrm{Tr}(\widehat{\rho}\sigma_z) = 2r - 1 = (2p-1)\cos 2\omega t, \qquad (4.94a)$$

$$\langle \sigma_x(t)\rangle = \mathrm{Tr}(\widehat{\rho}\sigma_x) = 2s = (2p-1)\sin 2\omega t. \qquad (4.94b)$$

☐ **Exercise 4.16** *Show that if we impose the condition $\widehat{\rho}^2 = \widehat{\rho}$ on the density matrix obtained, then it implies $p = 1$ or $p = 0$, either of which values corresponds to a pure state.*

☐ **Exercise 4.17** *Show that the eigenvalues are time-independent for the density matrix obtained through Eq. (4.93).*

☐ **Exercise 4.18** *A collection of atoms consists of an incoherent mixture of $3/4$ of the atoms in the up state (i.e., the state shown in Eq. (4.32)) and $1/4$ in the down state (i.e., the state orthogonal to the up state). Find its density matrix. What is the value of $\mathrm{Tr}(\rho^2)$ for this density matrix?*

☐ **Exercise 4.19** *A density matrix $\widehat{\rho}$ constituted out of a bunch of atoms in state $|1\rangle$ and $|2\rangle$ has elements $\rho_{11} = \langle 1|\widehat{\rho}|1\rangle = 1/2 = \rho_{22}$ and $\rho_{12} = \rho_{21} = 0$. At $t = 0$, one switches on a Hamiltonian $\widehat{H} = -\alpha\sigma_x$, where α is a parameter and σ_x is one of the Pauli matrices defined in Eq. (4.26). Find the components of $\widehat{\rho}$ as a function of time. Does it become diagonal for any $t \neq 0$?*

4.8   Coherence vector

There is another method of discussing the time evolution, which is really an offshoot of the density matrix formalism, and works well for systems whose state vectors belong to a very low-dimensional vector space.

We can outline the method by taking a 2D vector space of state vectors. The density matrix is a 2×2 matrix in this case. We can write this matrix in terms of four basis matrices, which we take to be the unit matrix and the three Pauli matrices introduced in Eq. (4.26), i.e., write

$$\widehat{\rho} = r_0 \mathbb{1} + \sum_{a=1}^{3} r_a \sigma_a, \tag{4.95}$$

where now, for the sake of convenience, we have called σ_x as σ_1, etc. Because ρ is Hermitian, it implies that the coefficient r_0 to r_3 are all real.

Take traces of both sides. Note that the Pauli matrices are all traceless, so that we obtain

$$\mathrm{Tr}\,\widehat{\rho} = 2r_0. \tag{4.96}$$

So, r_0 is fixed, and we can eliminate the symbol to write

$$\widehat{\rho} = \frac{1}{2}\mathbb{1} + \sum_{a=1}^{3} r_a \sigma_a. \tag{4.97}$$

All information in ρ is contained in the three real numbers r_a. Consider these to be components of a vector, called the *coherence vector*. The evolution of the system can then be inferred from the way the coherence vector changes with time.

Exactly similarly, we can write down the Hamiltonian using the Pauli matrices:

$$\widehat{H} = h_0 \mathbb{1} + \sum_{b=1}^{3} h_b \sigma_b. \tag{4.98}$$

Again, the coefficients h_0 through h_3 are real because the Hamiltonian is Hermitian.

Now substitute the forms of $\widehat{\rho}$ and H into the time evolution equation, Eq. (4.77). For r_0, there is no evolution: the derivative is zero. For the other elements of $\widehat{\rho}$, i.e., for the components of the coherence vector, we obtain

$$\sum_{a=1}^{3} \frac{dr_a}{dt} \sigma_a = \frac{i}{\hbar} \sum_{a=1}^{3} \sum_{b=1}^{3} r_a h_b [\sigma_a, \sigma_b]. \tag{4.99}$$

Note that h_0, the part in Hamiltonian that is proportional to the unit matrix, does not contribute to this evolution equation. This is true not only for a 2D vector space, but for any system, no matter how many dimensional its state vectors are.

For the case of 2D vector space, evaluating the commutator, we get

$$\sum_{c=1}^{3} \frac{dr_c}{dt} \sigma_c = \frac{i}{\hbar} \sum_{a=1}^{3} \sum_{b=1}^{3} \sum_{c=1}^{3} r_a h_b \cdot 2i\varepsilon_{abc} \sigma_c. \tag{4.100}$$

As explained in Ex. 4.20, this equation implies

$$\frac{dr_c}{dt} = -\frac{2}{\hbar} \sum_{a=1}^{3} \sum_{b=1}^{3} \varepsilon_{abc} r_a h_b. \tag{4.101}$$

Explicitly:

$$\frac{dr_1}{dt} = -\frac{2}{\hbar} \left(r_2 h_3 - r_3 h_2 \right), \tag{4.102a}$$

$$\frac{dr_2}{dt} = -\frac{2}{\hbar} \left(r_3 h_1 - r_1 h_3 \right), \tag{4.102b}$$

$$\frac{dr_3}{dt} = -\frac{2}{\hbar} \left(r_1 h_2 - r_2 h_1 \right). \tag{4.102c}$$

Solving these equations with appropriate initial conditions, one obtains the time evolution of the density matrix.

☐ **Exercise 4.20** *Verify that*

$$\text{Tr}\left(\sigma_a \sigma_b\right) = 2\delta_{ab}. \tag{4.103}$$

Use this equation to show that if $\sum_a A_a \sigma_a = \sum_a B_a \sigma_a$, then $A_a = B_a$ for each value of the index a.

As an example, let us redo the problem discussed in §4.7.3. We had $H = \hbar\omega\sigma_y$, which means

$$h_2 = \hbar\omega, \qquad h_1 = h_3 = 0. \tag{4.104}$$

Then,

$$\frac{dr_1}{dt} = 2\omega r_3, \qquad \frac{dr_2}{dt} = 0, \qquad \frac{dr_3}{dt} = -2\omega r_1. \tag{4.105}$$

Thus, r_2 is a constant, and the others follow the second-order differential equations

$$\frac{d^2 r_1}{dt^2} = -4\omega^2 r_1, \qquad \frac{d^2 r_3}{dt^2} = -4\omega^2 r_3. \tag{4.106}$$

The initial condition was given in Eq. (4.91), which means

$$r_3(0) = p - \frac{1}{2}, \tag{4.107}$$

whereas $r_1(0)$ and $r_2(0)$ vanish, and r_0 is not written because it is irrelevant. Thus, r_2 remains zero at all times, whereas for r_1 and r_3 we obtain the solutions

$$r_1(t) = (p - \frac{1}{2}) \sin 2\omega t,$$
$$r_3(t) = (p - \frac{1}{2}) \cos 2\omega t. \tag{4.108}$$

From this information, we can construct the density matrix and find the expectation values of any operator.

☐ **Exercise 4.21** *Check that the density matrix is the same as that obtained through Eq. (4.93).*

If we consider a system whose state vectors belong to an n-dimensional vector space, we can still make the decomposition of ρ and H into a multiple of the unit matrix plus a linear combination of n^2-1 traceless matrices. Denoting these traceless matrices by λ_a, if they satisfy the commutation relation

$$\left[\lambda_a, \lambda_b\right] = 2i \sum_c f_{abc}\lambda_c, \qquad (4.109)$$

then these f_{abc}'s would replace the ε_{abc} in Eq. (4.101).

The coherence vector can be thought of as just a method of solving the evolution equation for the density matrix. The only small advantage is that this method shows that the terms in $\widehat{\rho}$ and $\widehat{H}$ that are proportional to the unit matrix are irrelevant for the time-evolution of probabilities.

Chapter 5

Symmetry

Symmetries play a very important role in quantum mechanics, much more so than in classical mechanics. The reason for this is the vector space structure of the state vectors of a quantum system, as will be clear in the discussion of this chapter.

5.1 Symmetry and conservation

Our intuitive notion of symmetry is through geometrical objects. A square is more symmetric than a rectangle, whereas a circle is more symmetric than both. We can make the notion quantitative if we consider what are the operations on these geometrical objects that produce a result that is indistinguishable from the original one. For a rectangle, rotations in its plane by 180° or its multiples produce an identical figure. For a square, rotation by 90° and its multiples do the job, and this is why a square is more symmetric than a rectangle. For a circle, any rotation in its plane leaves the shape unchanged, and that is why the circle is more symmetric than either the square or the rectangle.

We can extend this notion to mathematical expressions as well. For example, suppose we are considering a particle in a 1D space, with a potential that obeys the condition

$$V(x) = V(-x). \tag{5.1}$$

In this case, we can say that the potential is symmetric under the transformation that changes the sign of x, something that we will denote by $x \rightarrow -x$. From the Heisenberg equation of the system, Eq. (4.4, p. 83), we see that this transformation on x would also imply that p changes sign: $p \rightarrow -p$.

Since the kinetic energy term in the Hamiltonian is quadratic in p, it does not change under this transformation. Thus, the entire Hamiltonian is invariant under the transformation $x \to -x$ if the potential obeys Eq. (5.1).

Let us now discuss what happens if we apply a transformation to all vectors in a vector space and find that the physical consequences of the transformed vectors are indistinguishable from those of the original ones. In that case, we would call the transformation a symmetry transformation.

Transformations on any vector $|\Psi\rangle$ are represented by operators on the vector space. Let us denote the operator corresponding to the symmetry transformation, or the *symmetry! operator* for short, by $\mathscr{S}$. All physical consequences in quantum mechanics rely on matrix elements and expectation values of various operators representing physical observables, i.e., on quantities of the form $\langle \Psi_1|\mathscr{O}|\Psi_2\rangle$. Consider now the same matrix element after the symmetry transformation has been applied. The transformed states will be denoted by

$$|\mathscr{S}\Psi_1\rangle \equiv \mathscr{S}\,|\Psi_1\rangle\,, \qquad |\mathscr{S}\Psi_2\rangle \equiv \mathscr{S}\,|\Psi_2\rangle\,, \tag{5.2}$$

and the transformed operator by $\mathscr{O}'$. If there is no discernible difference in any experiment, that must mean that the matrix elements with or without the transformation are equal, i.e.,

$$\langle \mathscr{S}\Psi_1|\mathscr{O}'|\mathscr{S}\Psi_2\rangle = \langle \Psi_1|\mathscr{O}|\Psi_2\rangle\,. \tag{5.3}$$

From Eq. (2.80, p. 38), we know that

$$\langle \mathscr{S}\Psi| = \langle \Psi|\,\mathscr{S}^\dagger, \tag{5.4}$$

so Eq. (5.3) can be written as

$$\left\langle \Psi_1\left|\mathscr{S}^\dagger\mathscr{O}'\mathscr{S}\right|\Psi_2\right\rangle = \langle \Psi_1|\mathscr{O}|\Psi_2\rangle\,. \tag{5.5}$$

This means that the operators on the vector spaces transform according to the rule

$$\mathscr{S}^\dagger\mathscr{O}'\mathscr{S} = \mathscr{O}, \tag{5.6}$$

or as

$$\mathscr{O}' = (\mathscr{S}^\dagger)^{-1}\mathscr{O}\mathscr{S}^{-1}. \tag{5.7}$$

Now consider that before and after the transformation, we are dealing only with normalized states. This means that

$$\langle \Psi \,|\, \Psi \rangle = 1 \tag{5.8}$$

where Ψ stands for either Ψ_1 or Ψ_2, and also that

$$\langle \mathscr{S}\Psi \,|\, \mathscr{S}\Psi \rangle = \langle \Psi \,|\, \mathscr{S}^\dagger \mathscr{S} \,|\, \Psi \rangle = 1. \tag{5.9}$$

Thus, we obtain the consequence that

$$\mathscr{S}^\dagger \mathscr{S} = \mathbb{1}. \tag{5.10}$$

This is the defining relation for a unitary operator introduced in §2.10. Alternatively, we can write

$$\mathscr{S}^{-1} = \mathscr{S}^\dagger \tag{5.11}$$

and rewrite Eq. (5.7) as

$$\mathscr{O}' = \mathscr{S}\mathscr{O}\mathscr{S}^{-1}. \tag{5.12}$$

The Hamiltonian operator is somewhat special because it governs the time evolution of the system, as discussed in Chapter 4. Consider a state $|\Psi\rangle$ being transformed to a state $\mathscr{S}\,|\Psi\rangle \equiv |\mathscr{S}\Psi\rangle$ by the action of the symmetry operator. After some small time δt, $|\Psi\rangle$ would change to $(1 - iH\delta t/\hbar)\,|\Psi\rangle$, whereas $|\mathscr{S}\Psi\rangle$ would change to $(1 - iH\delta t/\hbar)\,|\mathscr{S}\Psi\rangle$. Assuming that the symmetry operation does not depend on time explicitly, these two states after the time δt should also be related by symmetry transformation, i.e., we should have

$$\mathscr{S}\big(1 - iH\delta t/\hbar\big)\,|\Psi\rangle = \big(1 - iH\delta t/\hbar\big)\,|\mathscr{S}\Psi\rangle, \tag{5.13}$$

which gives

$$\mathscr{S}H = H\mathscr{S}. \tag{5.14}$$

We can write this equation as

$$\mathscr{S}H\mathscr{S}^{-1} = H. \tag{5.15}$$

Comparing this with Eq. (5.12), we conclude that the Hamiltonian must be invariant under the symmetry transformation.

There are important consequences of this statement. First, note that Eq. (5.14) means that

$$\left[\mathscr{S}, H \right] = 0, \tag{5.16}$$

i.e., the symmetry operator commutes with the Hamiltonian. According to the discussion of §2.7.3, we can say that the eigenstates of the Hamiltonian should also be eigenstates of the symmetry operator. Second, the Heisenberg equation, Eq. (3.20, p. 61), tells us that

$$\frac{d\mathscr{S}}{dt} = 0, \tag{5.17}$$

i.e., the operator is conserved. Thus, symmetry implies conserved quantities, and conservation laws always help in understanding the dynamics of a system.

5.2   Symmetry and degeneracy

We now elucidate an aspect of symmetry for which the vector space structure of quantum mechanics plays a crucial role. Consider a symmetry operator $\mathscr{S}$. As shown in §5.1, this means that the operator $\mathscr{S}$ commutes with the Hamiltonian. According to the theorem proved in §2.7.3, this implies that the eigenvectors of $\mathscr{S}$ and H are identical. In other words, an eigenstate of H can be taken to be an eigenstate of $\mathscr{S}$ as well.

Suppose now we consider an energy eigenstate $|\Psi\rangle$:

$$H |\Psi\rangle = E |\Psi\rangle. \tag{5.18}$$

The energy eigenvalue is E. Now consider the Hamiltonian operator acting on the state $\mathscr{S}|\Psi\rangle$, where $\mathscr{S}$ is a symmetry operator and therefore obeys Eq. (5.16). Therefore,

$$H\mathscr{S}|\Psi\rangle = \mathscr{S}H|\Psi\rangle = E\mathscr{S}|\Psi\rangle. \tag{5.19}$$

This shows that $\mathscr{S}|\Psi\rangle$ is also an eigenstate of the Hamiltonian with the same eigenvalue.

This can mean one of the following two things. First, if the eigenstate of the Hamiltonian is non-degenerate, it means that $\mathscr{S}|\Psi\rangle$ must be the same state as $|\Psi\rangle$, apart from a possible numerical factor:

$$\mathscr{S}|\Psi\rangle = s|\Psi\rangle. \tag{5.20}$$

This s is then the eigenvalue of the operator $\mathscr{S}$ for the same state $|\Psi\rangle$, and a state can be characterized by the eigenvalues of both H and $\mathscr{S}$. In Chapter 7, we will encounter examples of such eigenstates, where the symmetry operator will be the parity operator $x \to -x$, already invoked in §5.1.

Alternatively, if $\mathscr{S}|\Psi\rangle$ is not the same as $|\Psi\rangle$, the collection of states of the form $\mathscr{S}|\Psi\rangle$, for all different symmetry transformations of the Hamiltonian, form a set of states with degenerate energy eigenvalues. In this sense, symmetry implies degeneracy of energy eigenstates. We will see examples of such degeneracies in Chapter 9 and beyond.

5.3 Symmetry and group

5.3.1 Definition of a group

A group is a mathematical structure with a set of elements and a rule for combining any two of them. We will denote the elements of the set by g_1, g_2, etc., and the combination rule by $\circ$. The combination rule must have the following properties in order that the structure defined by the rule and the set of elements form a group.

1. For any g_1 and g_2, the combination $g_1 \circ g_2$ must also be a member of the set.

2. The combination rule must be associative, i.e., $(g_1 \circ g_2) \circ g_3 = g_1 \circ (g_2 \circ g_3)$ for any three elements of the set.

3. There must exist one element of the group, e, for which $g \circ e = e \circ g = g$. This element e is called the identity element of the group.

4. For any element g of the set, there must exist an element $\bar{g}$ such that $g \circ \bar{g} = \bar{g} \circ g = e$. The element $\bar{g}$ is called the inverse of the element g.

We give some examples.

- The set of all integers forms a group under the combination rule of addition.

- The set of all real numbers except 0 forms a group under the combination rule of multiplication.

□ **Exercise 5.1** *Show that the identity element is unique for any group.*

☐ **Exercise 5.2** *Verify that all integers form a group under addition. What is the identity element of this group? What is the inverse of the integer 2?*

☐ **Exercise 5.3** *Show that the set containing only two numbers, +1 and −1, form a group under multiplication.*

☐ **Exercise 5.4** *Consider the set consisting of two numbers, 0 and 1. If the combination rule is defined to be "addition mod 2", show that this defines a group.* [**Note:** *"addition mod 2" of two numbers means adding up the numbers, dividing the result by 2 and keeping the remainder only.*]

Note what the definition of a group does *not* imply. We do not say anywhere that the combination rule has to satisfy $g_1 \circ g_2 = g_2 \circ g_1$. For some specific group, this condition might hold. If it does, the group is called a *commutative group* or an *Abelian group*. However, not all groups are commutative. We will see examples of groups that are not.

The definition of a group also does not say that the inverse of an element has to be *different* from the element itself. In fact, for any group, it is obvious from the definition that the inverse of the identity element is the identity element itself.

☐ **Exercise 5.5** *Show that any group with three elements must be Abelian.*

☐ **Exercise 5.6** *Show that for any element x of a finite group G, there must exist a positive integer N such that $x^N = e$, where e is the identity element.*

☐ **Exercise 5.7** *Show that for all elements x, y of an Abelian group G, $(x \circ y)^n = x^n \circ y^n$ for all integers n.*

☐ **Exercise 5.8** *Show that the solutions of the equation $z^5 = 1$ form a group under ordinary multiplication whose elements satisfy $T_m \circ T_n = T_{(m+n) \bmod 5}$.*

5.3.2   Symmetry group

Suppose now we consider all symmetry transformations on a system as the elements of a set. Such a set can have finite number of elements, in which case the symmetry group is discrete, or it can be a continuous group with infinite number of elements. We decorate this set with the combination rule that $S_1 \circ S_2$ means applying the two transformations successively, first applying S_2 and then S_1. It is quite clear that this set, with this rule of combination, will form a group. Let us explain.

A symmetry transformation does not affect any physical property of the system. Therefore, two such transformations applied in succession will also not affect any physical property of the system. So, for S_1 and S_2 both belonging to the set of all symmetry transformations, it is clear that $S_1 \circ S_2$ does as well. The identity element is the operation of not doing anything, i.e., the identity transformation on the vector space. The inverse of any transformation is just performing the transformation backwards, i.e., if

$$\mathscr{S} \left| \Psi \right\rangle = \left| \mathscr{S} \Psi \right\rangle, \tag{5.21}$$

then we define the inverse $\mathscr{S}^{-1}$ by the rule

$$\mathscr{S}^{-1} \left| \mathscr{S} \Psi \right\rangle = \left| \Psi \right\rangle. \tag{5.22}$$

For example, if $\mathscr{S}$ is the transformation of rotating every vector by $30°$ about a certain axis, $\mathscr{S}^{-1}$ would imply a rotation of $-30°$ about the same axis, i.e., a rotation of $30°$ in the opposite direction. It is also easy to convince oneself that the symmetry operations are associative, because if we perform three symmetry operations in succession, it would not matter how we might think of bunching them together.

5.4 Some important discrete groups

5.4.1 Parity

For 1D problems, the *parity* transformation is defined to be the transformation

$$x \rightarrow -x. \tag{5.23}$$

In other words, the transformation reverses the spatial coordinate. For this reason, it is also called *space inversion*. For a 3D problem, the parity transformation can similarly be defined as

$$\boldsymbol{r} \rightarrow -\boldsymbol{r}, \tag{5.24}$$

or, in other words, by

$$x \rightarrow -x, \qquad y \rightarrow -y, \qquad z \rightarrow -z. \tag{5.25}$$

It is important to comment here that for 2D problems, parity is *not* defined by the transformation

$$x \rightarrow -x, \qquad y \rightarrow -y. \tag{5.26}$$

The reason is that this is not an inversion at all. It represents a rotation of the coordinate axes by 180°. Thus, for 2D problems, parity is still defined by Eq. (5.23), keeping the y coordinate unchanged. More generally, we can say that in a space of arbitrary number of dimensions, the parity transformation is defined to be a transformation that changes the sign of an odd number of coordinates. To be specific, in any number of spatial dimensions, we can define the parity transformation by Eq. (5.23), keeping all other coordinates fixed. Change of the other coordinates can be inflicted by some rotation. Thus, in three dimensions, the transformation in Eq. (5.24) is equivalent to that in Eq. (5.23) and a rotation in the y-z plane by 180°.

If we apply the parity operator $\mathscr{P}$ on a state vector $|\Psi\rangle$, the coordinates in the corresponding wavefunction change sign. It means that

$$\langle r \,|\mathscr{P}| \,\Psi\rangle = \langle -r \,|\, \Psi\rangle. \tag{5.27}$$

Either from this or from the definition given in Eq. (5.24), it is clear that we must have

$$\mathscr{P}^2 = \mathbb{1}. \tag{5.28}$$

Thus, the eigenvalues of the parity operator can be only $+1$ or -1. If parity is a symmetry of the system, the eigenstates can be either even or odd under parity. In Chapter 7, we will see examples of both kinds.

5.4.2 Exchange symmetry

If a system contains two (or more) particles of the same kind, the particles will be indistinguishable. If there is any mark, of whatever sort, to distinguish one particle from the other, that obviously means that the particles are not identical. On a macroscopic system, one can put such marks without altering their behavior to any discernible extent. For example, on a bowling alley one can paint the numbers 1,2, etc., on the individual pins and study something about a specified pin. But if we put such a tag on an electron, its behavior will dramatically change. It cannot be done. As a result, a statement like "we have electron 1 at a point r_1 and electron 2 at a point r_2" will make no sense. All we can say is that there is one electron at r_1 and another one at r_2. It is meaningless to say which one is which. Thus, the operation of changing two identical particles would have no effect on the probabilities associated with the system, meaning that it is a symmetry operation. Such symmetries are called *exchange symmetries*. We will explore the effects of exchange symmetry in Chapter 14.

5.4.3 Time reversal

Time-reversal symmetry in quantum mechanics is somewhat special since it needs to be implemented by antiunitary operators. This requires a slightly detailed discussion. But before embarking onto that, let us briefly discuss its role in classical theories.

In classical mechanics the existence of time-reversal symmetry implies invariance under $t \to -t$. It is clear that such a change keeps coordinates invariant, i.e., $r \to r$ under this transformation while the momentum (or velocity) changes sign: $p \to -p$. Newton's equations, for time-independent forces, are clearly invariant under time-reversal transformation. Moreover, from the force equation for a particle with mass m and charge q in an electromagnetic field given by

$$\frac{d^2 r}{dt^2} = \frac{q}{m} \left(E + v \times B \right), \qquad (5.29)$$

we find that invariance under time reversal implies that

$$E \to E, \qquad B \to -B. \qquad (5.30)$$

☐ **Exercise 5.9** *Check that the transformation rules of Eq. (5.30) keep the Maxwell's equations invariant, provided the charge density ρ and the current density j satisfy $\rho \to \rho$ and $j \to -j$ under time-reversal transformation.*

☐ **Exercise 5.10** *Check that the Maxwell equations, as well as Eq. (5.29), are invariant under parity transformation as well, where under parity the transformations of the electric and magnetic fields are given by*

$$E \to -E, \qquad B \to B. \qquad (5.31)$$

Coming back to quantum mechanics, we note that an essential difference between the classical Newtonian equation and the Schrödinger equation is that the former has second-order derivatives with respect to time whereas the latter has first-order. Thus, if $|\Psi(t)\rangle$ is a solution of the Schrödinger equation for a certain time-independent Hamiltonian, it is not guaranteed that $|\Psi(-t)\rangle$ will also be a solution. So we should be circumspect in identifying the time-reversal operator.

We would want the operator be such that the classical relations regarding position, momentum, and angular momentum hold good, i.e., we want

$$\mathcal{T}\widehat{r}\mathcal{T}^{-1} = \widehat{r}, \qquad (5.32a)$$

$$\mathcal{T}\widehat{p}\mathcal{T}^{-1} = -\widehat{p}, \qquad (5.32b)$$

$$\mathcal{T}\widehat{L}\mathcal{T}^{-1} = -\widehat{L}, \qquad (5.32c)$$

whether these are the time-dependent operators in the Heisenberg picture or time-independent ones in the Schrödinger picture. However, in quantum mechanics, it already creates a peculiarity. The commutator of position and momentum changes sign under time reversal, because momentum does. Thus, the only way to keep the commutation relation $[x_a, p_b] = i\hbar\delta_{ab}$ intact is to require that in the right side we should have

$$\mathcal{T} i \mathcal{T}^{-1} = -i, \tag{5.33}$$

i.e., there is a complex conjugation involved. This means that on any linear combination of two states, the time-reversal operator acts as follows:

$$\mathcal{T} |\alpha_1 \psi_1 + \alpha_2 \psi_2\rangle = \alpha_1^* \mathcal{T} |\psi_1\rangle + \alpha_2^* \mathcal{T} |\psi_2\rangle. \tag{5.34}$$

The time-reversal operator is therefore not a linear operator in the sense defined in Eq. (2.40, p. 28). It can be called *antilinear*. Mathematical properties of such operators should be examined carefully.

An additional peculiarity is encountered when one applies this operator on states. Usually in quantum mechanics a symmetry is implemented via a unitary operator U which takes a state $|\psi\rangle$ to $|\psi'\rangle = U|\psi\rangle$. Since U is unitary this ensures that $\langle\psi|\phi\rangle = \langle\psi'|\phi'\rangle$. For time reversal this is not quite the case, since the requirement of complex conjugation makes the operator implementing to be of the form

$$\mathcal{T} = U C_0, \tag{5.35}$$

where U is a unitary operator and C_0 denotes complex conjugation. Note that complex conjugation is definitely an operator on the Hilbert space. When it acts on a vector $|v\rangle$, the result is a vector whose components are complex conjugates of that of $|v\rangle$. Obviously, Eq. (5.35) gives an antilinear operator, since the definition implies $C_0(\alpha |v\rangle) = \alpha^* C_0 |v\rangle$.

Now consider states $|\psi\rangle$ and $|\phi\rangle$ such that

$$|\psi\rangle = \sum_n c_n |n\rangle, \quad |\phi\rangle = \sum_n d_n |n\rangle, \tag{5.36}$$

where $|n\rangle$ forms a complete basis and c_n and d_n are complex coefficients. The operation of the time-reversal operator on these states yields

$$\mathcal{T}|\psi\rangle = |\psi'\rangle = \sum_n c_n^* U |n\rangle,$$

$$\mathcal{T}|\phi\rangle = |\phi'\rangle = \sum_n d_n^* U |n\rangle. \tag{5.37}$$

From Eqs. (5.36) and (5.37) it is easy to see that

$$\langle \mathcal{T}\psi | \mathcal{T}\phi \rangle = \langle \phi | \psi \rangle = \langle \psi | \phi \rangle^{*}. \tag{5.38}$$

Operators with this property are called *antiunitary*. Thus, the time-reversal operator is both antilinear and antiunitary. This highlights the effect of complex conjugation.

To see one consequence of this antiunitarity condition, let us consider the case when the two states appearing there are related through

$$|\phi\rangle = \mathcal{T}^{-1}\widehat{O}\mathcal{T}|\psi\rangle \tag{5.39}$$

for some Hermitian operator $\widehat{O}$. This means that $\mathcal{T}|\phi\rangle = \widehat{O}\mathcal{T}|\psi\rangle$. Then the left side of Eq. (5.38) turns out to be $\langle \mathcal{T}\psi | \widehat{O}\mathcal{T}|\psi\rangle = \langle \psi' | \widehat{O} | \psi'\rangle$. The right side, with this form of $|\phi\rangle$, becomes $\langle \psi | \mathcal{T}^{-1}\widehat{O}\mathcal{T}|\psi\rangle^{*}$. However, the complex conjugation sign is irrelevant since the left side, being the expectation value of a Hermitian operator in a certain state, must be real. So we can write

$$\left\langle \psi' \left| \widehat{O} \right| \psi' \right\rangle = \left\langle \psi \left| \mathcal{T}^{-1}\widehat{O}\mathcal{T} \right| \psi \right\rangle. \tag{5.40}$$

This equation can be interpreted as an operator identity. An operator is called symmetric under time reversal if $\mathcal{T}^{-1}\widehat{O}\mathcal{T} = \widehat{O}$, and antisymmetric under time reversal if $\mathcal{T}^{-1}\widehat{O}\mathcal{T} = -\widehat{O}$. Some examples of symmetric operators are $\widehat{x}$ and $\widehat{H}$, while antisymmetric operators include $\widehat{p}$ and $\widehat{L}$.

The invariance of H, the generator of time translation, under time-reversal operation deserves a slightly more detailed discussion. To this end, let us consider a time-reversed state $|\psi'; t = 0\rangle \equiv \mathcal{T}|\psi; t = 0\rangle$ and evolve it forward in time through an infinitesimal time window δt. Then clearly, if H is the Hamiltonian generating this time translation, one has

$$|\psi'; t = \delta t\rangle = \left(1 - \frac{i}{\hbar}H\delta t\right)\mathcal{T}|\psi; t = 0\rangle \tag{5.41}$$

If the system is indeed time-reversal invariant, this state must also be obtained by evolving the state $|\psi; t = 0\rangle$ backward in time for $t = -\delta t$, and then applying time reversal on the final state. In other words, we must have

$$|\psi'; t = \delta t\rangle = \mathcal{T}\left(1 + \frac{i}{\hbar}H\delta t\right)|\psi; t = 0\rangle. \tag{5.42}$$

Comparing Eqs. (5.41) and (5.42), we find that they lead to the operator identity

$$\mathcal{T}iH = -iH\mathcal{T}. \tag{5.43}$$

There are two possible ways of satisfying Eq. (5.43). The first is to allow H to anticommute with $\mathcal{T}$. If we do this, we find that for an eigenstate $|n\rangle$ of H with energy ϵ_n, we have

$$H\mathcal{T}|n\rangle = -\epsilon_n \mathcal{T}|n\rangle. \tag{5.44}$$

This tells us that corresponding to any state of arbitrarily large energy, the time-reversed state will have a correspondingly low energy. This violates the idea of having a ground state of a system, and is therefore untenable. So one chooses

$$[\mathcal{T}, H] = 0 \tag{5.45}$$

and $\mathcal{T}i = -i\mathcal{T}$ which, once again, shows the necessity of complex conjugation for implementing time-reversal symmetry.

The time-reversal transformation, applied twice on a state vector, must result in the same state. In quantum mechanics, however, a state is defined only up to an overall phase, so this means that the time-reversal operator should obey the property

$$\mathcal{T}^2 = e^{i\alpha}\mathbb{1}. \tag{5.46}$$

Then,

$$\mathcal{T}^3 |\psi\rangle = \mathcal{T}\left(\mathcal{T}^2 |\psi\rangle\right) = \mathcal{T}\left(e^{i\alpha}|\psi\rangle\right) = e^{-i\alpha}\mathcal{T}|\psi\rangle, \tag{5.47}$$

utilizing the antilinear property in the middle. On the other hand, $\mathcal{T}|\psi\rangle$ is a state, so $\mathcal{T}^2$ acting on it should follow the rule of Eq. (5.46), i.e., we should have

$$\mathcal{T}^3 |\psi\rangle = \mathcal{T}^2\left(\mathcal{T}|\psi\rangle\right) = e^{i\alpha}\mathcal{T}|\psi\rangle. \tag{5.48}$$

Comparing the two expressions, we conclude that α can be 0 or π, so it is possible to have

$$\mathcal{T}^2 = \pm\mathbb{1}. \tag{5.49}$$

An interesting consequence of time reversal follows from Eq. (5.38). The equation is valid for arbitrary states $|\psi\rangle$ and $|\phi\rangle$. Let us, in particular, consider $|\phi\rangle = \mathcal{T}|\psi\rangle \equiv |\mathcal{T}\psi\rangle$. Then Eq. (5.38) gives

$$\langle \mathcal{T}\psi | \mathcal{T}^2\psi \rangle = \langle \mathcal{T}\psi | \psi \rangle. \tag{5.50}$$

This is an identity if the state $|\psi\rangle$ obeys $\mathcal{T}^2 |\psi\rangle = |\psi\rangle$. But, according to Eq. (5.49), $\mathcal{T}^2$ can also have opposite sign. If we take a state on which $\mathcal{T}^2 |\psi\rangle = - |\psi\rangle$, then for such a state Eq. (5.50) gives

$$\langle \mathcal{T}\psi | \psi \rangle = - \langle \mathcal{T}\psi | \psi \rangle. \tag{5.51}$$

It means that the indicated inner product is zero, i.e., the states $|\psi\rangle$ and $|\mathcal{T}\psi\rangle$ are orthogonal. However, if time reversal is a symmetry, we saw in Eq. (5.45) the Hamiltonian commutes with $\mathcal{T}$. In that case, for an energy eigenstate $|\Psi\rangle$,

$$H(\mathcal{T}|\Psi\rangle) = \mathcal{T}H |\Psi\rangle = \mathcal{T}E |\Psi\rangle = E(\mathcal{T}|\Psi\rangle). \tag{5.52}$$

This exercise uses the fact that the energy eigenvalues are real, and shows that $\mathcal{T}|\Psi\rangle$ and $|\Psi\rangle$ have the same energy eigenvalue. They cannot be the same state since they are orthogonal. Hence, the two states must be degenerate. This result is called *Kramers' degeneracy*, after the name of Hendrik A. Kramers who first pointed it out. The relationship with time reversal was shown later by Eugene Wigner.

□ **Exercise 5.11** *A particle having an electric dipole moment d interacts with the electric field through the Hamiltonian*

$$H_E = -d\boldsymbol{S} \cdot \boldsymbol{E}, \tag{5.53}$$

where $\boldsymbol{S}$ is the spin, whose behavior under any transformation is the same as that of the angular momentum $\boldsymbol{L}$. Show that this interaction would violate both parity and time-reversal symmetries.

□ **Exercise 5.12** *Argue that a similar interaction with the magnetic field*

$$H_M = -\mu\boldsymbol{S} \cdot \boldsymbol{B}, \tag{5.54}$$

where μ is the magnetic dipole moment, does not violate either parity or time-reversal symmetries.

5.5 Group representations

5.5.1 Some examples as starters

Consider an operation F that consists of a reflection on a plane mirror. We obtain a flipped image. If we do it twice, the result is the same as the original object, i.e., the result of identity transformation. We can write this statement as $F^2 = I$, where I stands for the identity transformation. These two operations, F and I, form a group that we can call the reflection group. The composition rule for this group can be written in the form of a table:

$$
\begin{array}{c|cc}
 & I & F \\
\hline
I & I & F \\
F & F & I
\end{array}
\tag{5.55}
$$

The row headings and the column headings have been indicated explicitly, and the table entry gives the result of "(row heading) ∘ (column heading)". However, note that if we substitute the operation I by the number $+1$ and F by -1, and use ordinary multiplication of numbers as the composition rule, we obtain the same table. We can thus say that the operations I and F can be represented by the numbers $+1$ and -1. This means that, no matter how different I and F may be from numbers, as far as the group composition property is concerned they behave the same way as ordinary numbers do under ordinary multiplication.

To cite a more trivial example, we can say that if we had replaced both F and I by the number $+1$ in the table of Eq. (5.55), that would also be a valid table for ordinary multiplication. Trivially valid, agreed, but valid nonetheless. Indeed, this also constitutes a representation of the reflection group.

So far we have been talking about representation through a number. A number can be thought of as a special case of a matrix, viz., a matrix with just one row and one column. More generally, a representation consists of setting up a correspondence between the group elements g and a set of square matrices $R(g)$,

$$
g \to R(g)
\tag{5.56}
$$

in such a way that the relation

$$
R(g_1)R(g_2) = R(g_1 \circ g_2)
\tag{5.57}
$$

holds for any elements g_1 and g_2 of the group. Here, on the left side, the two matrices $R(g_1)$ and $R(g_2)$ should be multiplied by the standard matrix multiplication rule. Such representations are called *matrix representations*. The number of rows (or columns) of the matrix corresponding to each element is called the *dimensionality* of the representation.

Note that there is nothing in our definition of representation that says that $R(g_1)$ and $R(g_2)$ will have to be *different* matrices if $g_1 \neq g_2$. Thus, in our example above, we have found two different 1D matrix representations (yes, a rather awkward way of saying *number representations*, but it generalizes easily to higher-dimensional matrices) of the reflection group:

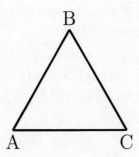

B

A   C

Figure 5.1 An equilateral triangle. The text discusses the symmetries of this triangle.

Representation serial	Representation of I	F
1	+1	+1
2	+1	−1

The second one can be called a *faithful representation* because it assigns a different matrix to each different element. The first one is *unfaithful*.

Clearly, a representation like the first one would work for any group. No matter what the group composition rule might be, it will always be satisfied if the elements entering into the composition as well as the result of the composition are all represented by the number +1. Thus, all groups have this trivial representation, which is 1D. In the case of the reflection group, the other representation that we had found was also 1D, but it is not trivial, i.e., it cannot be easily used for any other group.

Let us take another example of a group. Consider an equilateral triangle as shown in Figure 5.1 and consider which operations would produce a result that would be indistinguishable from the original figure. To make the argument clear, we have denoted the vertices of the triangle by the letters A, B, and C. We can now describe the symmetry transformations of this triangle as follows:

- An anticlockwise rotation by 120° about the center of the triangle, which takes the vertices A, B, and C respectively to C, A, and B. For the sake of brevity, we can call this operation simply by CAB, mentioning the final position of the three original vertices that were originally in alphabetical order.

- A rotation by 240° about the center of the triangle. In the notation we just introduced, it will be called BCA.

- A reflection about a mirror held perpendicular to the plane of the triangle along the line that passes through the left vertex and the mid-point of the opposite side. This operation will be denoted by ACB in the same notation.

- Again a reflection, but about a similar line that passes through the top vertex. This is operation CBA.

- Another reflection, this time about a line that passes through the right vertex. This is operation BAC.

- And finally, of course, the identity transformation, which does not do anything. We will write it as simply ABC in our notation.

With little trouble, it can be found that the group composition table has the following form:

	ABC	BCA	CAB	ACB	CBA	BAC
ABC	ABC	BCA	CAB	ACB	CBA	BAC
BCA	BCA	CAB	ABC	CBA	BAC	ACB
CAB	CAB	ABC	BCA	BAC	ACB	CBA
ACB	ACB	BAC	CBA	ABC	CAB	BCA
CBA	CBA	ACB	BAC	BCA	ABC	CAB
BAC	BAC	CBA	ACB	CAB	BCA	ABC

$$(5.58)$$

Each entry in the main 6×6 block of the table is obtained by the composition $R \circ C$, where R is the row heading given at the left, and C is the column heading appearing at the top.

☐ **Exercise 5.13** *Check and verify the group composition table in Eq. (5.58).*

For this group composition table, we can find the following representations:

1. All elements represented by the number +1. As we said earlier, this representation is valid for any group. This is a 1D representation.

2. Another 1D representation, in which ABC, BCA, and CAB are represented by +1 and the other three elements by −1.

3. A 2D representation with the following matrix assignments for the elements:

$$
\text{ABC}: \begin{pmatrix} 1 & 0 \\ 0 & 1 \end{pmatrix}, \qquad \text{BCA}: \begin{pmatrix} -\frac{1}{2} & -\frac{\sqrt{3}}{2} \\ \frac{\sqrt{3}}{2} & -\frac{1}{2} \end{pmatrix},
$$

$$
\text{CAB}: \begin{pmatrix} -\frac{1}{2} & \frac{\sqrt{3}}{2} \\ -\frac{\sqrt{3}}{2} & -\frac{1}{2} \end{pmatrix}, \qquad \text{ACB}: \begin{pmatrix} \frac{1}{2} & \frac{\sqrt{3}}{2} \\ \frac{\sqrt{3}}{2} & -\frac{1}{2} \end{pmatrix}, \tag{5.59}
$$

$$
\text{CBA}: \begin{pmatrix} -1 & 0 \\ 0 & 1 \end{pmatrix}, \qquad \text{BAC}: \begin{pmatrix} \frac{1}{2} & -\frac{\sqrt{3}}{2} \\ -\frac{\sqrt{3}}{2} & -\frac{1}{2} \end{pmatrix}.
$$

This assignment is not unique. If we denote the elements of the group g_i and the corresponding matrices by $R_i \equiv R(g_i)$, then another set given by

$$
R_i' = U R_i U^\dagger \tag{5.60}
$$

will work just as well, where U is any 2×2 unitary matrix. If we have two sets of matrices that are related as in Eq. (5.60), they represent the same operators in different basis, and we will call them as equivalent.

□ **Exercise 5.14** *In representation theory one can allow for an additional phase factor (called* cocycle*) while defining group representation. Such representations are called* projective *representations where Eq. (5.57) is modified to*

$$
D(R_1)D(R_2) = \omega(R_1, R_2)D(R_1 \circ R_2), \tag{5.61}
$$

where $\omega(R_1, R_2)$ is the phase factor or cocycle. Show that for such representations, the cocycles must satisfy

$$
\frac{\omega(R_1, R_2)}{\omega(R_2, R_3)} = \frac{\omega(R_1, R_2 \circ R_3)}{\omega(R_1 \circ R_2, R_3)}. \tag{5.62}
$$

□ **Exercise 5.15** *Consider the group with 8 elements which have a representation in the form of the matrices $\pm\mathbb{1}$, $\pm\sigma_z$, $\pm\sigma_x$, and $\pm i\sigma_y$, where $\mathbb{1}$ is the 2×2 identity matrix and the other ones involve the Pauli matrices defined in Eq. (4.26, p. 87). Find all cocycles for this representation.*

5.5.2   States and symmetry operators

The examples presented above help us in formulating the idea of representations in a more formal and mathematical language. In §2.6, we discussed that operators on a finite-dimensional vector space can be seen as matrices. The symmetry operators, represented by matrices, must be operators on some vector space. The elements of this vector space are the state vectors of the system under consideration. Thus, if we have an $n \times n$ matrix representation of a symmetry operator, it must act on an n-dimensional vector space. Any vector in this vector space, when operated by the symmetry operator, would produce a vector in the same vector space.

Let us now try to express these words in mathematical terms. Suppose we start with a state $|\psi\rangle$ and operate it with an operator $\mathcal{O}$. The result is a state $|\psi'\rangle$, i.e.,

$$|\psi'\rangle = \mathcal{O}\,|\psi\rangle. \tag{5.63}$$

We suppose that the state vectors are vectors in a finite-dimensional vector space. Then, using a complete set of basis vectors denoted by $|i\rangle$, we can write

$$\langle i\,|\,\psi'\rangle = \langle i\,|\mathcal{O}|\,\psi\rangle = \sum_k \langle i\,|\mathcal{O}|\,k\rangle\,\langle k\,|\,\psi\rangle. \tag{5.64}$$

The matrix element on the right side is called the element of the representation matrix for the operator $\mathcal{O}$. Denoting this matrix element by R_{ik}, we can say that if

$$\langle i\,|\mathcal{O}|\,\psi\rangle = \sum_k R_{ik}\,\langle k\,|\,\psi\rangle \tag{5.65}$$

for some matrix R, then R is a matrix representation of the operator $\mathcal{O}$.

5.5.3   Reducible and irreducible representations

Can we have a 2D representation for the reflection group discussed earlier? Sure, we can. For example, we can make an assignment like this:

$$I: \begin{pmatrix} 1 & 0 \\ 0 & 1 \end{pmatrix}, \qquad F: \begin{pmatrix} 1 & 0 \\ 0 & -1 \end{pmatrix}. \tag{5.66}$$

But this would be a joke! It is just taking the two different representations mentioned earlier and stacking them together. The first diagonal elements of these two matrices correspond to the numbers in the first representation, and the second diagonal elements correspond to the second representation. There is no connection between the first and the second rows (or columns), because the off-diagonal elements are zero.

On a similar note, we can ask whether we can construct a 3D representation of the symmetry group of an equilateral triangle. We can start with the 2D matrices R_i given in Eq. (5.59), and for each of them, write down the 3×3 matrices, which are of the form

$$\left(\begin{array}{c|cc} 1 & 0 & 0 \\ \hline 0 & & \\ 0 & & R_i \end{array} \right). \tag{5.67}$$

Obviously their multiplication would have also satisfied the group composition table, for the simple fact that the first 1×1 block corresponds to the trivial representation, and the next 2×2 block to the 2D representation shown in Eq. (5.59), and there is no communication between the two sectors. For this reason, representations like these are called *reducible representations*. On the other hand, the 2D representation of the triangle group shown in Eq. (5.59) cannot be written in such block diagonal form for all elements of the group, so it is an example of *irreducible representations*. All 1D representations are trivially irreducible.

It is therefore clear that, when talking about representations, it is enough to focus our attention on only irreducible representations. All reducible representations can be built from the irreducible ones. The irreducible ones are, in a sense, the basic representations.

5.5.4 Differential representations

So far in this section, we have been talking about finite-dimensional vector spaces, on which the group elements can be represented by matrices. Some quantum systems might also demand infinite-dimensional vector spaces. The definition of a representation in such spaces is obtained by extending the corresponding definition given for finite-dimensional spaces in Eq. (5.65). Suppose $|x\rangle$ denotes an eigenstate of the position operator, as discussed in Chapter 3. The exact analogy of Eq. (5.65) will then be obtained if we use these eigenstates and replace the summation by integration to write

$$\langle x | \mathcal{O} | \psi \rangle = \int dx' \, D(x, x') \, \langle x' | \psi \rangle, \tag{5.68}$$

where $D(x, x')$ acts on the wavefunction $\langle x' | \psi \rangle$. This $D(x, x')$ will then be a representation of the operator $\mathcal{O}$. We will look for representations of the form

$$D(x, x') = \delta(x - x') D_x, \tag{5.69}$$

so that we can write

$$\langle x | \mathscr{O} | \psi \rangle = D_x \langle x | \psi \rangle. \tag{5.70}$$

We will see that such objects are differential operators, and so representations like this are called *differential representations*.

Let us write Eq. (5.70) in a notation that will be easier to use. We use the notation of Eq. (5.63) to write

$$\langle x | \psi' \rangle = D_x \langle x | \psi \rangle. \tag{5.71}$$

The inner products appearing in this equation are just the wavefunctions, as explained in §3.5. Thus, we can write

$$\psi'(x) = D_x \psi(x). \tag{5.72}$$

The object D, therefore, shows how the functional form of a wavefunction changes when the states are operated by a symmetry operator. To be more precise, Eq. (5.70) implies that D_x is a differential representation of the operator $\mathscr{O}$ irrespective of whether the latter is a symmetry operator. The claim for the status of a symmetry operator will be settled only by considering the commutation with the Hamiltonian, as commented in Eq. (5.16). For specific symmetry operations, we will find the forms for these differential operators in §5.7.

5.6   Generators of group elements

5.6.1   Generators of discrete groups

While dealing with a specific group, it is not necessary to keep track of all elements separately or to think about the representation of each individual element. It is sufficient to keep track of a few elements and *generate* all other elements by performing the group multiplication operation on these chosen few. If we can identify a few elements from which all others can be obtained in this manner, the chosen few are called *generators* of the group.

For example, look at the group composition table of Eq. (5.58). The composition is a 6×6 table, as we can see. But all of these can be obtained if, for example, we start from the two elements

$$\langle \mathrm{BCA} \rangle \equiv a, \qquad \langle \mathrm{ACB} \rangle = b. \tag{5.73}$$

Note that

$$\langle \text{CAB} \rangle = \langle \text{BCA} \rangle \circ \langle \text{BCA} \rangle = a \circ a = a^2. \tag{5.74}$$

Therefore, we need not waste a new symbol for the element that we have denoted by CAB in Eq. (5.58): we can call it a^2, where the group composition symbol has been kept implicit. Next, note that we also have

$$\langle \text{CBA} \rangle = \langle \text{BCA} \rangle \circ \langle \text{ACB} \rangle = ab,$$
$$\langle \text{BAC} \rangle = \langle \text{CAB} \rangle \circ \langle \text{ACB} \rangle = a^2 b, \tag{5.75}$$

which shows that these elements can be expressed in terms of a and b. Finally, the identity element is identified by the relation

$$a^3 = b^2 = 1. \tag{5.76}$$

So we see that all elements have been written in terms of the two elements, which can be called the generators of the group. In fact, with this notation, it is not really important to remember the entire table given in Eq. (5.58). It is enough to give the information about only the generators. If we just give the information supplied in Eq. (5.76) and write

$$\langle a, b \,|\, a^3 = b^2 = 1 \rangle, \tag{5.77}$$

that would not be enough, because it does not yet contain the information about whether a and b commute. If we write

$$\langle a, b \,|\, a^3 = b^2 = 1, a^2 b = ba \rangle, \tag{5.78}$$

then that is enough: one can check that the table of Eq. (5.58) can be reproduced without any further information. This is called the *presentation* of the group. For symmetries such as space inversion, the group has only one generator, and the presentation is

$$\langle a \,|\, a^2 = 1 \rangle. \tag{5.79}$$

When we try to find representations for the group elements, it is enough to find the representation matrices for the generators. Representation of the other elements are obtained by matrix multiplication starting from the representation of the generators.

☐ **Exercise 5.16** *Check the relation $a^2 b = ba$ for the group given in Eq. (5.58) and construct the group composition table using the presentation.*

☐ **Exercise 5.17** *The presentation of a group is*

$$\langle a, b \, | \, a^n = b^2 = 1, (ab)^2 = 1 \rangle. \tag{5.80}$$

Show that the group has 2n elements.

☐ **Exercise 5.18** *Find the group generated by the matrices A and B under matrix multiplication where*

$$A = \begin{pmatrix} 1 & 0 & 0 \\ 0 & -1 & 0 \\ 0 & 0 & 1 \end{pmatrix}, \qquad B = \begin{pmatrix} 0 & 1 & 0 \\ 0 & 0 & 1 \\ 1 & 0 & 0 \end{pmatrix}. \tag{5.81}$$

5.6.2   Generators of Lie groups

In the department of continuous groups, generators are more of a necessity than a convenience. The reason is that one cannot use the group multiplication table like that shown in Eq. (5.58), simply because the number of elements is infinite. Therefore, one must seek some manner of writing the group elements in terms of a finite number of objects. Let us say that the group elements are functions of a finite number of parameters denoted by θ_a. For example, take the case of rotations in a 2D plane. The rotation can be by any angle, so the number of elements is infinite. However, any rotation can be specified by mentioning the angle of rotation, which is just one parameter that takes continuous values. For 3D rotations, one needs three such angles.

We will be concerned only with what are called *Lie groups*, for which the group elements are differentiable functions of these parameters. In that case, we can write an arbitrary group element g in the form

$$g = \exp(-i \sum_a \theta_a T_a), \tag{5.82}$$

where

$$T_a = i \left. \frac{\partial g}{\partial \theta_a} \right|_{\theta_a = 0 \, \forall a}. \tag{5.83}$$

These T_a's are called the generators of the group. They are operators on the vector space on which the elements of the group act. However, unlike the case of discrete groups, the T_a's in general are not group elements.

If we want to find the result of multiplication of two such elements, we need to multiply two exponential factors. Now, the catch is that the group elements, in general, do not commute. For non-commuting objects, the multiplication of two exponentials is given by the *Baker–Campbell–Hausdorff formula*:

$$e^A e^B = \exp\left(A + B + \frac{1}{2}[A, B] + \frac{1}{3!}\Big[A, [A, B]\Big] + \cdots\right), \qquad (5.84)$$

where

$$[A, B] = AB - BA, \qquad (5.85)$$

the *commutator* of the two operators A and B. This means that in order to be able to multiply two group elements, we need to know the commutators of their exponents, i.e., commutators of the generators. In particular, the commutator of any two generators must be a linear superposition of all generators so that the product denoted in Eq. (5.84) is also of the form of a group element. We can write this statement as

$$\Big[T_a, T_b\Big] = i \sum_c f_{abc} T_c, \qquad (5.86)$$

where the f_{abc}'s are constants, called *structure constants* of the Lie group. The collection of all commutators of the generators is called the *algebra* of the group.

When we talk about the representation of a Lie group, it is therefore enough to obtain representations of the generators. The representation of a group element can then be obtained through Eq. (5.82).

If the group combination happens to be commutative, the group is called an *Abelian group*. For such groups, all structure constants vanish. If there is at least one non-zero structure constant, the group is called a *non-Abelian* group. Already, in connection with Figure 2.1 (p. 31), we have argued that rotations in three dimensions do not commute. The algebra of this group will be discussed in §5.7.2, and again in Chapter 8.

□ **Exercise 5.19** *Prove the Baker–Campbell–Hausdorff formula.* [**Hint:** *Define* $g(s) = e^{sA} e^{sB}$. *Take* dg/ds *and use the Hadamard identity.*]

5.7 Important continuous groups

5.7.1 The translation group

To keep things simple, let us begin this discussion with the translation group in 1D. On the coordinate, the group operation means

$$x \to x' = x + \alpha. \qquad (5.87)$$

With the new coordinates x', one needs to use a new function to describe any physical situation. One needs to pick a function such that its value at a particular point remains unchanged. Thus, one should have

$$\psi'(x') = \psi(x), \qquad (5.88)$$

i.e., the value of the function should be the same, irrespective of whether one uses the old coordinate, x, or the new one, x'. Using Eq. (5.87), we therefore write

$$\psi'(x + \alpha) = \psi(x), \qquad (5.89)$$

which is equivalent to the statement

$$\psi'(x) = \psi(x - \alpha). \qquad (5.90)$$

As explained in Eq. (5.72), the differential operator should represent how this change will affect the wavefunctions. If we denote the generator by P, using Eq. (5.82) we should write

$$\exp(-i\alpha P)\psi(x) = \psi(x - \alpha). \qquad (5.91)$$

We can now use the power series expansion for the exponential on the left side and the Taylor series expansion on the right side. It is enough just to write down the expressions up to linear order in α:

$$(1 - i\alpha P)\psi(x) = \psi(x) - \alpha\frac{d}{dx}\psi(x). \qquad (5.92)$$

This shows that

$$P = -i\frac{d}{dx}, \qquad (5.93)$$

which is the differential representation of the generator.

This derivation is not restricted to quantum mechanics in any way. The differential representation of the translation operator is given by Eq. (5.93), no matter what kind of physics one might be doing. The specialty of quantum mechanics is the close similarity of this operator with the coordinate representation of the momentum operator. We find that

$$(\text{translation generator}) = \frac{(\text{momentum operator})}{\hbar}. \qquad (5.94)$$

This is the reason why we have denoted the translation generator by P, the momentum operator being called p.

For more than one dimension, one can define translation along each possible axes. The same argument would lead to the result

$$P_k = -i\frac{\partial}{\partial x_k} \tag{5.95}$$

for any orthogonal Cartesian coordinate x_k. One also sometimes writes it with a vector notation:

$$\boldsymbol{P} = -i\boldsymbol{\nabla}. \tag{5.96}$$

The algebra can be easily seen to be

$$\left[P_i, P_j\right] = 0, \tag{5.97}$$

since the partial derivatives commute. This algebra, and so the translation group, is Abelian.

Are there matrix representations of the translation group? There are, but the unitary representations are not faithful, and the faithful representations are not unitary. We are not interested in them for the purpose of quantum mechanics.

□ **Exercise 5.20** *Show that the matrix*

$$g_\alpha = \begin{pmatrix} 1 & \alpha \\ 0 & 1 \end{pmatrix} \tag{5.98}$$

constitutes a representation of translation by the amount α. Is this representation faithful? Is this matrix unitary?

□ **Exercise 5.21** *representation of the translation group, but it is not faithful.*

5.7.2 The rotation group

As in the case of the translation group, we start with the simple possible case: that of rotation in two dimensions. We denote the coordinates in this space by x_1 and x_2. A rotation by an angle θ inflicts the following changes in the coordinates:

$$\begin{aligned} x_1' &= x_1 \cos\theta - x_2 \sin\theta, \\ x_2' &= x_1 \sin\theta + x_2 \cos\theta. \end{aligned} \tag{5.99}$$

There is just one parameter that characterizes the rotation, so there should be one generator. Let us call it Λ. Taking only first-order terms in θ, the relation corresponding to Eq. (5.91) would be

$$\exp(-i\theta\Lambda)\psi(x_1, x_2) = \psi(x_1 + \theta x_2, x_2 - \theta x_1). \tag{5.100}$$

Expanding both sides as we did for the translation group, we obtain

$$\Lambda = -i\left(x_1\frac{\partial}{\partial x_2} - x_2\frac{\partial}{\partial x_1}\right). \tag{5.101}$$

For the 3D space, there will be three such generators, and they will be given by

$$\Lambda_j = -i\sum_{k,l}\varepsilon_{jkl}x_k\frac{\partial}{\partial x_l}, \tag{5.102}$$

where ε_{jkl} is the 3D Levi-Civita symbol, defined as something that is completely antisymmetric with respect to the interchange of any two indices. This leaves only one independent component, which is fixed conventionally by

$$\varepsilon_{123} = +1. \tag{5.103}$$

□ **Exercise 5.22** *Apply the differential operator for the 3D rotation generators on functions of three coordinates to show that the algebra of the generators is given by*

$$\left[\Lambda_1, \Lambda_2\right] = i\Lambda_3 \tag{5.104}$$

and its cyclic variants.

As in the case for the translation generators, it needs to be said that the differential representation of the rotation generators has nothing to do with quantum mechanics. The special aspect of quantum mechanics derives from the fact that the derivatives are equivalent to the momentum operator apart from a factor of $\hbar$. In fact, the angular momentum operators can be written as

$$\boldsymbol{L} = \boldsymbol{r} \times \boldsymbol{p}, \tag{5.105}$$

which in component form reads

$$L_j = -i\hbar\sum_{k,l}\varepsilon_{jkl}x_k\frac{\partial}{\partial x_l} \tag{5.106}$$

in the coordinate representation. Thus, we find the connection:

$$(\text{rotation generator}) = \frac{(\text{angular momentum})}{\hbar}. \tag{5.107}$$

These operators will satisfy the commutation relations

$$\left[L_i, L_j\right] = i\hbar \sum_k \varepsilon_{ijk} L_k. \tag{5.108}$$

More explicitly, it contains three independent commutators, given by

$$\left[L_x, L_y\right] = i\hbar L_z, \tag{5.109a}$$

$$\left[L_y, L_z\right] = i\hbar L_x, \tag{5.109b}$$

$$\left[L_z, L_x\right] = i\hbar L_y. \tag{5.109c}$$

There are big dissimilarities with the translation algebra. As Eq. (5.109) shows, the algebra is non-Abelian. Also, this algebra has matrix representations as well, which will be discussed in Chapter 8. The matrix representations are called *spin* operators, whereas the differential representation is called the *orbital angular momentum*.

Under rotation, all vectors transform the same way. It is therefore expected that the angular momentum operators shown above should have the same kind of commutation relations with the components of any vector operator. Indeed, that is true. It is easily seen that the commutators of the rotation generators with the components of the position vector are given by

$$\left[L_i, x_j\right] = i\hbar \sum_k \varepsilon_{ijk} x_k. \tag{5.110}$$

Also, the commutators with the momentum components are given by

$$\left[L_i, p_j\right] = i\hbar \sum_k \varepsilon_{ijk} p_k. \tag{5.111}$$

These commutators show the universal nature of the rotation generators.

☐ **Exercise 5.23** *Verify Eqs. (5.110) and (5.111), starting from the basic commutators involving position and momentum.*

☐ **Exercise 5.24** *Show that if any two vectors satisfy commutation rules like those in Eqs. (5.110) and (5.111), the commutators of L_j with their dot product vanishes, which is expected since scalars are unaffected by rotation.*

☐ **Exercise 5.25** *Practice some manipulations with the Levi-Civita symbols. In particular, verify the identity*

$$\varepsilon_{ijk}\varepsilon_{i'j'k'} = \det \begin{pmatrix} \delta_{ii'} & \delta_{ij'} & \delta_{ik'} \\ \delta_{ji'} & \delta_{jj'} & \delta_{jk'} \\ \delta_{ki'} & \delta_{kj'} & \delta_{kk'} \end{pmatrix}. \tag{5.112}$$

Use this to show that

$$\sum_i \varepsilon_{ijk}\varepsilon_{ij'k'} = \det \begin{pmatrix} \delta_{jj'} & \delta_{jk'} \\ \delta_{kj'} & \delta_{kk'} \end{pmatrix} = \delta_{jj'}\delta_{kk'} - \delta_{jk'}\delta_{kj'}, \tag{5.113a}$$

$$\sum_{i,j} \varepsilon_{ijk}\varepsilon_{ijk'} = 2\delta_{kk'}, \tag{5.113b}$$

$$\sum_{i,j,k} \varepsilon_{ijk}\varepsilon_{ijk} = 6. \tag{5.113c}$$

☐ **Exercise 5.26** *Every element of the Lie group SL(2,R) can be written in terms of real quantities a, b, and θ as*

$$T = \begin{pmatrix} \cos\theta & -\sin\theta \\ \sin\theta & \cos\theta \end{pmatrix} \exp\left[\begin{pmatrix} a & b \\ b & -a \end{pmatrix}\right]. \tag{5.114}$$

(a) *Show that* $\det T = 1$. [**Hint:** *You may use Eq. (2.159, p. 53).*]

(b) *Verify that these elements indeed form a group.*

(c) *Find the values of a, b, and θ for the element*

$$\begin{pmatrix} 2 & 1 \\ 1 & 1 \end{pmatrix}. \tag{5.115}$$

Part Two

Exact solutions

Chapter 6

The free particle

In this chapter, we describe the simplest possible system, viz., a free particle. To keep the notation simple, we will often pretend in this chapter that the particle moves only in one spatial dimension, so that there is only one coordinate x and its corresponding momentum p. However, the notation can be trivially modified to make the discussion applicable to motion of particles in 3D space.

6.1 Solution of free Schrödinger equation

We have introduced the Schrödinger equation in Chapter 3, which says that the time evolution of the state vector is governed by the Hamiltonian of the system:

$$i\hbar \frac{d}{dt} |\psi\rangle = H |\psi\rangle. \tag{6.1}$$

If the Hamiltonian does not have any explicit time dependence, the solution of this equation is simply

$$|\psi(t)\rangle = e^{-iHt/\hbar} |\psi(0)\rangle, \tag{6.2}$$

as given in Eq. (4.15, p. 85). This solution is valid irrespective of whether the particle is free. For a free particle, the extra simplification comes from the fact that the Hamiltonian involves only the momentum operator and not the position operator. Therefore, H commutes with all components of the momentum, and so the momentum eigenstates are also energy eigenstates. This means that we can use the eigenstates of the momentum operator as the energy eigenstates and write

$$H |p\rangle = E_p |p\rangle, \tag{6.3}$$

where E_p is the energy eigenvalue that depends on the momentum eigenvalue p. This feature can also be understood as a consequence of translational invariance. The application of the translation operator $T(a) = \exp[-i\widehat{p}a/\hbar]$, which takes $\psi(x)$ to $T(a)\psi(x) = \psi(x + a)$, does not change the energy of a free particle; consequently $\widehat{p}$ commutes with the Hamiltonian leading common eigenstates of both operators.

So we can take the inner product of both sides of Eq. (6.2) with a momentum eigenstate to write

$$\langle p\,|\,\psi(t)\rangle = \langle p\,|e^{-iHt/\hbar}|\,\psi(0)\rangle = e^{-iE_pt/\hbar}\,\langle p\,|\,\psi(0)\rangle, \qquad (6.4)$$

where we have used Eq. (6.3) to obtain the last step of Eq. (6.4). This shows how the momentum–space wavefunction evolves in time. In order to see the evolution of wavefunctions in the coordinate space, we take Eq. (6.4) and write

$$\int \frac{dp}{2\pi}\,\langle x\,|\,p\rangle\,\langle p\,|\,\psi(t)\rangle = \int \frac{dp}{2\pi}\,e^{-iE_pt/\hbar}\,\langle x\,|\,p\rangle\,\langle p\,|\,\psi(0)\rangle. \qquad (6.5)$$

Using the completeness relation of momentum eigenstates, Eq. (3.98, p. 78), we find that the left side of this equation is simply $\hbar\langle x|\psi(t)\rangle$. On the right side, using Eq. (3.72, p. 73), we obtain

$$\langle x\,|\,\psi(t)\rangle = \int \frac{dp}{2\pi\hbar}\,e^{-i(E_pt-px)/\hbar}\,\langle p\,|\,\psi(0)\rangle. \qquad (6.6)$$

The left side of this equation is the wavefunction at time t in the coordinate representation, which we write as $\psi(x, t)$. On the right side, the object $\langle p|\psi(0)\rangle$ is the momentum–space wavefunction at time $t = 0$, which we will write as $\widetilde{\psi}(p, 0)$. Eq. (6.6) represents the solution of the coordinate-space wavefunction $\psi(x, t)$ in terms of the initial condition in the form of the momentum–space wavefunction. The solution, of course, depends on this initial condition. Later in the chapter, we will examine the nature of the solution that results from specific initial conditions.

6.2   Wavepacket

We can certainly write

$$\langle p\,|\,\psi(0)\rangle = g(p)e^{i\alpha(p)}, \qquad (6.7)$$

where both $g(p)$ and $\alpha(p)$ are real functions, denoting the magnitude and the phase of $\widetilde{\psi}(p, 0)$. Let us consider a situation where the function $g(p)$ has the

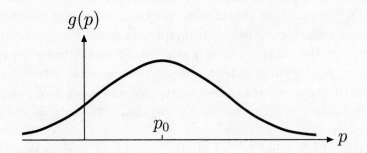

Figure 6.1 Schematic nature of the modulus function $g(p)$ appearing in the momentum-space wavefunction discussed in §6.2.

qualitative nature shown in Figure 6.1, i.e., it peaks around a certain value p_0 and dies out on both sides of the peak. Eq. (6.6) then tells us that the wavefunction in the coordinate representation is given by

$$\psi(x,t) = \int \frac{dp}{2\pi\hbar} \, \exp\left(-\frac{i}{\hbar}\left(E_p t - px - \hbar\alpha(p)\right)\right) g(p). \tag{6.8}$$

Such superpositions of momentum eigenfunctions over a range of values of the eigenvalue, which result in wavefunctions localized in some region of space, are called *wavepackets*.

The biggest contribution to the integral comes from values of p close to p_0, where $g(p)$ peaks. If, for a certain value of x, the phase factor varies wildly for values of p near p_0, contributions from different values of p will interfere destructively and we will not get a substantial contribution to $\psi(x,t)$. If, on the other hand, the variation of the phase factor is very small near p_0 for a certain value of x, the contributions will interfere constructively. The largest value of $|\psi(x,t)|$ will be obtained for $x = x_t$ where the variation of the phase factor vanishes, i.e., where

$$\frac{\partial}{\partial p}\left((E_p t - px_t - \hbar\alpha(p))\right)\Big|_{p=p_0} = 0. \tag{6.9}$$

This equation can be written in the form

$$x_t = vt + x_0 \tag{6.10}$$

where

$$v = \frac{\partial E_p}{\partial p}\Big|_{p=p_0}, \qquad x_0 = -\hbar\frac{d\alpha}{dp}\Big|_{p=p_0}. \tag{6.11}$$

Eq. (6.10) gives the time dependence of the coordinate of a classical particle moving in 1D. In quantum mechanics, we cannot specify the position of a particle because of the uncertainty principle. However, x_t is the point where the magnitude of the wavefunction peaks in the coordinate representation, which means it is the point where the particle is most likely to be found. Thus, Eq. (6.10) shows us that the position of the most likely occurrence of a free particle follows exactly the time dependence of the classical position of the particle.

We can also check what is the spatial extent of the wavepacket. As we move away from p_0 by an amount Δp, the change in the phase factor is given by

$$\Delta p \frac{\partial}{\partial p}\left(\frac{E_p t - px - \hbar\alpha(p)}{\hbar}\right)\bigg|_{p=p_0} = \frac{\Delta p (x - x_t)}{\hbar}, \tag{6.12}$$

using Eq. (6.10). Constructive interference will be obtained if this change in this phase factor is roughly within a quadrant, i.e., less than $\pi/2$. Thus, the width of the wavepacket in the coordinate representation will be given by

$$\Delta x \sim \frac{\pi\hbar}{2\Delta p}, \tag{6.13}$$

which is a manifestation of the uncertainty relation.

6.3   Spreading of wavepacket

In §6.2, we found how the peak of the wavepacket moves with time. We can also check how the spread of the wavepacket changes with time. In order to obtain analytical formulas, we assume a Gaussian form for $\langle p|\psi(0)\rangle$. We also take $p_0 = 0$ so that we can bring out the essential features of time evolution of a wavepacket without making the algebra overly cumbersome. Thus, we consider

$$\langle p\,|\,\psi(0)\rangle = N \exp\left(-a^2 p^2/\hbar^2\right), \tag{6.14}$$

where N is a normalization constant and a is also a constant. We have put an appropriate factor of $\hbar$ in the exponent so that a has the dimension of length, which will be convenient in the ensuing discussion. Eq. (6.8) now becomes

$$\psi(x,t) = N \int \frac{dp}{2\pi\hbar} \exp\left(-\frac{i(E_p t - px)}{\hbar} - \frac{a^2 p^2}{\hbar^2}\right). \tag{6.15}$$

For a non-relativistic particle of mass m, we have

$$E_p = \frac{p^2}{2m}. \tag{6.16}$$

Putting it in, the power of the exponential can be written in the form

$$\frac{1}{\hbar}\left[-bp^2 + ipx\right] = \frac{1}{\hbar}\left[-b\left(p - \frac{ix}{2b}\right)^2 - \frac{x^2}{4b}\right], \tag{6.17}$$

where $b = (a^2/\hbar) + (it/2m)$. The integration over momentum can be easily performed to produce the result

$$\psi(x,t) = N' \exp\left(-\frac{x^2}{4b\hbar}\right)$$

$$= N' \exp\left[-\frac{x^2\left(a^2 - (i\hbar t/2m)\right)}{4\left(a^4 + (\hbar^2 t^2/4m^2)\right)}\right], \tag{6.18}$$

where N' is different from N, but is still an overall normalizing factor that is not important. Thus, the probability of finding the particle at the point x at the time t is given by

$$\left|\psi(x,t)\right|^2 \propto \exp\left[-\frac{a^2 x^2}{2\left(a^4 + (\hbar^2 t^2/4m^2)\right)}\right]. \tag{6.19}$$

This is also a Gaussian function. It shows that the standard deviation of the position, Δx, is given by

$$(\Delta x)^2 = a^2 + \frac{\hbar^2 t^2}{4m^2 a^2} = \left(1 + \frac{\hbar^2 t^2}{4m^2 a^4}\right)(\Delta x)^2_{t=0}. \tag{6.20}$$

☐ **Exercise 6.1** *Note that the peak of the Gaussian of Eq. (6.19) always stays at $x = 0$. This is because we have assumed $p_0 = 0$ in the derivation. Show that if we take $\langle p|\psi(0)\rangle$ to be a Gaussian centered at $p = p_0$, the peak position would change with time in accordance with Eq. (6.10).*

☐ **Exercise 6.2** *The spread of a Gaussian wavepacket is such that its standard deviation becomes three times its initial value (at $t = 0$) at $t = 2m\ell^2/\hbar$. Find ℓ/a.* [**Hint:** *Use Eq. (6.20).*]

☐ **Exercise 6.3** *The amplitude at $x = a$ of a Gaussian wavepacket centered at origin changes to q times its initial value at $t = 2m\ell^2/\hbar$. Find the possible range of q as one varies ℓ. For what value of ℓ/a is $q = 3/2$?*

□ **Exercise 6.4** *The initial wavefunction of a free particle at $t = 0$ centered at x_0 is given by*

$$\psi(x,0) = \frac{1}{(2\pi a^2)^{1/4}} e^{-(x-x_0)^2/2a^2}. \tag{6.21}$$

Find $\psi(p,t)$, i.e., the momentum space representation of the wavefunction at time t.

There are a few important features of the result in Eq. (6.20). First, note that the standard deviation increases with time. It means that the wavepacket spreads with time. To put it differently, the uncertainty in position increases with time. This is intuitively expected: if you are not certain about the momentum at $t = 0$, you cannot be certain about the position at any later time. This uncertainty will increase monotonically with time.

Next, we note that the smaller the value of a, the bigger the spread factor. This means that if the spread of a wavefunction is smaller at an initial time, it spreads faster. This is because the small spread at the initial time implies a larger spread in momentum, which makes the wavefunction spread fast.

Another interesting feature of Eq. (6.20) is that the spread of the wave function depends on the mass of the particle. The spread for a heavy particle is smaller than that for a lighter particle in the same amount of time. This is one reason that such wavepacket spreading cannot be seen for macroscopic objects.

It would be instructive of have a quantitative idea about the magnitudes of the spreads involved. Since the position uncertainty at $t = 0$ is given by $(\Delta x)_{t=0}^2 = a\hbar$, we can write the time-dependent factor appearing in Eq. (6.20) in the form

$$1 + \frac{t^2}{4m^2 a^2} = 1 + \frac{\hbar^2 t^2}{4m^2 (\Delta x)_{t=0}^4}. \tag{6.22}$$

Consider how big the second term can be even for a very light particle, like a dust particle, of mass 10^{-3} g. Let us suppose that the position uncertainty at $t = 0$ to be of order 10^{-8} cm, about the size of an atom. Even then the extra term has the value of order $10^{-16} \times (t/1\,\mathrm{s})^2$. The position uncertainty changes appreciably when this term is comparable to or larger than 1, i.e., for $t \gtrsim 10^8$ s or about three years. For all ponderable periods of time of making any observation, the uncertainty therefore remains practically unchanged. In fact, we have taken a tiny particle of very small mass, and a ridiculously small position uncertainty at $t = 0$. With more reasonable values, it takes much more time for the wavefunction to have any appreciable spread for a macroscopic object.

We used a Gaussian wavepacket and demonstrated its spreading. However, it is not essential to make any assumption about the initial wavepacket. To see this, consider the Heisenberg equation of motion for the position operator

$\widehat{x}$. For the free Hamiltonian of a non-relativistic particle, it was shown in §4.1 that the solution is

$$\widehat{x}(t) = \widehat{x}(0) + \frac{\widehat{p}t}{m}. \tag{6.23}$$

This shows that the operator $\widehat{x}(t)$ does not commute with $\widehat{x}(0)$ for all $t \neq 0$. From Eq. (6.23), we obtain

$$\left[\widehat{x}(t), \widehat{x}(0)\right] = \frac{t}{m}\left[\widehat{p}, \widehat{x}(0)\right] = \frac{i\hbar t}{m}. \tag{6.24}$$

Both $\widehat{x}(t)$ and $\widehat{x}(0)$ are, of course, Hermitian operators. Therefore, we can use the theorem of Eq. (2.134, p. 48) to write

$$\Delta x(t)\,\Delta x(0) \geqslant \frac{\hbar t}{2m}. \tag{6.25}$$

Let us say

$$\Delta x(0) = a, \tag{6.26}$$

which is independent of time. Then we can rewrite this inequality in the form

$$\left(\Delta x(t)\right)^2 \geqslant \frac{\hbar^2 t^2}{4m^2 a^4}\left(\Delta x(0)\right)^2. \tag{6.27}$$

This is consistent with what we had obtained in Eq. (6.20) with a Gaussian wavepacket, and shows that any wavepacket should spread with time.

6.4 Propagator

Let us go back to Eq. (6.6). Using the expression for momentum eigenstates given in Eq. (3.72, p. 73), we can write this equation as

$$\langle x\,|\,\psi(t)\rangle = \int \frac{dp}{2\pi\hbar}\, e^{-iEt/\hbar}\,\langle x\,|\,p\rangle\,\langle p\,|\,\psi(0)\rangle\,. \tag{6.28}$$

More generally, if we put the momentum–space wavefunction at a time t' on the right side, we can write

$$\langle x\,|\,\psi(t)\rangle = \int \frac{dp}{2\pi\hbar}\, e^{-iE(t-t')/\hbar}\,\langle x\,|\,p\rangle\,\langle p\,|\,\psi(t')\rangle\,. \tag{6.29}$$

The momentum–space wavefunction can be written as

$$\langle p \,|\, \psi(t') \rangle = \int dx' \ \langle p \,|\, x' \rangle \langle x' \,|\, \psi(t') \rangle . \tag{6.30}$$

Putting Eq. (6.30) into Eq. (6.29), we obtain

$$\langle x \,|\, \psi(t) \rangle = \int dx' \ K(x, t; x', t') \langle x' \,|\, \psi(t') \rangle , \tag{6.31}$$

where

$$K(x, t; x', t') = \int \frac{dp}{2\pi\hbar} \ e^{-iE(t-t')/\hbar} \langle x \,|\, p \rangle \langle p \,|\, x' \rangle . \tag{6.32}$$

The function $K(x, t; x', t')$, through Eq. (6.31), expresses the coordinate space wavefunction at a certain time t in terms of the wavefunction at some other time t'. Said another way, it gives a connection between the coordinate space wavefunctions at two different times t and t'. This function is called the *propagator*.

☐ **Exercise 6.5** *More generally, i.e., even when the particle is not free, start from Eq. (6.2) to show that the propagator can be written as*

$$K(x, t; x', t') = \sum_n e^{-iE_n(t-t')/\hbar} \langle x \,|\, n \rangle \langle n \,|\, x' \rangle , \tag{6.33}$$

where the summation sign indicates a sum over discrete energy eigenstates $|n\rangle$ with energy eigenvalue E_n, or an integral over continuous energy eigenstates with a suitable integration measure.

The propagator can easily be calculated analytically for a free non-relativistic particle, for which

$$E = \frac{p^2}{2m}. \tag{6.34}$$

We then find that the propagator is of the form

$$K(x, t; x', t') = \frac{1}{2\pi\hbar} \int_{-\infty}^{+\infty} dp \ e^{i(-ap^2 + bp)}, \tag{6.35}$$

where

$$a = \frac{t - t'}{2m\hbar}, \qquad b = \frac{x - x'}{\hbar}. \tag{6.36}$$

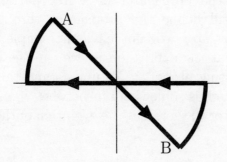

Figure 6.2 Contour in the complex plane for performing the integration that appears in Eq. (6.38).

For $a > 0$, i.e., $t' < t$, this integral can be evaluated. Changing the integration variable from p to $\xi = \sqrt{a}(p - (b/2a))$, we can write

$$K(x, t; x', t') = \frac{1}{2\pi\hbar\sqrt{a}} \exp(ib^2/4a) \int_{-\infty}^{+\infty} d\xi \, e^{-i\xi^2}. \qquad (6.37)$$

We now make a further change of the integration variable to $\xi' = \exp(i\pi/4)\xi = \sqrt{i}\,\xi$.

$$\int_{-\infty}^{+\infty} d\xi \, e^{-i\xi^2} = \frac{1}{\sqrt{i}} \int_{-\infty \exp(-i\pi/4)}^{+\infty \exp(-i\pi/4)} d\xi' \, e^{-\xi'^2}. \qquad (6.38)$$

The limits on the integration over ξ' mean that the integration has to be performed on the line AB of Figure 6.2, with a slope of $-\pi/4$ with the real axis.

In order to perform the remaining integration, we look at the closed contour in Figure 6.2. The integrand has no singularity anywhere on the complex plane, and therefore the integral around the contour should vanish. In the limit that the radius of the circular arcs are taken to infinity, the arcs do not contribute to the integral, and we obtain

$$\int_{-\infty \exp(-i\pi/4)}^{+\infty \exp(-i\pi/4)} d\xi' \, e^{-\xi'^2} = \int_{-\infty}^{+\infty} d\xi' \, e^{-\xi'^2} = \sqrt{\pi}. \qquad (6.39)$$

Putting in the results into Eq. (6.37), we obtain the propagator as

$$K(x, t; x', t') = \left(\frac{m}{2\pi\hbar i(t - t')}\right)^{1/2} \exp\left(\frac{im(x - x')^2}{2\hbar(t - t')}\right). \qquad (6.40)$$

This is, of course, the propagator of a free particle. In presence of interactions, we can still define the propagator through Eq. (6.31): the expression for the propagator will not have the simple form of Eq. (6.40). However, no matter what the expression is, we can see that the propagator must satisfy one interesting property. To understand this property, consider an instant of time t_1 which is somewhere in between t' and t, i.e., $t' < t_1 < t$. We can use the propagator to represent the evolution of the wavefunction from t_1 to t:

$$\langle x \,|\, \psi(t) \rangle = \int dx_1 \, K(x, t; x_1, t_1) \, \langle x_1 \,|\, \psi(t_1) \rangle. \tag{6.41}$$

The wavefunction at time t_1 appears in this equation, which can in turn be written as

$$\langle x_1 \,|\, \psi(t_1) \rangle = \int dx' \, K(x_1, t_1; x', t') \, \langle x' \,|\, \psi(t') \rangle. \tag{6.42}$$

Putting this back into the earlier equation, we obtain

$$\langle x \,|\, \psi(t) \rangle = \int dx_1 \, K(x, t; x_1, t_1) \int dx' \, K(x_1, t_1; x', t') \, \langle x' \,|\, \psi(t') \rangle. \tag{6.43}$$

Comparing this equation with Eq. (6.42), we obtain a transitive relation that has to be satisfied by any propagator:

$$K(x, t; x', t') = \int dx_1 \, K(x, t; x_1, t_1) K(x_1, t_1; x', t'). \tag{6.44}$$

The interpretation of this result is very simple. The particle goes from the position x' at a time t' to the position x at a time t. At any instant of time t_1 between t' and t, what will be the position of the particle? In absence of any measurement, we do not know the answer. This means that if we denote the position by x_1, we do not know what the value of x_1 is. Therefore we should sum over all such values, or in other words, integrate over x_1. This result has obvious extensions involving two or more intermediate values of time, and we will discuss this extension shortly.

6.5   Path integral formalism

Let us look at the exponential factor of Eq. (6.40). The velocity of the classical particle is given by

$$v = \frac{x - x'}{t - t'}. \tag{6.45}$$

So the exponent can be written as $\frac{1}{2}imv^2(t-t')/\hbar$. Note that the kinetic energy of the particle is $\frac{1}{2}mv^2$. For a free particle, this is also the same as its classical Lagrangian. Since velocity is constant for a free particle, the exponent can also be written as

$$\frac{i}{\hbar}\int_{t'}^{t} dt_1\ L(t_1) \equiv \frac{i}{\hbar}S_{\text{cl}}(t,t'). \qquad (6.46)$$

The symbol S appearing on the right side of this equation is known as the *action*, which is the time integral of the Lagrangian. The subscript 'cl' implies that the action has to be evaluated along the classical path taken by the particle.

We now use the generalization of Eq. (6.44) by introducing a number of intermediate time values to write

$$K(x,t;x',t') = \left(\prod_{i=1}^{n} \int dx_i\right) K(x,t;x_n t_n) \cdots K(x_2 t_2; x_1,t_1)$$
$$\times K(x_1,t_1;x',t'), \qquad (6.47)$$

such that every two successive moments of time are separated by a fixed amount ϵ. Using Eq. (6.40), we can write

$$K(x_{i+1}t_{i+1}; x_i t_i) = \frac{1}{F} \exp\left(\frac{i}{\hbar}\int_{t_i}^{t_{i+1}} d\tau\ L(\tau)\right), \qquad (6.48)$$

where

$$F = \left(\frac{2\pi\hbar i\epsilon}{m}\right)^{1/2}. \qquad (6.49)$$

Putting this back into Eq. (6.47), we obtain

$$K(x,t;x',t') = \frac{1}{F}\left(\prod_{i=1}^{n} \int \frac{dx_i}{F}\right) \exp\left(\frac{i}{\hbar}\int_{t'}^{t} d\tau\ L(\tau)\right). \qquad (6.50)$$

The notation has become a bit obscure here. There is integration over the x_i's, but the integrand does not show these variables explicitly. The point is that the action, appearing in the integrand, has to be calculated along the classical path that goes from the point x' at time t' to the point x at time t, through the points x_1 at time t_1, x_2 at time t_2, etc. But then, if there is an integration over all intermediate points, in the limit $n \to \infty$ it really means that we are really integrating over all possible paths. Thus, probably the content of Eq. (6.50) in this limit can be better expressed by writing

$$K(x, t; x', t') = \int [\mathcal{D}x_{\text{path}}] \, \exp\left(\frac{i}{\hbar} \mathcal{S}_{\text{path}}\right), \qquad (6.51)$$

where now the integration measure explicitly says that the integration is over all possible paths. This way, then, the evolution of the wavefunction is expressed as an integral over path, and this formulation of quantum mechanics is called the *path integral formulation*.

One can argue, of course, that from the discussion above, there is no way of convincingly concluding that the exponent contains the time integral of the Lagrangian. After all, the Lagrangian and the Hamiltonian of a free particle have the same classical value. So it is conceivable that the exponent contains the time integral of the Hamiltonian, and not of the Lagrangian.

Although a more rigorous treatment will be presented in §17.5, here we give two heuristic arguments to suggest that it is indeed the Lagrangian that appears in the exponent. The first is that it is the velocity and not the momentum that appears directly in the exponent. The Lagrangian is a function of the velocity, the Hamiltonian of the momentum.

The second thing is that, in the classical regime, motion of a system can be described by the minimization of the action. In other words, classical motion takes place on a path along which the action of the system is minimum. We now argue that in the classical limit, this result is recovered from the expression of Eq. (6.51).

In classical physics, dynamical variables are functions of time, which commute with one another. In quantum mechanics, as we have seen, the commutator of position and momentum contains the Planck constant $\hbar$. Thus, if we could tune this constant and take the limit $\hbar \to 0$, we should have recovered the classical results.

In the path integral of Eq. (6.51), all paths will contribute. If the magnitude of $\hbar$ is diminished indefinitely, i.e., if $\mathcal{S}/\hbar \to \infty$, even two nearby paths will contribute with very different phase factors. Contributions from many such paths will destructively interfere, and will cancel. However, near the minimum of the action, $\delta\mathcal{S} = 0$, so nearby paths will contribute with the same phase, and will therefore constructively interfere. Thus, contribution will come only from paths very close to the path of minimum action. The smaller $\hbar$ can be tuned, the more dominant the action along the classical path becomes. This is the *classical limit*.

In real life, one cannot really tune $\hbar$. So, physically it means that if the classical action corresponding to a system is very large compared to $\hbar$, quantum effects will be insignificant for this system, and the system can be described through the rules of classical mechanics. This statement, called the *correspondence principle*, acknowledges the fact that classical mechanics is a successful theory in its own regime, and ensures that its success is not spoiled by quantum mechanics.

In order to illustrate the applicability of classical mechanics, let us examine two systems, one for which quantum effects are negligible and another for which they are not. The first system is a dust particle of mass 10^{-3} g, moving a distance of $1\,\mathrm{mm}$ within a second. The action over the classical path can be easily calculated, and turns out to be $\frac{1}{2} \times 10^{-5}\,\mathrm{g\,cm^2\,s^{-1}}$. This is bigger than $\hbar$ by roughly a factor of 5×10^{21}. Quantum effects would be of no importance for such a particle. However, if an electron were traveling the same path in the same time, the action would be much smaller than $\hbar$, so the quantum effects would be pronounced.

☐ **Exercise 6.6** *Contemplate an atomic electron orbiting around the nucleus in a circular classical path of radius 10^{-8} cm. Calculate its speed. Estimate the action of the electron as it completes one revolution around the nucleus.*

6.6   Interference experiments

Although the path integral formulation is conceptually quite illuminating, it is usually very difficult to solve practical problems in this formulation. We shall discuss a few solvable cases in Chapter 17. Here we discuss interference experiments, where the application of the path integral formulation is simple and straightforward. We therefore go back to the interference experiments that we had outlined in §1.5.

In Figure 6.3, we reproduce the essential features of Figure 1.2 (p. 13): the source S to the left, the two slits A and B, and the screen on the right. We take an arbitrary point P on the screen and ask ourselves what would be the intensity of the particles at this point.

The easiest way of finding the answer utilizes the path integral approach. Since $\hbar$ is indeed small, we can take only the classical paths. The point of interest is that, because of the two slits, there are two classical paths. We should therefore add the contributions of these two paths to obtain the intensity of the particles at the point P. For the path passing through the slit A, the wavefunction of the particles reaching the point P is given by

$$\psi_{P|A} = \exp(iS_{\mathrm{SAP}}/\hbar)\psi_S, \qquad (6.52)$$

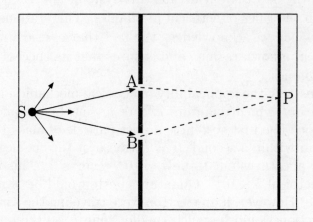

Figure 6.3 A two-slit interference experiment. The source is at S, the two slits at A and B, and P is an arbitrary point on the screen.

where $\mathcal{S}_{\mathrm{SAP}}$ denotes the classical action calculated along the path SAP. Similarly, for the path passing through the slit B, we obtain

$$\psi_{P|B} = \exp(i\mathcal{S}_{\mathrm{SBP}}/\hbar)\psi_S. \tag{6.53}$$

Thus, adding the two paths, we obtain

$$\psi_P = \psi_{P|A} + \psi_{P|B} = \left(\exp(\frac{i}{\hbar}\mathcal{S}_{\mathrm{SAP}}) + \exp(\frac{i}{\hbar}\mathcal{S}_{\mathrm{SBP}}) \right)\psi_S. \tag{6.54}$$

According to the discussion of §3.5.2, the probability of finding a particle at the point P is proportional to

$$|\psi_P|^2 = \left| \exp(\frac{i}{\hbar}\mathcal{S}_{\mathrm{SAP}}) + \exp(\frac{i}{\hbar}\mathcal{S}_{\mathrm{SBP}}) \right|^2 |\psi_S|^2$$

$$= 2\left(1 + \cos\frac{\mathcal{S}_{\mathrm{SAP}} - \mathcal{S}_{\mathrm{SBP}}}{\hbar} \right) |\psi_S|^2. \tag{6.55}$$

The interference pattern for this kind of experiments, where an initial beam is split and made to go through two different paths, depends therefore on the difference of the classical actions along the two paths. Loosely, this is summarized by saying that the interference pattern depends on the *path difference*.

For free particles not under the spell of any force, we can indeed interpret the result in terms of path difference. For free particles the Lagrangian is just the kinetic energy, which is conserved because there is no potential energy. Thus, the action will be given by the Lagrangian multiplied by time difference. Since the velocity of a free particle is constant, the time difference can be translated into path difference. This is the reason that we often calculate the path difference while analyzing interference experiments.

However, it is easy to contemplate systems where the difference in the lengths of two paths is not related to the difference of classical actions along the two paths. In fact, one can think of examples where two paths have the same lengths but the actions are different along the two paths. A beautiful experiment in this regard was performed by Colella, Overhauser, and Werner (COW), and reported in a paper published in 1975. They used a neutron beam, split it into two paths, and used strategically placed reflectors to make the two branches interfere. A schematic diagram of their set-up has been shown in Figure 6.4. In order to keep the geometry simple, we consider all four distances, SA, SB, AP, and BP, to be equal to ℓ. Obviously, the physical distances traversed by the two branches of the beam are the same, viz., 2ℓ. The two paths SAP and SBP form a rhombus, with one angle α as shown in the figure.

Suppose we now set up the whole system such that the plane of the rhombus is tilted with the horizontal. The arms SA and BP are both horizontal, but they are at different heights h_0 and h_1 from any arbitrarily fixed level. If the neutrons are released at the source with velocity v_0, the action along the path SA would be given by

$$\mathcal{S}_{SA} = (\tfrac{1}{2}mv_0^2 - mgh_0)\frac{\ell}{v_0}. \tag{6.56}$$

Action along the arm BP of the rhombus is given by a similar formula, with h_0 replaced by h_1, and also v_0 replaced by v_1, since the neutrons lose some speed in climbing the height within the earth's gravitational field. Therefore,

$$\mathcal{S}_{BP} - \mathcal{S}_{SA} = \frac{1}{2}m\ell(v_1 - v_0) - mg\ell\left(\frac{h_1}{v_1} - \frac{h_0}{v_0}\right). \tag{6.57}$$

In fact, this difference is also the action difference between the entire paths SAP and SBP, because the actions along the arms AP and SB are equal. To

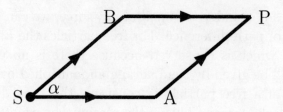

Figure 6.4 Schematic diagram for the neutron interference experiment described in the text. A neutral beam, originating at S, is split into two beams. One beam goes through A, the other through B. The splitter at S as well as the reflectors at A and B are not shown in the diagram.

write down the action difference in a compact way, we keep terms only in the first order in g, the acceleration due to gravity. Then

$$v_1 = \sqrt{v_0^2 - 2gh} \approx v_0 - \frac{gh}{v_0}, \tag{6.58}$$

where

$$h \equiv h_1 - h_0. \tag{6.59}$$

Therefore,

$$S_{\text{SBP}} - S_{\text{SAP}} = -\frac{3mg\ell h}{2v_0}. \tag{6.60}$$

The intensity of neutrons reaching the point P would therefore depend on this difference in action. The quantity termed h can be expressed in terms of ℓ and some angles. If the plane of the rhombus is vertical, clearly $h = \ell \tan \alpha$. If the plane of the rhombus subtends an angle θ with the vertical, we obtain

$$h = \ell \tan \alpha \cos \theta. \tag{6.61}$$

If we now vary θ, i.e., the inclination of the plane of the rhombus with the vertical, the intensity of neutrons at P would change in a well-defined way. The COW experiment measured this variation of intensity, thus providing a test of quantum mechanical effects induced by gravitational interactions.

□ **Exercise 6.7** *With $\theta = 0$, $\alpha = \pi/6$, and $\ell = 1\,\mathrm{m}$, estimate the distance between two successively maxima in the interference patterns for neutrons of energy $10\,\mathrm{MeV}$.*

□ **Exercise 6.8** *Suppose one carries out an Young's double-slit experiment with a particle of mass m, moving at a speed $0.1c$, where c is the speed of light in the vacuum. Find the linear distance between the nth maxima and minima in the interference pattern on a screen at a distance of $1\,\mathrm{m}$ from the slit. Estimate this distance for an electron, a proton, and a ball of mass $1\mathrm{g}$.*

Chapter 7

Exactly solvable problems in one dimension

In Chapter 6, we have discussed free particles. We carried out the discussion mostly in 1D space, although we mentioned that generalization to arbitrary dimensions is obvious and straightforward. We now start discussing particles in a potential. Problems of this kind and their solutions are quite sensitive to the number of spatial dimensions. In this chapter, we discuss particles subjected to a potential in a 1D space. Problems with higher-dimensional space will be discussed in later chapters of the book.

7.1 Stationary states

The Schrödinger equation, introduced in Chapter 3, admits particularly simple analytical solution in 1D for a variety of potentials $V(x)$. In this chapter, we are going to look at some of them to gain an insight into the methods of analytical solution of the Schrödinger equation in 1D.

From this chapter onward, we will work almost exclusively in the coordinate representation. From the discussion in §3.5.5, we know that the most general form of this equation for 1D is given by

$$i\hbar\frac{\partial\Psi(x,t)}{\partial t} = H[\widehat{p}, x; t]\Psi(x,t), \tag{7.1}$$

where $H[\widehat{p}, x; t]$ is the Hamiltonian, $\widehat{p} = -i\hbar\partial_x \equiv -i\hbar\partial/\partial x$ is the momentum operator, x is the position coordinate, and t denotes time. We will look for solutions of the form

$$\Psi(x,t) = e^{-iEt/\hbar}\psi(x), \tag{7.2}$$

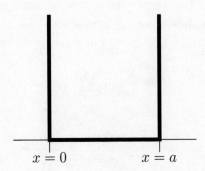

$x = 0$ $\quad\quad\quad\quad$ $x = a$

Figure 7.1 An infinite potential well in 1D. The thick line shows the plot of the potential as a function of x. The vertical lines indicate that the potential is infinite outside the lines.

where $\psi(x)$ is time-independent and can be interpreted as $\Psi(x,0)$, as hinted in Eq. (4.23, p. 86). It is easily seen, through Eq. (3.78, p. 74), that $\psi(x)$ satisfies the equation

$$-\frac{\hbar^2}{2m}\frac{d^2}{dx^2}\psi(x) + V(x)\psi(x) = E\psi(x) \tag{7.3}$$

for Hamiltonians of the general form

$$H = \frac{\widehat{p}^2}{2m} + V(\widehat{x}). \tag{7.4}$$

The solutions will be eigenstates of the Hamiltonian operator. We will denote the state corresponding to a fixed value of energy $E = E_n$ by $\psi_n(x)$, where the number n will be used to distinguish different states. Such numbers are called *quantum numbers*. The states obtained for all allowed values of n (whether discrete or continuous) are stationary states, since the expectation value of any operator in these states is independent of time. The central aim of this chapter is to find $\psi_n(x)$ and E_n corresponding to several specific potentials $V(x)$.

7.2 Particle in a box

We start with a very simple example that borders on the example of free particles described in Chapter 6. We consider a free particle in 1D, except that it can access only a finite part of the 1D space. In Figure 7.1, we show how this can happen. In the region $0 < x < a$, the particle is free to move because the potential vanishes there. Everywhere else, the potential is infinite. In summary, the Hamiltonian of a particle moving in this potential is

$$H = \frac{p^2}{2m} + V(x), \tag{7.5}$$

where

$$V(x) = \begin{cases} 0 & \text{for } 0 < x < a, \\ \infty & \text{elsewhere.} \end{cases} \tag{7.6}$$

The Schrödinger equation will therefore have two different forms: one inside the well and one outside. For the outside equation, the solution is very easy. Since $V(x)$ is infinite there, the only way that Eq. (7.3) can have a solution is to have

$$\psi_{\text{out}}(x) = 0, \tag{7.7}$$

where the subscript "out" implies the region outside the well, i.e., where $x < 0$ or $x > a$.

In the region inside where the potential is zero, Eq. (7.3) reduces to

$$\frac{d^2}{dx^2}\psi_{\text{in}}(x) + k^2 \psi_{\text{in}}(x) = 0, \tag{7.8}$$

where

$$k^2 \equiv \frac{2mE}{\hbar^2}. \tag{7.9}$$

This is a very familiar equation, and the general solution is of the form

$$\psi_{\text{in}}(x) = A\sin(kx + \delta), \tag{7.10}$$

for arbitrary constants A and δ.

We can obtain further information on the wavefunction from the fact that the wavefunction must be a continuous function of x. This means that we must impose

$$\psi_{\text{in}}(x) = \psi_{\text{out}}(x) \qquad \text{for } x = 0 \text{ and } x = a. \tag{7.11}$$

This implies that the constant δ in Eq. (7.10) must vanish, whereas the values of k must be such that

$$ka = n\pi \tag{7.12}$$

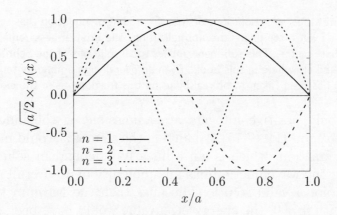

Figure 7.2 Nature of the wavefunctions for some energy eigenstates for the potential in Eq. (7.6). Outside the displayed range, the wavefunction is zero everywhere.

for some integer n. This is an interesting result. It shows that the particle cannot have any arbitrary value of energy. The only allowed values are of the form

$$E_n = \frac{\hbar^2 \pi^2 n^2}{2ma^2} \tag{7.13}$$

for integers n. Earlier, we have seen how the classical notion of the path of a motion becomes affected by the uncertainty principle. Now we encounter another hallmark of quantum mechanics, viz., the fact that the possible energy eigenvalues of a system need not be continuous. This phenomenon is called *quantization of energy*.

Coming back to the eigenstates, we find that the constant A of Eq. (7.10) can be determined only by imposing some kind of normalization condition on the wavefunction. If we use the usual one, i.e.,

$$\int_{-\infty}^{+\infty} dx\, |\psi(x)|^2 = 1, \tag{7.14}$$

then the constant A is determined, and the solutions can be written as

$$\psi_n(x) = \begin{cases} \sqrt{\dfrac{2}{a}} \sin \dfrac{n\pi x}{a} & \text{for } 0 < x < a, \\ 0 & \text{otherwise.} \end{cases} \tag{7.15}$$

In Figure 7.2, we show the nature of these solutions for some small values of the integer n.

> We emphasize that any solution of the Schrödinger equation has a multiplicative arbitrariness. Even after fixing the normalization, the arbitrariness remains in the form of a global phase factor. This comment applies to all solutions of the Schrödinger equation written here and elsewhere in this book. Even an overall sign ambiguity remains where the wavefunction has been taken to be real by using this freedom, as in the plot of Figure 7.2.

Note that only positive integers are admissible as the index n for the wavefunction given in Eq. (7.15). Changing the sign of n would merely change the sign of the function, which is not a linearly independent solution: in fact, any solution for the wavefunction has an ambiguity of an overall phase factor. Even $n = 0$ is not a valid solution, because then the wavefunction vanishes everywhere. With $n = 0$, the energy eigenvalue would have been zero. This is not possible in the present problem because of the uncertainty principle. The particle is constrained to stay within the region $0 < x < a$. Thus, the position uncertainty of the particle is given by $\Delta x \leqslant a$. The uncertainty relation then says that the momentum uncertainty obeys the relation $\Delta p \geqslant \hbar/a$. The expectation value of the magnitude of momentum must be at least as large as the uncertainty in the momentum, so the energy cannot vanish in any state. Also, note that the spacing between the successive energy levels $\delta E_n = E_{n+1} - E_n \sim 1/a^2$. Thus, the energy levels become more closely packed as the width of the well increases and the spectrum finally becomes continuous in the limiting case of $a \to \infty$.

☐ **Exercise 7.1** *For a particle in the well, find the position and momentum uncertainty in the state ψ_n and verify that these uncertainties obey the uncertainty relation.*

☐ **Exercise 7.2** *The wavefunctions inside the infinite well have been shown in Figure 7.2. Outside the well, the wavefunctions vanish, as stated in Eq. (7.7). This means that at $x = 0$ and $x = a$, the first derivative of $\psi(x)$ is discontinuous. By taking the integral of the Schrödinger equation from $x = -\epsilon$ to $x = +\epsilon$ where ϵ is a small number, show that this is expected.*

☐ **Exercise 7.3** *Take the walls to be at $-a/2$ and $+a/2$ and show that the same energies are obtained. What are the wavefunctions in this case?*

7.3 Some general features

Before trying to find solutions for other specific potentials, we make a few general remarks regarding the properties of ψ_n in 1D for generic $V(x)$.

The first of these is this: bound state wavefunctions for a 1D problem can never be degenerate. To see this, let us consider the Schrödinger equation for stationary state solutions, Eq. (7.3), which can be written as

$$\frac{d^2\psi(x)}{dx^2} + \frac{2m}{\hbar^2}\Big(E - V(x)\Big)\psi(x) = 0. \tag{7.16}$$

Let us also assume that there are two degenerate states $\psi_1(x)$ and $\psi_2(x)$ which are solutions of Eq. (7.16) with energy eigenvalue E_n. These eigenfunctions then must satisfy

$$\frac{1}{\psi_1(x)}\frac{d^2\psi_1(x)}{dx^2} = \frac{1}{\psi_2(x)}\frac{d^2\psi_2(x)}{dx^2} = \frac{2m}{\hbar^2}\Big(E_n - V(x)\Big). \tag{7.17}$$

This leads to

$$\psi_2(x)\frac{d^2\psi_1(x)}{dx^2} - \psi_1(x)\frac{d^2\psi_2(x)}{dx^2} = 0. \tag{7.18}$$

Integrating this relation with respect to x, one gets

$$\psi_2(x)\frac{d\psi_1(x)}{dx} - \psi_1(x)\frac{d\psi_2(x)}{dx} = c, \tag{7.19}$$

where c is a constant. We want to discuss bound state wavefunctions, which means that the wavefunctions should be normalizable. In Eq. (3.56, p. 70), we concluded that it requires the wavefunction to vanish at infinity. Thus, evaluating Eq. (7.19) at infinity, we find $c = 0$ so that

$$\psi_2(x)\frac{d\psi_1(x)}{dx} = \psi_1(x)\frac{d\psi_2(x)}{dx}. \tag{7.20}$$

One more integration leads to $\psi_1(x) = c'\psi_2(x)$ showing that ψ_1 and ψ_2 are same states up to an unimportant constant c'. Thus, our assumption of having degenerate states corresponding to a given energy must be at fault and there can be no such degeneracy.

The second statement regarding eigenfunctions corresponds to parity of the Hamiltonian and is as follows. If $V(x)$ is even under parity, the eigenfunctions of H must be either even or odd under parity. To see this, we note that if $V(x) = V(-x)$, then $\psi_n(x)$ and $\psi_n(-x)$ are both solutions of Eq. (7.16) with energy E_n. This can be seen by simply transforming $x \to -x$ in Eq. (7.16). Since we have shown that there can be no degeneracy, $\psi_n(x) = c_1\psi_n(-x)$ for some constant c_1. Changing the sign of x in the last relation once again, we find that $c_1^2 = 1$, which yields $c_1 = \pm 1$. Thus, one must have $\psi_n(x) = \pm \psi_n(-x)$, i.e., the eigenfunctions must have odd or even parity.

Finally, it can be shown that for 1D potentials which lead to discrete spectrum with eigenvalues E_n (higher energies indicate larger integer n and the ground state correspond to $n = 0$), the nth eigenfunction has n nodes (points where $\psi_n(x) = 0$) between $-\infty < x < \infty$. This statement is often called the *oscillation theorem*. We are not going to provide a rigorous proof of this theorem here. However, the qualitative logic behind the theorem can be stated as follows. From the structure of Eq. (7.16), it is easy to see that a higher curvature (i.e, higher value of the second derivative) of the wavefunction leads to higher energy. Now let us assume that the ground state is non-degenerate with minimum possible curvature, *i.e*, zero nodes. Then, from the orthogonality requirement of the ground and the first excited states, one finds that the first excited state must have an odd number of nodes. Since higher number of nodes leads to higher curvature and energy, we expect it to have one node. Similarly, the least number of nodes which allows the second excited state to be orthogonal to both the ground and the first excited states is two, and so on.

We can check these general comments against the solution of the problem in §7.2. In order to check parity, we first note that the potential given in Eq. (7.6) does not have the property $V(x) = V(-x)$. However, this can be easily remedied by taking the coordinate variable to be $x' = x - \frac{1}{2}a$ instead of x. Our statement about parity would then imply that the wavefunctions should be symmetric or antisymmetric for x' replaced by $-x'$, i.e., for x replaced by $a - x$. Indeed, from the plots of Figure 7.1 (p. 153) or even directly from the formula of Eq. (7.15), this feature can be seen very easily. As regards the oscillation theorem, again we need to look at Eq. (7.15) and remember that since the ground state has $n = 1$ for the solution of Eq. (7.15), we need to redefine n in order to check the oscillation theorem, which has been stated with a definition of n such that the ground corresponds to $n = 0$. Thus, in terms of the n that appears in Eq. (7.15), the nth eigenstate should have $n - 1$ nodes, which is what Figure 7.2 (p. 155) shows.

In the remaining part of this chapter, we focus on solutions of the Schrödinger equation for a few more specific forms of $V(x)$.

☐ **Exercise 7.4** *For a particle of mass m in an arbitrary potential with energy eigenvalues E_n, the lowest of which represents a bound state, prove*

$$\sum_{n=1}^{\infty} \frac{2m(E_n - E_0)}{\hbar^2} \left| \langle n \left| x \right| 0 \rangle \right|^2 = 1, \qquad (7.21)$$

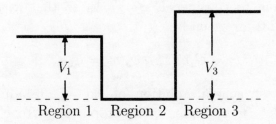

Figure 7.3 A potential well showing three regions 1, 2, and 3. The potential assumes finite constant values V_1 and V_3 in regions 1 and 3 while it vanishes in region 2.

> where $|n\rangle$ denotes the eigenstate corresponding to E_n. *This is known as the* Thomas–Reiche–Kuhn *sum rule.* [**Hint:** *First show that* $[x, [x, H]] = -\hbar^2/m$. *Then take the expectation value of both sides in the ground state* $|0\rangle$. *Finally, insert Eq. (2.33, p. 26) judicially to reach the result.*]

7.4 Square-well potential

The most general square-well potential has been shown in Figure 7.3:

$$V(x) = \begin{cases} V_1 & \text{for } x \leq 0, \\ 0 & \text{for } 0 < x < a, \\ V_3 & \text{for } x \geq a, \end{cases} \qquad (7.22)$$

where a is the width of the well, and V_1 and V_3 are parameters which determine its depth. For $V_1 = V_3 = V$, the well is said to be symmetric.

We call the different regions 1, 2, and 3, as shown in Figure 7.3, and outline the solution to the general problem. For finite V_1 and V_3, the wavefunction does not vanish for $x < 0$ and $x > a$ but are determined by the Schrödinger equations in regions 1 and 3 given by

$$\left(\frac{d^2}{dx^2} - \kappa_1^2 \right) \psi_1(x) = 0, \qquad (7.23a)$$

$$\left(\frac{d^2}{dx^2} - \kappa_3^2 \right) \psi_3(x) = 0, \qquad (7.23b)$$

where

$$\kappa_1 = \sqrt{2m(V_1 - E)/\hbar^2}, \qquad (7.24)$$

and κ_3 is defined the same way, using V_3 instead of V_1. In the rest of this section, we are going to consider $E \leq V_1, V_3$ so that $\kappa_{1(2)}$ are real. The solutions of Eq. (7.23) are given by

$$\psi_1(x) = C_1 e^{\kappa_1 x}, \quad \psi_3(x) = C_2 e^{-\kappa_3 x}. \tag{7.25}$$

Notice that the most general solution of Eq. (7.23) admits also a term proportional to $e^{-\kappa_1 x}$ in ψ_1 and a term proportional to $e^{+\kappa_3 x}$ in ψ_3. But these terms are not allowed because then the wavefunction diverges at infinity, and will not be square-integrable.

The solution in region 2 is still given by Eq. (7.10), except that the constants δ and A appearing there, as well as the constants C_1 and C_2 appearing in Eq. (7.25), are now to be fixed by matching boundary conditions at $x = 0$ and $x = a$ that are appropriate for this problem. These conditions are

$$\psi_1(x = 0) = \psi_2(x = 0), \quad \left.\frac{d\psi_1(x)}{dx}\right|_{x=0} = \left.\frac{d\psi_2(x)}{dx}\right|_{x=0},$$

$$\psi_3(x = a) = \psi_2(x = a), \quad \left.\frac{d\psi_3(x)}{dx}\right|_{x=a} = \left.\frac{d\psi_2(x)}{dx}\right|_{x=a}. \tag{7.26}$$

Substituting Eqs. (7.25) and (7.10) into Eq. (7.26), one obtains, after some straightforward algebra,

$$k \cot \delta = \kappa_1, \quad k \cot(ka + \delta) = -\kappa_3. \tag{7.27}$$

Eliminating δ from Eq. (7.27), we get

$$ka = n\pi - \sin^{-1}\left(\frac{ka}{\sqrt{V_1/E_\star}}\right) - \sin^{-1}\left(\frac{ka}{\sqrt{V_3/E_\star}}\right), \tag{7.28}$$

where we have used the expressions of κ_1 and κ_3 as defined through Eq. (7.24), n is an integer, and we have defined the shorthand

$$E_\star = \frac{\hbar^2}{2ma^2} \tag{7.29}$$

which can be called the typical kinetic energy of a particle confined within a box of length a. It is understood that the trigonometric inverses appearing in Eq. (7.28) are evaluated within the interval 0 and $\pi/2$. The bound states for the asymmetric square well can be obtained by numerical solution of the transcendental equation yielding discrete values of $k \equiv k_n$ and energies

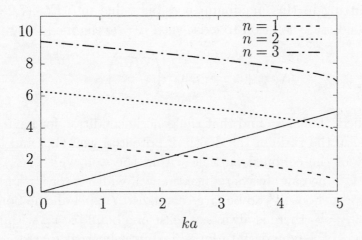

Figure 7.4 Graphical solution of Eq. (7.28) for $V_1 = 25E_\star$ and $V_3 = 36E_\star$. The straight line is the graph for the left side of the equation and the lines marked with different values of n represent the right side. Note that the range of ka is chosen in accordance with V_3 and V_1. In this case, there are two solutions, and therefore two bound states.

$E_n = \hbar^2 k_n^2 / 2m$ as shown in Figure 7.4 for representative values of V_1 and V_3. Note that in the limit of $V_1/E_\star$, $V_3/E_\star \to \infty$, one obtains $k_n = n\pi/a$; thus, we reproduce the infinite-well result as a special case. Also, for $V_1 = V_3 = V$, one obtains the result

$$ka = n\pi - 2\sin^{-1}\left(\frac{ka}{\sqrt{V/E_\star}}\right), \qquad (7.30)$$

where V is the common value of V_1 and V_3.

☐ **Exercise 7.5** *Deduce Eq. (7.28), starting from Eq. (7.27) and using the definitions given in Eq. (7.24).*

We close the section with a discussion of the minimal well depth required (for a given well width a) to have at least one bound state within the well. For the sake of this discussion, we assume that $V_1 \leqslant V_3$. (Although that is how the things were drawn in Figure 7.4, notice that in our discussion so far, we have never assumed any relation between V_1 and V_3.) Certainly the solution has to satisfy $ka \leqslant \sqrt{V_1/E_\star}$, because otherwise the inverse sine function does not exist. Putting in this maximum possible value into Eq. (7.28), we find that for a bound state solution to exist with $n = 1$, the inequality

$$\sqrt{\frac{V_1}{E_\star}} \geqslant \frac{\pi}{2} - \sin^{-1}\left(\sqrt{\frac{V_1}{V_3}}\right) \tag{7.31}$$

must be satisfied. Thus, we find that the generic condition for having a bound state depends on the ratio of the depth of potential well (V_1) and the typical kinetic energy of the confined particle within the well width ($E_\star$). We note that for $V_1 \neq V_3$, one can always reduce the well width a (or for a given width a, one can always choose V_3 to be large enough) to expel all the bound states. However, for $V_1 = V_3$ there is always at least one bound state within the well. Such a presence of a bound state for an arbitrarily weak/narrow symmetric square well is a property of 1D square wells; for higher dimensions, one would need a minimum well depth to have at least one bound state. This has been indicated in Ex. 9.4 (p. 229) later.

☐ **Exercise 7.6** *With the numerical values of $V_1/E_\star$ and $V_3/E_\star$ used for plotting Figure 7.4, show without the help of the graph that not more than two bound states can arise.*

7.5   Square potential barrier

The motion of a particle in the presence of a potential barrier presents one of the most striking distinguishing features between classical and quantum mechanics. In order to discuss the quantum mechanical problem, we employ a completely different strategy compared to what we did for potential wells. For wells, we solved the energy eigenvalues by solving the Schrödinger eigenvalue equation. For a barrier, we consider a particle coming with a given energy E and find what happens after it hits the barrier. If the height of the barrier is V, classical mechanics would tell us that the particle only gets reflected off the barrier if $E < V$. However, for a quantum particle, there is a probability of tunneling through the barrier even if $E < V$. To elucidate this, let us consider a square potential barrier defined in Figure 7.5:

$$V(x) = \begin{cases} V_0 & \text{for } 0 \leq x \leq a, \\ 0 & \text{otherwise.} \end{cases} \tag{7.32}$$

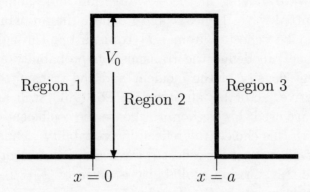

Figure 7.5 A square potential barrier showing three regions 1, 2, and 3. The potential is a finite constant V_0 in region 2 while it vanishes in regions 1 and 3.

As in the case of a square well, we divide the space into three regions 1, 2, and 3 as shown in Figure 7.5 and consider a particle approaching the barrier from the left in region 1. First, we consider the case $E < V_0$. Defining

$$k_1^2 = \frac{2mE}{\hbar^2}, \qquad k_2^2 = \frac{2m(V_0 - E)}{\hbar^2}, \qquad (7.33)$$

we find that the Schrödinger equations in these regions are given by

$$\left(\frac{d^2}{dx^2} + k_1^2\right)\psi_{1,3}(x) = 0, \qquad \left(\frac{d^2}{dx^2} - k_2^2\right)\psi_2(x) = 0. \qquad (7.34)$$

The solutions to these equations consist of left and right moving waves and can be written as

$$\psi_1(x) = e^{ik_1x} + re^{-ik_1x}, \qquad (7.35a)$$
$$\psi_2(x) = pe^{k_2x} + qe^{-k_2x}, \qquad (7.35b)$$
$$\psi_3(x) = te^{ik_1x}. \qquad (7.35c)$$

Remember that the full solution, including the time dependence of the wavefunction, is of the form shown in Eq. (7.2). Thus, for example, the first term in $\psi_1(x)$ has an exponential factor of $\exp i(k_1x - \omega t)$ where $\omega = E/\hbar$. This represents a wave moving along the $+x$ direction. This is the incident wave. The other part in ψ_1, by a similar argument, represents a wave moving in the $-x$ direction and therefore should be the reflected wave. Similarly, ψ_3 represents a wave moving in the $+x$ direction and should therefore be called the transmitted wave. Hence, r and t denote the probability amplitude of the particle to be reflected from and transmitted through the barrier. More precisely, one can define the transmission probability as the ratio of probability current density of the incident and the transmitted waves. An overall multiplicative factor for all terms in Eq. (7.35) will leave this ratio unaffected. We have used this freedom to choose the coefficient of the incident wave to be 1. With this choice, the reflection probability, defined as the ratio of the probability densities of the reflected wave and the incident wave, is given by $R = |r|^2$. The transmission probability is given by $T = 1 - |r|^2$, so that these quantities always satisfy $R + T = 1$.

As mentioned earlier, classically the particle will be reflected if $E < V_0$, so that we must have $R = 1$ and $T = 0$. However, we will see that the quantum mechanical treatment gives a non-zero value of T, implying that a particle can penetrate through a barrier that is higher than its energy.

To compute the transmission amplitude, we impose continuity of the wavefunction and its derivative at $x = 0$ and $x = a$ as in the case of a square well. This yields

$$1 + r = p + q,$$
$$ik_1(1 - r) = k_2(p - q),$$
$$pe^{k_2 a} + qe^{-k_2 a} = te^{ik_1 a},$$
$$k_2(pe^{k_2 a} - qe^{-k_2 a}) = k_1 te^{ik_1 a}. \tag{7.36}$$

Note that these equations do not depend on the normalization of the wavefunctions given in Eq. (7.35). Eliminating p and q from the above expressions, some straightforward algebra leads to

$$T = |t|^2 = \frac{4k_1^2 k_2^2}{4k_1^2 k_2^2 + (k_1^2 + k_2^2)^2 \sinh^2(k_2 a)},$$
$$= \left(1 + \frac{V_0^2}{4E(V_0 - E)} \sinh^2\left(\sqrt{(V_0 - E)/E_\star}\right)\right)^{-1}, \tag{7.37}$$

where $E_\star = \hbar^2/2ma^2$ as defined in Eq. (7.29).

We note that the function $\sinh x$ satisfies

$$\sinh x = \frac{e^x - e^{-x}}{2} \simeq \frac{e^x}{2} \text{ for } x \gg 1. \tag{7.38}$$

Thus, from Eq. (7.37), we find that when the energy difference $V_0 - E$ becomes large compared to the characteristic energy $E_\star$ that is related to the width of the barrier, T can be replaced by

$$T \simeq \left(\frac{16E(V_0 - E)}{V_0^2} \right) \exp \left(-2\sqrt{(V_0 - E)/E_\star} \right). \qquad (7.39)$$

In this regime, the tunneling probability therefore falls off exponentially with the barrier height.

In contrast, for $x \ll 1$ we have $\sinh x \simeq x$. Therefore, for $E \simeq V_0$, the transmission probability is given by

$$T \simeq \left(1 + \frac{V_0^2}{4EE_\star} \right)^{-1}. \qquad (7.40)$$

Next, let us consider the case $E > V_0$. A straightforward analysis exactly in the same line as the $E < V_0$ case yields

$$T' = \frac{4k_1^2 k_2'^2}{4k_1^2 k_2'^2 + (k_1^2 - k_2'^2)^2 \sin^2(k_2' a)} \qquad (7.41)$$

where $k_2'^2 = 2m(E - V_0)/\hbar^2$. In fact, Eq. (7.41) can be easily obtained from Eq. (7.37) by replacing k_2 in that equation by ik_2'. Note that $T' < 1$ as long as V_0/E is finite, which demonstrates that a quantum particle has a non-zero probability of being reflected even when its energy E is larger than the barrier height V_0. Finally, we point out that for $k_2' a = n\pi$, $T' = 1$ and the barrier becomes transparent in this limit. This phenomenon is called *transmission resonance*.

□ **Exercise 7.7** *Consider the square well problem in §7.4 but now with $E > V_1, V_3$.*

  (a) *By matching appropriate boundary conditions on the wavefunction and its derivative, show that the energy eigenvalues are continuous.*

  (b) *Find the transmission probability of the particle in the presence of the well.*

  (c) *Compare your result with the potential barrier problem addressed in this section for $V_1 = V_3$.*

□ **Exercise 7.8** *For a symmetric well, show that the eigenfunctions are either symmetric or antisymmetric under the transformation $x \to a - x$. Explain why it is so.*

□ **Exercise 7.9** *Consider a system with two potential barriers, each of height V_0 and width a_1, separated by a distance a_2. Find the transmission probability of a quantum particle of energy $E \le V_0$ which incidents on the left barrier from the left.*

□ **Exercise 7.10** *Consider an array of N square wells with $V = -V_0$ of equal width a which are separated by potential-free regions of width b. Such a model Hamiltonian is often called the Kronig–Penney model. Let the wavefunction in the nth region be denoted by*

$$\psi_n(x) = A_n e^{ik_n x} + B_n e^{-ik_n x}. \tag{7.42}$$

Then the relation between the coefficients appearing here and the ones appearing in the previous potential region would be of the form

$$\begin{pmatrix} A_n \\ B_n \end{pmatrix} = \begin{pmatrix} 1/t_n^* & -r_n^*/t_n^* \\ -r_n/t_n & 1/t_n \end{pmatrix} \begin{pmatrix} A_{n-1} \\ B_{n-1} \end{pmatrix} \tag{7.43}$$

where t_n and r_n are the transmission and reflection coefficients. The connecting matrix in this equation is called transfer matrix. Note that in this language propagation of a particle through free space by distance a leads to a diagonal transfer matrix with diagonal elements $\exp[\pm ika]$ where k's are the wavenumbers.

(a) *By matching appropriate boundary conditions on the wavefunction and its derivative, compute the transfer matrix T_n between wavefunctions across the nth well. Hence show that the eigenvalues of T_n are roots of the equation*

$$\lambda^2 + 2\lambda[\cosh(\alpha b)\cos(\beta a) + t_1 \sinh(\alpha b)\sin(\beta a)] + 1 = 0, \tag{7.44}$$

where $\alpha^2 = -2mE/\hbar^2$, $\beta^2 = 2m(E+V_0)/\hbar^2$, and $t_1 = (\alpha^2 - \beta^2)/(2\alpha\beta)$.

(b) *For $N \to \infty$, argue that the finiteness of the eigenvalues of the transfer matrix demands that $|\lambda| = 1$. Using this, write $\lambda = e^{i\theta}$. Hence show that the energies are given by the equation $\cos\theta = \cosh\alpha b \cos\beta a + t_1 \sinh\alpha b \sin\beta a$.*

(c) *Analyzing the left and right sides of the equation for energies derived earlier, show that there are allowed/forbidden regions of energies in which eigenvalues can/cannot exist. This demonstrates the formation of energy bands in infinite 1D periodic potentials.*

7.6   Spike potential

7.6.1   General features

In this section, we consider the effect of a very strong and very localized potential in 1D, idealizing it by a Dirac delta function:

$$V(x) = W\delta(x). \tag{7.45}$$

The energy eigenfunctions will satisfy the equation

$$-\frac{\hbar^2}{2m}\frac{d^2\psi(x)}{dx^2} = \left(E - W\delta(x)\right)\psi(x),\tag{7.46}$$

where E is the eigenvalue. We can divide the 1D space into two regions, one for $x > 0$ and one for $x < 0$. On either side, the potential is zero, so that the solutions are of the general form

$$\psi(x) = \begin{cases} A_1 e^{ikx} + B_1 e^{-ikx} & \text{for } x < 0, \\ A_2 e^{ikx} + B_2 e^{-ikx} & \text{for } x > 0, \end{cases}\tag{7.47}$$

where

$$k = \sqrt{\frac{2mE}{\hbar^2}}.\tag{7.48}$$

The two sides must match at $x = 0$ with appropriate conditions. First, the wavefunction must be continuous, which implies

$$A_1 + B_1 = A_2 + B_2.\tag{7.49}$$

The derivatives, on the other hand, would not be continuous. This can be seen by integrating Eq. (7.46) in a very small region containing the position of the potential. This gives

$$-\frac{\hbar^2}{2m}\left(\frac{d\psi}{dx}\bigg|_{x=+\epsilon} - \frac{d\psi}{dx}\bigg|_{x=-\epsilon}\right) = \int_{-\epsilon}^{+\epsilon} dx\,\left[E - W\delta(x)\right]\psi(x).\tag{7.50}$$

If we now take the limit of ϵ going to zero, the right side will not vanish because of the Dirac delta function. We will obtain

$$\lim_{\epsilon \to 0}\left(\frac{d\psi}{dx}\bigg|_{x=+\epsilon} - \frac{d\psi}{dx}\bigg|_{x=-\epsilon}\right) = \frac{2mW}{\hbar^2}\psi(0).\tag{7.51}$$

Using the general solution of Eq. (7.47), we see that this condition implies

$$(A_2 - B_2 - A_1 + B_1)ik = \frac{2mW}{\hbar^2}(A_1 + B_1).\tag{7.52}$$

The rest of the analysis would depend on the sign of W. If W is negative, the potential is a well. For positive W, we are talking of a potential barrier. These cases will have to be discussed separately.

☐ **Exercise 7.11** *What is the dimension of the constant W? Use it to check the dimensional consistency of Eq. (7.52).*

7.6.2   Bound state in a well

For $W < 0$, i.e., for an attractive potential, we want to check whether there can be a bound state. For a bound state, we write

$$E = -\mathscr{E}, \tag{7.53}$$

where $\mathscr{E}$ is positive. The quantity k defined in Eq. (7.48) is purely imaginary. The solutions shown in Eq. (7.47) will then have exponentially growing and falling parts. For a bound state, the wavefunction must vanish at infinity, so we will have solutions of the form

$$\psi(x) = \begin{cases} Ae^{\kappa x} & \text{for } x < 0, \\ Ae^{-\kappa x} & \text{for } x > 0, \end{cases} \tag{7.54}$$

where

$$\kappa = \sqrt{\frac{2m\mathscr{E}}{\hbar^2}}, \tag{7.55}$$

and we have used the condition for the continuity of the wavefunction at $x = 0$ to make the coefficients equal. Using now Eq. (7.51) which gives the discontinuity of the derivative, we obtain

$$-2\kappa A = \frac{2mW}{\hbar^2} A. \tag{7.56}$$

Plugging in the definition of κ, we obtain

$$\mathscr{E} = \frac{mW^2}{2\hbar^2}. \tag{7.57}$$

This is the bound state energy. There is only one bound state. The wavefunction for this state is given in Eq. (7.54).

7.6.3   Reflection and transmission at a barrier

We now consider the case $W > 0$. The potential is now a barrier, and we can discuss transmission and reflection through it, much like what we did in §7.5. Now k, defined in Eq. (7.48), is real, so we have plane wave solutions in Eq. (7.47). We will consider a plane wave coming from the left, part of which is reflected at, and partly transmitted through, the barrier at $x = 0$. Recalling now the arguments given in connection with Eq. (7.35), we now take $B_2 = 0$ and fix the normalization with $A_1 = 1$. Then Eq. (7.49) gives

$$1 + B_1 = A_2, \tag{7.58}$$

whereas Eq. (7.51) regarding the discontinuity of the derivatives gives the relation

$$(A_2 - 1 + B_1)ik = \frac{2mW}{\hbar^2}A_2. \tag{7.59}$$

Solving these equations, one obtains

$$B_1 = \frac{1}{\dfrac{i\hbar^2 k}{mW} - 1}, \qquad A_2 = \frac{1}{1 + \dfrac{imW}{\hbar^2 k}}. \tag{7.60}$$

Therefore, the reflection probability is

$$R = |B_1|^2 = \frac{1}{1 + \dfrac{2E\hbar^2}{mW^2}}, \tag{7.61}$$

and the transmission probability is $1 - R$.

□ **Exercise 7.12** *Show that these expressions for reflection and transmission probability follow from Eq. (7.37) in the limit $V_0 \to \infty$, $a \to 0$ with $V_0 a = W$.*

7.7 Simple harmonic oscillator

The Hamiltonian for a simple harmonic oscillator is

$$H = \frac{p^2}{2m} + \frac{1}{2}m\omega_c^2 x^2, \tag{7.62}$$

where m is the mass of the oscillator and ω_c is a constant. In this section, we will find the energy eigenvalues and eigenfunctions of this system, using two different methods.

The solution of eigenvalues and eigenvectors of this Hamiltonian is of immense practical importance since almost any complicated potential can be reduced, near its stable points, to a parabolic potential. Therefore, the low-energy spectrum of several complicated quantum systems near the stable points of their potentials is characterized by the solution of this Hamiltonian.

7.7.1 Coordinate space solution

In the coordinate representation, the stationary states of a simple harmonic oscillator will have wavefunctions of the form $\Psi(x,t) = \psi(x)\exp(-iEt/\hbar)$, where $\psi(x)$ satisfies the equation

$$E\psi(x) = \left(-\frac{\hbar^2}{2m}\frac{d^2}{dx^2} + \frac{1}{2}m\omega_c^2 x^2\right)\psi(x). \tag{7.63}$$

To solve this problem we scale all coordinates by the length scale

$$\ell_0 = \sqrt{\frac{\hbar}{m\omega_c}}. \tag{7.64}$$

Note that ℓ_0 is the only length scale that can be formed out of the parameters in the Hamiltonian (and the universal constant $\hbar$) and is therefore the natural choice for rescaling coordinates. Defining dimensionless coordinates and energies through the relations

$$\xi = \frac{x}{\ell_0}, \qquad \epsilon = \frac{2E}{\hbar\omega_c}, \tag{7.65}$$

we can rewrite Eq. (7.63) as

$$\epsilon\psi(\xi) = \left(-\frac{d^2}{d\xi^2} + \xi^2\right)\psi(\xi). \tag{7.66}$$

To solve this equation, we first note that, for very large ξ, the solution $\psi(x) = A\exp(-\xi^2/2)$ works, for an arbitrary constant A, if we consider only the leading power of ξ. This prompts us to try a general solution of the form

$$\psi(x) = A\exp(-\xi^2/2)v(\xi). \tag{7.67}$$

Substituting this solution into Eq. (7.66), we find

$$\frac{d^2v(\xi)}{d\xi^2} - 2\xi\frac{dv(\xi)}{d\xi} + (\epsilon - 1)v(\xi) = 0. \tag{7.68}$$

We compare Eq. (7.68) with the well-known Hermite differential equation discussed in detail in Appendix C. We see that the series solutions converge only if $\epsilon - 1$ is an even integer $2n$, and the solution in this case is called the *Hermite polynomial* $H_n(\xi)$. Thus, we write the solutions as

$$\epsilon_n = 2n + 1, \tag{7.69a}$$

$$\psi_n(\xi) = A_n e^{-\xi^2/2} H_n(\xi). \tag{7.69b}$$

Eq. (7.69a) readily gives

$$E_n = (n + \frac{1}{2})\hbar\omega_c. \tag{7.70a}$$

The coefficient A_n can now be fixed by demanding $\int dx |\psi_n(x)|^2 = 1$. This integral can be easily evaluated using the orthonormality relation of the Hermite polynomials (see Appendix D) and we finally obtain

$$\psi_n(x) = \left(\frac{1}{\ell_0\sqrt{\pi}2^n n!}\right)^{1/2} e^{-x^2/2\ell_0^2} H_n(x/\ell_0). \tag{7.70b}$$

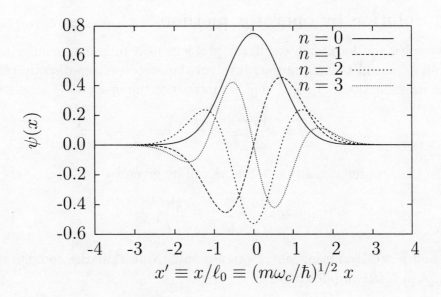

Figure 7.6 Plot of energy eigenfunctions for several energy levels of the simple harmonic oscillator. Notice the symmetry of each of these under the transformation $x \to -x$.

Note that these eigenvalues form an infinite tower of equally spaced energies with non-degenerate eigenstates. The eigenfunctions $\psi_n(x)$ can be easily shown, using the relation $H_n(x) = (-1)^n H_n(-x)$, to be odd under parity for odd integers n and even for even integers n. These eigenfunctions are plotted in Figure 7.6 as a function of ξ for several representative values of n. It is easy to see from these plots that $\psi_n(x)$ has n nodes between $\pm\infty$. This is because $H_n(x)$ is a polynomial of degree n, which has n zeros.

We note here that Eq. (7.68) is a second-order differential equation, and has therefore two linearly independent solutions. As discussed in Appendix B, the second solution can even be found by using the first solution, through Eq. (B.45, p. 437). In the present context, the other solution is irrelevant since it is not normalizable, and therefore cannot pertain to bound states.

□ **Exercise 7.13** *Consider the simple harmonic oscillator in the presence of an electric field $\mathcal{E}$ which adds a term $V(x) = -e\mathcal{E}x$ in the Schrödinger equation. Find the energy eigenvalue and eigenfunctions for such an oscillator.*

7.7.2   Solution by operator method

Next, we solve the harmonic oscillator problem in a different manner using a much more useful technique that involves the state vectors directly rather than the wavefunctions. To this end, we introduce the operator

$$a = \frac{1}{\sqrt{2}} \left(\frac{\widehat{x}}{\ell_0} + i \frac{\widehat{p}\ell_0}{\hbar} \right). \tag{7.71}$$

The Hermitian conjugate of this operator will be given by

$$a^\dagger = \frac{1}{\sqrt{2}} \left(\frac{\widehat{x}}{\ell_0} - i \frac{\widehat{p}\ell_0}{\hbar} \right), \tag{7.72}$$

since $\widehat{x}$ and $\widehat{p}$ are Hermitian operators anyway. Note that the commutation relation $[\widehat{x}, \widehat{p}] = i\hbar\mathbb{1}$ implies that

$$\left[a, a^\dagger\right] = \mathbb{1}. \tag{7.73}$$

In what follows, we are going to use the state vectors $|n\rangle$ that are related to the wavefunctions by $\langle x|n\rangle = \psi_n(x)$. To understand the action of a and $a^\dagger$ on these state vectors, we note that the Hamiltonian of the simple harmonic oscillators can be represented in terms of a and $a^\dagger$ as

$$H = \hbar\omega_c (a^\dagger a + \frac{1}{2}). \tag{7.74}$$

This statement can be directly verified by computing $a^\dagger a$ in terms of $\widehat{x}$ and $\widehat{p}$ using the definitions of a and $a^\dagger$ in Eqs. (7.71) and (7.72). It is now easy to check, using Eq. (7.73), the following relations:

$$\left[H, a\right] = -\hbar\omega_c a, \tag{7.75a}$$

$$\left[H, a^\dagger\right] = +\hbar\omega_c a^\dagger. \tag{7.75b}$$

Suppose now there is an energy eigenstate $|\psi\rangle$ with eigenvalue E:

$$H|\psi\rangle = E|\psi\rangle. \tag{7.76}$$

Then, using Eq. (7.75a), we find

$$Ha\,|\psi\rangle = (aH - \hbar\omega_c a)\,|\psi\rangle = (E - \hbar\omega_c)a\,|\psi\rangle\,, \qquad (7.77)$$

meaning that $a\,|\psi\rangle$ is also an energy eigenstate whose eigenvalue is lower by an amount $\hbar\omega_c$. This leads to the name *lowering operator* for the operator a. A similar argument shows that the state $a^\dagger\,|\psi\rangle$ is higher than that of the state $|\psi\rangle$ by an amount $\hbar\omega_c$, and so $a^\dagger$ is called the *raising operator*. Together, they are termed the *ladder operators*. Alternatively, the operation of $a^\dagger$ on an energy eigenstate is seen as the creation of a packet of energy $\hbar\omega_c$ in the state and therefore $a^\dagger$ is called the *creation operator*. Similarly, a is called the annihilation operator because its action seems to annihilate a packet of energy $\hbar\omega_c$.

It might seem that by applying the operator a, one can obtain states of arbitrarily low values of energy. That would be a disaster, because the system will not have any ground state. The avalanche can be stopped only if there is an energy eigenstate $|0\rangle$ with the property that

$$a\,|0\rangle = |\Omega\rangle\,, \qquad (7.78)$$

where we have the null vector on the right side. If that happens, applying further factors of the operator a will not produce any new state: the result will always be the null vector. Thus, $|0\rangle$ will be the ground state, i.e., the lowest energy state. Using Eq. (7.74), it is now easy to see that

$$H\,|0\rangle = \frac{1}{2}\hbar\omega_c\,|0\rangle\,, \qquad (7.79)$$

i.e., the ground state energy is

$$E_0 = \frac{1}{2}\hbar\omega_c. \qquad (7.80)$$

Quite often, Eq. (7.78) is written as $a\,|0\rangle = 0$. Strictly speaking, this is incorrect, since an operator acting on a state produces a state, not a number. However, this sloppy notation does not cause any harm, because ultimately one has to deal with matrix elements, and any matrix element involving the null vector is always zero.

The higher states will have energies in steps of $\hbar\omega_c$, as indicated by Eq. (7.75b). The state obtained by applying the operator $(a^\dagger)^n$ will have the energy shown in Eq. (7.70a). The normalized state, to be denoted by $|n\rangle$, will involve some numerical constants which will be given below, in Eq. (7.85).

From the eigenvalue equation for H and the eigenvalues just described, it follows that

$$a^\dagger a\,|n\rangle = n\,|n\rangle\,. \qquad (7.81)$$

Thus, energy eigenstates are also the eigenstates of $a^\dagger a$ with integer eigenvalues n. Thus, the energy of any state can be thought of as the sum of a ground state energy $\frac{1}{2}\hbar\omega$ plus the sum of n quanta, each of which has an energy $\hbar\omega$. Since the number of such quanta is the eigenvalue of $a^\dagger a$, the operator $a^\dagger a$ is called the *number operator*.

Next, let us examine the effect of a and $a^\dagger$ on the eigenstates $|n\rangle$. To this end, we operate on the state $a^\dagger |n\rangle$ with $a^\dagger a$ from the left to obtain $(a^\dagger a)a^\dagger |n\rangle$. Using the commutation relation of a and $a^\dagger$, we find that

$$(a^\dagger a)a^\dagger|n\rangle = a^\dagger(aa^\dagger)|n\rangle = a^\dagger(a^\dagger a + \mathbb{1})|n\rangle = (n+1)a^\dagger|n\rangle. \qquad (7.82)$$

Thus, the state $a^\dagger|n\rangle$ is an eigenstate of $a^\dagger a$ with eigenvalue $n+1$. This leads us to conclude that $a^\dagger|n\rangle = c_n|n+1\rangle$, where c_n is a normalization constant. To fix this constant, we compute the normalization of the state $a^\dagger |n\rangle = c_n |n+1\rangle$. Note that $(a^\dagger |n\rangle)^\dagger = \langle n|a$, so we find

$$|c_n|^2 = \|a|n\rangle\|^2 = \langle n|aa^\dagger|n\rangle = n+1, \qquad (7.83)$$

where we have used the commutation relation of Eq. (7.73). An identical analysis, which we urge the reader to carry out, shows that $a|n\rangle = c'_n |n-1\rangle$ with $c'_n = \sqrt{n}$. To summarize,

$$a^\dagger |n\rangle = \sqrt{n+1}\,|n+1\rangle\,, \qquad (7.84a)$$
$$a\,|n\rangle = \sqrt{n}\,|n-1\rangle\,. \qquad (7.84b)$$

Of course, there can be factors of extra phases depending on the phase convention, since Eq. (7.83) and the similar equation for c'_n determine only the modulus squares of the constants.

Eq. (7.84) can be used to write the state $|n\rangle$ directly in terms of the ground state ket $|0\rangle$. Using $a^\dagger |0\rangle = |1\rangle$, then $(a^\dagger)^2 |0\rangle = a^\dagger |1\rangle = \sqrt{2}\,|2\rangle$, and so on, we can see that

$$|n\rangle = \frac{(a^\dagger)^n}{\sqrt{n!}}\,|0\rangle \qquad (7.85)$$

are the states with unit norm. The connection with the coordinate space wavefunctions obtained earlier can be established through the definition of wavefunctions: $\psi_n(x) = \langle x|n\rangle$. The procedure is indicated in Ex. 7.16.

We end this section by noting that the discrete equally spaced eigenvalues and the corresponding eigenstates which we have found for the quantum harmonic oscillator do not reduce to the continuous spectrum of a classical oscillator in any simple way; in other words, there is no simple way to take the classical limit of this problem using the energy eigenstates. We shall see how to do this using the so-called coherent states, i.e., eigenstates of the lowering operator. This will be discussed in Chapter 17.

□ **Exercise 7.14** *Show that*

$$\left[a, (a^\dagger)^n\right] = n(a^\dagger)^{n-1}. \tag{7.86}$$

Use this to show directly Eq. (7.81).

□ **Exercise 7.15** *Show, by induction or otherwise, that the expectation value of the Hamiltonian of Eq. (7.74) in the state $|n\rangle$ defined in Eq. (7.85) is indeed $(n + \frac{1}{2})\hbar\omega_c$.*

□ **Exercise 7.16** *Eq. (7.78) implies $\langle x|a|0\rangle = 0$. Starting from the definition of the lowering operator given in Eq. (7.71), set up a differential equation for the ground state wavefunction $\psi_0(x) = \langle x|0\rangle$ by using Eq. (3.65, p. 72). Then solve this equation to show that one obtains the wavefunction indicated through Eq. (7.70b).*

□ **Exercise 7.17** *Wavefunctions for higher levels can be found in a similar manner. Using Eq. (7.84a) with $n = 0$ and the form of $\psi_0(x)$, find $\psi_1(x)$ and check with Eq. (7.70b). [**Note:** Of course, an overall phase can be arbitrary.]*

□ **Exercise 7.18** *Write the matrix elements $\langle m|a|n\rangle$ and $\langle m|a^\dagger|n\rangle$. Use them to obtain $\langle m|a^2|n\rangle$ and $\langle m|(a^\dagger)^2|n\rangle$.*

□ **Exercise 7.19** *Consider the simple harmonic oscillator discussed in this section.*

(a) *Find the uncertainty in position Δx and momentum Δp of the oscillator in its ground state. Hence show that the ground state of oscillator satisfies $\Delta x \Delta p = \hbar/2$.*

(b) *Show that $\langle m|(a^\dagger)^p a^q|n\rangle = 0$ unless $p - q = m - n$.*

(c) *Using the above result or otherwise, find $\langle m|\hat{x}^4|n\rangle$.*

7.8 Particle on a lattice

In this section, we discuss the problem of a particle whose kinetic energy is due to hopping between sites of a 1D lattice. We shall mostly discuss the case when these hops are between neighboring sites of a lattice; however, its generalization to cases where the hopping extends to other neighbors is straightforward.

To discuss these hopping processes, we define an operator h_η by the rule

$$h_\eta^\dagger |x\rangle = -\mathcal{J}_0 |x + \eta a\rangle, \tag{7.87}$$

where $\mathcal{J}_0 > 0$ is the amplitude of the *hopping* process, the minus sign ensures that the hopping leads to reduction of energy (otherwise the particle will be localized on a site in the ground state), a denotes the lattice spacing, and $\eta = \pm 1$ for nearest neighbor hopping. Taking Eq. (7.87) in the form $\langle x|h_\eta = -\mathcal{J}_0\langle x + \eta a|$ and taking an inner product with a state vector $|\psi\rangle$, we find

$$\langle x |h_\eta| \psi \rangle = -\mathcal{J}_0 \psi(x + \eta a). \tag{7.88}$$

Just as the kinetic energy term, due to the presence of the momentum operator which acts as a derivative and therefore connects $\psi(x)$ to the values of ψ at nearby points, the operator h_η connects the wavefunction at different lattice sites. In this sense, it can contribute to the kinetic energy term on the lattice. We can therefore contemplate a Hamiltonian of the form

$$H = \sum_{\eta = \pm 1} h_\eta. \tag{7.89}$$

□ **Exercise 7.20** *Argue that the operator h_η is not Hermitian, but the operator defined in Eq. (7.89) is.*

The energy eigenvalue equation that follows from the Schrödinger equation for such particles can be written as

$$E |\psi\rangle = H |\psi\rangle = h_+ |\psi\rangle + h_- |\psi\rangle. \tag{7.90}$$

Taking the inner product with $\langle x|$ and using Eq. (7.88), we can write this equation in terms of the wavefunction:

$$E\psi(x) = -\mathcal{J}_0 \left[\psi(x + a) + \psi(x - a)\right]. \tag{7.91}$$

To solve for E, we express $\psi(x)$ in terms of its Fourier components:

$$\psi(x) = \int_{-\pi/a}^{\pi/a} \frac{dk}{2\pi} e^{ikx} \psi_k \tag{7.92}$$

where the presence of the lattice spacing a leads to a finite upper cutoff for the momenta of particles. Substituting Eq. (7.91) into Eq. (7.92), we find

$$E_k = -2\mathcal{J}_0 \cos ka. \tag{7.93}$$

We note that in the long-wavelength $(k \to 0)$ limit the lattice details are expected to be unimportant and we should recover the standard continuum result. To check this, we note that in this limit, a straightforward expansion of the cosine term in Eq. (7.93) leads to $E_k \simeq -2\mathcal{J}_0 + \mathcal{J}_0 a^2 k^2$. The constant term can always be absorbed in the definition of energy while the second term, in real space, corresponds to a Hamiltonian $\hat{p}^2/2m$ with the identification $k = -id/dx$ and $m = \hbar^2/(2\mathcal{J}_0 a^2)$.

Next, we address the problem of a particle in a 1D lattice in the presence of a static electric field. This is generally known as the *Wannier–Stark problem*. In addition to the hopping term discussed earlier in this section, the quantum particle is subjected to a constant electric field $\mathcal{E}$ which leads to a potential $V_0 = -e\mathcal{E}x$. We note that on an infinite lattice, this potential is not bounded; we shall get back to this point in detail later. In what follows, we shall express the coordinates in units of lattice spacing a by writing $x = na$ where n is an integer.

The Schrödinger equation in the presence of such a static field is given by

$$E\psi_n = -\mathcal{J}_0\Big(\psi_{n+1} + \psi_{n-1}\Big) - e\mathcal{E}an\psi_n, \qquad (7.94)$$

where ψ_n denotes, for the rest of this section, the value of the wavefunction ψ at the point $x = na$. Let us now write the energy eigenvalues in the form

$$E_\ell = -e\mathcal{E}a\ell \qquad (7.95)$$

where possible values of ℓ will be specified shortly. Then Eq. (7.94) can be rewritten in the form

$$\frac{e\mathcal{E}a}{\mathcal{J}_0}(\ell - n)\psi_n^{(\ell)} = \psi_{n+1}^{(\ell)} + \psi_{n-1}^{(\ell)}, \qquad (7.96)$$

where $\psi_n^{(\ell)}$ is the wavefunction for the eigenvalue E_ℓ. Note the remarkable analogy of this equation with an identity involving the Bessel functions:

$$(2p/y)J_p(y) = J_{p+1}(y) + J_{p-1}(y) \qquad (7.97)$$

which holds for integer p and any real y, and has been proved in Appendix D, in Eq. (D.39, p. 479). This analogy shows two things. First, we can identify the wavefunctions as Bessel functions:

$$\psi_n^{(\ell)} = J_{\ell-n}\left(\frac{2\mathcal{J}_0}{e\mathcal{E}a}\right). \qquad (7.98)$$

Second, since both p of Eq. (7.97) and n of Eq. (7.96) are integers, ℓ must be an integer. Eq. (7.95) then tells us that the energy eigenvalues form an infinite ladder with equal spacing controlled by the electric field. This is called the *Wannier–Stark ladder*. The energy spectrum is not bounded from below for an infinite chain since the potential itself is not.

One interesting feature of the wavefunctions is that they are very strongly localized. This can be seen from the series expansion of Bessel functions given in Appendix C, in Eq. (C.47, p. 461). It shows that for integral positive ν,

$$J_\nu(y) = \frac{1}{\nu!} \left(\frac{y}{2}\right)^\nu \left[1 - \frac{y^2}{4}\frac{1}{\nu+1} + \cdots\right]. \tag{7.99}$$

Clearly, for large ν, the term with the lowest power of y dominates because of factors like $1/(\nu+1)$ in the denominator of the other terms. Since $\nu! \sim (\nu/e)^\nu$ for $\nu \gg 1$, we can write

$$J_\nu(y) \sim (y/\nu)^\nu, \tag{7.100}$$

ignoring some other factors which are of no interest. Because the Bessel functions with negative indices are simply related to those with positive indices, as shown in Eq. (C.53, p. 462), we can write more generally,

$$J_{\ell-n}(y) \sim \left(\frac{y}{|\ell - n|}\right)^{|\ell-n|}, \tag{7.101}$$

which shows the sharp fall with increasing values of $|\ell - n|$.

The localization is stronger for stronger electric fields. Thus, a particle starting on a site n would not race out to infinity (or to the extremities of the chain). This is a purely quantum phenomenon, since classical intuition suggests that the applied electric field should accelerate the particle to the end of the chain. Moreover, it should be remembered that the solution of the time-evolution equation also contains a time-dependent factor $\exp(-iE_\ell t/\hbar)$, so that the wavefunction $\psi_n^{(\ell)}(t)$ comes back to the initial state periodically, independent of the value of $\mathcal{J}_0$. It can be shown that this indicates a semi-classical oscillatory motion of the particle which is to be contrasted to the classically expected acceleration in the presence of an electric field. This is known as *Bloch oscillation*. Finally, we point out that the presence of other terms (such as other bands in solids) may destabilize such a metastable state causing the particle to move; this phenomenon is called *electric breakdown*. We shall not discuss this further in this text.

□ **Exercise 7.21** *Consider the potential* $V(x) = V_0[\exp(-2ax) - 2\exp(-ax)]$.

(a) Write down the Schrödinger equation for this potential.

(b) Introduce a variable $y = 2\exp(-ax)\sqrt{2mV_0}/(\hbar a) = 2y_0\exp(-ax)$. Write down the Schrödinger equation in terms of y.

(c) Now make the transformation $\psi = \exp(-y/2)y^s w(y)$ where $s = \sqrt{2mE}/(a\hbar)$, and write down the equation for $w(y)$. Show that the solution of this equation is the confluent hypergeometric function $F(p, q, y)$ discussed in Appendix C. Identify p and q in terms of parameters of the problem.

(d) Finally, show that the energy eigenvalue corresponds to $E_n = -V_0\left(1 - (n + 1/2)/y_0\right)$.

□ **Exercise 7.22** Consider the potential $V(x) = V_0/\cosh^2(ax)$.

(a) Write down the Schrödinger equation for this potential.

(b) Make the transformation $\psi(x) = w(x)/\cosh^s(ax)$ and write down the equation for $w(x)$. Here

$$s = \frac{1}{2}\left(-1 + \sqrt{1 + \frac{8mV_0}{\hbar^2 a^2}}\right). \tag{7.102}$$

(c) Introduce a variable $y = \sinh^2(ax)$. Write down the Schrödinger equation in terms of y. Show that the solution of this equation is the confluent hypergeometric function $F(p, q, r, y)$. Identify p, q, and r in terms of parameters of the problem.

(d) Finally, show that the energy eigenvalue corresponds to $E_n = -\frac{\hbar^2 a^2}{4m}(s - n)$. Find the number of levels from the condition of positivity of the energy.

□ **Exercise 7.23** Consider the double oscillator potential $V(x) = \frac{1}{2}m\omega^2(|x| - a)^2$. Note that we have two length scales in the problem, viz., ℓ_0 (defined in Eq. (7.64)), and a.

(a) Write down the Schrödinger equation for this potential. Show by substituting $z = \sqrt{2}(x/\ell_0)(1 - a/|x|)$ that the solution in both these regions are parabolic cylinder functions $D_\nu(z)$ (also called Weber functions) which satisfy the differential equation

$$\frac{d^2 D_\nu(y)}{dz^2} + \left(\nu + \frac{1}{2} - \frac{z^2}{4}\right)D_\nu(z) = 0. \tag{7.103}$$

(b) By matching boundary conditions at $x = 0$, show that the energy eigenvalues are given as solutions of the transcendental equations $D_\nu(\sqrt{2}a/\ell_0) = 0$ for odd parity eigenvalues (where $E_\nu = \hbar\omega(\nu + 1/2)$). What is the condition for even parity eigenvalues?

(c) Show from your solution that $\nu = n$ is recovered for $a = 0$.

Chapter 8

Angular momentum

One of the main differences between 1D problems discussed in Chapter 7 and their higher-dimensional counterparts is the presence of rotational degrees of freedom. For reasons elaborated in Chapter 5, the description of such rotational motion necessitates the introduction of the concept of angular momentum, which will be the main focus of this chapter.

8.1   General considerations

It was mentioned in §5.7.2 that the rotation algebra admits both differential and matrix representations. In order to discuss some general properties of both kinds of representations, it is customary to denote the angular momentum generators by J. The letter L would then apply only for the differential generators or orbital angular momenta, and the letter S will be used for spin if necessary.

To reiterate, the angular momentum algebra is the set of the following commutator relations:

$$\left[J_x, J_y\right] = i\hbar J_z, \tag{8.1a}$$

$$\left[J_y, J_z\right] = i\hbar J_x, \tag{8.1b}$$

$$\left[J_z, J_x\right] = i\hbar J_y. \tag{8.1c}$$

In a compact notation, we can write all three relations as

$$\left[J_\alpha, J_\beta\right] = i\hbar \sum_\gamma \varepsilon_{\alpha\beta\gamma} J_\gamma, \tag{8.2}$$

where the indices take the 'values' x, y, z, and $\varepsilon_{\alpha\beta\gamma}$ is the completely antisymmetric Levi-Civita symbol defined earlier in §5.7.

In order to discuss the angular momentum states, we have to first set up a basis on the vector space. Earlier in Chapter 2, we said that the eigenstates of a Hermitian operator can be used as a basis. We can choose, for this purpose, the eigenvectors of one of the three components of the angular momentum. These states will not be eigenstates of the other two components, since the operators for different components of $\boldsymbol{J}$ do not commute, as seen in Eq. (8.2). Conventionally, one chooses the eigenstates of J_z. This is, of course, no special assumption: we can always consider the basis as the eigenvectors of angular momentum component in one direction and call this direction to be the z direction.

But there is a problem. There is a lot of degeneracy among the eigenstates of J_z. Whenever there is degeneracy, the eigenstates cannot be uniquely defined, because any combination of states in a degenerate subspace of eigenvectors also qualifies as an eigenvector. For a basis, we need well-defined vectors, with no ambiguity in the defintion.

Whenever one encounters such a situation with an operator A, one looks for some other operator that commutes with A. It is guaranteed that such an operator can be found. Let us call one such operator B. Then, according to the result we derived in §2.7.3, A and B have simultaneous eigenstates. If these eigenstates are such that each of them has different combinations of eigenvalues of the two operators, then they can be unambiguously defined in terms of the combination, and can be used as a basis. If degeneracy still remains, one will have to invoke other operators that commute with both A and B, until one finds a set of commuting operators whose simultaneous eigenvectors are uniquely defined.

To see that degeneracy implies the existence of a commuting operator, consider that the operator A has an n-fold degeneracy in an eigenvalue a_1. This means that if we write the matrix for A in the basis of its eigenvectors, it would be of the form

$$A = \mathrm{diag}\,(\underbrace{a_1, \cdots, a_1}_{n \text{ times}}, a_2, a_3, \cdots). \tag{8.3}$$

In the first $n \times n$ block, it is the multiple of the unit matrix. Thus, if we take another matrix that is anything but a multiple of the unit matrix in this block and a multiple of the unit matrix in the remaining block, it would commute with A.

We therefore have to look for an operator that commutes with J_z. Although J_x and J_y do not, there is a combination of the components that in fact commutes with J_z, or for that matter with any other component. This combination is

$$J^2 = J_x^2 + J_y^2 + J_z^2 = \sum_\alpha J_\alpha J_\alpha. \tag{8.4}$$

Intuitively, it is obvious that it should commute with any component of $\boldsymbol{J}$ because J^2 is a scalar which should not be affected by rotations. Explicitly, we can check this statement as follows:

$$
\begin{aligned}
\left[J^2, J_\alpha \right] &= \sum_{\beta = x,y,z} \left[J_\beta J_\beta, J_\alpha \right] \\
&= \sum_{\beta = x,y,z} \left(J_\beta \left[J_\beta, J_\alpha \right] + \left[J_\beta, J_\alpha \right] J_\beta \right) \\
&= \sum_{\beta,\gamma = x,y,z} i\hbar \epsilon_{\beta\alpha\gamma} \left(J_\beta J_\gamma + J_\gamma J_\beta \right) = 0.
\end{aligned}
\tag{8.5}
$$

The expression contains a product of two terms, one of which is antisymmetric under the $\beta \leftrightarrow \gamma$ interchange while the other is symmetric. The sum of such product vanishes.

This means that there will be vectors in the Hilbert space which are eigenvectors of both J^2 and J_z, and we can use them as our basis states. Let us denote these states by $|\lambda, m\rangle$, where λ and m characterize the eigenvalues of J^2 and J_z respectively. In other words, these states satisfy the relations

$$
J^2 |\lambda, m\rangle = \lambda \hbar^2 |\lambda, m\rangle,
\tag{8.6a}
$$
$$
J_z |\lambda, m\rangle = m\hbar |\lambda, m\rangle,
\tag{8.6b}
$$

where we have put factors of $\hbar$ to make λ and m dimensionless.

To learn about the possible eigenvalues, we need to understand the effect of J_x and J_y on the states $|\lambda, m\rangle$. To this end, we construct a specific linear combination of these two operators:

$$
J_\pm = J_x \pm iJ_y.
\tag{8.7}
$$

The utility of such a construction can be understood by inspecting the effect of $J_\pm$ on the states $|\lambda, m\rangle$. To do this, we first note that Eq. (8.2) leads to the following commutation relations:

$$
\left[J_z, J_\pm \right] = \pm \hbar J_\pm,
$$
$$
\left[J_+, J_- \right] = 2\hbar J_z.
\tag{8.8}
$$

Using these relations, it is easy to see that

$$
\begin{aligned}
J_z \Big(J_\pm |\lambda, m\rangle \Big) &= \Big(J_\pm J_z + [J_z, J_\pm] \Big) |\lambda, m\rangle \\
&= (m \pm 1)\hbar J_\pm |\lambda, m\rangle,
\end{aligned}
\tag{8.9}
$$

where we have used Eq. (8.6b) to obtain the last step. Also, $J_\pm$ obviously commute with J^2 since J_x and J_y do, which tells us that

$$J^2 J_\pm \left|\lambda, m\right\rangle = J_\pm J^2 \left|\lambda, m\right\rangle = \lambda \hbar^2 J_\pm \left|\lambda, m\right\rangle. \tag{8.10}$$

Eqs. (8.9) and (8.10) imply that if we apply the operator J_+ or J_- on the eigenvector $\left|\lambda, m\right\rangle$, we obtain a state which is an eigenstate of J^2 with the same eigenvalue, and also an eigenstate of J_z with an eigenvalue that is shifted by one unit. In other words,

$$J_\pm \left|\lambda, m\right\rangle = N_\pm(\lambda, m) \left|\lambda, m \pm 1\right\rangle, \tag{8.11}$$

where $N_\pm(\lambda, m)$ denote normalization constants that obviously can depend on λ and m. Since the operators $J_\pm$ connect one state with another with a shifted value of m, they are called *ladder operators*. Sometimes, the name *raising operator* is used for J_+ and *lowering operator* for J_-. Recall that we had encountered such ladder operators while discussing the simple harmonic oscillator in §7.7.2.

In fact, the magnitudes of these normalization constants can be determined easily. For this, note that

$$J_- J_+ = (J_x - iJ_y)(J_x + iJ_y) = J_x^2 + J_y^2 + i\Big[J_x, J_y\Big]. \tag{8.12}$$

Using Eq. (8.4) and the commutation relation from Eq. (8.1a), we can write this relation in the form

$$J_- J_+ = J^2 - J_z^2 - \hbar J_z. \tag{8.13}$$

Taking the expectation values of both sides in the state $\left|\lambda, m\right\rangle$, we obtain

$$\begin{aligned}
\left\langle \lambda, m \left| J_- J_+ \right| \lambda, m \right\rangle &= \left\langle \lambda, m \left| J^2 - J_z^2 - \hbar J_z \right| \lambda, m \right\rangle \\
&= \Big(\lambda - m(m+1)\Big)\hbar^2,
\end{aligned} \tag{8.14}$$

using Eq. (8.6) in the last step. However, notice that J_+ and J_- are Hermitian conjugates of each other, i.e.,

$$J_+^\dagger = J_-, \qquad J_-^\dagger = J_+. \tag{8.15}$$

Therefore,

$$\left\langle \lambda, m \left| J_- J_+ \right| \lambda, m \right\rangle = \Big\| J_+ \left|\lambda, m\right\rangle \Big\|^2. \tag{8.16}$$

Eq. (8.14) therefore gives the absolute value of the normalization factor N_+ introduced in Eq. (8.11). The phase of this normalization factor cannot be determined: it has to be fixed by convention. We take the convention that this factor is equal to unity, and write

$$J_+\,|\lambda,m\rangle = \hbar\sqrt{\lambda - m(m+1)}\,|\lambda,m+1\rangle . \qquad (8.17a)$$

Similarly, evaluating the expectation value of the operator J_+J_- in the state $|\lambda,m\rangle$, we can find the relation

$$J_-\,|\lambda,m\rangle = \hbar\sqrt{\lambda - m(m-1)}\,|\lambda,m-1\rangle . \qquad (8.17b)$$

Finally, there is a connection between λ and the possible values of m. This can be seen from Eq. (8.14). Since the left side is the norm squared of a vector, as shown in Eq. (8.16), the expression obtained for it must be a positive real number. This means that we must have

$$\lambda \geqslant m(m+1) \qquad \forall m. \qquad (8.18)$$

For a given value of λ, how can this inequality be ensured if every time we operate an eigenstate with J_+, we produce a new state with a higher value of m? The only way this can happen is if there exists a value of m for which the coefficient on the right side of Eq. (8.17a) is zero, so that the right side equals the null vector. Any operator acting on it will give back the null vector, so no higher state will be produced by operating further with J_+. It means that for a given value of λ, there must be a maximum value of m, to be denoted by $m_{\max}$, satisfying the condition

$$\lambda = m_{\max}(m_{\max}+1). \qquad (8.19)$$

Similarly, in order that states with indefinitely low values of m are not produced by the action of J_-, we must have a minimum, $m_{\min}$, given by

$$\lambda = m_{\min}(m_{\min}-1). \qquad (8.20)$$

Equating the right sides of Eqs. (8.19) and (8.20), we obtain

$$(m_{\max}+m_{\min})(m_{\max}-m_{\min}+1) = 0. \qquad (8.21)$$

This must mean

$$m_{\min} = -m_{\max}, \qquad (8.22)$$

because the other factor cannot be zero since $m_{\min}$ cannot be greater than $m_{\max}$.

Let us then use the notation

$$j \equiv m_{\max} = -m_{\min}. \tag{8.23}$$

Instead of using λ as the eigenvalue of J^2, we can use j, in which case Eq. (8.19) can be used to write the eigenvalue of J^2 in the form

$$\lambda = j(j+1). \tag{8.24}$$

The possible values of m, for a given value of j, range from $-j$ to $+j$ in steps of unity. This leads to $2j + 1$ states for different values of m corresponding to every j and consequently to a $2j + 1$ dimensional Hilbert space. Therefore, $2j + 1$ must be a positive integer, which means that the possible values of j are given by

$$j = 0, \frac{1}{2}, 1, \frac{3}{2}, 2, \cdots, \tag{8.25}$$

integers and half-integers. We will henceforth label the eigenstates by j and m instead of by λ and m, i.e., rewrite the eigenvalue equations of Eq. (8.6) in the form

$$J^2 |j, m\rangle = j(j+1)\hbar^2 |j, m\rangle , \tag{8.26a}$$
$$J_z |j, m\rangle = m\hbar |j, m\rangle , \tag{8.26b}$$

with possible values of m given by

$$m = -j, -j+1, \cdots, j-1, j. \tag{8.27}$$

In view of the notation for the maximum eigenvalue of J_z, the result of the ladder operators should be written as

$$J_\pm |j, m\rangle = a_\pm^{(j,m)} \hbar |j, m \pm 1\rangle , \tag{8.28}$$

where we introduce the shorthand

$$a_\pm^{(j,m)} = \sqrt{j(j+1) - m(m \pm 1)}. \tag{8.29}$$

We will see later that all possible values for the eigenvalue of J^2, as indicated in Eq. (8.25), are not allowed for orbital angular momentum: only the integral values are allowed. For spin, however, both integral and half-integral values are permissible.

□ **Exercise 8.1** *Show that*

$$\boldsymbol{J}^2 = \frac{1}{2}\Big(J_+ J_- + J_- J_+ \Big) + J_z^2. \tag{8.30}$$

8.2   Coordinate representation

In this section, we shall find the coordinate space representation of the angular momentum operators in the form of differential operators, and their corresponding eigenfunctions. Already from §5.7.2, we know that one can read off the definition of angular momentum in coordinate space as

$$\widehat{\boldsymbol{L}} = -i\hbar \boldsymbol{r} \times \boldsymbol{\nabla} = \boldsymbol{r} \times \widehat{\boldsymbol{p}}. \tag{8.31}$$

Note that this coincides exactly with the classical definition of angular momentum with the reinterpretation of the classical momentum $\boldsymbol{p}$ as the operator $\widehat{\boldsymbol{p}}$. We do not use the operator mark on $\boldsymbol{r}$ since in the coordinate representation, this operator acts multiplicatively.

To obtain the coordinate space representation of the eigenfunctions of angular momentum, we begin with Eq. (8.31). In coordinate representation, the operator $\widehat{L}^2$ can be written as

$$\widehat{L}^2 = (\boldsymbol{r} \times \widehat{\boldsymbol{p}}) \cdot (\boldsymbol{r} \times \widehat{\boldsymbol{p}}) = r^2 \widehat{p}^2 - (\boldsymbol{r} \cdot \widehat{\boldsymbol{p}})^2 + i\hbar \boldsymbol{r} \cdot \widehat{\boldsymbol{p}}, \tag{8.32}$$

where $r = |\boldsymbol{r}|$. To simplify this expression, we note that

$$\boldsymbol{r} \cdot \boldsymbol{\nabla} = r\partial_r, \tag{8.33}$$

where the shorthand

$$\partial_r = \frac{\partial}{\partial r} \tag{8.34}$$

has been used here, and similar ones will be used in what follows. Using it, one can write

$$\begin{aligned} \widehat{L}^2 &= -\hbar^2 \left[r^2 \nabla^2 - (r\partial_r)^2 - r\partial_r \right] \\ &= -\hbar^2 \left[r^2 \nabla^2 - \partial_r(r^2\partial_r) \right]. \end{aligned} \tag{8.35}$$

In spherical polar coordinates where $\boldsymbol{r} = (r, \theta, \phi)$, ∇^2 can be expressed as

$$\nabla^2 = \frac{1}{r^2} \left[\partial_r(r^2\partial_r) + \frac{1}{\sin^2\theta} \left(\sin\theta\, \partial_\theta \left(\sin\theta\, \partial_\theta \right) + \partial_\phi^2 \right) \right]. \tag{8.36}$$

Note that the first term in Eq. (8.36) exactly cancels the last term in the right side of Eq. (8.35). Thus, we finally obtain

$$\widehat{L}^2 = -\frac{\hbar^2}{\sin^2\theta} \left(\sin\theta\, \partial_\theta \left(\sin\theta\, \partial_\theta \right) + \partial_\phi^2 \right). \tag{8.37}$$

The reason for expressing $\widehat{L}^2$ in terms of spherical polar coordinates instead of the usual Cartesian coordinates becomes apparent if we try to solve Eq. (8.26) using the latter. Clearly, in polar coordinates we need to solve a partial differential equation involving only the two angular variables θ and ϕ; the radial coordinate r drops out of the picture. Furthermore, as we shall see momentarily, in this coordinate system Eq. (8.26) allows for easy solution via separation of variables. Both of these simplifying features are absent in Cartesian coordinate representation, as the reader can easily verify.

□ **Exercise 8.2** *Verify Eq. (8.32), remembering that the components of $\boldsymbol{r}$ and $\boldsymbol{p}$ do not commute.*

Similarly, we can also find expressions for the Cartesian components of orbital angular momentum in the spherical coordinates. For this, we recall that, in spherical coordinates,

$$\boldsymbol{\nabla} = \mathring{\boldsymbol{r}}\partial_r + \mathring{\boldsymbol{\theta}}\frac{1}{r}\partial_\theta + \mathring{\boldsymbol{\phi}}\frac{1}{r\sin\theta}\partial_\phi, \tag{8.38}$$

where $\mathring{\boldsymbol{r}}$, etc., are the unit vectors along the directions of increasing values of the spherical coordinate components. Using the orthogonality of the unit vectors, we have

$$\mathring{\boldsymbol{r}} \times \mathring{\boldsymbol{\theta}} = \mathring{\boldsymbol{\phi}}, \qquad \mathring{\boldsymbol{r}} \times \mathring{\boldsymbol{\phi}} = -\mathring{\boldsymbol{\theta}}. \tag{8.39}$$

Putting these results into Eq. (8.31), we obtain

$$\widehat{\boldsymbol{L}} = -i\hbar r\mathring{\boldsymbol{r}} \times \boldsymbol{\nabla} = -i\hbar\left[\mathring{\boldsymbol{\phi}}\partial_\theta - \frac{\mathring{\boldsymbol{\theta}}}{\sin\theta}\partial_\phi\right]. \tag{8.40}$$

The different Cartesian components of $\widehat{\boldsymbol{L}}$ can then be read off from the relations

$$\begin{aligned}
\mathring{\boldsymbol{r}} &= \sin\theta\cos\phi\,\mathring{\boldsymbol{x}} + \sin\theta\sin\phi\,\mathring{\boldsymbol{y}} + \cos\theta\,\mathring{\boldsymbol{z}}, \\
\mathring{\boldsymbol{\theta}} &= \cos\theta\cos\phi\,\mathring{\boldsymbol{x}} + \cos\theta\sin\phi\,\mathring{\boldsymbol{y}} - \sin\theta\,\mathring{\boldsymbol{z}}, \\
\mathring{\boldsymbol{\phi}} &= -\sin\phi\,\mathring{\boldsymbol{x}} + \cos\phi\,\mathring{\boldsymbol{y}},
\end{aligned} \tag{8.41}$$

which give

$$L_x = -i\hbar\left[-\sin\phi\partial_\theta - \cot\theta\cos\phi\partial_\phi\right], \tag{8.42a}$$

$$L_y = -i\hbar\left[\cos\phi\partial_\theta - \cot\theta\sin\phi\partial_\phi\right], \tag{8.42b}$$

$$L_z = -i\hbar\partial_\phi. \tag{8.42c}$$

☐ **Exercise 8.3** *The relation between Cartesian and spherical coordinates is given by*

$$x = r \sin\theta \cos\phi,$$
$$y = r \sin\theta \sin\phi,$$
$$x = r \cos\theta. \tag{8.43}$$

Use these relations, and the chain rule of differentiation, to convert the Cartesian results of Eq. (5.106, p. 130) to the results given in Eq. (8.42).

Our aim is to solve the eigenfunctions from Eq. (8.26). The eigenfunctions are defined in terms of the Hilbert space vectors as

$$Y_{\ell,m}(\theta,\phi) = \langle \theta;\phi \,|\, \ell,m \rangle. \tag{8.44}$$

Similar to the definition of Eq. (3.65, p. 72), the differential operators of Eqs. (8.37) and (8.42) imply the following relations in terms of the eigenfunctions:

$$\left\langle \theta,\phi \left| \widehat{L}_z \right| \ell,m \right\rangle = -i\hbar\partial_\phi \langle \theta;\phi \,|\, \ell,m \rangle, \tag{8.45}$$

and other similar equations with $\widehat{L}_x$, $\widehat{L}_y$, and $\widehat{L}^2$. These are the differential equations we need to solve.

From Eqs. (8.44) and (8.45), we see that in the coordinate representation, Eq. (8.26b) takes the form

$$-i\partial_\phi Y_{\ell m}(\theta,\phi) = m Y_{\ell m}(\theta,\phi). \tag{8.46}$$

The solution of this equation is of the form

$$Y_{\ell m}(\theta,\phi) = \Theta(\theta)\exp(im\phi)/\sqrt{2\pi}, \tag{8.47}$$

where $\Theta(\theta)$ is a function of only θ. Since m is thus connected to the azimuthal angle ϕ, it is sometimes called the *azimuthal quantum number*.

Although the ϕ-dependence of this solution is similar to the x-dependence of the free particle solution in 1D, there is one crucial difference. Since the azimuthal angle ϕ is periodic in 2π, any single-valued wavefunction must come back to itself when ϕ is increased by any multiple of 2π. This imposes an importance constraint on m:

$$m = 0, \pm 1, \pm 2, \cdots \tag{8.48}$$

Accordingly, ℓ can also be only integer. The half-integers values, which are allowed in general for angular momentum eigenvalues, are inaccessible to the orbital angular momentum.

Table 8.1 Spherical harmonics for values of ℓ up to 2, as obtained from Eq. (8.51).

ℓ	m	$Y_{\ell m}(\theta, \phi)$
0	0	$\sqrt{\frac{1}{4\pi}}$
1	0	$\sqrt{\frac{3}{4\pi}} \cos\theta$
1	± 1	$\sqrt{\frac{3}{8\pi}} \sin\theta \, e^{\pm i\phi}$
2	0	$\sqrt{\frac{5}{16\pi}} (3\cos^2\theta - 1)$
2	± 1	$\sqrt{\frac{15}{32\pi}} \sin 2\theta \, e^{\pm i\phi}$
2	± 2	$\sqrt{\frac{15}{32\pi}} \sin^2\theta \, e^{\pm 2i\phi}$

To obtain the θ-dependence of the eigenfunctions, we now use Eq. (8.26a). From Eq. (8.37), we see that the differential equation is

$$-\frac{1}{\sin\theta}\left[\partial_\theta\left(\sin\theta\partial_\theta\right) + \frac{1}{\sin\theta}\partial_\phi^2\right]Y_{\ell m}(\theta,\phi) = \ell(\ell+1)Y_{\ell m}(\theta,\phi). \qquad (8.49)$$

Putting the form of the solution already obtained in Eq. (8.47), we find that the differential equation for $\Theta(\theta)$ is given by

$$\frac{1}{\sin\theta}\partial_\theta\left[\sin\theta\partial_\theta\Theta(\theta)\right] + \left[\ell(\ell+1) - \frac{m^2}{\sin^2\theta}\right]\Theta(\theta) = 0. \qquad (8.50)$$

This shows that the θ-dependent part of the solution can depend on both ℓ and m, and therefore can be called $\Theta_{\ell m}(\theta)$. In fact, after consultation with the material of Appendix C.3, we realize that these solutions are the associated Legendre functions denoted by $P_\ell^m(\cos\theta)$. In terms of them, we can write the solutions of Eq. (8.49) as

$$Y_{\ell m}(\theta,\phi) = \sqrt{\frac{(2\ell+1)(\ell-m)!}{4\pi(\ell+m)!}} P_\ell^m(\cos\theta)e^{im\phi}. \qquad (8.51)$$

These functions are called the *spherical harmonics*. They have been normalized by the condition

$$\int_0^{2\pi} d\phi \int_0^\pi d\theta \, \sin\theta \left|Y_{\ell m}(\theta,\phi)\right|^2 = 1. \qquad (8.52)$$

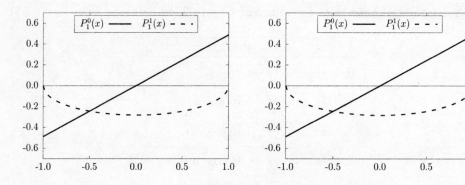

Figure 8.1 Plot of $P_\ell^m(x)$ for all possible values of m with $\ell = 1, 2$. The argument of the functions is $\cos\theta$, and therefore ranges from -1 to $+1$.

However, it will have to be remembered that an overall phase is purely conventional. In fact, different phase conventions are used in different contexts.

We list the normalized expressions of the first few $Y_{\ell m}$'s in Table 8.1. Also, the plots of the P_ℓ^m's as a function of the argument are shown in Figure 8.1. This concludes our discussion on coordinate representation of the angular momentum operator.

☐ **Exercise 8.4** *Consider the raising and the lowering operators $\widehat{L}_\pm$.*

(a) *Starting from the definition of $\widehat{\boldsymbol{L}}$ in Eq. (8.31), construct the representation of $\widehat{L}_\pm$ in terms of θ, ϕ, ∂_θ, and ∂_ϕ.*

(b) *Show that $\widehat{L}_+ Y_{\ell\ell}(\theta, \phi) = 0$ leads to the differential equation*

$$[\partial_\theta - \ell \cot\theta]\Theta_{\ell\ell}(\theta) = 0. \tag{8.53}$$

(c) *Solve this equation to confirm that*

$$Y_{\ell\ell}(\theta, \phi) \propto \sin^\ell\theta \, e^{i\ell\phi}. \tag{8.54}$$

Normalize this expression for $\ell = 1, 2$ and check with the solutions given in Table 8.1.

☐ **Exercise 8.5** *As a continuation, apply the lowering operators $\widehat{L}_-$ on the expression for $Y_{\ell\ell}$ to obtain the expressions for $Y_{\ell m}$ with $m < \ell$, and check the results against Table 8.1.*

☐ **Exercise 8.6** *Show that the orbital angular momentum operator $\widehat{\boldsymbol{L}}$ and the momentum operator $\widehat{\boldsymbol{p}}$ satisfy the relation*

$$\widehat{\boldsymbol{L}} \times \widehat{\boldsymbol{p}} + \widehat{\boldsymbol{p}} \times \widehat{\boldsymbol{L}} = 2i\hbar\widehat{\boldsymbol{p}}. \tag{8.55}$$

□ **Exercise 8.7** *For any operator $\widehat{\boldsymbol{u}}$ which transforms as a vector under rotation ($\widehat{\boldsymbol{r}}$ and $\widehat{\boldsymbol{p}}$ are examples of such vectors) show that*

$$\left[\widehat{L}_i, \widehat{u}_j\right] = i\hbar\epsilon_{ijk}\widehat{u}_k. \tag{8.56}$$

Using this or otherwise show that for a constant unit vector $\overset{\circ}{\boldsymbol{n}}$,

$$\left[\overset{\circ}{\boldsymbol{n}} \cdot \widehat{\boldsymbol{L}}, \widehat{\boldsymbol{u}}\right] = -i\hbar\overset{\circ}{\boldsymbol{n}} \times \widehat{\boldsymbol{u}}. \tag{8.57}$$

8.3 Spin $1/2$ algebra

In this section, we are going to provide a detailed discussion of the simplest possible case of half-integer angular momentum, viz., $j = \frac{1}{2}$. As explained in §8.2, orbital angular momentum cannot have this value. It can only refer to the spin of a system. Therefore, we will use the notation $\widehat{S}$ for $\widehat{J}$, and accordingly modify the notations for the eigenvalues.

The study of this value of spin is very important because subatomic particles like the electron, the proton, and the neutron all have this value of spin. There are many other important particles as well, and also the substructures of the proton and the neutron called *quarks*, which have this value of the spin angular momentum.

8.3.1 Basic formalism

For $s = \frac{1}{2}$, the dimensionality of the Hilbert space is 2. Such a Hilbert space can be spanned by two state vectors which we choose to be eigenstates of S_z. These are given by

$$|s, m_s\rangle = |1/2, \pm 1/2\rangle. \tag{8.58}$$

We note from our discussion in Chapter 2 that any state vector in this 2D Hilbert space can be written as a linear combination of these two basis vectors. In what follows, we choose these vectors to be

$$|1/2, 1/2\rangle = \begin{pmatrix} 1 \\ 0 \end{pmatrix}, \quad |1/2, -1/2\rangle = \begin{pmatrix} 0 \\ 1 \end{pmatrix}. \tag{8.59}$$

Such a representation of the state vectors allows for a straightforward matrix representation of the $s = 1/2$ angular momentum operators. To see this, let us first recall Eqs. (8.26b) and (8.28) to write

$$\widehat{S}_z|1/2, \pm 1/2\rangle = \pm\frac{1}{2}\hbar|1/2, \pm 1/2\rangle, \tag{8.60a}$$

$$\widehat{S}_\pm|1/2, \mp 1/2\rangle = \hbar|1/2, \pm 1/2\rangle. \tag{8.60b}$$

Eq. (8.60a) implies that, in the basis defined in Eq. (8.59), the representation of S_z will be of the form

$$\widehat{S}_z = \frac{1}{2}\hbar \begin{pmatrix} 1 & 0 \\ 0 & -1 \end{pmatrix}. \tag{8.61}$$

Similarly, Eq. (8.60b) yields

$$\widehat{S}_+ = \hbar \begin{pmatrix} 0 & 1 \\ 0 & 0 \end{pmatrix}, \quad \widehat{S}_- = \hbar \begin{pmatrix} 0 & 0 \\ 1 & 0 \end{pmatrix}. \tag{8.62}$$

Since $\widehat{S}_\pm = \widehat{S}_x \pm i\widehat{S}_y$, one obtains

$$\widehat{S}_x = \frac{1}{2}\hbar \begin{pmatrix} 0 & 1 \\ 1 & 0 \end{pmatrix}, \quad \widehat{S}_y = \frac{1}{2}\hbar \begin{pmatrix} 0 & -i \\ i & 0 \end{pmatrix}. \tag{8.63}$$

From Eqs. (8.61) and (8.63), we find that these angular momentum operators can be written as

$$\widehat{S}_i = \frac{1}{2}\hbar\sigma_i, \tag{8.64}$$

where $i = x, y, z$ and σ_i are the 2×2 traceless matrices given by

$$\sigma_x = \begin{pmatrix} 0 & 1 \\ 1 & 0 \end{pmatrix}, \quad \sigma_y = \begin{pmatrix} 0 & -i \\ i & 0 \end{pmatrix}, \quad \sigma_z = \begin{pmatrix} 1 & 0 \\ 0 & -1 \end{pmatrix}. \tag{8.65}$$

These matrices were first used by Pauli for describing properties of electron spin and are named after him as Pauli matrices. Recall that they were used earlier, in §4.4 for example, to serve as a basis for 2×2 Hermitian matrices, along with the unit matrix. Here we see that they reappear as the generators of rotation.

The properties of these matrices will be useful in our subsequent discussion and so we chart them out here. First, we note that

$$(\sigma_i)^2 = \mathbb{1}, \tag{8.66a}$$
$$\det \sigma_i = -1, \tag{8.66b}$$

for all the three Pauli matrices. Second, these matrices satisfy the following anticommutation and commutation relations:

$$\{\sigma_i, \sigma_j\} = \sigma_i\sigma_j + \sigma_j\sigma_i = 2\delta_{ij}, \tag{8.67a}$$
$$[\sigma_i, \sigma_j] = \sigma_i\sigma_j - \sigma_j\sigma_i = 2i\sum_k \epsilon_{ijk}\sigma_k, \tag{8.67b}$$

where δ_{ij} denotes the Kronecker delta function, ϵ_{ijk} is the complete antisymmetric tensor with $\epsilon_{xyz} = 1$, and the indices i, j, and k can take values x, y, and z. Third, for arbitrary 3D vectors $\boldsymbol{A} = (A_x, A_y, A_z)$, one can construct matrix $\boldsymbol{\sigma} \cdot \boldsymbol{A}$ which is given by

$$\boldsymbol{\sigma} \cdot \boldsymbol{A} = \sum_{i=x,y,z} \sigma_i A_i = \begin{pmatrix} A_z & A_x - iA_y \\ A_x + iA_y & -A_z \end{pmatrix}, \tag{8.68}$$

where $\boldsymbol{\sigma} = (\sigma_x, \sigma_y, \sigma_z)$. Such dot products lead to the useful identity

$$\begin{aligned}(\boldsymbol{\sigma} \cdot \boldsymbol{A})(\boldsymbol{\sigma} \cdot \boldsymbol{B}) &= \sum_{i,j=x,y,z} \sigma_i \sigma_j A_i B_j \\ &= \sum_{i,j} \left(\delta_{ij} + i\sum_k \epsilon_{ijk}\sigma_k\right) A_i B_j = \boldsymbol{A} \cdot \boldsymbol{B} + i\boldsymbol{\sigma} \cdot (\boldsymbol{A} \times \boldsymbol{B}).\end{aligned}$$
$$\tag{8.69}$$

Note that for $\boldsymbol{A} = \boldsymbol{B}$, this yields $(\boldsymbol{\sigma} \cdot \boldsymbol{A})^2 = |\boldsymbol{A}|^2$. Finally, using this relation, we note that for any unit vector $\mathring{\boldsymbol{n}}$, we have $(\boldsymbol{\sigma} \cdot \mathring{\boldsymbol{n}})^{2n} = 1$ for any integer n, which leads to the identity

$$\begin{aligned}e^{i\boldsymbol{\sigma}\cdot\boldsymbol{n}\theta} &= \left(1 - \frac{\theta^2}{2!} + \frac{\theta^4}{4!} - \cdots\right) + i(\boldsymbol{\sigma} \cdot \boldsymbol{n})\left(\theta - \frac{\theta^3}{3!} + \cdots\right) \\ &= \cos\theta + i(\boldsymbol{\sigma} \cdot \mathring{\boldsymbol{n}})\sin\theta.\end{aligned} \tag{8.70}$$

Special cases of this equation were used earlier, for example, in Eq. (4.34, p. 89).

Using the relations mentioned earlier, we now focus on several physical properties of spin-$\frac{1}{2}$ particles in quantum mechanics. To this end, let us first consider the rotation operator. Recall that the angular momenta are the generators of rotation, and for spin-$\frac{1}{2}$ particles the angular momenta are given by Eq. (8.64). Therefore, the rotation operator about an arbitrary axis by an angle α can be written as

$$U_{\text{spin}} = \exp(-i\widehat{\boldsymbol{S}} \cdot \mathring{\boldsymbol{n}}\alpha/\hbar) = \exp(-i\boldsymbol{\sigma} \cdot \mathring{\boldsymbol{n}}\alpha/2), \tag{8.71}$$

where $\mathring{\boldsymbol{n}} = (n_x, n_y, n_z)$ is the unit vector along the axis. The exponentiation can be easily done, as shown in Eq. (8.70), yielding the result

$$\begin{aligned}U_{\text{spin}} &= \cos(\alpha/2)\mathbb{1} - i(\boldsymbol{\sigma} \cdot \mathring{\boldsymbol{n}})\sin(\alpha/2) \\ &= \begin{pmatrix} \cos\frac{\alpha}{2} - in_z\sin\frac{\alpha}{2} & -i(n_x - in_y)\sin\frac{\alpha}{2} \\ -i(n_x + in_y)\sin\frac{\alpha}{2} & \cos\frac{\alpha}{2} + in_z\sin\frac{\alpha}{2} \end{pmatrix}.\end{aligned} \tag{8.72}$$

Thus, any two-component ket $|\psi\rangle$ under such a rotation transforms to

$$|\psi'\rangle = U_{\text{spin}}|\psi\rangle. \tag{8.73}$$

The key point we note here is that such rotations are fundamentally in contrast to rotations of state vectors by action of integer angular momentum operators. This is easily seen by noting $U_{\text{spin}} \to -U_{\text{spin}}$ for $\alpha \to \alpha + 2\pi$, and hence $|\psi\rangle \to -|\psi\rangle$ under such transformation. Thus, the state ket requires a rotation by 4π and not 2π to come back to itself and hence are double-valued functions. Note that, in contrast, the Pauli matrices σ_i do not acquire a negative sign. For example, a rotation around $\mathring{z}$ changes the Pauli matrix σ_x as

$$
\begin{aligned}
e^{i\sigma_z\alpha/2}\sigma_x e^{-i\sigma_z\alpha/2} &= \left[\cos\frac{\alpha}{2}\mathbb{1} + i\sigma_z\sin\frac{\alpha}{2}\right]\sigma_x\left[\cos\frac{\alpha}{2}\mathbb{1} - i\sigma_z\sin\frac{\alpha}{2}\right] \\
&\dot{=} \sigma_x\cos\alpha - \sigma_y\sin\alpha,
\end{aligned} \tag{8.74}
$$

which clearly comes back to itself after 2π rotation. One can analogously show that all expectation values of the form $\langle\psi|\sigma_i|\psi\rangle$ do come back to themselves under such transformations. As a result, the negative sign acquired by a spinor wavefunction under rotation by 2π cannot be directly measured in experiments involving a single spinor particle.

☐ **Exercise 8.8** *Verify Eq. (8.67).*

☐ **Exercise 8.9** *Consider the unit vector $\mathring{n} = (\sin\theta\cos\phi, \sin\theta\sin\phi, \cos\theta)$ in spin space. For a spin pointing along $\mathring{n}$, consider the spin operator $S_n = \hbar(\boldsymbol{\sigma}\cdot\mathring{n})/2$. Find the 2×2 matrix corresponding to this operator and its eigenvalues and eigenvectors.*

☐ **Exercise 8.10** *Find the eigenvalues and eigenvectors of σ_x and σ_y.*

☐ **Exercise 8.11** *The Pauli matrices are all traceless. Use the commutation relations to argue that they must be so.*

8.3.2   Some applications

Next, we consider the application of the formalism developed above to a few simple problems. More complicated problems in this regard will be treated in later chapters. The first problem that we will consider is a spin-$\frac{1}{2}$ particle in the presence of a magnetic field. Without loss of generality, we take the field to be in the x-z plane. The Hamiltonian of such a particle is given by

$$H_0 = -\frac{e\hbar}{m}(\boldsymbol{\sigma}\cdot\boldsymbol{B}) = -b(\cos\theta\,\sigma_z + \sin\theta\,\sigma_x), \tag{8.75}$$

where θ is the angle between the magnetic field and the z-axis, and $b = e\hbar|\boldsymbol{B}|/m$. The Hamiltonian H_0, whose origin shall be discussed in detail in Chapter 9, can be roughly understood as follows. As a simplest possibility, we will seek an interaction that is linear in the magnetic field. Since the field strength is a vector, we need another vector, pertaining to the particle, to produce a dot product that will be a scalar. For a particle, the only inherent vector available is its spin $\boldsymbol{S}$, so we should have the dot product $\boldsymbol{S} \cdot \boldsymbol{B}$ in the Hamiltonian. H_0 now follows from this reasoning using the definition $\boldsymbol{S} = \frac{1}{2}\hbar\boldsymbol{\sigma}$, and the fact that e/m, involving the charge e and the mass m of the particle, has the proper dimension to make H_0 have dimension of energy if the SI units are used for the charge and the field. To obtain the energy eigenvalues and eigenstates of this Hamiltonian, we need to solve the time-independent Schrödinger equation $H_0\chi_0 = E\chi_0$, where χ_0 is a two-component spinor. It is easy to see that this problem amounts to finding eigenvectors and eigenvalues of a 2×2 matrix. Using Eq. (8.65), one can write the Schrödinger equation as

$$-b \begin{pmatrix} \cos\theta & \sin\theta \\ \sin\theta & -\cos\theta \end{pmatrix} \chi_0 = E\chi_0, \tag{8.76}$$

which leads to the energy eigenvalues and eigenfunctions:

$$E_+ = -b, \qquad \chi_0^{(+)} = \begin{pmatrix} \cos\frac{\theta}{2} \\ \sin\frac{\theta}{2} \end{pmatrix},$$

$$E_- = +b, \qquad \chi_0^{(+)} = \begin{pmatrix} \sin\frac{\theta}{2} \\ -\cos\frac{\theta}{2} \end{pmatrix}. \tag{8.77}$$

If the particles have some other contribution to the energy that is independent of spin, this extras spin-dependent contribution will add to it, and the result will be different for different spin states. This splitting of energy levels due to interaction of the spin with magnetic field is called the *Zeeman effect*.

Another class of problems involving Pauli spin matrices which has important implications for condensed matter physics involves coupling of spin and momenta of electrons through a special type of interaction called spin–orbit coupling. While the details of the origin of such interaction is beyond the scope of this book, we may borrow some of the Hamiltonians that it gives rise to and solve for their energy eigenvalues and eigenfunctions. A class of such Hamiltonians, applicable to electrons confined in the x-y plane, turns out to have the form

$$H_1 = c_1\hat{\boldsymbol{z}} \cdot (\boldsymbol{\sigma} \times \widehat{\boldsymbol{p}}) = c_1(\sigma_x\widehat{p}_y - \sigma_y\widehat{p}_x)$$

$$= c_1 \begin{pmatrix} 0 & \widehat{p}_y + i\widehat{p}_x \\ \widehat{p}_y - i\widehat{p}_x & 0 \end{pmatrix}. \tag{8.78}$$

The energy eigenvalues and eigenvectors for the Schrödinger equation $H_1\chi_1 = E\chi_1$ can be easily obtained by switching to momentum space. Writing

$$\chi_1 = \begin{pmatrix} u(k_x, k_y) \\ v(k_x, k_y) \end{pmatrix} e^{i(k_x x + k_y y)}, \tag{8.79}$$

we obtain the Schrödinger equation in momentum space

$$\hbar c_1 \begin{pmatrix} 0 & k_y + ik_x \\ k_y - ik_x & 0 \end{pmatrix} \begin{pmatrix} u(k_x, k_y) \\ v(k_x, k_y) \end{pmatrix} = E \begin{pmatrix} u(k_x, k_y) \\ v(k_x, k_y) \end{pmatrix}. \tag{8.80}$$

This equation has the solution

$$E_\pm = \pm \hbar c_1 \sqrt{k_x^2 + k_y^2}, \tag{8.81a}$$

$$\begin{pmatrix} u^\pm(k_x, k_y) \\ v^\pm(k_x, k_y) \end{pmatrix} = \frac{1}{\sqrt{2}} \begin{pmatrix} 1 \\ \mp i e^{i \arctan(k_y/k_x)} \end{pmatrix}. \tag{8.81b}$$

We note that the wavefunction in Eq. (8.81) shows that the direction of spin of the electron described by H_1 is tied to its direction of motion — a property which goes under the name of *helicity* in particle physics and is applicable to massless relativistic particles in that context. This can be easily seen by considering the case when the electron moves along y so that $k_x = 0$. In this case one has $(u^\pm, v^\pm) = (1, \pm 1)$, which are eigenstates of σ_x with eigenvalues ± 1, indicating that the electron spin described by these wavefunctions points along $\pm x$. Thus, in this case, the spin of the particle turns out to be orthogonal to its direction of motion. The reader is urged to explicitly verify this property for other values of (k_x, k_y). In §17.4, we have further discussion on Hamiltonians such as that in Eq. (8.78).

□ **Exercise 8.12** *Consider applying a constant magnetic field B along x to the 2D electron whose Hamiltonian is given by Eq. (8.78). The Hamiltonian of the system in the presence of the magnetic field is $H = H_1 - \mu B \sigma_x$, where μ is a constant called the* magnetic moment *of the particle. Show that the energy spectrum of H is identical to H_1 except that k_x has to be modified by a B-dependent term in Eq. (8.81).*

8.4   Higher matrix representations

We can just as easily talk about spins with values higher than $\frac{1}{2}$. For spin equal to s, the matrix representations of the angular momentum operators are $(2s + 1)$-dimensional. The task is then to find higher-dimensional matrix representation of the angular momentum operators. We will also outline how the rotation matrices can be obtained from them.

8.4.1 Angular momentum matrices

The groundwork was laid in §8.1. In particular, the operator S_z was chosen to be diagonal. Eq. (8.26b) implies that the matrix elements of this operator are given by

$$\langle s, m' | S_z | s, m \rangle = m\hbar\delta_{m'm}. \tag{8.82}$$

To obtain the representations of S_x and S_y, we first obtain the representations of S_+ and S_-. These are found through Eq. (8.17). With the phase convention chosen for the normalization constants of the states, we obtain

$$\langle s, m' | S_+ | s, m \rangle = \hbar a_+^{(s,m)} \delta_{m',m+1}, \tag{8.83a}$$

$$\langle s, m' | S_- | s, m \rangle = \hbar a_-^{(s,m)} \delta_{m',m-1}, \tag{8.83b}$$

with $a_\pm^{(s,m)}$ defined in Eq. (8.29). Once these are known, it is trivial to obtain S_x and S_y through the definitions in Eq. (8.7). A few examples of the results are given below. First, the matrices for $s = 1$:

$$S_x = \frac{\hbar}{\sqrt{2}} \begin{pmatrix} 0 & 1 & 0 \\ 1 & 0 & 1 \\ 0 & 1 & 0 \end{pmatrix}, \qquad S_y = \frac{\hbar}{\sqrt{2}} \begin{pmatrix} 0 & -i & 0 \\ i & 0 & -i \\ 0 & i & 0 \end{pmatrix},$$

$$S_z = \hbar \begin{pmatrix} 1 & 0 & 0 \\ 0 & 0 & 0 \\ 0 & 0 & -1 \end{pmatrix}, \qquad S^2 = 2\hbar^2 \mathbb{1}. \tag{8.84}$$

Then we give the matrices for $s = \frac{3}{2}$:

$$S_x = \frac{1}{2}\hbar \begin{pmatrix} 0 & \sqrt{3} & 0 & 0 \\ \sqrt{3} & 0 & 2 & 0 \\ 0 & 2 & 0 & \sqrt{3} \\ 0 & 0 & \sqrt{3} & 0 \end{pmatrix},$$

$$S_y = \frac{1}{2}\hbar \begin{pmatrix} 0 & -\sqrt{3}i & 0 & 0 \\ \sqrt{3}i & 0 & -2i & 0 \\ 0 & 2i & 0 & \sqrt{3}i \\ 0 & 0 & -\sqrt{3}i & 0 \end{pmatrix},$$

$$S_z = \frac{1}{2}\hbar \begin{pmatrix} 3 & 0 & 0 & 0 \\ 0 & 1 & 0 & 0 \\ 0 & 0 & -1 & 0 \\ 0 & 0 & 0 & -3 \end{pmatrix},$$

$$S^2 = \frac{15}{4}\hbar^2 \mathbb{1}. \tag{8.85}$$

In these equations, $\mathbb{1}$ stands for the unit matrix of appropriate size. Matrices for representations with higher values of s can be constructed as easily. Note that S_z is diagonal by our convention, and the matrices for S_x and S_y have non-zero elements only in the next-to-diagonal places. This follows from the structure of the operators S_+ and S_-.

□ **Exercise 8.13** *Find the eigenvectors of S_x, S_y, and S_z for spin-1 and spin-3/2 matrices. Using these or otherwise, construct the corresponding raising and lowering operators.*

8.4.2 Wigner matrices

In order to obtain the rotation matrices, we need to exponentiate the generators, i.e., the spin matrices. For the 2D representation of the spin matrices, this exponentiation can be easily done, and the result was shown in Eq. (8.72). Here, we outline how one can obtain the rotation matrices from higher-dimensional spin matrices.

In Eq. (5.82, p. 126), we mentioned how the group elements are related to the generators. If we follow that prescription, we would write an arbitrary element of the rotation group as

$$g = \exp(-i\theta_x J_x - i\theta_y J_y - i\theta_z J_z). \tag{8.86}$$

But this is not the unique way. For the rotation group, for example, we can also write a general element in the form

$$g = e^{-i\gamma J_z} e^{-i\beta J_y} e^{-i\alpha J_z}. \tag{8.87}$$

This means that a general rotation is seen as a combination of a rotation around the z-axis, followed by a rotation around the y-axis, and finally another rotation around the z-axis. When the exponents are combined by using the Baker–Campbell–Hausdorff formula, one obtains an expression like that in Eq. (8.86). The parameters α, β, and γ which appear in Eq. (8.87) are called the *Euler angles*.

There is an advantage of this parametrization of a general rotation. Suppose we take the basis consisting of the angular momentum eigenstates. Then the matrix elements of the two outlying factors of the right side of Eq. (8.87) can easily be written down, since J_z is diagonal in this basis. Thus, for a particular representation of the generators characterized by a specific value of j, we can write the matrix elements as

$$\langle m' | g | m \rangle = e^{-i\alpha m} e^{-i\gamma m'} \left[d(\beta) \right]_{m'm}, \tag{8.88}$$

with

$$\left[d(\beta) \right]_{m'm} = \langle m' | e^{-i\beta J_y} | m \rangle. \tag{8.89}$$

This matrix is called *Wigner's d-matrix*. Sometimes, the name is also used for the matrix involving the other exponential factors appearing on the right side of Eq. (8.88).

We now outline how the d-matrix is evaluated by giving an example with the 3D representation. Although the exponential involves infinite number of terms, high powers of J_y can be reduced to lower powers by using the *Cayley–Hamilton theorem*, which says that a matrix must satisfy the polynomial equation satisfied by its eigenvalues. Thus, J_y must satisfy the equation

$$J_y^3 = J_y \tag{8.90}$$

in the 3D representation. Using this, we can write

$$e^{-i\beta J_y} = 1 - i J_y \left(\beta - \frac{\beta^3}{3!} + \cdots \right) + J_y^2 \left(-\frac{\beta^2}{2!} + \frac{\beta^4}{4!} + \cdots \right)$$
$$= 1 - i J_y \sin \beta - J_y^2 (1 - \cos \beta). \tag{8.91}$$

Using the explicit form for J_y from Eq. (8.84), we obtain

$$d = \begin{pmatrix} \frac{1}{2}(1 + \cos \beta) & -\frac{1}{\sqrt{2}} \sin \beta & \frac{1}{2}(1 - \cos \beta) \\ \frac{1}{\sqrt{2}} \sin \beta & \cos \beta & -\frac{1}{\sqrt{2}} \sin \beta \\ \frac{1}{2}(1 - \cos \beta) & \frac{1}{\sqrt{2}} \sin \beta & \frac{1}{2}(1 + \cos \beta) \end{pmatrix}. \tag{8.92}$$

Similarly, one can find the rotation matrices for any other matrix representation.

8.5 Angular momentum and time-reversal transformation

In §5.4.3, we introduced the general properties of the time-reversal operator. Here, we will discuss the action of the time-reversal operator on angular momentum eigenstates.

8.5.1 Explicit form of the time-reversal operator

We start with the simple example of constructing an explicit representation of the operator for spin-$\frac{1}{2}$ particles. It was stated that the time-reversal operator is of the form

$$\mathcal{T} = UC_0, \tag{8.93}$$

where U is a unitary operator on the Hilbert space and C_0 is the complex conjugation operator on the same. Our task is to find the form of U applicable for spin-$\frac{1}{2}$ particles.

Under time-reversal transformation, an operator transforms as

$$\mathcal{O}' = \mathcal{T}\mathcal{O}\mathcal{T}^{-1} = UC_0\mathcal{O}C_0^{-1}U^{-1}. \tag{8.94}$$

We mentioned in §5.4.3 that the orbital angular momentum operator changes sign under time-reversal transformation. Spin angular momentum must do the same. So, we want

$$UC_0\sigma_i C_0^{-1}U^{-1} = -\sigma_i. \tag{8.95}$$

From the explicit form of the Pauli matrices shown in Eq. (8.65), we see that σ_y changes sign under complex conjugation whereas the others do not.

$$C_0\sigma_x C_0^{-1} = \sigma_x, \quad C_0\sigma_y C_0^{-1} = -\sigma_y, \quad C_0\sigma_z C_0^{-1} = \sigma_z. \tag{8.96}$$

Looking at Eq. (8.95), we see that we need a U that satisfies the relations

$$U\sigma_x U^{-1} = -\sigma_x, \quad U\sigma_y U^{-1} = \sigma_y, \quad U\sigma_z U^{-1} = -\sigma_z. \tag{8.97}$$

In view of the fact that the Pauli matrices anticommute among themselves, the solution of these equations is easy to find:

$$U = e^{i\delta}\sigma_y, \tag{8.98}$$

for an arbitrary real phase δ. It is convenient to choose the phase factor to be i so that U is real, and write

$$\mathcal{T} = i\sigma_y C_0 \tag{8.99}$$

in the 2D Hilbert space of vectors whose elements are the states of a spin-$\frac{1}{2}$ particle.

There is one important property to note for this operator.

$$\mathcal{T}^2 = i\sigma_y C_0 i\sigma_y C_0 = i\sigma_y C_0 i\sigma_y C_0^{-1} = -\sigma_y^2 = -\mathbb{1}, \tag{8.100}$$

using the fact that $i\sigma_y$ is real. In §5.4.3, it was mentioned that it was indeed a possibility. We now see that the possibility is realized on the space of states of a spin-$\frac{1}{2}$ particle. An implication is Kramers' degeneracy, i.e., existence of doubly degenerate energy eigenstates for Hamiltonians which are invariant under time-reversal transformation.

□ **Exercise 8.14** *Find the eigenvectors of the operator σ_x. Call them $|\chi_\pm\rangle$, where the subscript indicates the sign of the eigenvalue. Show that $\mathcal{T}|\chi_\pm\rangle \propto \chi_\mp$ with $\mathcal{T}$ given in Eq. (8.99), as is expected since the angular momentum components change sign under time-reversal transformation.*

□ **Exercise 8.15** *Repeat the same exercise with the eigenvectors of σ_y.*

□ **Exercise 8.16** *Argue that the Hamiltonian of Eq. (8.78) is invariant under time-reversal transformation, and explain how Kramers' degeneracy is realized in the states.*

It is to be noted that in a fixed external magnetic field $\boldsymbol{B}$, the Hamiltonian of Eq. (8.75) does not commute with the time-reversal operator given in Eq. (8.99). Therefore, no degeneracy is expected in the energy levels through Kramers' theorem. However, if we include the transformation of $\boldsymbol{B}$ as well under time-reversal transformation, as shown in Eq. (5.30, p. 113), then the interaction $\boldsymbol{\sigma} \cdot \boldsymbol{B}$ is invariant under time-reversal transformation. In this case, Kramers' degeneracy is realized by the fact that the state with $S_z = +m$ in magnetic field B is degenerate with the state with $S_z = -m$ in magnetic field $-B$.

Generalization to higher matrix representations is not difficult. As noted before, one should have

$$\mathcal{T}\widehat{\boldsymbol{S}}\mathcal{T}^{-1} = -\widehat{\boldsymbol{S}} \tag{8.101}$$

for any representation of the spin operators. Note that, with the convention adopted in Eq. (8.83), the matrices $S_\pm$ are real. This means that the matrix for S_x is real and S_y is imaginary. In addition, S_z is real because it is diagonal in our representation. To summarize, the effect of the complex conjugation operator is as follows:

$$C_0 S_x C_0^{-1} = S_x, \quad C_0 S_y C_0^{-1} = -S_y, \quad C_0 S_z C_0^{-1} = S_z. \tag{8.102}$$

Therefore, Eq. (8.101) demands that the matrix U defined in Eq. (8.93) satisfies the relations

$$U S_x U^{-1} = -S_x, \quad U S_y U^{-1} = S_y, \quad U S_z U^{-1} = -S_z. \tag{8.103}$$

Changing the signs of the x and z components of a vector while keeping the y component unchanged: that is exactly what is done by rotation of π about the y-axis. In other words,

$$U = e^{-i\pi S_y/\hbar}. \tag{8.104}$$

Putting it into Eq. (8.93), we obtain the representation of the time-reversal operator for states with any value of spin. Note that the form obtained in Eq. (8.99) also conforms to this general rule, except that the exponentiation was performed through Eq. (8.70). For $s = 1$, the result of the exponentiation was given in Eq. (8.92), which shows that, for this representation,

$$\mathcal{T} = \begin{pmatrix} 0 & 0 & 1 \\ 0 & -1 & 0 \\ 1 & 0 & 0 \end{pmatrix} C_0. \tag{8.105}$$

The representation of the square of the time-reversal operator can also be found easily. Since S_y is purely imaginary, the matrix U appearing in Eq. (8.104) is real, so $C_0 U C_0^{-1} = U$. Hence,

$$\mathcal{T}^2 = U C_0 U C_0 = U C_0 U C_0^{-1} = U^2 = e^{-2i\pi S_y/\hbar}. \tag{8.106}$$

It was argued earlier, in the context of spin-$\frac{1}{2}$ particles, that for particles with half-integral spin, a 2π rotation changes the state vectors by a sign. Hence, we can succinctly say that, on angular momentum eigenstates $|\psi\rangle$,

$$\mathcal{T}^2 |\psi\rangle = (-1)^{2s} |\psi\rangle. \tag{8.107}$$

8.5.2 Phases

We find from Eq. (8.101) that the time-reversal operator acting on an angular momentum eigenstate $|s, m\rangle$ must produce the state $|s, -m\rangle$. But there may be a phase factor in the relation, i.e., we should write

$$\mathcal{T} |s, m\rangle = e^{i\delta(s,m)} |s, -m\rangle. \tag{8.108}$$

The phase can be chosen arbitrarily if, instead of the definition given in Eq. (8.104), we use the freedom of putting an extra arbitrary phase on the right side. However, that phase would be related to the phase $\delta(s, m)$, which is what we show now. Of course, the entire deduction assumes that there is no phase in the definition of the ladder operators, Eq. (8.83).

For this, let us first note that Eq. (8.101) implies

$$\mathcal{T} S_\pm \mathcal{T}^{-1} = -S_\mp. \tag{8.109}$$

Now,

$$
\begin{aligned}
\mathcal{T} S_- |s, m\rangle &= \mathcal{T} \hbar a_-^{(s,m)} |s, m - 1\rangle \\
&= \hbar a_-^{(s,m)} e^{i\delta(s,m-1)} |s, -m + 1\rangle,
\end{aligned} \tag{8.110}
$$

according to the definition of Eq. (8.108). But we can write the same expression in a different form:

$$
\begin{aligned}
\mathcal{T} S_- |s, m\rangle &= (\mathcal{T} S_- \mathcal{T}^{-1}) \mathcal{T} |s, m\rangle = -S_+ e^{i\delta(s,m)} |s, -m\rangle \\
&= -e^{i\delta(s,m)} \hbar a_+^{(s,-m)} |s, -m + 1\rangle.
\end{aligned} \tag{8.111}
$$

Equating the two expressions and noting that $a_-^{(s,m)} = a_+^{(s,-m)}$ from their definitions, we find

$$e^{i\delta(s,m-1)} = -e^{i\delta(s,m)}. \tag{8.112}$$

Applying this at each step of going down the ladder, we obtain

$$e^{i\delta(s,m)} = (-1)^{s-m} e^{i\delta(s,s)}. \tag{8.113}$$

Thus, once the phase is chosen by adjusting an extra allowed phase factor in Eq. (8.104), the time-reversal phases of all other states are determined. There are two dominant conventions in this regard. One of them chooses $e^{i\delta(s,s)} = 1$ so that $e^{i\delta(s,m)} = (-1)^{s-m}$, and the other takes $e^{i\delta(s,s)} = (-1)^s$ so that $e^{i\delta(s,m)} = (-1)^m$.

☐ **Exercise 8.17** *We made our choice of the time-reversal operator for spin-$\frac{1}{2}$ particles in Eq. (8.99). What value of $\delta(s, s)$ does this choice correspond to? Verify that $\exp\{i\delta(\frac{1}{2}, -\frac{1}{2})\} = -\exp\{i\delta(\frac{1}{2}, \frac{1}{2})\}$.*

8.6   Addition of angular momenta

In this section, we are going to consider the rule of adding two angular momenta. More precisely, we would like to know the eigenstates and eigenvalues of the angular momentum operator

$$\widehat{\boldsymbol{J}} = \widehat{\boldsymbol{J}}_1 + \widehat{\boldsymbol{J}}_2, \tag{8.114}$$

where $\widehat{\boldsymbol{J}_1}$ and $\widehat{\boldsymbol{J}_2}$ are two angular momentum operators which commute:

$$\left[\widehat{\boldsymbol{J}}_1, \widehat{\boldsymbol{J}}_2\right] = 0. \tag{8.115}$$

The two individual angular momenta might be the angular momenta of two different particles, in which case $\widehat{\boldsymbol{J}}$ would be the angular momentum of the system comprising the two particles. Alternatively, $\widehat{\boldsymbol{J}}_1$ and $\widehat{\boldsymbol{J}}_2$ might be the orbital and spin angular momenta of the same particle, in which case $\widehat{\boldsymbol{J}}$ would be the total angular momentum of the particle. For the formalism that we now unfold, it does not matter which option we have in mind.

We will tackle this problem in two steps. First, in §8.6.1, we shall find the allowed eigenvalues of $\widehat{\boldsymbol{J}}$ in terms of those of $\widehat{\boldsymbol{J}_1}$ and $\widehat{\boldsymbol{J}_2}$. This will be followed in §8.6.2 by obtaining the eigenstates of the total angular momentum, i.e., of $\widehat{J}^2$ and $\widehat{J}_z$, in terms of the eigenstates of the individual angular momenta.

8.6.1   Relation between the states in the two bases

The eigenstates of the first angular momentum are given by

$$\begin{aligned}
\widehat{J}_1^2 \left|j_1, m_1\right\rangle &= j_1(j_1 + 1)\hbar^2 \left|j_1, m_1\right\rangle, \\
\widehat{J}_{1z} \left|j_1, m_1\right\rangle &= m_1\hbar \left|j_1, m_1\right\rangle,
\end{aligned} \tag{8.116}$$

whereas the states for the second one can be obtained by replacing the subscript 1 by 2 in these equations. Thus, for the combined system, we can take the basis states in the form $\left|j_1, m_1; j_2, m_2\right\rangle$, which are eigenstates of the mutually commuting operators $\widehat{\boldsymbol{J}}_1^2, \widehat{J}_{1z}, \widehat{\boldsymbol{J}}_2^2, \widehat{J}_{2z}$. Another basis can be constructed by using eigenstates of the total angular momentum. Since $\widehat{\boldsymbol{J}}^2$ does not commute with $\widehat{J}_{1z}$ and $\widehat{J}_{2z}$, we need to consider eigenstates of the mutually commuting operators $\widehat{\boldsymbol{J}}_1^2, \widehat{\boldsymbol{J}}_2^2, \widehat{\boldsymbol{J}}^2$, and $\widehat{J}_z$, which will be denoted by $\left|j_1, j_2; j, m\right\rangle$. Our task is to find the relation between these two sets of basis states. Since j_1 and j_2 appear in the specification of both sets of states, we will omit them for the sake of brevity, and write the first set of states as $\left|m_1; m_2\right\rangle$

and the second set of states as $|j, m\rangle$. Where numbers are used to denote any specific state, the two kinds will be distinguished by the presence of a comma in one notation and of a semicolon in the other.

□ **Exercise 8.18** *Find the commutators $[J^2, J_{1z}]$ and $[J^2, J_{2z}]$.*

To obtain the eigenvalues of $\widehat{\boldsymbol{J}}$ in terms of those of $\widehat{\boldsymbol{J}_1}$ and $\widehat{\boldsymbol{J}_2}$, we first note the relation between the z components of these operators: $\widehat{J}_z = \widehat{J}_{1z} + \widehat{J}_{2z}$. Operating on the state $|j; m\rangle$, we find

$$m\hbar|j, m\rangle = (\widehat{J}_{1z} + \widehat{J}_{2z})|j, m\rangle. \tag{8.117}$$

To make further progress, we need to express the state $|j, m\rangle$ in terms of the states in the $|m_1; m_2\rangle$ basis. Since $\widehat{\boldsymbol{J}_1}$ leaves the eigenstates of $\widehat{\boldsymbol{J}_2}$ unchanged, and vice versa, we can write

$$\widehat{\boldsymbol{J}}_1^2|m_1; m_2\rangle = j_1(j_1 + 1)\hbar^2 |m_1; m_2\rangle,$$
$$\widehat{J}_{1z}|m_1; m_2\rangle = m_1\hbar|m_1; m_2\rangle, \tag{8.118}$$

and similar ones for $\widehat{\boldsymbol{J}}_2^2$ and $\widehat{J}_{2z}$ acting on $|m_1; m_2\rangle$. Using this notation and the fact that the $|m_1; m_2\rangle$ states form a complete set of basis states, we can write

$$|j, m\rangle = \sum_{m_1, m_2} |m_1; m_2\rangle \langle m_1; m_2 | j, m\rangle. \tag{8.119}$$

This relation can be written as

$$|j, m\rangle = \sum_{m_1, m_2} C_{j,m}^{m_1; m_2}|m_1; m_2\rangle, \tag{8.120}$$

where

$$C_{j,m}^{m_1; m_2} = \langle m_1; m_2 | j, m\rangle. \tag{8.121}$$

These numbers, which could be complex in general, are called the *Clebsch–Gordan coefficients*, and we will presently discuss how to determine them. It has to be remembered that these coefficients depend on j_1 and j_2 as well. We have suppressed them only to avoid clutter in the notation. Using Eq. (8.118) along with Eq. (8.120), one can write Eq. (8.117) as

$$m|j, m\rangle = \sum_{m_1, m_2} (m_1 + m_2)C_{j,m}^{m_1; m_2}|m_1; m_2\rangle. \tag{8.122}$$

It is easy to see that the only possible consistent outcome of Eqs. (8.120) and (8.122) is

$$m = m_1 + m_2. \tag{8.123a}$$

In other words,

$$C_{j,m}^{m_1;m_2} = 0 \qquad \text{if } m \neq m_1 + m_2. \tag{8.123b}$$

Using this, we find that the states $|j, m\rangle$ can be expressed in terms of the states for the individual angular momenta $\widehat{J}_1$ and $\widehat{J}_2$ as

$$|j, m\rangle = \sum_{m_1} C_{j,m}^{m_1;m-m_1} |m_1; m - m_1\rangle. \tag{8.124}$$

Once we establish this relation, the only thing left is to find the maximum and minimum possible values of j. The maximum value of j can be easily found out by noting that the maximal values of m_1 and m_2 are j_1 and j_2 respectively. Thus, Eq. (8.123a) tells us that the maximal value of m is $j_1 + j_2$. Since the maximal value of m is also the maximum possible value of j, we find that

$$j_{\text{max}} = j_1 + j_2. \tag{8.125}$$

To find the minimum value of j one now notes that for a given j, there are total $2j + 1$ states. The number of states in the $|m_1; m_2\rangle$ basis is $(2j_1 + 1)(2j_2 + 1)$. Since the total number of states in both the bases must be the same, we find

$$\sum_{j=j_{\text{min}}}^{j_1+j_2} (2j + 1) = (2j_1 + 1)(2j_2 + 1), \tag{8.126}$$

where we have used Eq. (8.125) as the upper limit of the sum. This equation provides an unique solution for j_{min} given by

$$j_{\text{min}} = |j_1 - j_2|. \tag{8.127}$$

Thus, we find that j can take values $j_1 + j_2, j_1 + j_2 - 1, \cdots, |j_1 - j_2|$. Note that the maximum and the minimum values of j are identical to those obtained by adding two classical vectors $\boldsymbol{J}_1$ and $\boldsymbol{J}_2$ with magnitudes j_1 and j_2 respectively. However, for classical vectors, the magnitude of $\boldsymbol{J} = \boldsymbol{J}_1 + \boldsymbol{J}_2$ can take any value between $j_1 + j_2$ and $|j_1 - j_2|$; in contrast, for quantum operators, the allowed values are discrete.

☐ **Exercise 8.19** *Prove Eq. (8.127).*

☐ **Exercise 8.20** *If one has to add three or more mutually commuting angular momenta, one adopts exactly the strategy for adding multiple numbers, viz., add two at one time, and then add the next one to the sum, and go on like this. What are the possible values of the total spin if a system consists of three spin-1 particles?* [**Note:** *There may be more than one state with the same values of total j and m. Check that the total number of $|j, m\rangle$ states agree with the number of $|j_1, m_1; j_2, m_2; j_3, m_3\rangle$ states.*]

8.6.2   Clebsch–Gordan coefficients

We now see how the Clebsch–Gordan coefficients can be determined, i.e., how the $|j, m\rangle$ states can be constructed in terms of the eigenstates of the individual angular momenta.

We start with the maximum possible angular momentum, i.e., $j = j_1 + j_2$. The maximum m value is $j_1 + j_2$, and there is only one possible combination of m_1 and m_2 that can yield this value, viz., $m_1 = j_1$ and $m_2 = j_2$. Therefore, it is trivial to write this state in terms of the individual states:

$$|j_1 + j_2, j_1 + j_2\rangle = |j_1; j_2\rangle. \qquad (8.128)$$

Equivalently, we can say that the Clebsch–Gordan coefficients for the state on the left side are

$$C^{m_1; m_2}_{j_1 + j_2, j_1 + j_2} = \delta_{j_1, m_1} \delta_{j_2, m_2}, \qquad (8.129)$$

making use of the Kronecker deltas.

Let us now try to construct the state with the next lower value of the total m, i.e., $m = j_1 + j_2 - 1$. As shown in Eq. (8.11), this can be obtained by applying the lowering operator J_- on the state given in Eq. (8.128). However, the linear relation that was used for the z-component applies for any other component as well, so that we can write

$$\widehat{J}_- = \widehat{J}_{1-} + \widehat{J}_{2-}. \qquad (8.130)$$

We can apply $\widehat{J}_-$ on the left side of Eq. (8.128) and $\widehat{J}_{1-} + \widehat{J}_{2-}$ on the right side. On the left side, we get

$$\widehat{J}_- |j_1 + j_2, j_1 + j_2\rangle = \hbar\sqrt{2(j_1 + j_2)}\, |j_1 + j_2, j_1 + j_2 - 1\rangle, \qquad (8.131)$$

using Eq. (8.28). On the right side, the operator J_{1-} affects only the state with j_1, leaving the other undisturbed:

$$\widehat{J}_{1-} |j_1 + j_2, j_1 + j_2\rangle = \widehat{J}_{1-} |j_1; j_2\rangle = \hbar\sqrt{2j_1}\, |j_1 - 1; j_2\rangle. \qquad (8.132)$$

Similarly, J_{2-} acts only on j_2. Thus, we obtain

$$|j_1 + j_2, j_1 + j_2 - 1\rangle = \sqrt{\frac{j_1}{j_1 + j_2}}\, |j_1 - 1; j_2\rangle$$

$$+ \sqrt{\frac{j_2}{j_1 + j_2}}\, |j_1; j_2 - 1\rangle. \qquad (8.133)$$

This is the state corresponding to the next-to-maximum value of m for the total angular momentum $j_1 + j_2$. We can now apply Eq. (8.130) on this state to obtain the next lower state, and continue doing so until we reach the bottom of the ladder. The state with the lowest possible value of m will be obtained as

$$|j_1 + j_2, -j_1 - j_2\rangle = |-j_1; -j_2\rangle, \qquad (8.134)$$

because there is only one way of obtaining the lowest possible m value.

Having exhausted the ways for obtaining all states with the highest possible angular momentum, we can now look out for the states with the next highest value of the same. For $j = j_1 + j_2 - 1$, the highest possible value of m will be $j_1 + j_2 - 1$. We have already constructed one state with this value of m, shown in Eq. (8.133). That state belongs to the total angular momentum equal to $j_1 + j_2$. For $j = j_1 + j_2 - 1$, the state with the same value of m must be orthogonal to the state shown in Eq. (8.133). That alone is enough to obtain the state, apart from an overall possible phase, which we will assign arbitrarily, and write

$$|j_1 + j_2 - 1, j_1 + j_2 - 1\rangle = \sqrt{\frac{j_2}{j_1 + j_2}}\, |j_1 - 1; j_2\rangle$$

$$- \sqrt{\frac{j_1}{j_1 + j_2}}\, |j_1; j_2 - 1\rangle. \qquad (8.135)$$

We can now apply Eq. (8.130) on both sides to obtain all states corresponding to this value of j. Once that is done, we can start looking at the states with $j = j_1 + j_2 - 2$. Here, like before, we find the state with the highest m by using orthogonality with the states with the same value of m obtained for higher values of j. Then the lowering operator can be used to find states with lower values of m. The process continues until we find all states, i.e., all Clebsch–Gordan coefficients.

a) Example with $j_1 = j_2 = \frac{1}{2}$

For $j_1 = j_2 = \frac{1}{2}$, the possible values of the total angular momentum j are 1 and 0. As described above, we start with the largest value of j, i.e., 1. Clearly,

$$|1,1\rangle = \left|\frac{1}{2};\frac{1}{2}\right\rangle, \tag{8.136}$$

as shown in Eq. (8.128). Applying the lowering operator on both sides, we obtain

$$|1,0\rangle = \frac{1}{\sqrt{2}}\left(\left|-\frac{1}{2};\frac{1}{2}\right\rangle + \left|\frac{1}{2};-\frac{1}{2}\right\rangle\right). \tag{8.137}$$

Continuing the process, we obtain

$$|1,-1\rangle = \left|-\frac{1}{2};-\frac{1}{2}\right\rangle, \tag{8.138}$$

which can be checked easily.

The state with $j = 0$ can be obtained by using the fact that it should be orthogonal to the state shown in Eq. (8.137). The result is

$$|0,0\rangle = \frac{1}{\sqrt{2}}\left(\left|-\frac{1}{2};\frac{1}{2}\right\rangle - \left|\frac{1}{2};-\frac{1}{2}\right\rangle\right). \tag{8.139}$$

These results have been summarized in Table 8.2 in the form of Clebsch–Gordan coefficients.

b) Example with $j_1 = 1$, $j_2 = \frac{1}{2}$

As another example, we outline the case of $j_1 = 1$, $j_2 = \frac{1}{2}$, for which the possible values of j are $\frac{3}{2}$ and $\frac{1}{2}$. Here,

$$\left|\frac{3}{2},\frac{3}{2}\right\rangle = \left|1;\frac{1}{2}\right\rangle. \tag{8.140}$$

Applying the lowering operator, we obtain

$$\left|\frac{3}{2},\frac{1}{2}\right\rangle = \sqrt{\frac{2}{3}}\left|0;\frac{1}{2}\right\rangle + \sqrt{\frac{1}{3}}\left|1;-\frac{1}{2}\right\rangle. \tag{8.141}$$

The state with $j = \frac{1}{2}$ and $m = \frac{1}{2}$ should be orthogonal to this state:

$$\left|\frac{1}{2},\frac{1}{2}\right\rangle = \sqrt{\frac{1}{3}}\left|0;\frac{1}{2}\right\rangle - \sqrt{\frac{2}{3}}\left|1;-\frac{1}{2}\right\rangle. \tag{8.142}$$

All states with lower values of m can be obtained by using the lowering operator, repeatedly if necessary.

□ **Exercise 8.21** *Complete the exercise of finding the states and check the results against Table 8.2.*

8.6.3   Orthogonality and recursion relations

There are some relations between different Clebsch–Gordan coefficients, which we show now. Some of these relations come from the orthogonality and completeness of the angular momentum eigenstates. For example, for fixed j_1 and j_2, the $|m_1; m_2\rangle$ states form a complete set, so that we can write

$$\sum_{m_1, m_2} \langle j, m \,|\, m_1; m_2 \rangle \langle m_1; m_2 \,|\, j', m' \rangle = \langle j, m \,|\, j', m' \rangle . \qquad (8.143)$$

Using now the orthogonality of the angular momentum eigenstates, and using the definition of Eq. (8.121), we can write this equation as

$$\sum_{m_1, m_2} \left(C_{j,m}^{m_1; m_2} \right)^* C_{j', m'}^{m_1; m_2} = \delta_{jj'} \delta_{mm'} . \qquad (8.144)$$

In a similar way, we can prove that

$$\sum_{j, m} \left(C_{j,m}^{m_1; m_2} \right)^* C_{j,m}^{m_1'; m_2'} = \delta_{m_1 m_1'} \delta_{m_2 m_2'} . \qquad (8.145)$$

We have been using real Clebsch–Gordan coefficients, so the complex conjugation occurring in these formulas can be ignored. These relations can be summarized by saying that the matrix of entries in Table 8.2 is an orthogonal matrix, or, if complex phases are also used, then a unitary matrix.

□ **Exercise 8.22** *Check the orthogonality of the rows and columns of the entries in Table 8.2.*

There are also recursion relations, which relate the Clebsch–Gordan coefficients for different values of m, for the same j. To derive them, we recall the definition of Eq. (8.124) and apply Eq. (8.130). This gives

$$a_-^{(j,m)} |j, m - 1\rangle = \sum_{m_1} C_{j,m}^{m_1; m - m_1} \left(a_-^{(j_1, m_1)} |m_1 - 1; m - m_1\rangle \right.$$

$$\left. + a_-^{(j_2, m - m_1)} |m_1; m - m_1 - 1\rangle \right). \qquad (8.146)$$

Table 8.2 Clebsch–Gordan coefficients for several choices of j_1 and j_2. Blank spaces correspond to zeroes.

$j_1 = \frac{1}{2}$ $j_2 = \frac{1}{2}$	Possible $|j, m\rangle$ states			
$m_1\ m_2$	$|1,1\rangle$	$|1,0\rangle$	$|0,0\rangle$	$|1,-1\rangle$
$\frac{1}{2}\ \frac{1}{2}$	1			
$\frac{1}{2}\ \frac{1}{2}$		$\sqrt{1/2}$	$\sqrt{1/2}$	
$\frac{1}{2}\ -\frac{1}{2}$		$\sqrt{1/2}$	$-\sqrt{1/2}$	
$-\frac{1}{2}\ -\frac{1}{2}$				1

$j_1 = 1$ $j_2 = \frac{1}{2}$	Possible $|j, m\rangle$ states					
$m_1\ m_2$	$\left|\frac{3}{2},\frac{3}{2}\right\rangle$	$\left|\frac{3}{2},\frac{1}{2}\right\rangle$	$\left|\frac{1}{2},\frac{1}{2}\right\rangle$	$\left|\frac{3}{2},-\frac{1}{2}\right\rangle$	$\left|\frac{1}{2},-\frac{1}{2}\right\rangle$	$\left|\frac{3}{2},-\frac{3}{2}\right\rangle$
$1\ \frac{1}{2}$	1					
$1\ -\frac{1}{2}$		$\sqrt{1/3}$	$\sqrt{2/3}$			
$0\ \frac{1}{2}$		$\sqrt{2/3}$	$-\sqrt{1/3}$			
$0\ -\frac{1}{2}$				$\sqrt{2/3}$	$\sqrt{1/3}$	
$-1\ \frac{1}{2}$				$\sqrt{1/3}$	$-\sqrt{2/3}$	
$-1\ -\frac{1}{2}$						1

$j_1 = 1$ $j_2 = 1$	Possible $|j, m\rangle$ states								
$m_1\ m_2$	$|2,2\rangle$	$|2,1\rangle$	$|1,1\rangle$	$|2,0\rangle$	$|1,0\rangle$	$|0,0\rangle$	$|2,-1\rangle$	$|1,-1\rangle$	$|2,-2\rangle$
$1\ 1$	1								
$1\ 0$		$\sqrt{1/2}$	$\sqrt{1/2}$						
$0\ 1$		$\sqrt{1/2}$	$-\sqrt{1/2}$						
$1\ -1$				$\sqrt{1/6}$	$\sqrt{1/2}$	$\sqrt{1/3}$			
$0\ 0$				$\sqrt{2/3}$	0	$-\sqrt{1/3}$			
$-1\ 1$				$\sqrt{1/6}$	$-\sqrt{1/2}$	$\sqrt{1/3}$			
$-1\ 0$							$\sqrt{1/2}$	$\sqrt{1/2}$	
$0\ -1$							$\sqrt{1/2}$	$-\sqrt{1/2}$	
$-1\ -1$									1

Using Eq. (8.124) for the left side as well, we obtain

$$a_-^{(j,m)} \sum_{m_1} C_{j,m-1}^{m_1;m-m_1-1} |m_1; m - m_1 - 1\rangle$$

$$= \sum_{m_1} C_{j,m}^{m_1;m-m_1} \left(a_-^{(j_1,m_1)} |m_1 - 1; m - m_1\rangle \right.$$

$$\left. + a_-^{(j_2,m-m_1)} |m_1; m - m_1 - 1\rangle \right). \qquad (8.147)$$

In each sum, the m_1 values range from $-j_1$ to $+j_1$. In the first term on the right side, let us introduce a new variable $m_1' = m_1 - 1$. The range of the sum on m_1' can still be taken from $-j_1$ to $+j_1$, because the term with $m_1' = -j_1 - 1$ is zero anyway.

$$a_-^{(j,m)} \sum_{m_1} C_{j,m-1}^{m_1;m-m_1-1} |m_1; m - m_1 - 1\rangle$$

$$= \sum_{m_1'} C_{j,m}^{m_1'+1;m-m_1'-1} a_-^{(j_1,m_1'+1)} |m_1'; m - m_1' - 1\rangle$$

$$+ \sum_{m_1} C_{j,m}^{m_1;m-m_1} a_-^{(j_2,m-m_1)} |m_1; m - m_1 - 1\rangle. \qquad (8.148)$$

The primes on m_1 can now be dropped since the variable is summed over anyway. Since the angular momentum eigenstates are orthogonal, the sums can be equal only if each individual term is equal on both sides. This gives the relation

$$a_-^{(j,m)} C_{j,m-1}^{m_1;m-m_1-1} = a_-^{(j_1,m_1+1)} C_{j,m}^{m_1+1;m-m_1-1}$$

$$+ a_-^{(j_2,m-m_1)} C_{j,m}^{m_1;m-m_1}. \qquad (8.149a)$$

Similarly, if we apply $\widehat{J}_+$ instead of $\widehat{J}_-$, we obtain

$$a_+^{(j,m)} C_{j,m+1}^{m_1;m-m_1+1} = a_+^{(j_1,m_1-1)} C_{j,m}^{m_1-1;m-m_1+1}$$

$$+ a_+^{(j_2,m-m_1)} C_{j,m}^{m_1;m-m_1}. \qquad (8.149b)$$

8.7   Tensor operators

Techniques for adding angular momenta are also helpful in finding matrix elements of a special set of operators whose transformation properties under rotation are similar to those of the angular momentum eigenstates. In this section, we first define these operators and then discuss the kind of manipulations where their transformation properties play a key role.

8.7.1   Definition and properties

Rotation matrices are representations of the rotation group, as described in Chapter 5. A state $|\psi\rangle$ changes by the rule

$$|\psi\rangle \longrightarrow R\,|\psi\rangle\,, \tag{8.150}$$

where R is the appropriate matrix. In terms of the generators, for infinitesimal angles θ_a, we can write

$$R = 1 - \frac{i}{\hbar}\sum_a \theta_a \widehat{J}_a, \tag{8.151}$$

where the summation over a runs on the three generators. Thus, a state changes as

$$|\psi\rangle \longrightarrow |\psi\rangle - \frac{i}{\hbar}\sum_a \theta_a \widehat{J}_a\,|\psi\rangle\,. \tag{8.152}$$

To determine the effect of an arbitrary rotation, it is therefore sufficient to know the objects $\widehat{J}_a\,|\psi\rangle$, i.e., the results of the operation of the generators on the states. The angular momentum eigenstates have a special status in this regard. These are the states $|j,m\rangle$ defined such that the effect of the operation of the angular momentum operators are given by

$$\widehat{J}_z\,|j,m\rangle = m\hbar\,|j,m\rangle\,, \tag{8.153a}$$
$$\widehat{J}_\pm\,|j,m\rangle = a_\pm^{(j,m)}\hbar\,|j,m\pm1\rangle\,, \tag{8.153b}$$

as discussed earlier in this chapter, with $a_\pm^{(j,m)}$ defined in Eq. (8.29).

What happens to an operator under the same transformation? As explained in §2.11, any operator A must transform as

$$A' = RAR^{-1}. \tag{8.154}$$

Let us see what this implies in terms of the rotation generators, i.e., the angular momentum operators. Using the form of R from Eq. (8.151), we obtain, to first order in the infinitesimal angles,

$$RAR^{-1} = A - \frac{i}{\hbar}\sum_a \theta_a\Big(J_a A - A J_a\Big)$$
$$= A - \frac{i}{\hbar}\sum_a \theta_a\Big[J_a, A\Big]. \tag{8.155}$$

While this rule is valid for any operator, a special class of operators are called *tensor operators*, which have a very special rule of transformation. A tensor operator of rank-k is a set of $2k+1$ operators, to be denoted by $T_q^{(k)}$ with q ranging from $-k$ to $+k$, whose change under infinitesimal rotations has the same form as the change of the angular momentum eigenstates. Comparing Eqs. (8.152) and (8.155), we see that this would require that their commutators with the angular momentum operators be as follows:

$$\left[J_z, T_q^{(k)}\right] = q\hbar T_q^{(k)}, \tag{8.156a}$$

$$\left[J_+, T_q^{(k)}\right] = a_+^{(k,q)}\hbar T_{q+1}^{(k)}, \tag{8.156b}$$

$$\left[J_-, T_q^{(k)}\right] = a_-^{(k,q)} T_{q-1}^{(k)}. \tag{8.156c}$$

Of course, these commutators also imply the relation

$$\left[J^2, T_q^{(k)}\right] = k(k+1)\hbar^2 T_q^{(k)}. \tag{8.157}$$

□ **Exercise 8.23** *Derive Eq. (8.157) from Eq. (8.156).*

Because of these very specialized transformation properties, the tensor operators have a lot of similarities with the angular momentum eigenstates. One of the most important properties is that if we form a product of two tensor operators of rank k_1 and k_2, and make the combination

$$T_q^{(k)} = \sum_{q_1} C_{k,q}^{q_1; q-q_1} T_{q_1}^{(k_1)} T_{q-q_1}^{(k_2)} \tag{8.158}$$

in analogy with Eq. (8.124), then the quantity on the left side behaves like a tensor operator of rank k.

To show that this is true, let us first check Eq. (8.156a). One can use the identity for commutators given in Eq. (3.16, p. 60) to write

$$\left[\widehat{J}_z, T_{q_1}^{(k_1)} T_{q-q_1}^{(k_2)}\right] = \left[\widehat{J}_z, T_{q_1}^{(k_1)}\right] T_{q-q_1}^{(k_2)} + T_{q_1}^{(k_1)}\left[\widehat{J}_z, T_{q-q_1}^{(k_2)}\right]. \tag{8.159}$$

Applying Eq. (8.156a) for each commutator on the right side, the desired commutation rule with $\widehat{J}_z$ can be obtained trivially. Next, using the similar relation with $\widehat{J}_+$ in place of $\widehat{J}_z$, we obtain

$$\left[\widehat{J}_+, T_q^{(k)}\right] = \hbar \sum_{q_1} C_{k,q}^{q_1; q-q_1} \left(a_+^{(k_1,q_1)} T_{q_1+1}^{(k_1)} T_{q-q_1}^{(k_2)}\right.$$
$$\left. + a_+^{(k_2, q-q_1)} T_{q_1}^{(k_1)} T_{q-q_1+1}^{(k_2)}\right). \tag{8.160}$$

We now replace q_1 in the first term on the right side by $q_1' = q_1 + 1$ to obtain

$$\left[\widehat{J}_+, T_q^{(k)}\right] = \hbar \sum_{q_1'} C_{k,q}^{q_1'-1;q-q_1'+1} a_+^{(k_1,q_1'-1)} T_{q_1'}^{(k_1)} T_{q-q_1'+1}^{(k_2)}$$
$$+\hbar \sum_{q_1} C_{k,q}^{q_1;q-q_1} a_+^{(k_2,q-q_1)} T_{q_1}^{(k_1)} T_{q-q_1+1}^{(k_2)}. \qquad (8.161)$$

The limits on the sum do not change, for reasons discussed while deriving Eq. (8.149). The prime on q_1 can be removed since it is summed over, so that we obtain

$$\left[\widehat{J}_+, T_q^{(k)}\right] = \hbar \sum_{q_1} \left(C_{k,q}^{q_1-1;q-q_1+1} a_+^{(k_1,q_1-1)} \right.$$
$$\left. +C_{k,q}^{q_1;q-q_1} a_+^{(k_2,q-q_1)} \right) T_{q_1}^{(k_1)} T_{q-q_1+1}^{(k_2)}. \qquad (8.162)$$

Applying the recursion formula Eq. (8.149b) now, we see that $T_q^{(k)}$ has the desired commutation relation with $\widehat{J}_+$. The proof for the commutation relation with $\widehat{J}_-$ is similar, and is not repeated. In summary, we have shown that $T_q^{(k)}$ is indeed a tensor operator of rank k.

8.7.2 Connection with Cartesian components of tensors

The tensor operators $T_q^{(k)}$, for a fixed value of k, are very simply related to the Cartesian components of a rank-k tensor. To see this connection, let us consider tensors involving only the position operator. The position operator can be replaced by any other vector operator, with obvious changes in the notation, to include other operators.

Tensors with rank 0 are scalars, and they need not be discussed. Let us consider rank-1 tensors. The Cartesian components of a rank-1 tensor $T^{(1)}$ involving the position vector will be

$$T_x^{(1)} = f(r)x, \qquad T_y^{(1)} = f(r)y, \qquad T_z^{(1)} = f(r)z, \qquad (8.163)$$

where $f(r)$ is an arbitrary scalar function of r. Let us now write the expressions in terms of the spherical coordinates. We see that

$$T_z^{(1)} = f(r)r\cos\theta = K_1 f(r)r Y_1^0(\theta, \phi), \qquad (8.164)$$

where Y_1^0 is the spherical harmonic listed in Table 8.1 (p. 189), and K_1 is a numerical constant which can be easily read from that table.

As for the x- and y-components, we note that we can make linear combinations:

$$T_x^{(1)} \pm iT_y^{(1)} = f(r)(x \pm iy) = f(r)r\sin\theta e^{\pm i\phi}$$
$$= \sqrt{2}K_1 f(r)rY_1^{\pm 1}(\theta, \phi). \tag{8.165}$$

Since the functions of r are invariant under rotation, we see that the Cartesian components can be expressed as linear superpositions of tensor operators with $k = 1$, as proposed.

We now turn to tensors of rank 2. We can write any such tensor as a sum of a symmetric tensor and an antisymmetric one. Components of the antisymmetric tensor can be suitably multiplied with the Levi-Civita symbol, the result being a rank-1 tensor which has already been discussed in this context. The symmetric one has 6 Cartesian components. Leaving aside the combination

$$T^{(0)} = \sum_i T_{ii}^{(2)} \tag{8.166}$$

which transforms like a scalar, we have 5 components to worry about. It can be easily seen that one can make linear combinations of these 5 components which transform like spherical harmonics with $\ell = 2$. For example,

$$T_{xz}^{(2)} \pm iT_{yz}^{(2)} = f(r)(x \pm iy)z = f(r)r^2 \sin\theta\cos\theta\, e^{\pm i\phi}$$
$$= K_2 f(r)r^2 Y_2^{\pm 1}. \tag{8.167}$$

□ **Exercise 8.24** *Find the suitable combinations of spherical harmonics for other components of a rank-2 tensor operator.*

The exercise can be continued. The point to note is that whenever any two indices of a tensor appear in an antisymmetric combination, we can reduce it in terms of some lower-rank tensor by using the Levi-Civita symbol. Therefore, only completely symmetric tensors of each rank need to be considered. For symmetric rank-3 tensors, the number of independent components is 10. They can be written in terms of Y_3^m and T_1^m, using all admissible values of m for each kind. The first kind has 7 spherical components, and the second kind has three, which add up to 10.

□ **Exercise 8.25** *By counting, show that rank-4 symmetric tensors can be expressed in terms of spherical harmonics with the ℓ value equal to 4, 2, and 0.*

□ **Exercise 8.26** *More generally, show that rank-n symmetric tensors can be expressed in terms of: (a) rank-$2k$ tensors with $k = 0, ..., m$ if $n = 2m$, (b) rank-$(2k+1)$ tensors with $k = 0, ..., m$ if $n = 2m + 1$.*

8.7.3 Wigner–Eckart theorem

The similarity of the structure of Eq. (8.124) for states and Eq. (8.158) for operators is not surprising. In fact, as stated, both equations bank on the fact that the quantities behave in a certain way under rotation.

Clearly, then, we can even write same kind of relations involving one tensor operator and one angular momentum eigenstate, i.e., combinations of the form

$$\sum_q C_{j,m}^{q;m_2} T_q^{(k)} \left| \alpha, j_2, m_2 \right\rangle, \tag{8.168}$$

where α denotes, collectively, all other quantum numbers on which the specification of the state might depend. With $m_2 = m - q$, these objects would transform like the state $\left| \alpha, j, m \right\rangle$. However, the sum need not be normalized properly, so we can write

$$\sum_q C_{j,m}^{q;m_2} T_q^{(k)} \left| \alpha, j_2, m_2 \right\rangle = T^{(k)} \left| \alpha, j, m \right\rangle, \tag{8.169}$$

where $T^{(k)}$, without any subscript, denotes a quantity that commutes with all angular momentum components.

Now, recall the definition of the Clebsch–Gordan coefficients given in Eq. (8.121) to write this expression in the form

$$T^{(k)} \left| \alpha, j, m \right\rangle = \sum_q T_q^{(k)} \left| \alpha, j_2, m_2 \right\rangle \left\langle q; m - q \mid j, m \right\rangle. \tag{8.170}$$

To invert this set of equations, we recall the statement that the Clebsch–Gordan coefficients form a unitary matrix. Therefore,

$$T_q^{(k)} \left| \alpha, j_2, m_2 \right\rangle = \sum_{j,m} T^{(k)} \left| \alpha, j, m \right\rangle \left\langle j, m \mid q; m_2 \right\rangle. \tag{8.171}$$

The sum would obtain contributions only for the values of j from $|k - j_2|$ to $k + j_2$ and for $m = q + m_2$.

Now suppose we want to find the matrix element of the tensor operator $T_q^{(k)}$ between two angular momentum eigenstates. The *other* quantum numbers may or may not be the same. This is easily done now:

$$\left\langle \alpha', j_1, m_1 \left| T_q^{(k)} \right| \alpha, j_2, m_2 \right\rangle = \sum_{j,m} \left\langle \alpha', j_1, m_1 \left| T^{(k)} \right| \alpha, j, m \right\rangle$$

$$\times \left\langle j, m \mid q; m_2 \right\rangle. \tag{8.172}$$

The first factor on the right side contains the operator $T^{(k)}$ in the middle. Since it commutes with all angular momentum generators, the factors containing the azimuthal angle ϕ must integrate to zero if $m \neq m_1$. And, when $m = m_1$, the factor becomes unity, so there is no ϕ-dependence, i.e., no dependence on the azimuthal quantum number. Next, we note that the second factor on the right side is the complex conjugate of a Clebsch–Gordan coefficient. Thus, we can write

$$\langle \alpha', j_1, m_1 | T_q^{(k)} | \alpha, j_2, m_2 \rangle = \sum_{j, m_1} \langle \alpha', j_1 | T^{(k)} | \alpha, j \rangle \left(C_{j, m_1}^{q; m_2} \right)^* . \quad (8.173)$$

The sum on the right side does not really do much. Because the angular momentum eigenstates are orthogonal, the matrix element of $T^{(k)}$ will vanish unless $j = j_1$. Also, the Clebsch–Gordan coefficients will vanish unless $m_1 = q + m_2$. Thus, for the matrix element, there is only one term that contributes to the sum, and we can write

$$\langle \alpha', j, m | T_q^{(k)} | \alpha, j_2, m_2 \rangle = \langle \alpha', j | T^{(k)} | \alpha, j \rangle \left(C_{j, m}^{q; m_2} \right)^* . \quad (8.174)$$

This statement is called the *Wigner–Eckart theorem*. Since we have chosen a convention in which the Clebsch–Gordan coefficients are all real, we can disregard the complex conjugation in its statement. Put simply, this theorem says that the matrix element of a tensor operator between two angular momentum eigenstates is equal to an invariant matrix element times a Clebsch–Gordan coefficient. The invariant part is often called the *reduced matrix element*.

Chapter 9

Exactly solvable problems in three dimensions

We have discussed 1D problems in Chapter 7. In this chapter, we discuss mainly 3D problems, although some of the generalities might apply to spaces whose dimensions are different.

9.1 Generalities

The stationary state form of the Schrödinger equation for a single particle system was given in Eq. (4.24, p. 86). For non-relativistic systems with velocity-independent potential energies, the general form of the Hamiltonian would be

$$H = \frac{p^2}{2M} + V(r) \tag{9.1}$$

for a particle of mass M. Accordingly, the Schrödinger equation for stationary states has the form

$$\nabla^2 \psi(r) + \frac{2M}{\hbar^2} \Big(E - V(r) \Big) \psi(r) = 0. \tag{9.2}$$

The Laplacian operator, ∇^2, involves second-order derivatives with respect to the coordinates. This is the differential equation we want to solve in this chapter for various choices of the potential.

9.2 Particle in a 3D box

The problem of a particle in an infinite rectangular-shaped 3D potential well is just an extension of the same problem in 1D. So, the solution can be borrowed

from what we have already done in §7.2, but there will be some new features
of the solutions which need some attention.

The potential is infinite except within the box defined by

$$0 \leqslant x \leqslant \ell_1, \qquad 0 \leqslant y \leqslant \ell_2, \qquad 0 \leqslant z \leqslant \ell_3. \tag{9.3}$$

Inside the box, the potential vanishes. The wavefunctions for the stationary
states can be written in the form

$$\psi(x, y, z) = X(x)Y(y)Z(z), \tag{9.4}$$

and the differential equation corresponding to each factor will be the same as
the corresponding 1D problem. The energy eigenvalues will be a sum of three
terms of the form obtained for the 1D problem, i.e., they will be

$$E_{n_1,n_2,n_3} = \frac{\pi^2\hbar^2}{2M}\left(\frac{n_1^2}{\ell_1^2} + \frac{n_2^2}{\ell_2^2} + \frac{n_3^2}{\ell_3^2}\right). \tag{9.5}$$

The energy eigenfunctions should be the product, as shown in Eq. (9.4):

$$\psi_{n_1,n_2,n_3} = N \sin\left(\frac{n_1\pi x}{\ell_1}\right)\sin\left(\frac{n_2\pi y}{\ell_2}\right)\sin\left(\frac{n_3\pi z}{\ell_3}\right), \tag{9.6}$$

where N is a normalization constant. As for the case of the 1D problem, we
see that the solution requires that each of the three integers n_1, n_2, and n_3
should be strictly positive. If any of them is zero, the wavefunction vanishes.
The negative integers give the same wavefunction as the positive integers once
the normalization factor is adjusted.

☐ **Exercise 9.1** *Show that*

$$N = \sqrt{8/\ell_1\ell_2\ell_3}. \tag{9.7}$$

in Eq. (9.6) for a normalized wavefunction.

The only important difference with the 1D problem is that if more than one
side of the box are equal, the wavefunctions will have degeneracy. For example,
consider the case of a cubical box, for which all sides are equal, of length ℓ.
Then the energy eigenvalues will depend on the combination $n_1^2 + n_2^2 + n_3^2$. The
degeneracy of any energy eigenvalue will be equal to the number of ways that
this sum can be the same. For example, for $n_1^2 + n_2^2 + n_3^2 = 6$, there will be a
three fold degeneracy, because any one of the three integers can be equal to 2
and the rest equal to 1.

☐ **Exercise 9.2** *A particle is in a rectangular box with sides L, $2L$, and $2L$.
Find the degeneracy of the first and the second excited states.*

☐ **Exercise 9.3** *The energy of an eigenstate of a particle in a cubic box with
sides L is $E_{n_1,n_2,n_3} = 11\hbar^2\pi^2/(2ML^2)$. Find the possible values of n_1, n_2, and
n_3 and hence the degeneracy of the state.*

9.3   3D problems with central potentials

In this section, we outline some aspects of the solution of Eq. (9.2) when the potential energy is a function of only the radial coordinate, i.e.,

$$V(\boldsymbol{r}) = V(r). \tag{9.8}$$

Such potentials are called *central potentials*. Solutions for specific forms of $V(r)$ will be taken up in some later sections. In this section, we want to discuss the general nature of the solutions that are independent of the r-dependence of the central potential.

9.3.1   The eigenvalue equation

Obviously, it is convenient to attempt the solution in spherical polar coordinates, for which the Laplacian operator is given by

$$\boldsymbol{\nabla}^2 \psi = \frac{1}{r^2} \frac{\partial}{\partial r} \left(r^2 \frac{\partial \psi}{\partial r} \right) + \frac{1}{r^2 \sin\theta} \frac{\partial}{\partial \theta} \left(\sin\theta \frac{\partial \psi}{\partial \theta} \right) + \frac{1}{r^2 \sin^2\theta} \frac{\partial^2 \psi}{\partial \phi^2}. \tag{9.9}$$

Although the potential is independent of the angular coordinates θ and ϕ, the angular coordinates appear in the differential equation, Eq. (9.2). The solution $\psi(\boldsymbol{r})$ will be a function of all the coordinates r, θ, ϕ.

In order to find the solution, let us try an ansatz that the solution is separable in the three coordinates, i.e., the solution is of the form

$$\psi(\boldsymbol{r}) = R(r)\Theta(\theta)\Phi(\phi). \tag{9.10}$$

Putting this form into Eq. (9.2) and using the expression for the Laplacian operator given in Eq. (9.9), we obtain

$$\frac{1}{R} \frac{d}{dr} \left(r^2 \frac{dR}{dr} \right) + \frac{1}{\Theta \sin\theta} \frac{d}{d\theta} \left(\sin\theta \frac{d\Theta}{d\theta} \right)$$

$$+ \frac{1}{\Phi \sin^2\theta} \frac{d^2\Phi}{d\phi^2} + \frac{2Mr^2}{\hbar^2} \left(E - V(r) \right) = 0, \tag{9.11}$$

where we have used ordinary derivatives instead of partial derivatives because each of the functions R, Θ, and Φ is a function of only one variable. Let us write this equation in the form

$$\frac{1}{R} \frac{d}{dr} \left(r^2 \frac{dR}{dr} \right) + \frac{2Mr^2}{\hbar^2} \left(E - V(r) \right)$$

$$= -\frac{1}{\Theta \sin\theta} \frac{d}{d\theta} \left(\sin\theta \frac{d\Theta}{d\theta} \right) - \frac{1}{\Phi \sin^2\theta} \frac{d^2\Phi}{d\phi^2}. \tag{9.12}$$

The remarkable thing about this form of the equation is that the left side is a function of only the radial coordinate r, whereas the right side is a function of the angular variables θ and ϕ. This is not possible unless both sides are equal to some constant. So let us write

$$\frac{1}{R}\frac{d}{dr}\left(r^2\frac{dR}{dr}\right) + \frac{2Mr^2}{\hbar^2}\left(E - V(r)\right) = K_1, \tag{9.13}$$

$$\frac{1}{\Theta\sin\theta}\frac{d}{d\theta}\left(\sin\theta\frac{d\Theta}{d\theta}\right) + \frac{1}{\Phi\sin^2\theta}\frac{d^2\Phi}{d\phi^2} = -K_1, \tag{9.14}$$

where K_1 is a constant. Further, we note that Eq. (9.14) can again be written as

$$\frac{\sin\theta}{\Theta}\frac{d}{d\theta}\left(\sin\theta\frac{d\Theta}{d\theta}\right) + K_1\sin^2\theta = -\frac{1}{\Phi}\frac{d^2\Phi}{d\phi^2}, \tag{9.15}$$

where the left side is a function of only θ, and the right side of only ϕ. Thus, we can use the same argument again and say that for a solution of the form given in Eq. (9.10), there must exist a constant K_2 such that

$$\frac{\sin\theta}{\Theta}\frac{d}{d\theta}\left(\sin\theta\frac{d\Theta}{d\theta}\right) + K_1\sin^2\theta = K_2, \tag{9.16}$$

$$\frac{1}{\Phi}\frac{d^2\Phi}{d\phi^2} = -K_2. \tag{9.17}$$

Of these two equations, let us consider Eq. (9.17) first. The solutions of this equation are of the form $\Phi = \exp(\pm i\sqrt{K_2}\,\phi)$. If K_2 is negative, these are exponentially growing and falling solutions, both monotonic in ϕ. However, remember that ϕ is an angular variable with period 2π. Thus, for equal values of r and θ, the values ϕ and $\phi + 2\pi$ denote the same physical point. Hence, we should demand that

$$\Phi(\phi) = \Phi(\phi + 2\pi). \tag{9.18}$$

This means that the constant K_2 has to be non-negative. Let us therefore write

$$K_2 = m^2 \tag{9.19}$$

for a real number m. The solutions are then of the form

$$\Phi(\phi) = e^{\pm im\phi}, \tag{9.20}$$

and Eq. (9.18) implies that this m must be an integer. Remember that the mass of the particle was denoted by M, so there is no conflict of notations here.

With the knowledge of the nature of ϕ-dependence, let us now look back at Eq. (9.16). We write this equation by introducing a new variable

$$\xi = \cos\theta. \tag{9.21}$$

Using Eq. (9.19) for the constant K_2, we obtain

$$\frac{d}{d\xi}\left((1-\xi^2)\frac{d\Theta}{d\xi}\right) + \left[K_1 - \frac{m^2}{1-\xi^2}\right]\Theta = 0. \tag{9.22}$$

This equation has been discussed in detail in Appendix C, in particular in §C.3. Solution for this equation are obtained when

$$K_1 = l(l+1) \tag{9.23}$$

for some integer $l \geqslant |m|$ and is called the *associated Legendre function*, denoted by $P_l^m(\xi)$. Thus, we find that the angular dependence of the solution given in Eq. (9.10) is of the form

$$Y_l^m(\theta,\phi) = P_l^m(\cos\theta)e^{im\phi}, \tag{9.24}$$

where m can take positive and negative values in the range

$$|m| \leqslant l, \tag{9.25}$$

implying that l can be zero or a positive integer. The functions defined in Eq. (9.24) are called *spherical harmonics*.

These are the eigenfunctions of angular momentum, encountered in Chapter 8. There is no surprise that they reappeared here. In fact, if we take the value of K_1 from Eq. (9.23) and put it into Eq. (9.14), the resulting equation is

$$\widehat{L}^2\Theta(\theta)\Phi(\phi) = l(l+1)\hbar^2\Theta(\theta)\Phi(\phi), \tag{9.26}$$

with $\widehat{L}^2$ given in Eq. (8.37, p. 186). The solutions are

$$\Theta_{lm}(\theta) = P_l^m(\cos\theta), \qquad \Phi_m(\phi) = e^{im\phi}. \tag{9.27}$$

In summary, solutions of the energy eigenfunctions in a central potential are of the form

$$\psi(r,\theta,\phi) = R(r)Y_l^m(\theta,\phi). \tag{9.28}$$

The angular dependence of the solution is the same, irrespective of the form of the central potential. The radial dependence, understandably, cannot be obtained without the knowledge of the function $V(r)$. However, inserting the forms for the constant K_1 from Eq. (9.23) into Eq. (9.13), we find that the radial function $R(r)$ should satisfy a differential equation of the form

$$\frac{d}{dr}\left(r^2\frac{dR}{dr}\right) + \left[\frac{2Mr^2}{\hbar^2}\left(E - V(r)\right) - l(l+1)\right]R = 0. \qquad (9.29)$$

In the rest of this chapter, we will solve this radial equation for several forms of the central potential $V(r)$.

9.3.2   Degeneracy and conservation

Even without the explicit knowledge about $V(r)$, there are two striking things that have emerged from the discussion, which are worth some scrutiny.

The first question is about the eigenfunctions. The wavefunctions depend on the polar and azimuthal angles, θ and ϕ. Polar angle is the angle measured with respect to a fixed coordinate axis. The Hamiltonian does not make any reference to any direction in the 3D space. So, how to define this axis from which the polar angle will be measured?

The second question is about the eigenvalues. Clearly, the eigenvalues come out from the solution of Eq. (9.29). This differential equation does not contain the quantum number m that appears in the angular part of the solution. Therefore, the eigenvalues would be independent of m. In other words, all eigenfunctions corresponding to different values of m but same values of all other quantum numbers (like l) will be degenerate. In §5.1, we showed that conserved quantities cause degeneracy. Is there a conservation law that is responsible for this degeneracy?

An easy hint to this question is available from the knowledge of classical mechanics. According to Newton's laws of motion, the equation of motion of a particle is given by

$$\frac{d\boldsymbol{p}}{dt} = \boldsymbol{F}, \qquad (9.30)$$

where $\boldsymbol{p}$ is the momentum and $\boldsymbol{F}$ is the force applied. From this, it easily follows that

$$\frac{d}{dt}(\boldsymbol{r} \times \boldsymbol{p}) = \boldsymbol{r} \times \boldsymbol{F}. \qquad (9.31)$$

For any system where the potential does not have any directional dependence, the force is in the radial direction, and therefore the right side vanishes. This implies that the orbital angular momentum,

$$\boldsymbol{L} = \boldsymbol{r} \times \boldsymbol{p}, \tag{9.32}$$

is conserved.

Coming back to quantum mechanics, we can then try to see whether the orbital angular momentum commutes with the Hamiltonian. In particular, let us take any one component of the angular momentum, say,

$$L_z = xp_y - yp_x, \tag{9.33}$$

and see whether it commutes. It is relatively easy to see that it commutes with the kinetic energy part of the Hamiltonian, using the basic commutation relations between the components of the coordinate and the momentum given in Eq. (3.15, p. 60). For the potential energy part, let us check the result of the operation of $[L_z, V(r)]$ on an arbitrary function of the coordinates. Using the coordinate space representation of the momentum operator, we see that

$$\left[xp_y, V(r) \right] \psi(\boldsymbol{r}) = -i\hbar \left[x \frac{\partial}{\partial y} \Big(V(r)\psi(\boldsymbol{r}) \Big) - V(r)x \frac{\partial}{\partial y} \psi(\boldsymbol{r}) \right]$$
$$= -i\hbar x \frac{\partial V(r)}{\partial y} \psi(\boldsymbol{r}), \tag{9.34}$$

and similarly

$$\left[yp_x, V(r) \right] \psi(\boldsymbol{r}) = -i\hbar y \frac{\partial V(r)}{\partial x} \psi(\boldsymbol{r}). \tag{9.35}$$

Since

$$\frac{\partial V(r)}{\partial x_i} = \frac{\partial V(r)}{\partial r} \frac{x_i}{r}, \tag{9.36}$$

it follows that L_z commutes with any central potential. The exercise can also be carried out for other components of the angular momentum, giving the same result.

We therefore conclude that the Hamiltonian involving a central potential commutes with all components of angular momentum. Therefore, we can choose the eigenstates of the Hamiltonian in such a way that they are eigenstates of angular momentum. Now, the different components of angular momentum do not commute among themselves, as discussed in earlier chapters. The operator L^2 commutes with all components, and therefore we can say that the operators H, L^2, and any one component of $\boldsymbol{L}$ form a set of mutually commuting operators. We can pick this one component in this set to be L_z, for any arbitrary choice of the direction of the z-axis. So, we can take the eigenstates of the Hamiltonian to be also eigenstates of L^2 and L_z. That is exactly what we have done. In the wavefunctions, the quantum number l corresponds to the eigenvalue of L^2 being $l(l+1)\hbar^2$, and the quantum number m corresponds to the eigenvalue of L_z being $m\hbar$.

This discussion already answers one of two issues of discomfort mentioned earlier. We have chosen a particular axis when we decided that the eigenfunctions of the Hamiltonian will also be eigenfunctions of the component of angular momentum along that axis. The polar angle θ that appears in the wavefunctions is the angle measured from this axis, and the azimuthal angle ϕ is the angle in the plane perpendicular to this axis.

The question of degeneracy can also now be answered. Since the Hamiltonian commutes with each component of the angular momentum, it also commutes with the ladder operators L_+ and L_-. Recall that L_+ acting on an eigenstate of L_z produces another eigenstate of L_z with a higher eigenvalue. Thus, all states with different values of m_l must be degenerate. This is a general feature of particles in a central potential. Since the angular momentum operators are generators of rotations, we can say that this degeneracy results from the rotational symmetry of the Hamiltonian.

We want to add an additional comment. As mentioned, the energy eigenvalues are obtained from the solution of Eq. (9.29), which contains the quantum number l. This means that, in general, the energy eigenvalues can depend on l. However, there is no guarantee. For very specific kinds of the central potential, it might happen that the energy eigenvalues turn out to be independent of l. That will then indicate some symmetry beyond the rotational symmetry in the Hamiltonian. We will see an example when we discuss the hydrogen atom in §9.6.

9.4   Particle in a spherical potential well

In this section, we will discuss bound state energies of a particle in a spherical potential well. The zero of energy will be taken in such a way that the potential energy is zero within a spherical region of radius a, whereas it has some positive constant value everywhere outside. We will consider two cases, one in which the potential outside is infinite, and another in which it is finite.

9.4.1   Infinite well

First, we take up the case when the potential is given by

$$V(r) = \begin{cases} 0 & \text{for } r \leqslant a, \\ \infty & \text{otherwise.} \end{cases} \tag{9.37}$$

This is similar to the 1D potential discussed in §7.2. In this 3D incarnation, it is the problem of a particle moving freely in a spherical cavity of radius a and not allowed to go outside this region.

As discussed in case of the 1D problem, the wavefunction will vanish outside the spherical cavity. Therefore, the wavefunction inside the cavity must satisfy the continuity condition

$$R(a) = 0. \tag{9.38}$$

In order to solve the second-order differential equation of Eq. (9.29), we will need another boundary condition. This will be invoked later.

The equation for the radial part of the wavefunction inside the cavity is given by

$$r^2 \frac{d^2 R}{dr^2} + 2r \frac{dR}{dr} + \left[\frac{2Mr^2 E}{\hbar^2} - l(l+1) \right] R = 0. \tag{9.39}$$

We now define a shorthand

$$k \equiv \sqrt{\frac{2ME}{\hbar^2}} \tag{9.40}$$

which has the dimension of inverse length. Using the dimensionless radial variable

$$\rho \equiv kr, \tag{9.41}$$

Eq. (9.39) can be transformed into the form

$$\rho^2 \frac{d^2 R}{d\rho^2} + 2\rho \frac{dR}{d\rho} + \left[\rho^2 - l(l+1) \right] R = 0. \tag{9.42}$$

This is the spherical Bessel equation, and its solution has been discussed in Appendix C. As is usual for second-order differential equations, there are two independent solutions for the spherical Bessel equation. They are usually denoted by $j_l(\rho)$ and $\eta_l(\rho)$. In order to choose between them, we need to invoke a boundary condition that we have not discussed yet. The point at the center of the cavity, i.e., at $r = 0$, is an ordinary point, and the wavefunction must be finite there. This implies that the solutions to be used must be spherical Bessel functions of the first kind, i.e, the ones denoted by $j_l(\rho)$. Expressions for these functions in Eq. (C.61, p. 463) confirm that these functions are regular at the origin, whereas those of the other kind, shown in Eq. (C.65, p. 464), are not. Thus, we have, up to an irrelevant normalization factor,

$$R(\rho) = j_l(\rho). \tag{9.43}$$

Note that the solution will also have to satisfy the boundary condition of Eq. (9.38), i.e., we must have

$$j_l(ka) = 0. \tag{9.44}$$

The solution of this equation will give the allowed values of k, and hence the energy eigenvalues through Eq. (9.40). Thus, if a solution of the equation occurs at $ka = \nu_l$, the energy eigenvalues, computed from Eq. (9.40), will be

$$E_l = \frac{\hbar^2}{2Ma^2}\nu_l^2. \tag{9.45}$$

The lowest eigenvalue will correspond to the smallest value of ka for which Eq. (9.44) has a solution. This occurs for $l = 0$. In fact, by directly plugging into the differential equation of Eq. (9.42), it can be seen that for $l = 0$ the solution is

$$j_0(kr) = \frac{\sin kr}{kr}. \tag{9.46}$$

The smallest value of k for which Eq. (9.44) is satisfied is therefore given by

$$ka = \pi, \tag{9.47}$$

and so the ground state energy of the system, obtained from Eq. (9.40), is

$$E_0 = \frac{\pi^2\hbar^2}{2Ma^2}. \tag{9.48}$$

9.4.2 Finite well

In this case, the potential is

$$V(r) = \begin{cases} 0 & \text{for } r \leqslant a, \\ V_0 & \text{otherwise.} \end{cases} \tag{9.49}$$

The differential equation for the radial wavefunction is the same in the region inside the well, and so the solution is the same, as given in Eq. (9.43). The difference is that now the wavefunction need not be zero outside the well, so Eq. (9.38) does not apply, and therefore the bound states are not obtained by solving Eq. (9.44). Instead, we need to set up the differential equation for the wavefunction in the outside region, which would be

$$r^2 \frac{d^2 R}{dr^2} + 2r \frac{dR}{dr} + \left[\frac{2Mr^2(E - V_0)}{\hbar^2} - l(l+1) \right] R = 0. \tag{9.50}$$

Bound states will correspond to the solutions with $E < V_0$. Thus, we can define the real parameter

$$\kappa = \sqrt{\frac{2M(V_0 - E)}{\hbar^2}} \tag{9.51}$$

and rewrite the differential equation in the form

$$\rho'^2 \frac{d^2 R}{d\rho'^2} + 2\rho' \frac{dR}{d\rho'} + \left[\rho'^2 - l(l+1) \right] R = 0. \tag{9.52}$$

with

$$\rho' = i\kappa r \tag{9.53}$$

to be used for the outside region. This is still the spherical Bessel differential equation. So the solutions outside the well would also be spherical Bessel functions. However, this time we should choose the combination of the two independent solutions that vanishes at infinity. This combination, described in Appendix C, is given by

$$h_l^{(1)}(\rho') = j_l(\rho') + i\eta_l(\rho'), \tag{9.54}$$

and is called Hankel function of the first kind. The solution inside the well and outside can then be matched at the edge of the well by demanding that the wavefunction should be continuous. This is left as an exercise for the reader.

☐ **Exercise 9.4** *For the finite well, find the ground state eigenvalue and eigenfunctions by matching boundary condition at the edge of the well (at $r = a$). Hence show that the lowest bound state energy is given by the condition $\kappa = -k \cot ka$ where k and κ are defined in Eqs. (9.40) and (9.51) respectively. Find the critical radius of the well so that at least one bound state exists.*

☐ **Exercise 9.5** *Repeat the exercise for a circular well in 2D.*

9.5   Two-particle systems

Suppose we have a system of two particles, of masses m_1 and m_2. The Hamiltonian of this system can be written as

$$H = \frac{\boldsymbol{p}_1^2}{2m_1} + \frac{\boldsymbol{p}_2^2}{2m_2} + V(\boldsymbol{r}_1, \boldsymbol{r}_2). \qquad (9.55)$$

The wavefunction is now a function of the coordinates of both particles, i.e., we should write it as $\Psi(\boldsymbol{r}_1, \boldsymbol{r}_2)$. Thus, the Schrödinger equation should be

$$i\hbar \frac{\partial}{\partial t} \Psi(\boldsymbol{r}_1, \boldsymbol{r}_2) = \left[-\frac{\hbar^2}{2m_1} \boldsymbol{\nabla}_1^2 - \frac{\hbar^2}{2m_2} \boldsymbol{\nabla}_2^2 + V(\boldsymbol{r}_1, \boldsymbol{r}_2) \right] \Psi(\boldsymbol{r}_1, \boldsymbol{r}_2), \qquad (9.56)$$

where $\boldsymbol{\nabla}_1$ and $\boldsymbol{\nabla}_2$ involve derivatives with respect to $\boldsymbol{r}_1$ and $\boldsymbol{r}_2$ respectively.

As in the case of classical two-body problem, it is convenient to rewrite the equation in terms of the center-of-mass coordinate of the two particles,

$$\boldsymbol{r}_{\mathrm{CM}} = \frac{m_1 \boldsymbol{r}_1 + m_2 \boldsymbol{r}_2}{m_1 + m_2}, \qquad (9.57\mathrm{a})$$

and the relative coordinate,

$$\boldsymbol{r} = \boldsymbol{r}_1 - \boldsymbol{r}_2. \qquad (9.57\mathrm{b})$$

It is then a trivial exercise to check that

$$\frac{1}{m_1} \boldsymbol{\nabla}_1^2 + \frac{1}{m_2} \boldsymbol{\nabla}_2^2 = \frac{1}{m_1 + m_2} \boldsymbol{\nabla}_{\mathrm{CM}}^2 + \left(\frac{1}{m_1} + \frac{1}{m_2} \right) \boldsymbol{\nabla}^2. \qquad (9.58)$$

Suppose further that the potential energy of the two-particle system depends *only* on the relative coordinate. This is expected if the system is not subjected to any external force, and the potential energy results only from the mutual interaction of the two particles in the system. In this case, we can write the potential energy as $V(\boldsymbol{r})$, and the Schrödinger equation becomes

$$i\hbar \frac{\partial}{\partial t} \Psi(\boldsymbol{r}_1, \boldsymbol{r}_2) = \left[-\frac{\hbar^2}{2M} \boldsymbol{\nabla}_{\mathrm{CM}}^2 - \frac{\hbar^2}{2\mu} \boldsymbol{\nabla}^2 + V(\boldsymbol{r}) \right] \Psi(\boldsymbol{r}_1, \boldsymbol{r}_2), \qquad (9.59)$$

where M is the total mass, and μ is called the *reduced mass*, defined by

$$\mu \equiv \frac{m_1 m_2}{m_1 + m_2}. \qquad (9.60)$$

Clearly, in this case, the wavefunction can be written as a product of two functions,

$$\Psi(\boldsymbol{r}_1, \boldsymbol{r}_2) = \Phi(\boldsymbol{r}_{\mathrm{CM}})\psi(\boldsymbol{r}). \tag{9.61}$$

Substituting this form of the solution into Eq. (9.59), we find that the part that depends on the center-of-mass coordinate satisfies the differential equation

$$i\hbar \frac{\partial}{\partial t} \Phi(\boldsymbol{r}_{\mathrm{CM}}) = -\frac{\hbar^2}{2M} \boldsymbol{\nabla}^2_{\mathrm{CM}} \Phi(\boldsymbol{r}_{\mathrm{CM}}), \tag{9.62a}$$

whereas the other part satisfies the differential equation

$$i\hbar \frac{\partial}{\partial t} \psi(\boldsymbol{r}) = \left[-\frac{\hbar^2}{2\mu} \boldsymbol{\nabla}^2 + V(\boldsymbol{r}) \right] \psi(\boldsymbol{r}). \tag{9.62b}$$

We see that this is a huge simplification. The two-particle problem is essentially reduced to two single-particle problems. The first equation, Eq. (9.62a), is the Schrödinger equation of a free particle of mass M, and the second one is also a single-particle Schrödinger equation for the function of the relative coordinate that involves the mutual potential of the two-particle system.

9.6 Hydrogen atom

The hydrogen atom is a bound state of a proton and an electron. It is a two-particle system, and therefore we can use the results obtained in §9.5. This is what we are going to do in this section.

9.6.1 The potential

The potential energy of this two-particle system depends only on the relative separation between the proton and the electron, and is given by

$$V(r) = -\frac{\alpha\hbar c}{r}. \tag{9.63}$$

Here c is the speed of light which has been put along with $\hbar$ to obtain the correct dimension of the potential, and α is a dimensionless constant whose value is about $1/137$; it is called the *fine-structure constant*. We have to put this potential into Eq. (9.62b) in order to obtain the radial solutions and to obtain ideas about the energy eigenvalues.

Surely, it is more conventional to write the numerator of Eq. (9.63) in terms of the electric charge. But the problem is that there are many different ways that the unit of electric charge is chosen, and therefore the relation between the electric charge and the potential depends on the system of units chosen. If we denote the magnitude of the electron's charge by e, its relation to the constant α is the following:

$$\alpha = \begin{cases} \dfrac{e^2}{4\pi\epsilon_0\hbar c} & \text{in SI units,} \\ \dfrac{e^2}{\hbar c} & \text{in Gaussian units,} \\ \dfrac{e^2}{4\pi\hbar c} & \text{in Heaviside--Lorentz units,} \end{cases} \tag{9.64}$$

and so on. Of course, the interaction potential has to be the same irrespective of the units, which means that the systems of units differ in their definition of the quantity e. We have no intention of getting into this mess, so we remain non-committal about the units and let the reader substitute his or her favorite units to this discussion by using Eq. (9.64). In fact, we will use just one single letter for the numerator,

$$\kappa \equiv \alpha\hbar c, \tag{9.65}$$

so that the equations look less cumbersome.

Strictly speaking, we should follow the formalism of two-particle systems developed in §9.5. However, the proton is more than 1800 times heavier than the electron. The center-of-mass of this system is, to a very good approximation, the position of the proton, and the reduced mass is roughly equal to the electron mass. Since we are not concerned about the center-of-mass motion of the system, it is a good approximation to pretend that the proton does not move at all, and the relative coordinate merely gives the position of the electron in a coordinate system where the position of the proton is taken as the origin. Thus, instead of using the notation μ for the reduced mass, we will use the electron mass m in what follows in this section. To avoid confusion, the azimuthal quantum number, which was denoted by m in Chapter 8, would be denoted by m_l here.

9.6.2   Solution of the differential equation

We now write down the Schrödinger equation corresponding to stationary states for the potential of Eq. (9.63). This is

$$\left[-\frac{\hbar^2}{2m}\boldsymbol{\nabla}^2 - \frac{\kappa}{r} \right] \psi(\boldsymbol{r}) = E\psi(\boldsymbol{r}). \tag{9.66}$$

Since the potential depends only on the magnitude of the separation vector, we can use the analysis of §9.3 to infer that the angular dependence of any wavefunction is in the form of a spherical harmonic, and the radial part of the solution, $R(r)$, satisfies the differential equation

$$\frac{d^2 R}{dr^2} + \frac{2}{r}\frac{dR}{dr} + \left[\frac{2m}{\hbar^2}\left(E + \frac{\kappa}{r}\right) - \frac{l(l+1)}{r^2}\right] R = 0, \tag{9.67}$$

which is obtained from Eq. (9.29). This is the equation that needs to be solved to obtain the energy eigenvalues and also the radial part of the energy eigenfunctions.

As a first step, we try to get rid of the first derivative term. The motivation and the technique for this exercise has been explained in §B.3 of Appendix B. For the equation at hand, the procedure dictates that we define a function

$$u(r) = rR(r). \tag{9.68}$$

Writing Eq. (9.67) using $u(r)$ instead of $R(r)$, we obtain

$$\frac{d^2 u}{dr^2} + \left[\frac{2m}{\hbar^2}\left(E + \frac{\kappa}{r}\right) - \frac{l(l+1)}{r^2}\right] u = 0. \tag{9.69}$$

We can now check how the solution is supposed to behave for very small and very large values of r. For $r \to 0$, the term with the $1/r^2$ factor dominates over the other terms that multiply $u(r)$, so the equation is effectively

$$\frac{d^2 u}{dr^2} = \frac{l(l+1)}{r^2} u. \tag{9.70}$$

The only solution of this equation that does not diverge at $r = 0$ is

$$u(r) \propto r^{l+1}. \tag{9.71}$$

On the other hand, for $r \to \infty$, terms with r in the denominator can be neglected, so the equation reduces to

$$\frac{d^2 u}{dr^2} + \frac{2mE}{\hbar^2} u = 0. \tag{9.72}$$

We are looking for bound state solutions for which

$$E = -|E|. \tag{9.73}$$

Thus, the solutions of Eq. (9.72) are in the form of rising and falling exponentials. The rising exponential is physically unacceptable since it becomes infinite for $r \to \infty$, and therefore cannot be normalized. Hence, the acceptable solution of Eq. (9.72) can be written as

$$u(r) \propto \exp(-r/r_E), \tag{9.74}$$

where r_E is defined as

$$r_E \equiv \sqrt{\frac{\hbar^2}{2m|E|}}, \tag{9.75}$$

a quantity that has the dimension of length. While we are doing this, we might also want to note that the quantity κ has dimensions of energy times length, so if we define

$$a_B = \frac{\hbar^2}{m\kappa} = \frac{\hbar}{mc\alpha}, \tag{9.76}$$

this a_B will also have the dimension of length. This quantity is called the *Bohr radius*, and its value is

$$a_B \simeq 0.529 \text{ Å} = 0.0529 \text{ nm}, \tag{9.77}$$

where 'Å' means Angstrom (10^{-10} m) and 'nm' means nanometer (10^{-9} m). Using these newly defined quantities, we can rewrite Eq. (9.69) in the form

$$\frac{d^2u}{dr^2} + \left[-\frac{1}{r_E^2} + \frac{2}{ra_B} - \frac{l(l+1)}{r^2} \right] u = 0. \tag{9.78}$$

In order to obtain the solution valid for all values of r, it is convenient to rewrite this equation using a dimensionless independent variable, and we choose this variable to be

$$\rho = 2r/r_E. \tag{9.79}$$

In terms of this variable, the differential equation is

$$\frac{d^2u}{d\rho^2} + \left[-\frac{1}{4} + \frac{r_E}{\rho a_B} - \frac{l(l+1)}{\rho^2} \right] u = 0. \tag{9.80}$$

Combining the wisdom gathered from Eqs. (9.71) and (9.74), we now write the general solution of this equation as

$$u(\rho) = L(\rho)\rho^{l+1} \exp(-\rho/2), \tag{9.81}$$

which should be valid everywhere. The task is now to determine the function $L(\rho)$. Clearly, for $r \to 0$ the exponential factor becomes equal to unity, so we need $L(\rho)$ to approach a constant in order that Eq. (9.71) is valid in that limit. On the other hand, the behavior at $r \to \infty$ is dominated by the exponential factor. Any power of ρ adds, to the simplified form of Eq. (9.72), terms which are suppressed by inverse powers of ρ. We therefore have to make sure that $L(\rho)$ goes like some power of ρ for large ρ.

We now put the solution of the form given in Eq. (9.81) into Eq. (9.80). After some straightforward calculation, this gives the following differential equation for $L(\rho)$:

$$\rho \frac{d^2 L}{d\rho^2} + \left(2(l+1) - \rho\right)\frac{dL}{d\rho} + \left(\frac{r_E}{a_B} - (l+1)\right) L = 0. \tag{9.82}$$

In Eq. (C.91, p. 468), we have presented the differential equation for associated Laguerre polynomials. Let us write it here, altering the notations a bit, which will make the comparison easier. The equation is

$$x \frac{d^2 y}{dx^2} + (\lambda + 1 - x)\frac{dy}{dx} + \nu y = 0. \tag{9.83}$$

It has been shown in Appendix C that polynomial solutions exist when both λ and ν are non-negative integers, and the solutions are called the associated Laguerre polynomial $L_\nu^\lambda(x)$. Comparing the two equations, we see that Eq. (9.82) can be put into the form of Eq. (9.83) if we take

$$\lambda = 2l + 1, \qquad \nu = n - l - 1, \tag{9.84}$$

with

$$\frac{r_E}{a_B} = n. \tag{9.85}$$

Because ν must be an integer, n must be an integer as well. Further, since ν is non-negative, we conclude that we must have

$$n \geqslant l + 1. \tag{9.86}$$

Thus, the allowed values of l are the integers in the range

$$0 \leqslant l \leqslant n - 1. \tag{9.87}$$

The energy eigenvalues can now be obtained from Eq. (9.85). Using the definitions of r_E and a_B from Eqs. (9.75) and (9.76), we obtain

$$E_n = -\frac{m\kappa^2}{2n^2\hbar^2} = -mc^2 \cdot \frac{\alpha^2}{2n^2}. \tag{9.88}$$

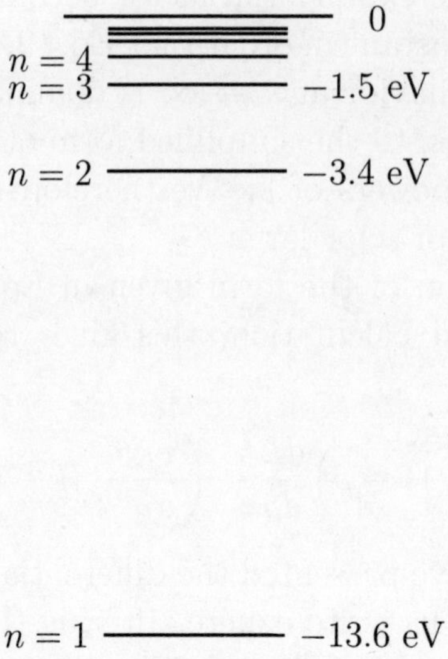

$$\underline{}\ 0$$

$n = 4$
$n = 3$ $\underline{}$ -1.5 eV

$n = 2$ $\underline{}$ -3.4 eV

$n = 1$ $\underline{}$ -13.6 eV

Figure 9.1 A few low-lying energy levels of the hydrogen atom. The longer horizontal line at the top denotes zero energy, which must be higher than the energy of any bound state.

One should be careful if one wants to write this equation using the electric charge of the electron rather than the fine-structure constant, because the relation between the two depends on the system of electromagnetic units used, as seen from Eq. (9.64). In SI units, for example, one obtains

$$E_n = -me^4 \cdot \frac{1}{2(4\pi\epsilon_0\hbar)^2 n^2}. \tag{9.89}$$

In Gaussian units, the factor $4\pi\epsilon_0$ would be absent, because it is absorbed in the definition of e. In writing the expression for the energy eigenvalues, we have put a subscript on the energy eigenvalue to indicate its dependence on the integer n, and also recalled that the energy must be negative, as mentioned in Eq. (9.73). The number n is called the *principal quantum number*.

It is interesting to note that the eigenfunctions depend on other quantum numbers. The solution of the angular part of the differential equation contains two integer parameters, l and m_l. However, the eigenvalues are independent of these quantum numbers. The significance of this result will be discussed later.

9.6.3 Spectral lines of hydrogen

The energy eigenvalues of a hydrogen atom are therefore characterized by a positive integer n through Eq. (9.88). Putting $\alpha = 1/137$, it is easy to check that the ground state energy, corresponding to $n = 1$, is

$$E_1 = -13.6 \, \text{eV}. \tag{9.90}$$

The energy eigenvalue for any higher value of n can be obtained by dividing this value by n^2. For very large values of n, the energy is still negative but very close to zero, which means that the electron is very loosely bound. The eigenvalues, for a few values of n, have been shown in Figure 9.1.

If an electron in a state n_1 jumps to a lower state n_2 by emitting a photon, the energy difference between the states will be equal to the energy of the resulting photon, E_γ:

$$E_\gamma = \frac{1}{2}mc^2\alpha^2 \left(\frac{1}{n_2^2} - \frac{1}{n_1^2} \right). \tag{9.91}$$

For $n_2 = 1$, there can be photons of many energies, depending on the value of n_1. The wavelengths of these radiations will be close together, and together they are called the *Lyman series* of spectral lines. The lines corresponding to $n_2 = 2$ are collectively called the *Balmer series*, those with $n_2 = 3$ are called the *Paschen series*, and so on. Only a few lines of the Balmer series fall within the visible region, as seen in Figure 9.2.

□ **Exercise 9.6** *Calculate the wavelengths to verify that the Lyman series lines are in the ultraviolet region ($\lambda < 400$ nm) and those in the Paschen series are in the infrared region ($\lambda > 700$ nm). Only a few of the Balmer series lines are in the visible region, 400 nm $< \lambda < 700$ nm. Identify what the values of n_1 for these are.*

9.6.4 Nature of the wavefunctions

Let us summarize all information about the wavefunctions of the stationary states. The functions are of the form

$$\Psi_{nlm_l}(\boldsymbol{r}, t) = \psi_{nlm_l}(\boldsymbol{r})e^{-iE_n t/\hbar}, \tag{9.92}$$

where

$$\psi_{nlm_l}(\boldsymbol{r}) = Y_{lm_l}(\theta, \phi)R_{nl}(r), \tag{9.93}$$

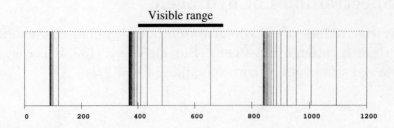

Figure 9.2 Atomic spectrum of hydrogen. The numbers at the bottom indicate wavelengths in nanometers. The range of visible light is marked. The three regions of closely bunched lines represent the lower-wavelength limits of Lyman, Balmer, and Paschen series respectively.

with

$$R_{nl}(r) = N_{nl} L_{n-l-1}^{2l+1}(\rho) \rho^l \exp(-\rho/2), \qquad (9.94)$$

where N_{nl} is a normalization constant. The angular parts of the wavefunctions are the spherical harmonics, $Y_{lm}(\theta, \phi)$. Their dependence with the polar and azimuthal angles have been discussed in §8.2, with plots of the θ-dependence given in Figure 8.1 (p. 190). As commented earlier, this feature is common to all 3D systems with a central potential.

> One apology for the notation. In Eq. (9.94), there is a factor ρ^l where the superscript is a power, but there is another factor L_{n-l-1}^{2l+1} where the superscript indicates just a parameter in the function. Unfortunately, the use is so conventional that we could not avoid it.

Let us discuss here only the radial parts of the wavefunctions. Recall that the variable ρ appearing in Eq. (9.94) is given by Eq. (9.79), which can be written as

$$\rho = \frac{2r}{na_B}. \qquad (9.95)$$

Therefore, plugging into Eq. (9.94), we can write

$$R_{10}(r) = N_{10} L_0^1(2r/a_B) e^{-r/a_B}, \qquad (9.96a)$$

$$R_{20}(r) = N_{20} L_1^1(r/a_B) e^{-r/2a_B}, \qquad (9.96b)$$

$$R_{21}(r) = N_{21} L_0^3(r/a_B) \cdot \left(\frac{r}{a_B}\right) e^{-r/2a_B}, \qquad (9.96c)$$

Table 9.1 Radial eigenfunctions of the hydrogen atom for the first three values of the principal quantum number, $n = 1, 2, 3$.

n	l	R_{nl}
1	0	$\dfrac{2}{\sqrt{a_B^3}} \exp\left(-\dfrac{r}{a_B}\right)$
2	0	$\dfrac{1}{2\sqrt{2a_B^3}} \left(2 - \dfrac{r}{a_B}\right) \exp\left(-\dfrac{r}{2a_B}\right)$
2	1	$\dfrac{1}{2\sqrt{6a_B^3}} \left(\dfrac{r}{a_B}\right) \exp\left(-\dfrac{r}{2a_B}\right)$
3	0	$\dfrac{2}{81\sqrt{3a_B^3}} \left(27 - 18\dfrac{r}{a_B} + 2\dfrac{r^2}{a_B^2}\right) \exp\left(-\dfrac{r}{3a_B}\right)$
3	1	$\dfrac{4}{81\sqrt{6a_B^3}} \left(6 - \dfrac{r}{a_B}\right)\left(\dfrac{r}{a_B}\right) \exp\left(-\dfrac{r}{3a_B}\right)$
3	2	$\dfrac{4}{81\sqrt{30a_B^3}} \left(\dfrac{r^2}{a_B^2}\right) \exp\left(-\dfrac{r}{3a_B}\right)$

and so on. The associated Laguerre polynomials have been listed in Eq. (C.96, p. 469). The normalization constants are to be obtained from the condition

$$\int_0^\infty dr \, r^2 \Big[R_{nl}(r)\Big]^2 = 1. \tag{9.97}$$

We have tabulated the integrand of this equation in Table 9.1 for a few low values of n, with the proper normalization constant obtained from this condition.

Looking at Table 9.1 we see that a factor of $\sqrt{a_B^3}$ appears in the denominator in each of the eigenfunctions. It is easy to understand the reason for this factor. Eq. (9.97) tells us that the dimension of the radial wavefunction R_{nl} should be equal to $L^{-3/2}$, where L denotes the dimension of length. The Hamiltonian of the system contains the constants m, $c\alpha$, and $\hbar$, and with them one can form only one quantity that has the dimension of length, which is the Bohr radius given in Eq. (9.76). Therefore, the normalized wavefunctions must contain the factor $a_B^{-3/2}$.

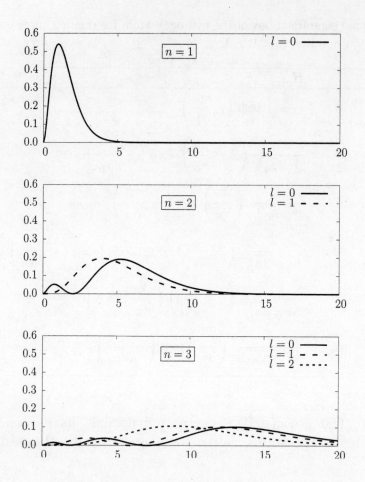

Figure 9.3 Probability density distributions of radial wavefunctions of the hydrogen atom.

One feature of Figure 9.3 is that the radial probability density, denoted by the integrand of Eq. (9.97), vanishes for some values of r. Of course, it vanishes at $r = 0$ because of the presence of the factor r^2 in the volume element, and also vanishes at infinity because of the exponentially decaying factor in Eq. (9.94). But it also vanishes at some other points for many of the combinations of n and l. The reason for this is that the associated Laguerre polynomial L_ν^λ is a polynomial of degree ν and therefore has ν zeros. For the solution of Eq. (9.94), we therefore should expect $n - l - 1$ zeros in the probability density distribution, which is what we see in Figure 9.3.

Since the radial probability density depends only on the quantum numbers n and l, for many practical purposes it is convenient to refer to the state of the electron in terms of only these two numbers. In such nomenclature, the l values are replaced by a letter code, s for $l = 0$, p for $l = 1$, d for $l = 2$, etc. Thus, for example, the ground state with $n = 1$ and $l = 0$ is called the $1s$ state, whereas the excited states with $n = 2$ and $l = 1$, irrespective of the value of m_l, are called the $2p$ states.

Another striking aspect of Figure 9.3 is that, as n increases, the electron has a larger probability of reaching regions farther from the proton. This means that if we calculate the expectation value of r, it will be larger for states with higher values of n. There is an easy way of checking this. Instead of calculating $\langle r \rangle$, we can calculate $\langle 1/r \rangle$ and see how it varies with n. This can be easily done through the *Hellmann–Feynman theorem*, which says that if β is any parameter in the Hamiltonian,

$$\left\langle \psi \left| \frac{\partial H}{\partial \beta} \right| \psi \right\rangle = \frac{\partial E}{\partial \beta}, \tag{9.98}$$

where $|\psi\rangle$ is an eigenstate of H with eigenvalue E. The hydrogen atom Hamiltonian contains the parameter κ, and we note that

$$\frac{\partial H}{\partial \kappa} = -\frac{1}{r}. \tag{9.99}$$

Therefore, using the Hellmann–Feynman theorem for the energy eigenstates, we find

$$\left\langle \frac{1}{r} \right\rangle = \frac{m\kappa}{n^2\hbar^2} = \frac{1}{n^2 a_B}, \tag{9.100}$$

using the expression for the energy eigenvalue from Eq. (9.88).

The expectation value of r is not the inverse of the expectation value of $1/r$. However, it shows at least that the expectation value of r should be bigger for larger values of n.

☐ **Exercise 9.7** *Find the expectation value of r for the ground state, using the ground state wavefunction from Table 9.1.*

☐ **Exercise 9.8** *Prove the Hellmann–Feynman theorem, Eq. (9.98).* [**Hint:** *Start from the right side and note that for a normalized $|\psi\rangle$, irrespective of how it has been normalized, $\langle\psi|\partial\psi/\partial\beta\rangle = 0$.*]

9.6.5   Degeneracy of eigenvalues

We pointed out that the energy eigenvalues depend only on the principal quantum number n, and not on the other quantum numbers l and m. For each l, the value of m_l has a range from $-l$ to $+l$, which contains $2l+1$ values in total. And for each n, the value of l can be any integer from 0 to $n-1$. Therefore, the degeneracy of an energy eigenvalue seems to be

$$\sum_{l=0}^{n-1}(2l+1) = n^2. \tag{9.101}$$

In fact, the degeneracy is twice as much. This is because the electron is a spin-$\frac{1}{2}$ particle. The Hamiltonian is independent of spin and therefore commutes with the spin operators. Thus, in the set of commuting operators that we use to write the eigenstates, we can also include S^2 and S_z. The eigenvalue of S^2 is not a variable: it is the same for all states, equal to $\frac{1}{2}(\frac{1}{2}+1)\hbar^2$. But there are two possible eigenvalues of S_z. So, a complete specification of the energy eigenstate of the electron in the hydrogen atom should have the values of n, l, m_l, and m_s, corresponding to the eigenvalues of the operators H, L^2, L_z, and S_z.

The degeneracy of the m_l values owes its origin to rotational symmetry, as was argued in §9.3. The degeneracy with respect to the eigenvalues of S_z can also be seen as a consequence of the fact that the Hamiltonian is independent of spin. But why do the energy eigenvalues not depend on l? There must be some additional symmetry in the Hamiltonian that forces this to happen. We will now search for that additional symmetry.

9.6.6   A hint from classical physics

We can obtain some hint of this additional symmetry by considering the classical problem. Classically, under the action of the central attractive potential, a particle will be orbiting the center of the force in elliptical orbits in general. To make the problem simple, let us consider a circular orbit, which is a limiting case of an ellipse. In Figure 9.4, we have shown the orbit of a particle, indicating the direction of momentum at a certain position, which is tangential to the circle. The angular momentum, $\boldsymbol{r} \times \boldsymbol{p}$, is perpendicular to the plane of the figure. Thus, the vector $\boldsymbol{p} \times \boldsymbol{L}$ will be along the radial direction. The magnitude of this vector can be easily calculated. Since $\boldsymbol{p}$ and $\boldsymbol{L}$ are perpendicular to each other, we obtain

$$\left| \boldsymbol{p} \times \boldsymbol{L} \right| = pL = p^2 r, \tag{9.102}$$

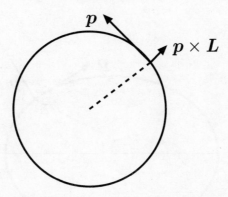

Figure 9.4 Motivating the additional conserved quantity for a particle in a circular orbit.

where r is the radius of the circular orbit. This means that we will have

$$\boldsymbol{p} \times \boldsymbol{L} - p^2 \boldsymbol{r} = 0 \qquad (9.103)$$

at any point on the orbit of the particle. This is certainly an additional conservation law, claiming that the vector on the left side vanishes at all times.

We can dress this conservation law in a little different manner, which will help us to go over to the general case of elliptical orbits. The centripetal force is supplied by the Coulomb attraction, which means we should have

$$\frac{mv^2}{r} = \frac{\kappa}{r^2}, \qquad (9.104)$$

where κ has been defined in Eq. (9.65). Thus, if we define a vector by

$$\boldsymbol{M} = \frac{1}{m\kappa} \boldsymbol{p} \times \boldsymbol{L} - \frac{\boldsymbol{r}}{r}, \qquad (9.105)$$

that would vanish for a circular orbit, according to Eq. (9.103).

Let us now see what can be said about this vector for the general case of elliptical orbits. Now the vector $\boldsymbol{M}$ will not vanish, but it is straightforward to show that it is conserved:

$$\frac{d\boldsymbol{M}}{dt} = 0. \qquad (9.106)$$

The vector $\boldsymbol{M}$ is called the *Runge–Lenz vector*.

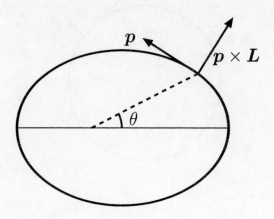

Figure 9.5 Interpretation of the Runge–Lenz vector.

□ **Exercise 9.9** *Verify Eq. (9.106).*

   To have an interpretation of the vector, we look at Figure 9.5. The origin, or the center of the force, is at a focus of the ellipse. If we consider the point at the end of the major axis of the ellipse, both $\boldsymbol{p} \times \boldsymbol{L}$ and $\boldsymbol{r}$ are along the major axis. Considering another point elsewhere on the ellipse, it is easy to convince oneself that the direction of $\boldsymbol{M}$ would be to the left of the figure. Thus, we conclude that the conserved Runge–Lenz vector is parallel to the major axis at all times, pointing in the direction from the focus to the closest point on the ellipse.

   As for the magnitude of this vector, we note that

$$\boldsymbol{r} \cdot \boldsymbol{M} = \frac{1}{m\kappa} \boldsymbol{r} \cdot (\boldsymbol{p} \times \boldsymbol{L}) - r. \qquad (9.107)$$

Noting that

$$\boldsymbol{r} \cdot (\boldsymbol{p} \times \boldsymbol{L}) = \boldsymbol{L} \cdot (\boldsymbol{r} \times \boldsymbol{p}) = L^2, \qquad (9.108)$$

we obtain

$$r(1 - M\cos\theta) = \frac{L^2}{m\kappa}, \qquad (9.109)$$

where θ denotes the polar angle on the plane. The right side is constant, and this is the equation in plane polar coordinates of an ellipse with eccentricity

M. Thus, the conservation of $\boldsymbol{M}$ means that the major axis of the elliptical orbit does not change its direction with time, and the magnitude, i.e., the eccentricity of the orbit, is fixed as well.

9.6.7 Alternative method for finding energy eigenvalues

Let us now see the quantum mechanical implication of the conserved Runge–Lenz vector. First, we notice that the definition in Eq. (9.105) does not give a Hermitian operator. So we add its Hermitian conjugate and divide by 2 to obtain the quantum mechanical version of the Runge–Lenz vector:

$$\boldsymbol{M} = \frac{1}{2m\kappa}\left(\boldsymbol{p} \times \boldsymbol{L} - \boldsymbol{L} \times \boldsymbol{p}\right) - \frac{\boldsymbol{r}}{r}. \tag{9.110}$$

Clearly, the conservation laws will now manifest as vanishing commutators with the Hamiltonian. Each of the components of $\boldsymbol{L}$ and $\boldsymbol{M}$ commutes with the Hamiltonian.

□ **Exercise 9.10** *In support of Eq. (9.110), show that*

$$\left(\boldsymbol{p} \times \boldsymbol{L}\right)^{\dagger} = -\boldsymbol{L} \times \boldsymbol{p}. \tag{9.111}$$

[**Hint:** *It is easiest to show it at purely operator level, i.e., recognizing that*

$$\left(\boldsymbol{p} \times \boldsymbol{L}\right)_i = \sum_{j,k}\varepsilon_{ijk}p_j L_k = \sum_{j,k}\sum_{r,s}\varepsilon_{ijk}\varepsilon_{krs}p_j x_r p_s, \tag{9.112}$$

and taking the Hermitian conjugate of this operator.]

□ **Exercise 9.11** *Show that the operator $\boldsymbol{M}$, as defined in Eq. (9.110), indeed commutes with the Hamiltonian.*

The symmetry of the system will be expressed by the commutation relations of the components of $\boldsymbol{L}$ and $\boldsymbol{M}$ among themselves. This algebra is not self-contained. Although the commutation of the components of $\boldsymbol{L}$ gives the standard angular momentum algebra as given in Eq. (5.108, p. 131), the commutation of the components of $\boldsymbol{M}$ involves the Hamiltonian. However, we can work in a subspace of the entire Hilbert space that corresponds to a particular energy value, say E. In this subspace, the Hamiltonian can be replaced by its eigenvalue. We then define

$$\boldsymbol{M}' = \sqrt{\frac{-m\kappa^2}{2E}}\,\boldsymbol{M}. \tag{9.113}$$

Because of the extra factor, the dimension of $\boldsymbol{M}'$ is the same as the dimension of angular momentum, which makes it easier to keep track of the commutators. The commutators are in fact

$$\left[L_i, L_j\right] = i\hbar \sum_k \varepsilon_{ijk} L_k, \tag{9.114a}$$

$$\left[M_i', M_j'\right] = i\hbar \sum_k \varepsilon_{ijk} L_k, \tag{9.114b}$$

$$\left[L_i, M_j'\right] = -i\hbar \sum_k \varepsilon_{ijk} M_k'. \tag{9.114c}$$

These commutation relations look much neater if we define the combinations

$$A_i = \frac{1}{2}(L_i + M_i'), \qquad D_i = \frac{1}{2}(L_i - M_i'). \tag{9.115}$$

In fact, the relations in Eq. (9.114) take the form

$$\left[A_i, A_j\right] = i\hbar \sum_k \varepsilon_{ijk} A_k, \tag{9.116a}$$

$$\left[D_i, D_j\right] = i\hbar \sum_k \varepsilon_{ijk} D_k, \tag{9.116b}$$

$$\left[A_i, D_j\right] = 0. \tag{9.116c}$$

We have two sectors now, one with the A generators and the other with the D generators, which are mutually commuting. We also note that each sector satisfies the rotation algebra, so we know what the possible representations are. Moreover, from the definitions it can be shown that

$$\boldsymbol{L} \cdot \boldsymbol{M} = \boldsymbol{M} \cdot \boldsymbol{L} = 0, \tag{9.117}$$

which translates to the relation

$$\boldsymbol{A}^2 = \boldsymbol{D}^2 = \frac{1}{4}(\boldsymbol{L}^2 + \boldsymbol{M}'^2). \tag{9.118}$$

From our knowledge of the rotation algebra obtained in Chapter 8, we know how the eigenvalues of $\boldsymbol{A}^2$ and $\boldsymbol{D}^2$ would look like. So we conclude that

$$\boldsymbol{L}^2 + \boldsymbol{M}'^2 \xrightarrow{\text{eigenvalue}} 4a(a+1)\hbar^2, \tag{9.119}$$

where a can be any non-negative integer or half-integer.

The connection to the energy eigenvalues can be established through the operator $\boldsymbol{M}^2$. Writing the definition of Eq. (9.110) in component notation, we obtain

$$\sum_i M_i M_i = \frac{1}{(2m\kappa)^2} \sum_i \sum_{j,k} \sum_{m,n} \varepsilon_{ijk} \varepsilon_{imn} (p_j L_k - L_j p_k)(p_m L_n - L_m p_n)$$

$$-\frac{1}{2m\kappa} \sum_i \sum_{j,k} \varepsilon_{ijk} \left((p_j L_k - L_j p_k)\frac{x_i}{r} + \frac{x_i}{r}(p_j L_k - L_j p_k) \right)$$

$$+1. \tag{9.120}$$

The expression looks cumbersome, but it simplifies a lot after we use relations of the type shown in Eq. (5.110, p. 131) and Eq. (5.111). For example, let us consider the four terms which contain an $1/r$ factor. The very first term contains the combination

$$\sum_{i,j,k} \varepsilon_{ijk} p_j L_k x_i = \sum_{i,j,k} \varepsilon_{ijk} \left(L_k p_j - i\hbar \sum_l \varepsilon_{kjl} p_l \right) x_i. \tag{9.121}$$

Now note that we can write $p_j x_i = x_i p_j - i\hbar \delta_{ij}$, so after the summation over all indices, we obtain $\sum \varepsilon_{ijk} L_k p_j x_i = \boldsymbol{L}^2$. For the other term, we use a summation identity involving the Levi-Civita symbols, given in Eq. (5.113b, p. 132), to obtain $\sum \varepsilon_{ijk} \varepsilon_{kjl} p_l x_i = \boldsymbol{p} \cdot \boldsymbol{r}$. Thus,

$$\sum_i \sum_{j,k} \varepsilon_{ijk} p_j L_k x_i = \boldsymbol{L}^2 + 2i\hbar \boldsymbol{p} \cdot \boldsymbol{r}. \tag{9.122}$$

Similarly, the last term containing $1/r$ can be simplified to

$$\sum_i \sum_{j,k} \varepsilon_{ijk} x_i L_j p_k = -\boldsymbol{L}^2 + 2i\hbar \boldsymbol{r} \cdot \boldsymbol{p}. \tag{9.123}$$

The other two terms are simpler to tackle. They give

$$\sum_i \sum_{j,k} \varepsilon_{ijk} L_j p_k x_i = -\boldsymbol{L}^2,$$

$$\sum_i \sum_{j,k} \varepsilon_{ijk} x_i p_j L_k = \boldsymbol{L}^2. \tag{9.124}$$

So, in Eq. (9.120), the terms containing $1/r$ sum up to

$$-\frac{1}{2m\kappa} \left(\frac{4}{r}\boldsymbol{L}^2 + 2i\hbar \sum_k \left[\frac{x_k}{r}, p_k \right] \right) = \frac{2}{m\kappa^2} \left(-\frac{\kappa}{r} \right) (\boldsymbol{L}^2 + \hbar^2)$$

$$= \frac{2}{m\kappa^2} V(r)(\boldsymbol{L}^2 + \hbar^2). \tag{9.125}$$

The remaining terms in Eq. (9.120) can also be simplified in a similar manner, yielding the final result

$$\boldsymbol{M}^2 = 1 + \frac{2}{m\kappa^2} H(\boldsymbol{L}^2 + \hbar^2). \tag{9.126}$$

In the subspace of a fixed energy eigenvalue, we can therefore write

$$\boldsymbol{M'}^2 = -\frac{m\kappa^2}{2E} - \boldsymbol{L}^2 - \hbar^2. \tag{9.127}$$

Using Eq. (9.119) now, we see that the energy eigenvalues should satisfy the relation

$$\frac{m\kappa^2}{2E} = -4a(a+1)\hbar^2 - \hbar^2, \tag{9.128}$$

or

$$E = -\frac{m\kappa^2}{2\hbar^2(2a+1)^2}. \tag{9.129}$$

This is the same as Eq. (9.88) where we identify the principal quantum number n to be equal to $2a+1$, so that it can be any integer equal to or greater than 1.

☐ **Exercise 9.12** *In writing Eq. (9.125), we have used the following commutators:*

$$\left[L^2, f(r)\right] = 0,$$
$$\sum_k \left[\frac{x_k}{r}, p_k\right] = \frac{2i\hbar}{r}. \tag{9.130}$$

Prove these results.

☐ **Exercise 9.13** *Simplify the terms containing two powers of momentum to show that ultimately if gives the contribution*

$$\frac{1}{m^2\kappa^2} \boldsymbol{p}^2(\boldsymbol{L}^2 + \hbar^2), \tag{9.131}$$

so that Eq. (9.126) is obtained after adding the other contributions.

9.7   3D isotropic oscillator

We have discussed 1D harmonic oscillators earlier in §7.7. The potential is proportional to the square of the coordinate. The 3D analog of the harmonic oscillator will have the potential

$$V(r) = \frac{1}{2}M\omega^2 r^2, \tag{9.132}$$

where the mass of the particle is denoted by M in this case, to avoid conflict with the integration constant that appears in Eq. (9.19).

The potential is a central potential, so the separation of Eq. (9.28) should apply for this problem, which means that the θ and ϕ dependence of the stationary wavefunctions are in the form of spherical harmonics. The radial part of the wavefunction obeys the differential equation

$$\frac{d}{dr}\left(r^2\frac{dR}{dr}\right) + \left[\frac{2Mr^2}{\hbar^2}\left(E - \frac{1}{2}M\omega^2 r^2\right) - l(l+1)\right]R = 0. \tag{9.133}$$

It really makes little sense to try to solve the energy eigenvalues from here. The equations are separable in Cartesian coordinates. In fact, it is separable even if the frequencies are different along the different axes. The energy eigenvalues will be the sum of the eigenvalues of three linear oscillators. For the isotropic case where the frequencies are the same, these eigenvalues are of the form

$$E_n = (n + \frac{3}{2})\hbar\omega. \tag{9.134}$$

All eigenvalues, except for the lowest one, will have degeneracy. As for the case of the hydrogen atom, part of this degeneracy is due to the rotational symmetry, and part of it is due to some additional symmetry.

☐ **Exercise 9.14** *Show that the degeneracy of the energy E_n is $\frac{1}{2}(n+1)(n+2)$. Check your results for $n = 1, 2$ by explicitly writing the states in terms of the states of the three different oscillators.*

☐ **Exercise 9.15** *Show that the ground and the first excited states of the 3D harmonic oscillators have mutually orthogonal wavefunctions.*

9.8   Cooper problem

In this section, we shall discuss the Cooper problem which pertains to the formation of bound states between two electrons on top of a filled Fermi surface of electrons. This problem was first solved by Leon Cooper in 1956, and it provided the much-needed insight into the theory of superconductivity subsequently developed by Bardeen, Cooper, and Schrieffer.

To understand this problem, let us consider N non-interacting electrons in a metal. The situation is similar to the electrons being in a spherical well and energy $E_{\boldsymbol{k}} = \hbar^2 \boldsymbol{k}^2/(2m)$, where m denotes the mass of the electron. Here $\boldsymbol{k}$ is almost continuous since L, the dimension of the sample (which is the radius of the well), is large compared to the lattice spacing a. In what follows, we shall take $\boldsymbol{k}$ to be a continuous variable. Each of these $\boldsymbol{k}$ represents a free-electron eigenstate. The ground (lowest energy) state of the system therefore amounts to putting N electrons in these states so that their energy is minimum. Since two electrons cannot occupy the same state, a property known as the *Pauli exclusion principle*, the electrons start filling the states starting from the lowest values of the momentum, and this leads to a sphere of filled electronic states whose radius, k_F, depends on N. All states with $|\boldsymbol{k}| < k_F$ are therefore filled; the rest are empty. This sphere is called the *Fermi sphere*, and k_F is known as the *Fermi momentum*.

For the Cooper problem, we consider the situation where two additional electrons are put at momenta $\boldsymbol{k}_F + \boldsymbol{\delta k}$ and $-(\boldsymbol{k}_F + \boldsymbol{\delta k})$, where $|\boldsymbol{\delta k}| \ll k_F$. Here $\boldsymbol{k}_F$ refers to a momentum whose direction is along the radius of the Fermi sphere and whose magnitude is fixed to k_F. This constitutes putting two electrons atop the surface of the Fermi sphere. In addition, one imagines having a small attractive interaction between the two electrons. This attractive interaction is a caricature of the effective interaction between the electrons arising from their interaction with lattice vibrations (or phonons) inside the metal. We are not going to bother about the origin of such interaction here. Instead, we merely assume that, in the momentum space, the form of such interaction is given by

$$V(\boldsymbol{k} - \boldsymbol{k}') = \begin{cases} -g, & \text{for } k_F \leqslant |\boldsymbol{k}|, |\boldsymbol{k}'| \leqslant k_F + k_D, \\ 0 & \text{otherwise,} \end{cases} \qquad (9.135)$$

where $\epsilon(|\boldsymbol{k}_D|) = \hbar\omega_D$ is the typical phonon-related scale (ω_D is in fact called the *Debye frequency*). Thus, we are led to the two-particle Hamiltonian

$$H_2 = -\frac{h^2}{2m}(\nabla_1^2 + \nabla_2^2) + V(\boldsymbol{r}_1 - \boldsymbol{r}_2). \qquad (9.136)$$

Since V depends on the relative coordinates of the two particles, it is natural to move to the center-of-mass and relative coordinate description. Writing the wavefunction as $\Psi(\boldsymbol{r}_1, \boldsymbol{r}_2) = \Phi(r_{\text{CM}})\psi(\boldsymbol{r})$ and following the analysis leading to Eq. (9.62b), we obtain the equation for the relative coordinate $\boldsymbol{r}$ as

$$\left[-\frac{\hbar^2}{m}\boldsymbol{\nabla}^2 + V(\boldsymbol{r}) \right] \psi(\boldsymbol{r}) = (2E_F + E)\psi(\boldsymbol{r}), \qquad (9.137)$$

where we have written the energy eigenvalue as $2E_F + E$, meaning that E is the energy measured from $2E_F = \hbar^2 k_F^2/m$.

We now express this equation in the momentum space. We obtain

$$(2\epsilon_k - E)\psi_{\boldsymbol{k}} = g \int \frac{d^3k'}{(2\pi)^3}\psi_{\boldsymbol{k}'} \qquad (9.138)$$

where the integral extends over the region $k_F \leqslant |\boldsymbol{k}| \leqslant |\boldsymbol{k}_D|$ in the momentum space, and $\epsilon_k = \hbar^2 k_F k/m$, where k is the component of $\boldsymbol{\delta k}$ along $\boldsymbol{k}_F$. Here we have used the fact $k \leq |\boldsymbol{\delta k}| \ll k_F$ so that one can ignore the quadratic term in δk. Further, we have used the fact that the center-of-mass momentum of the two electrons vanishes by construction. In our notation, $E = 0$ corresponds to the energy of two non-interacting particles atop the Fermi sphere; thus, the bound state solution corresponds to $E < 0$.

To solve this equation, we note that the right side is independent of $\boldsymbol{k}$, which requires $\psi_{\boldsymbol{k}} = C/(2\epsilon_{\boldsymbol{k}} - E)$, where C is a constant. Putting this solution in Eq. (9.138), one finds the consistency condition

$$g \int \frac{d^3k}{(2\pi)^3}\frac{1}{2\epsilon_k - E} = 1. \qquad (9.139)$$

After performing the angular integrations, this can be written in the form

$$gN(0) \int_0^{\hbar\omega_D} d\epsilon \frac{1}{2\epsilon - E} = 1, \qquad (9.140)$$

where in the last line we have used $(1/(2\pi)^3) \int d^3k' = N(0) \int d\epsilon$, where $N(0)$ is the Jacobian of the transformation from integral over k to that over energy (it is commonly referred to as *density of states*). This Jacobian is assumed to be a constant over the range of the integration. Note that this is only reasonable if $\hbar\omega_D \ll E_F = \hbar^2 k_F^2/(2m)$; it turns out that this condition holds very well in standard normal metals.

With these approximations, the integral in Eq. (9.140) becomes analytically tractable and yields

$$E = -\frac{2\hbar\omega_D}{e^{2/(N(0)g)} - 1} \simeq -2\hbar\omega_D e^{-2/(N(0)g)}, \qquad (9.141)$$

where the last expression is obtained under the assumption that $N(0)g$ is small. Thus, we find that one has bound states of two electrons due to attractive interaction between them.

The above problem demonstrates that it is possible to have a bound state for arbitrarily small g. This is in contradiction with the standard expectation obtained from the analysis of a potential well where we find that one needs a minimal well depth to have a bound state. The resolution of this conundrum comes from noting that the presence of the filled Fermi sphere makes the problem effectively 1D; this is reflected from the fact that the Schrödinger equation for the pair, Eq. (9.138), can be reduced to a 1D equation involving a single variable ϵ, as has been done in Eq. (9.140). This observation provided the first clue to a possible mechanism of pair formation between electrons in a 3D metallic system due to weak attractive interaction between them.

Chapter 10

Particles in magnetic fields

In earlier chapters, we have analyzed the quantum behavior of particles in many different kinds of Hamiltonian. Some of these Hamiltonians are quite contrived and used for exemplary purposes only. Some are realistic cases, like the case of the hydrogen atom in Chapter 9. In this chapter, we continue discussing realistic interactions, viz., interactions of particles with magnetic fields.

10.1   The Hamiltonian

Classical electrodynamics is described by Maxwell equations, which connect the derivatives of the electric field E and magnetic field B to the sources of the electromagnetic field, the charge density ρ and the current density J. All the quantities mentioned are in general functions of position and time. The behavior of a particle of charge q in such a field is described by the Lorentz force law. In the SI units, this force is given by

$$F = q\Big(E + v \times B\Big). \tag{10.1}$$

If we want to solve the classical problem of the path of a particle in an electromagnetic field, we can just set up Newton's equation of motion with the force shown above.

Clearly, this method does not help us solve the corresponding quantum mechanical problem, because the concept of force is not used in the formulation of quantum mechanics. We need to set up the Hamiltonian of the particle in an electromagnetic field. This cannot be done by using the field strength vectors E and B. Instead, we need to use the potentials A and φ, which are related to the fields by the relations

$$\boldsymbol{B} = \boldsymbol{\nabla} \times \boldsymbol{A}, \tag{10.2a}$$

$$\boldsymbol{E} = -\boldsymbol{\nabla}\varphi - \frac{\partial \boldsymbol{A}}{\partial t}. \tag{10.2b}$$

With these potentials, there is a very simple prescription for finding the Hamiltonian of a charged particle in an electromagnetic field. If the charge of the particle is q, here it is:

1. Take the Hamiltonian without any electromagnetic field, and in it replace the momentum operator by $\boldsymbol{p} - q\boldsymbol{A}$.

2. Add the term $q\varphi$ to the resulting expression.

In other words, for a non-relativistic particle with charge q, the Hamiltonian in the electromagnetic field would be

$$H = \frac{1}{2m}(\boldsymbol{p} - q\boldsymbol{A})^2 + q\varphi. \tag{10.3}$$

□ **Exercise 10.1** *Show that the Lagrangian corresponding to the Hamiltonian given above is*

$$L = \frac{1}{2}mv^2 - q\left(\varphi - \boldsymbol{v} \cdot \boldsymbol{A}\right). \tag{10.4}$$

Derive the Lagrange's equation of motion.

Some explanation is in order. Notice that we now have a velocity-dependent force law in Eq. (10.1). This necessitates a velocity-dependent potential energy term in the Lagrangian, as we see in Eq. (10.4). Thus, the canonical momentum turns out to be

$$p_i = \frac{\partial L}{\partial v_i} = mv_i + qA_i. \tag{10.5}$$

The quantity $m\boldsymbol{v}$ is called the *kinetic momentum* or *mechanical momentum*,

$$\boldsymbol{\Pi} = m\boldsymbol{v}, \tag{10.6}$$

whose relation to the canonical momentum is given by

$$\boldsymbol{\Pi} = \boldsymbol{p} - q\boldsymbol{A}. \tag{10.7}$$

This is the reason that the momentum operator of the free Lagrangian is replaced by $\boldsymbol{p} - q\boldsymbol{A}$, as stated above.

It is straightforward to see that this choice leads to the correct force law. One of the two Hamilton's equations of motion,

$$\frac{dx_i}{dt} = \frac{\partial H}{\partial p_i}, \tag{10.8}$$

reproduces Eq. (10.5). The other one is

$$\frac{dp_i}{dt} = -\frac{\partial H}{\partial x_i}, \tag{10.9}$$

which gives

$$\frac{d\Pi_i}{dt} + q\frac{dA_i}{dt} = \frac{q}{m}\sum_k (p_k - qA_k)\frac{\partial A_k}{\partial x_i} - q\frac{\partial \varphi}{\partial x_i}$$

$$= q\sum_k v_k \frac{\partial A_k}{\partial x_i} - q\frac{\partial \varphi}{\partial x_i}. \tag{10.10}$$

Using the fact that on the path of the particle,

$$\frac{dA_i}{dt} = \frac{\partial A_i}{\partial t} + \sum_k v_k \frac{\partial A_i}{\partial x_k}, \tag{10.11}$$

we see that the equation of motion is

$$\frac{d\Pi_i}{dt} = F_i, \tag{10.12}$$

with the force given in Eq. (10.1), using the expressions for electric and magnetic fields given in Eq. (10.2).

When we try to do quantum mechanics, we need to find the correct commutation laws between various operators. One question that arises is this: is it the kinetic momentum or the canonical momentum that conforms to the commutation rules of Eq. (3.15, p. 60)? As for the commutation between coordinates and momenta, it does not really matter which one we choose, because the difference between the two is $q\boldsymbol{A}$, which is a function of only position, and therefore commutes with the position coordinates. But what about the commutation between different components of momentum? Here, we can again fall back on the classical Poisson brackets, which are defined in terms of the canonical momentum. We conclude that the correct commutation relations are

$$\left[\widehat{x}_a, \widehat{x}_b\right] = 0, \qquad \left[\widehat{p}_a, \widehat{p}_b\right] = 0, \qquad \left[\widehat{x}_a, \widehat{p}_b\right] = i\hbar\delta_{ab}, \tag{10.13}$$

Table 10.1 Comparison between different electromagnetic systems of units. The speed of light in the vacuum is given by $c = 1/\sqrt{\epsilon_0 \mu_0}$.

Quantity in SI	Multiply by the following to change over to	
	Heaviside–Lorentz units	Gaussian units
q, ρ	$\sqrt{\epsilon_0}$	$\sqrt{4\pi\epsilon_0}$
$\boldsymbol{J}$	$1/\sqrt{\mu_0}$	$\sqrt{4\pi\epsilon_0}$
$\boldsymbol{E}, \varphi$	$1/\sqrt{\epsilon_0}$	$1/\sqrt{4\pi\epsilon_0}$
$\boldsymbol{B}, \boldsymbol{A}$	$\sqrt{\mu_0}$	$\sqrt{\mu_0/4\pi}$

where p_a are components of the canonical momentum. This means that, in the coordinate space representation, we can use

$$\boldsymbol{p} \longrightarrow -i\hbar\boldsymbol{\nabla} \tag{10.14}$$

for the canonical momentum.

The relations of Eq. (10.13) also mean that the components of the kinetic momentum do not commute among themselves. In fact, using Eqs. (10.5) and (10.13), it is easy to show that

$$\left[\Pi_a, \Pi_b\right] = iq\hbar \sum_c \varepsilon_{abc} B_c, \tag{10.15}$$

where B_c denotes components of the magnetic field.

☐ **Exercise 10.2** *Verify Eq. (10.15).*

We have been using the SI units in the discussion of this chapter. The other commonly used units are Gaussian and Heaviside–Lorentz. Conversion between these three systems of units have been summarized in Table 10.1.

10.2 Landau levels

We will now use the Hamiltonian encountered in §10.1 to find the energy eigenvalues for the quantum mechanical problem of a particle in a magnetic field $\boldsymbol{B}$ that is constant in both space and time. We can choose our axes such that the z-axis points towards $\boldsymbol{B}$, so that we can write

$$\boldsymbol{B} = B\hat{\boldsymbol{z}}. \tag{10.16}$$

We will solve the eigenvalue problem in two ways. The different procedures will give us some insight into the choice of the potentials, a matter that will be clarified in §10.3.

10.2.1 Energy eigenvalues and eigenfunctions

In order to set up the Hamiltonian, we first need to find an expression for the potentials that give rise to that magnetic field. Since there is no electric field to be reckoned, we can set the scalar potential φ to be equal to zero. As for the vector potential, we can choose

$$A_y = xB, \qquad A_x = A_z = 0. \tag{10.17}$$

Note that this choice is not unique. We will comment about some other choices later. Right now, using this choice, we want to write down the Hamiltonian, which would be

$$H = \frac{1}{2m}\left(\widehat{p}_x^2 + (\widehat{p}_y - qxB)^2 + \widehat{p}_z^2\right). \tag{10.18}$$

This Hamiltonian commutes with p_y and p_z, because the coordinates y and z do not appear in it. The energy eigenfunctions therefore must also be eigenfunctions of p_y and p_z, and we can write them as

$$\psi(x, y, z) = g(x)e^{ik_y y}e^{ik_z z}. \tag{10.19}$$

Substituting this form in the eigenvalue equation

$$H\psi = E\psi \tag{10.20}$$

and using Eq. (10.18), we obtain a differential equation for $g(x)$:

$$\frac{1}{2m}\left(\widehat{p}_x^2 + (\hbar k_y - qxB)^2 + \hbar^2 k_z^2\right)g(x) = Eg(x). \tag{10.21}$$

This is equivalent to the equation

$$\frac{1}{2m}\left(\widehat{p}_{x'}^2 + q^2 B^2 x'^2\right)G(x') = \left(E - \frac{\hbar^2 k_z^2}{2m}\right)G(x'), \tag{10.22}$$

where

$$G(x') = g(x) \tag{10.23}$$

and

$$x' = x - k_y \ell_0^2, \tag{10.24}$$

with

$$\ell_0 = \sqrt{\frac{\hbar}{|qB|}}, \tag{10.25}$$

which is the typical length scale that can be formed out of the charge, the magnetic field, and $\hbar$, and is called the *magnetic length*.

The equation for $G(x')$ is therefore just the equation for a simple harmonic oscillator. From the expression for energy eigenvalues of harmonic oscillators, we conclude that for a particle in a magnetic field in the z direction, the energy eigenvalues are given by

$$E = (n + \frac{1}{2})\hbar\omega_c + \frac{\hbar^2 k_z^2}{2m}, \tag{10.26}$$

with

$$\omega_c = \frac{|qB|}{m}. \tag{10.27}$$

This quantity is called the *cyclotron frequency*.

> The name *cyclotron frequency* has nothing to do with quantum mechanics. It comes from the fact that in a cyclotron, charged particles are forced to run into circular paths under the influence of a magnetic field perpendicular to the plane of the path. If the radius of the path is r, the magnitude of the centripetal force supplied by the magnetic field B will be given by
>
> $$\frac{mv^2}{r} = |qvB|. \tag{10.28}$$
>
> Thus, the time taken for the particle to go once around the circular path would be given by
>
> $$T = \frac{2\pi r}{|v|} = \frac{2\pi m}{|qB|} = \frac{2\pi}{\omega_c}. \tag{10.29}$$

The eigenvalues shown in Eq. (10.26) contain two different contributions. One of them, the part proportional to k_z^2, produces a continuum of energies, starting from zero as the lowest value. The other part contains discrete contributions. Together, it means that for each value of k_z, there is a base value $\hbar^2 k_z^2/2m$, and on top of that there is a ladder of harmonic oscillator energies. These rungs of harmonic oscillator levels are called *Landau levels*.

The eigenfunctions can easily be written down, taking the cue from Eq. (10.19) and using the harmonic oscillator wavefunctions from §7.7. For the nth Landau level, the wavefunction is

$$\psi(x, y, z) \propto \exp\left(-\frac{1}{2}\left(\frac{x}{\ell_0} - k_y\ell_0\right)^2 + ik_y y + ik_z z\right) H_n\left(\frac{x}{\ell_0} - k_y\ell_0\right), \tag{10.30}$$

where H_n denotes the nth Hermite polynomial. The normalization factor has not been shown in the wavefunction, which is why the proportionality sign has been used instead of the equality sign.

10.2.2   An alternative method

As commented in connection with Eq. (10.17), the choice of gauge is not unique. Let us therefore revisit the problem to make sure that the energy eigenvalues that we obtained are independent of our particular choice of gauge.

We will stick to $A_z = 0$, but will not specify what A_x and A_y are. They can be functions of x and y. All we care about is that the functions will give the magnetic field of Eq. (10.16). Since there is no z-dependence in the Hamiltonian, let us write

$$H = H' + \frac{p_z^2}{2m}, \tag{10.31}$$

where H' contains only the $x-$ and y-components of position and canonical momentum. Now, looking at Eqs. (10.3) and (10.7), we see that the combination of quantities appearing in the Hamiltonian are really the components of the kinetic momentum. Hence, we write

$$H' = \frac{1}{2m}\left(\Pi_x^2 + \Pi_y^2\right). \tag{10.32}$$

The operators appearing in H' are the components of the kinetic momentum, and they depend on $\boldsymbol{B}$, as can be seen from Eq. (10.15). It will be convenient for us if we use dimensionless operators defined by

$$P_i = \frac{\Pi_i}{\sqrt{|qB|\hbar}} = \frac{\Pi_i \ell_0}{\hbar}, \tag{10.33}$$

where the index i can denote either the x or the y directions. In terms of these operators, we obtain

$$H' = \frac{1}{2}\hbar\omega_c\left(P_x^2 + P_y^2\right). \tag{10.34}$$

This is vaguely reminiscent of the Hamiltonian of a simple harmonic oscillator. The connection will be clearer if we write the harmonic oscillator Hamiltonian in the form

$$H_{\text{ho}} = \frac{1}{2}\hbar\omega\left(\widehat{P}^2 + \widehat{X}^2\right), \tag{10.35}$$

using the dimensionless operators

$$\widehat{P} = \frac{1}{\sqrt{m\hbar\omega}}\,\widehat{p}, \qquad \widehat{X} = \sqrt{\frac{m\omega}{\hbar}}\,\widehat{x}. \tag{10.36}$$

The similarity between Eqs. (10.32) and (10.35) is now obvious. In fact, the similarity is even deeper. For the harmonic oscillator Hamiltonian, we have

$$\left[\widehat{X}, \widehat{P}\right] = \frac{1}{\hbar}\left[\widehat{x}, \widehat{p}\right] = i. \tag{10.37}$$

We commented that the components of the kinetic momentum do not commute in an electromagnetic field, but rather satisfy Eq. (10.15). The dimensionless operators defined in Eq. (10.33) therefore satisfy the commutation rule

$$\left[P_x, P_y\right] = i, \tag{10.38}$$

the same as in Eq. (10.37). Thus, the two Hamiltonians are completely equivalent. The eigenvalues of H' are therefore the harmonic oscillator eigenvalues with a frequency ω_c. Adding the contribution of the z-component of momentum, we therefore obtain the energy eigenvalues to be

$$E = (n + \frac{1}{2})\hbar\omega_c + \frac{\hbar^2 k_z^2}{2m}, \tag{10.39}$$

exactly what we obtained earlier.

Although the energy eigenvalues are gauge independent, the eigenfunctions are not. This should not be a surprise, because the eigenfunctions are not measurable things. The comment about eigenfunctions is the subject of §10.3.

10.2.3   2D particle in orthogonal magnetic field

Next, we comment on the behavior of a particle constrained in a two-dimensional (2D) plane in the presence of a magnetic field orthogonal to the plane. We take the orthogonal direction along the z-axis. The energy and the wavefunction of such a system can be directly read off from Eqs. (10.26) and (10.30). To see this, consider a 3D free particle in the presence of a magnetic field inside a box whose width in the z-direction is L_z. It is easy to see from the analysis of §7.2 that the effect of this would be to quantize k_z in units of $2\pi/L_z$, i.e., to ensure that $k_z = 2\pi r/L_z$ where r is a positive integer. As argued in §7.2, negative r values do not give independent states, and $r = 0$ is not a state at all. Thus, the energy of such a particle would be given by

$$E_{n,r} = (n + \frac{1}{2})\hbar\omega_c + \frac{2\pi^2\hbar^2 r^2}{mL_z^2}. \tag{10.40}$$

Now if we reduce the length of the box, the second term on the right side grows in energy and in the limit of $L_z \to 0$ (strictly speaking the dimensionless ratio L_z/ℓ_0 is taken to zero), all states except those with the lowest value of r, i.e., $r = 1$, become irrelevant because they are infinitely separated from the other states and do not play any role in the low-energy behavior of the system. The energy corresponding to $r = 1$ is a constant that attaches to all energies obtained from the first term of Eq. (10.40), and we can adjust the zero of energy in such a way that this term is eliminated, obtaining

$$E_n = (n + \frac{1}{2})\hbar\omega_c. \tag{10.41}$$

This yields the energy of a 2D particle in the presence of a magnetic field. The wavefunction can also be obtained from Eq. (10.30) by putting $k_z = 0$.

Before ending this section, we briefly comment about the degeneracy of the Landau levels in the 2D limit. We note that the number of allowed states in any given Landau level is essentially the number of allowed transverse momenta modes. So if the dimension of the sample is $L_x = L_y = L$, then $k_y = 2\pi\nu/L$ where ν is an integer. To find the maximum allowed value of ν, we note that the shift of the center of the Gaussian wavefunction (for $n = 0$) for any given k_y is $k_y\ell_0^2$. Assuming that the center of the wavefunction is located within the sample, we find the requirement $k_y\ell_0^2 < L$, or, in other words,

$$2\pi\nu_{\max}/L = L/\ell_0^2. \tag{10.42}$$

This leads to the maximum allowed value of ν given by

$$\nu_{\max} = \frac{L^2}{2\pi\ell_0^2} \propto B, \tag{10.43}$$

using the definition of ℓ_0 from Eq. (10.25). Thus, the degeneracy is macroscopic. Classically, if we imagine that an electron is performing a cyclotron orbit with area $a_0 = \pi\ell_0^2$, then the degeneracy is (approximately) given by the number of such cyclotron orbits that one can fit in the sample. Also, we note that the degeneracy increases linearly with the strength of the applied field B. Thus, one may, in principle, in a many-body 2D system with N electrons, increase this magnetic field where the number of available states within one Landau level is larger than N. In this situation, the kinetic energy of all the electrons is reduced to $\hbar\omega_c/2$ since all of them occupy the lowest Landau level. The ground state of the system becomes macroscopically degenerate. The mechanism by which such a degeneracy is lifted by interaction between electrons is key to understanding the physics of fractional quantum Hall effect; we are not going to discuss this any further here.

10.3 Gauge invariance

Eq. (10.2) implies that there is some arbitrariness in the definitions of the potentials. In particular, if we change $\boldsymbol{A}$ and φ to their primed avatars, defined by

$$\boldsymbol{A}' = \boldsymbol{A} + \boldsymbol{\nabla} f, \qquad \varphi' = \varphi - \frac{\partial f}{\partial t}, \tag{10.44}$$

we will obtain the same electric and magnetic fields. Since the classical equation of motion of a particle involves the force given by Eq. (10.1) that contains only the electric and magnetic fields, the equation is unaffected by the changes in the potential given in Eq. (10.44). This statement is called *gauge invariance*, and the redefinition of the form given in Eq. (10.44) is called a *gauge transformation*.

This statement, as it is, will not be true for quantum mechanics. The reason lies in a profound difference in the way the Hamiltonian appears in the equation of motion. For classical physics, the Hamiltonian appears *only* through its derivatives with respect to position or momentum, as seen in Eqs. (10.8) and (10.9). In quantum mechanics, however, the Hamiltonian appears directly in the Schrödinger equation. No derivatives are involved. The Hamiltonian, as we see in Eq. (10.3), contains the vector and scalar potentials. Hence, one would naively expect that the changes of Eq. (10.44) will affect the equation of motion. We will demonstrate that this is indeed true, and will show that a modified prescription of gauge transformations will be valid in the quantum mechanical case.

To begin this discussion, we start by writing down the Schrödinger equation corresponding to the Hamiltonian of Eq. (10.3). In the coordinate representation, this equation would be

$$i\hbar\frac{\partial \psi}{\partial t} = \frac{1}{2m}\left(-i\hbar\boldsymbol{\nabla} - q\boldsymbol{A}\right)^2 \psi + q\varphi\psi. \tag{10.45}$$

Here, the potentials have the most general form, i.e., they can depend on both coordinate and time. The wavefunction ψ also depends on both. Let us reorganize the terms a bit and write

$$\left(i\hbar\frac{\partial}{\partial t} - q\varphi\right)\psi = \frac{1}{2m}\left(-i\hbar\boldsymbol{\nabla} - q\boldsymbol{A}\right)^2\psi. \tag{10.46}$$

We will now see whether this equation can be invariant if we change the potentials as in Eq. (10.44). Let us also keep open the possibility that the wavefunction might also change in the process by the rule

$$\psi' = e^{i\theta}\psi, \tag{10.47}$$

where θ can be a function of space and time as well. Using Eqs. (10.44) and (10.47), it is easy to see that

$$\left(i\hbar\frac{\partial}{\partial t} - q\varphi'\right)\psi' = e^{i\theta}\left[\left(i\hbar\frac{\partial}{\partial t} - q\varphi\right)\psi + \left(-\hbar\frac{\partial\theta}{\partial t} + q\frac{\partial f}{\partial t}\right)\psi\right]. \tag{10.48}$$

The expression on the right side does not have any simple relation with the corresponding quantity with the unprimed variables, since θ is an arbitrary function. However, if we now fix θ by the relation

$$\theta(\boldsymbol{r}, t) = \frac{q}{\hbar}f(\boldsymbol{r}, t), \tag{10.49}$$

then we see that the expression with the primed quantities is related by a phase factor with the same expression involving the unprimed quantities. Keeping that in mind, we now look at the right side of Eq. (10.46). We find that

$$\left(-i\hbar\boldsymbol{\nabla} - q\boldsymbol{A}'\right)\psi' = e^{i\theta}\left[\left(-i\hbar\boldsymbol{\nabla} - q\boldsymbol{A}\right)\psi + \left(\hbar\boldsymbol{\nabla}\theta - q\boldsymbol{\nabla}f\right)\psi\right]. \tag{10.50}$$

Here also, if we use Eq. (10.49), the primed expression is just a phase factor times the unprimed expression. Applying the differential operator $(-i\hbar\boldsymbol{\nabla} - q\boldsymbol{A})$ one more time does not change this fact. Thus, we see that if we use the primed potentials and the primed wavefunction, each side of Eq. (10.46) is equal to $e^{i\theta}$ times the corresponding expression with the unprimed variables. So the factor of $e^{i\theta}$ cancels from both sides, and the form of the equation remains invariant.

The conclusion can be summarized now. The Schrödinger equation is gauge invariant, provided we augment the transformations of the potentials, Eq. (10.44), with a change of phase of the wavefunction given by Eqs. (10.47) and (10.49). If the phase is not changed, i.e., if one puts $\theta = 0$, the gauge invariance is not there unless the function f is a constant, which means the potentials are not changed at all.

To explore the implication of this statement on physical observables, it is more convenient to use the vector space notation. Eqs. (10.47) and (10.49) imply that, under gauge transformation, the state vectors in the Hilbert space change by the rule

$$|\psi'\rangle = U |\psi\rangle \tag{10.51}$$

with

$$U = \exp\left(\frac{iq}{\hbar} f(\widehat{\boldsymbol{r}}, t)\right). \tag{10.52}$$

The matrix elements of any gauge invariant operator $\mathscr{O}$ should not change under this transformation, i.e., we should have

$$\langle \psi'_1 | \mathscr{O} | \psi'_2 \rangle = \langle \psi_1 | \mathscr{O} | \psi_2 \rangle. \tag{10.53}$$

An example of a gauge-invariant operator is the position operator $\widehat{\boldsymbol{r}}$, for which Eq. (10.53) is clearly true since U is unitary, and $f(\widehat{\boldsymbol{r}}, t)$ commutes with $\widehat{\boldsymbol{r}}$. The canonical momentum operator, $\widehat{\boldsymbol{p}}$, is not gauge invariant. But the kinetic momentum, defined through Eq. (10.7), is gauge invariant, and it is easily seen that

$$\left\langle \psi'_1 \left| \widehat{\boldsymbol{p}} - q\widehat{\boldsymbol{A}'} \right| \psi'_2 \right\rangle = \left\langle \psi_1 \left| \widehat{\boldsymbol{p}} - q\widehat{\boldsymbol{A}} \right| \psi_2 \right\rangle. \tag{10.54}$$

What about the Hamiltonian, given in Eq. (10.3)? Clearly, the term that depends on $\boldsymbol{p} - q\boldsymbol{A}$ is gauge invariant. The other part, containing the scalar potential, is not. As a result, the Hamiltonian is not gauge invariant. However, if we consider the gauge transforming function f to be time-independent, Eq. (10.44) clearly shows that the scalar potential does not change under such a transformation. As a result, the Hamiltonian is also invariant, and Eq. (10.53) applies when $\mathscr{O}$ is substituted by the Hamiltonian. In particular, it implies that the expectation values of the Hamiltonian do not change, i.e., the energy eigenvalues are unchanged.

If $f(\boldsymbol{r}, t)$ is a linear function of time, φ' and φ differ by a constant. This means that the Hamiltonian obtains a constant shift because of the gauge transformation. This is like a shift of the zero of energy. All eigenvalues show the same shift.

For more complicated time dependence of $f(\boldsymbol{r}, t)$, the transformed Hamiltonian becomes time-dependent. We have not discussed such Hamiltonians anywhere yet, and will not do so until in Chapter 12, so we do not comment on such gauge transformations here.

☐ **Exercise 10.3** *The definition of Eq. (10.47) implies that θ should be dimensionless. Show that the right side of Eq. (10.49) indeed defines a dimensionless combination.* [**Hint:** *Find the dimension of $q\boldsymbol{B}$ from Eq. (10.1), and then of $q\boldsymbol{A}$ from Eq. (10.2) to arrive at the dimension of qf from Eq. (10.44).*]

☐ **Exercise 10.4** *For the uniform and static magnetic field discussed in §10.2, we chose the potentials through Eq. (10.17). Another possible choice is*

$$A_x = -yB, \qquad A_y = 0, \qquad A_z = 0. \tag{10.55}$$

(a) *Find the function f, defined in Eq. (10.44), that connects this gauge with the one in Eq. (10.17).*

(b) *Find the ground state wavefunctions for the choices of Eqs. (10.17) and (10.55), and show that they are related by Eqs. (10.47) and (10.49).*

☐ **Exercise 10.5** *Verify that Eq. (10.54) is true by evaluating the relevant matrix elements in the coordinate representation.*

We can restate this point, starting from the opposite side. If we change the phase of a wavefunction by a constant, the phase cancels from both sides of the free Schrödinger equation, and is therefore inconsequential. The story is different if the phase depends on position and time, i.e., if we perform a *local phase transformation*. For these, the free Hamiltonian does not produce a gauge invariant Schrödinger equation: the derivatives of the phase spoil the invariance. However, if the Schrödinger equation contained the electromagnetic potentials, this disturbance is inconsequential, because the offending terms can be absorbed in the definitions of the potentials. In this way, local phase transformations can be seen to be connected with the electromagnetic interactions. This viewpoint is utilized in *gauge theories* for building up theories of other interactions as well.

The presence of a vector potential $\boldsymbol{A}(\boldsymbol{x})$ is equivalent to a local phase transformation of the wavefunction

$$\psi(\boldsymbol{x}) \to \psi'(\boldsymbol{x}) = \psi(\boldsymbol{x}) \exp\left[-\frac{iq}{\hbar}\int^{\boldsymbol{x}} d\boldsymbol{x}' \cdot \boldsymbol{A}(\boldsymbol{x}')\right] \tag{10.56}$$

in the sense that

$$\left(\widehat{\boldsymbol{p}} - q\boldsymbol{A}(\boldsymbol{x})\right)\psi(\boldsymbol{x}) = \widehat{\boldsymbol{p}}\psi'(\boldsymbol{x}). \tag{10.57}$$

In the phase factor in Eq. (10.56), $\boldsymbol{x}'$ spans an arbitrary curve in three-dimensional space from some fixed reference point (usually chosen to be at infinity) to $\boldsymbol{x}$. Note that such a phase is generally path-dependent, i.e., it depends on the choice of $d\boldsymbol{x}'$.

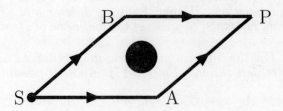

Figure 10.1 Schematic diagram for the interference experiment demonstrating the Aharonov–Bohm effect. The magnetic field, perpendicular to the plane of the paper, is confined to the region denoted by a big blob. A charged particle beam is divided to pass through two paths, avoiding the magnetic field altogether.

☐ **Exercise 10.6** *Verify Eq. (10.57).*

☐ **Exercise 10.7** *Consider a transformation to the rotating frame which changes the spatial coordinates*

$$\boldsymbol{x} = (x, y, z) \to \boldsymbol{x}' = (x', y', z'), \tag{10.58}$$

where the axis of rotation is $\hat{\boldsymbol{z}}$ and the rotation frequency is ω, so that $z' = z$, and

$$\begin{pmatrix} x' \\ y' \end{pmatrix} = \begin{pmatrix} \cos\omega t & -\sin\omega t \\ \sin\omega t & \cos\omega t \end{pmatrix} \begin{pmatrix} x \\ y \end{pmatrix}. \tag{10.59}$$

Find the Hamiltonian of a classical free particle in the new frame.

☐ **Exercise 10.8** *Using the Hamiltonian obtained in the last problem or otherwise, show that for a quantum particle obeying Schrödinger equation, such a transformation is equivalent to a local phase transformation with $\boldsymbol{A} = -(m/q)\boldsymbol{\omega} \times \boldsymbol{x}$. [**Note:** This shows that rotations can generate artificial vector potentials.]*

10.4   Aharonov–Bohm effect

We have discussed that the Hamiltonian depends on the scalar and vector potentials. As a result, the potentials appear in the Schrödinger equation. This fact has a direct experimental consequence that was pointed out by Yakir Aharonov and David Bohm and is known by their names.

The idea has been schematically represented in Figure 10.1. A beam of charged particles is split into two paths and then designed to fall on a screen, where they interfere. A magnetic field threads the region between the two paths. It is localized, meaning that the magnitude is negligibly small everywhere outside a small region. This region has been marked by a blob at the center of the diagram.

In the path integral formalism developed in §6.5, we argued that the phase difference of the two waves arriving at the point P would be equal to the difference in action along the two classical paths divided by $\hbar$. In this picture, there are two paths, and we will call them "Path 1" and "Path 2". Given that the Lagrangian in a magnetic field is given by Eq. (10.4), the phase difference will depend on the factor

$$\Delta\phi = \frac{q}{\hbar} \int dt \Big(\boldsymbol{v} \cdot \boldsymbol{A}(1) - \boldsymbol{v} \cdot \boldsymbol{A}(2) \Big), \tag{10.60}$$

where $\boldsymbol{A}(1)$ and $\boldsymbol{A}(2)$ denote the values of the vector potential encountered along paths 1 and 2 respectively. Since $dt\,\boldsymbol{v}$ is the infinitesimal line element along the paths, we can write this equation as

$$\Delta\phi = \frac{q}{\hbar} \int d\boldsymbol{\ell} \cdot \Big(\boldsymbol{A}(1) - \boldsymbol{A}(2) \Big). \tag{10.61}$$

The expression on the right side is just the line integral along the closed path comprising path 1 followed by path 2 in the backward direction. Using a well-known theorem of vector calculus, we can write

$$\Delta\phi = \frac{q}{\hbar} \oint d\boldsymbol{\ell} \cdot \boldsymbol{A} = \frac{q}{\hbar} \int d\boldsymbol{S} \cdot \boldsymbol{B}, \tag{10.62}$$

using the relation between $\boldsymbol{A}$ and $\boldsymbol{B}$ given in Eq. (10.2). The integral on the right side now is the flux of magnetic field through the area enclosed by the two paths. This will be non-zero because $\boldsymbol{B}$ is non-zero in a region inside this area. The exact geometry of the paths is not important. What is important is the magnetic field enclosed in the area, which produces a phase difference between the two paths, which can be changed by changing the value of the magnetic field. For example, if the magnetic field is provided by a long solenoid, we can tune the current in the solenoid, which will affect the phase difference, which can be observable by changes in interference fringes.

The Aharonov–Bohm effect makes a very important point. In classical physics, the potentials $\boldsymbol{A}$ and φ are introduced only to help solve the field equations. There is no observable effect of the potentials. The fields $\boldsymbol{E}$

and $\boldsymbol{B}$ determine all physical effects. In an Aharonov–Bohm setup, however, the magnetic field is zero everywhere on the paths of the charged particles, although the vector potential cannot be zero because it must produce the correct magnetic flux through the area enclosed. Particles interact with the local fields, i.e., the fields at the position where the particle is at any given time. However, here we see that although the local magnetic field is zero everywhere on the paths of the particle, there is an observable effect obtained by changing the field. This means that the particle recognizes the vector potential at its position. The vector potential thus is a part of the physical reality, not just a figment of imagination devised for obtaining solutions of some equations.

10.5   Interactions with spin

10.5.1   Interaction Hamiltonian

In §10.3, we discussed that the free Hamiltonian is not invariant under a local phase transformation. We also discussed, in §10.1, how to add the electromagnetic interactions in order to produce a Schrödinger equation that is gauge invariant. While that prescription produces the minimal Hamiltonian that can produce a gauge invariant Schrödinger equation, it does not preclude the possibility that there might be additional terms in the Hamiltonian which are gauge invariant by themselves. Even in classical electrodynamics, such terms arise when one considers, instead of a point charge, extended charge distributions whose multipole moments interact with the magnetic and electric fields and their derivatives. For a point charge q moving in a circle, the interaction energy derived from Maxwell equations is equal to

$$E_{\text{loop}} = -\frac{q}{2m} \boldsymbol{L} \cdot \boldsymbol{B}, \tag{10.63}$$

where $\boldsymbol{L}$ is the orbital angular momentum of the particle.

Quantum mechanics allows us to think of the spin angular momentum as well, which means we can contemplate a similar interaction with the spin of a particle. This means that we can talk about an interaction with the magnetic field for a particle with $\boldsymbol{L} = 0$, i.e., a particle not having magnetic moment from orbital angular momentum arising from being in the influence of other objects. Writing the spin operator as $\boldsymbol{S}$, we can consider the interaction of the magnetic field with spin:

$$H_{\text{spin}} = -\boldsymbol{\mu} \cdot \boldsymbol{B}, \tag{10.64}$$

where

$$\boldsymbol{\mu} = g\frac{q}{2m}\boldsymbol{S} \tag{10.65}$$

is the *magnetic moment* of the particle. The factor g is a numerical factor, called the *Landé g-factor*. For the electron, for example, this factor is very close to 2. For the proton, it is approximately equal to 5.585. The ratio $|q|\hbar/(2m)$ has the dimensions of the magnetic moment and is often given a special name for specific particles. For the electron, the ratio is called the *Bohr magneton* and denoted by μ_B. For the proton, the ratio is called the *nuclear magneton* and denoted by μ_N.

Because of these interactions with the magnetic field, levels that were otherwise degenerate with energy E_0 would split into states with energies dependent on the eigenvalue of the component of the angular momentum along $\boldsymbol{B}$. For example, the $L = 1$ states, which are degenerate in the hydrogen atom, would split into three different levels corresponding to three different eigenvalues of L_z. This is called the *Zeeman effect*. The same kind of thing will happen because of the interaction with spin. Since electrons have spin equal to half, there will be a two fold splitting due to the interaction of the spin with the magnetic field. This phenomenon was historically called the *anomalous Zeeman effect*, although in modern times the name *Zeeman effect* is applied for both.

□ **Exercise 10.9** *Show that the quantity g defined in Eq. (10.65) is dimensionless, with the dimension of qB given by the Lorentz force law, Eq. (10.1).*

□ **Exercise 10.10** *If one still uses Eq. (10.64) as the definition of magnetic moment in the Gaussian units of electromagnetism, how will Eqs. (10.63) and (10.65) be modified?*

10.5.2   Spin precession

If we consider the classical motion of a particle with spin in a magnetic field $\boldsymbol{B}$, the direction of the spin vector precesses about the direction of the field. This phenomenon is called *spin precession*.

Such a classical description of spin precession has no place in quantum mechanics. The reason is that the different components of spin do not commute among themselves, so they cannot be measured simultaneously, and therefore one cannot even define anything called the direction of the spin vector at a given time. However, a precessional behavior is seen for the expectation values of the spin components, which we demonstrate now.

We suppose that the spin of the particle is $\frac{1}{2}$, in units of $\hbar$ of course. Then the spin operators are given by the Pauli matrices:

$$S = \frac{1}{2}\hbar\boldsymbol{\sigma}. \tag{10.66}$$

We also take, without loss of generality, the magnetic field B to be in the y direction. Then the interaction Hamiltonian of Eq. (10.64) reduces to the form

$$H = -\hbar\omega\sigma_y, \quad \omega = \frac{gqB}{2m}. \tag{10.67}$$

Suppose now the spin was initially in a state with $S_z = +\frac{1}{2}$, i.e., the state vector at the initial time was

$$|\chi(0)\rangle = \begin{pmatrix} 1 \\ 0 \end{pmatrix}. \tag{10.68}$$

The time evolution of this system has in fact been solved earlier in this book, in §4.4. With minor adjustments in notation, we can use the results obtained there and write the state vector at time t as

$$|\chi(t)\rangle = \begin{pmatrix} \cos\omega t \\ -\sin\omega t \end{pmatrix}. \tag{10.69}$$

Let us now see what is the probability, at time t, that we will find the spin in the original state. If we denote the original state by $|\chi_+\rangle$ and the orthogonal state to be $|\chi_-\rangle$, these two will form an orthonormal basis in the 2D vector space. The state vector of Eq. (10.69) can be written as

$$|\chi(t)\rangle = \cos\omega t \,|\chi_+\rangle - \sin\omega t \,|\chi_-\rangle. \tag{10.70}$$

The probability for being in the state $|\chi_+\rangle$ is therefore

$$|\langle\chi_+|\chi(t)\rangle|^2 = \cos^2\omega t. \tag{10.71}$$

This sinusoidal variation is typical of the classical picture of precession of the spin vector.

We can make the comparison clearer by considering the expectation value of the z-component of the spin vector at any time t. This expectation value was obtained in §4.4:

$$\langle\chi(t)\,|\sigma_z|\,\chi(t)\rangle = \cos 2\omega t. \tag{10.72}$$

If we consider the classical motion of a vector rotating with an angular frequency 2ω, this is what a component should be equal to at any time t. Thus, although one cannot define the *spin vector* because of the non-commuting nature of its components, the expectation value of spin behaves like a classical spin vector.

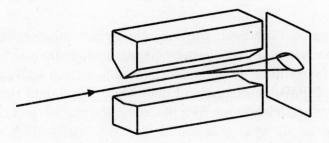

Figure 10.2 Schematic diagram of the apparatus for the Stern–Gerlach experiment. From the left side of the figure, a collimated beam of spin-$\frac{1}{2}$ particle comes. In the middle portion, we see the two poles of a magnet. In the screen on the right end of the figure, we see the beam split into two beams.

☐ **Exercise 10.11** *Solve the more general problem in which the initial state of the spin is not an eigenstate of σ_z but of $\sigma_z \cos\alpha + \sigma_y \sin\alpha$.*

10.6 Stern–Gerlach experiment

A milestone experiment was conceived by Otto Stern and performed by Walther Gerlach in 1921–22. It aimed at measuring the magnetic moment of atoms. We omit the details and present the basic idea here, which involves the passage of particles with spin through an inhomogeneous magnetic field.

To understand the idea, let us first discuss what is expected from the classical theory. Suppose a particle has a magnetic moment $\boldsymbol{\mu}$, and it passes through a magnetic field B in the z direction. It will interact with the magnetic field through the interaction Hamiltonian

$$H = -\mu_z B, \qquad (10.73)$$

where μ_z is the z-component of the magnetic moment. If the magnetic field has a variation in the z direction, the gradient of this Hamiltonian will be non-zero, which means that the particle would experience a force in the z direction and will be deflected. The force will be given by

$$\boldsymbol{F} = \mu_z \frac{\partial B}{\partial z} \mathring{\boldsymbol{z}}, \qquad (10.74)$$

so that, for a given field gradient, the deflection will be proportional to μ_z.

Classically, in a beam of particles, the magnetic moments will be randomly oriented. Thus, there will be all possible values of μ_z within a given maximum, and accordingly all possible amounts of deflection. We should therefore see a continuous band of particles on a screen placed on the exit path of the particles.

That was not at all what was found in the Stern–Gerlach experiment. What they saw were discrete lines where the particles arrived on the screen, implying that only discrete values of μ_z were available in the beam.

This is a beautiful vindication of quantum theory. According to the quantum theory, the magnetic moment is proportional to the spin, as shown in Eq. (10.65), and the possible values of S_z are discrete, as discussed in Chapter 8. For example, if we send a stream of spin-$\frac{1}{2}$ particles, there can be two possible values of S_z, viz., $+\frac{1}{2}\hbar$ and $-\frac{1}{2}\hbar$, and accordingly two possible paths of the outgoing particles. If the stream consists of spin-1 particles, there will be three possible paths, corresponding to the S_z values of $+1$, 0, and -1.

Let us see what we can do further with the different output streams. Suppose we put a blockade on all but one output paths. It means that we have singled out the stream of particles with a specific value of S_z. If we send it through another Stern–Gerlach apparatus, what will happen?

If the second apparatus is identical, meaning that the direction and gradient of the magnetic field is the same as in the first one, there will be no further splitting of the beam. The beam will, of course, suffer more deflection, but the entire beam has the same S_z, so all particles will be deflected by the same amount. This is reminiscent of Newton's experiment with the colors of light. Once you pass a collimated beam of white light through a prism, the colors separate. But if now you block all colors except one and pass the beam through a second prism, there is no further separation; the beam will only be deflected further.

However, there is a difference. The difference will show if we pass one outcoming stream of a Stern–Gerlach experiment through another apparatus, but this time with a field gradient in a different direction. To see the effect, let us take the simplest case in which the particles have spin $\frac{1}{2}$. Let us say that after the first Stern–Gerlach apparatus, we have blocked the path of the $s_z = -\frac{1}{2}$ particles. Only the $s_z = +\frac{1}{2}$ ray comes out, and we let it pass through another Stern–Gerlach apparatus in which the magnetic field, as well as its gradient, is in the x direction. The incoming beam, which is the eigenstate of S_z as given in Eq. (10.68), now experiences a force that involves the operator S_x. The incoming state can be written as

$$|\chi_+\rangle = \frac{1}{\sqrt{2}}|\xi_+\rangle + \frac{1}{\sqrt{2}}|\xi_-\rangle, \tag{10.75}$$

where

$$\xi_\pm = \frac{1}{\sqrt{2}}\begin{pmatrix} 1 \\ \pm 1 \end{pmatrix} \tag{10.76}$$

are the eigenstates of S_x. The two eigenstates would now separate in x direction.

One can continue this game, putting a third Stern–Gerlach apparatus on the way, which would have the magnetic field and field gradient in the z direction again. This apparatus would intake the $S_x = +\frac{1}{2}$ part of the beam, and will split it into two components with different S_z eigenvalues, and so on. The process cannot continue forever because every time we block the path of some particles, we are cutting down on the number of particles that reaches the other end. After a number of steps, the intensity will decrease to the level that the experiment cannot be performed.

10.7 Effect of magnetic field for particles in a 2D lattice

In §7.8, we have studied the Schrödinger equation for a free particle on a 1D lattice whose kinetic energy originates from hopping between neighboring lattice sites. We have seen that the energy eigenvalue equation for such a particle is given by

$$E\psi(x) = -t\left[\psi(x+a) + \psi(x-a)\right], \tag{10.77}$$

where t denotes the strength of the hopping and a denotes the lattice spacing. In this section, we shall consider the 2D case, where a particle hops on a 2D square lattice in the presence of a magnetic field perpendicular to the plane.

To this end, let us first consider the simple case, where the magnetic field is set to zero. In this case, the Schrödinger equation of the particle is easy to write down. If t_x and t_y denote hopping amplitudes in x- and y- directions, the

Schrödinger equation takes the same form as Eq. (10.77) but with additional terms due to hopping in the y-direction. This is given by

$$E\psi(x,y) = -t_x \left[\psi(x+a,y) + \psi(x-a,y) \right]$$
$$-t_y \left[\psi(x,y+a) + \psi(x,y-a) \right]. \qquad (10.78)$$

To solve for the energy dispersion, we proceed in a similar manner as in 1D. The translational invariance of the problem suggests that we move to the Fourier space. This is done by writing

$$\psi(x,y) = \int_{-\pi/a}^{\pi/a} \int_{-\pi/a}^{\pi/a} \frac{d^2 k}{(2\pi)^2} e^{i(k_x x + k_y y)} \psi(k_x, k_y). \qquad (10.79)$$

Note that the limits of the integration are finite. This can be understood as follows. The allowed positions are the lattice sites, whose coordinates are multiples of the lattice spacing a. Thus, changing Fourier space variables k_x and k_y by $2\pi/a$ would not produce any difference on the wavefunction which is defined only on the lattice sites. The Fourier variables can therefore be taken in the region $-\pi/a \leq k_x, k_y \leq \pi/a$. This region in momentum space is referred to as the first *Brillouin zone*. We shall restrict ourselves to momenta defined within this zone.

Substituting Eq. (10.79) into Eq. (10.78) and performing some straightforward manipulations similar to those carried out in §7.8, we find

$$E \equiv E(k_x, k_y) = -2t_x (\cos k_x a + \beta \cos k_y a), \qquad (10.80)$$

where $\beta = t_y/t_x$. We note that the energy is a periodic function of momenta, and the spectrum, within the Brillouin zone, has a single minimum at $\mathbf{k} = (k_x, k_y) = (0,0)$.

Next we consider a magnetic field $\mathbf{B} = B\hat{\mathbf{z}}$. Throughout our discussion, we shall choose the gauge $\mathbf{A} = xB\hat{\mathbf{y}}$. We have seen in §10.3 that such a field leads to an additional phase of the wavefunction. In the present case, this phase is given by

$$\psi(x,y) \rightarrow \psi'(x,y) = \psi(x,y) \exp\left[-\frac{iq}{\hbar} \int_{y_0}^{y} dy\, A_y \right], \qquad (10.81)$$

where y_0 is the y coordinate of the reference site. This is kept arbitrary at the moment, and we shall see that it does not enter the Schrödinger equation and is hence unimportant. Substituting Eq. (10.81) into Eq. (10.78), we find

$$E\psi(x, y) = -\sum_{\eta=\pm1} \Bigg[t_x\psi(x + \eta a, y)$$

$$+t_y \exp\left(-\frac{iq}{\hbar} \int_{y}^{y+\eta a} dy\, A_y \right) \psi(x, y + \eta a) \Bigg], \qquad (10.82)$$

where we have multiplied both sides of the equation by $\exp[-(iq/\hbar)\int_{y_0}^y A_y dy]$. Note that for the present gauge choice, the relative phase between $\psi(x_1, y)$ and $\psi(x_2, y)$ vanishes since $dy = 0$ between them. Thus, the modification of the Schrödinger equation constitutes inclusion of a relative phase for t_y:

$$t_y \to t_y' = t_y \exp\left[-\frac{iq}{\hbar} \int_{y}^{y+a} dy\, A_y \right]$$

$$= t_y \exp\left[-2\pi i \frac{\Phi}{\Phi_0} x/a \right] = t_y e^{-i\alpha_0}, \qquad (10.83)$$

where $\Phi = Ba^2$ is the flux through a plaquette of the square lattice and $\Phi_0 = 2\pi\hbar/q$ is a universal constant that has the dimension of magnetic flux. It is referred to as the flux quanta. The relative phase $\alpha_0 = 2\pi(\Phi/\Phi_0)x/a$ is known as the *Peierls phase*. In terms of this phase, the Schrödinger equation can be written as

$$E\psi(x, y) = -\sum_{\eta=\pm1} \left[t_x\psi(x + \eta a, y) + t_y e^{-i\eta\alpha_0}\psi(x, y + \eta a) \right]. \qquad (10.84)$$

Next, we make a Fourier transformation on the wavefunction. Using Eq. (10.79) and putting $t_x = t_y = t$ for simplicity, we find that the Schrödinger equation, in momentum space, can be written as

$$\frac{E}{2t}\psi(k_x, k_y) = -\cos k_x a\, \psi(k_x, k_y) - \cos k_y a\, \psi\left(k_x - \frac{2\pi\Phi}{a\Phi_0}, k_y \right). \quad (10.85)$$

Here we note several points about the problem. First, in contrast to the continuum problem, the magnetic field does not preclude the usage of momentum labels k_x and k_y. Second, the effect of the magnetic field is to provide a coupling between otherwise independent momentum modes k_x and $k_x - (2\pi/a)\Phi/\Phi_0$; the strength of the magnetic field determines which modes would be coupled through the factor Φ/Φ_0. Finally, a magnetic field which corresponds to $B = n\Phi_0/a^2$, where n is an integer, is equivalent to having no magnetic field. This follows from the fact that such a magnetic field couples $\psi(k_x, k_y)$ to $\psi(k_x - 2\pi n/a, k_y) \equiv \psi(k_x, k_y)$, since k_x has a periodicity of $2\pi/a$. Thus, the periodicity of the lattice ensures that Eq. (10.85) reduces to the momentum space equivalent of Eq. (10.78) and leads to Eq. (10.80). Hence, the effect of the magnetic field vanishes in this case.

The solution of Eq. (10.85) requires numerical endeavor for arbitrary $\Phi < \Phi_0$. For example, when Φ/Φ_0 is an irrational number, Eq. (10.85) involves L coupled equations (where L denotes the system dimension along x) whose solutions are to be numerically obtained. The energy spectrum, plotted as a function of Φ/Φ_0, takes the shape of a butterfly and is called the *Hofstadter butterfly*.

In contrast, Eq. (10.85) admits analytic solutions when $\Phi/\Phi_0 = p/q$, where p and q are co-prime integers. In these cases, a momentum mode k_x is connected to q other k_x modes; this results in q coupled equations in momentum space which can be analytically solved for small q. Here we shall analyze the problem for $p = 1$ and $q = 2$. In this case, the mode k_x is coupled to $k_x - \pi/a$, and Eq. (10.85) can be written as

$$\frac{E}{2t}\psi(k_x, k_y) = -\cos(k_x a)\psi(k_x, k_y) - \cos(k_y a)\psi(k_x - \frac{\pi}{a}, k_y),$$

$$\frac{E}{2t}\psi(k_x - \frac{\pi}{a}, k_y) = \cos(k_x a)\psi(k_x - \frac{\pi}{a}, k_y) - \cos(k_y a)\psi(k_x, k_y),$$

$$(10.86)$$

where we have used the fact $k_x \equiv k_x - 2\pi/a$. Note that in Eq. (10.86) the range of k_x is reduced: $-\pi/(2a) \leq k_x \leq \pi/(2a)$. Thus, the effect of the magnetic field is to reduce the width of the Brillouin zone along x. Such a reduced Brillouin zone is often call the *magnetic Brillouin zone*. Of course, the fact that the width is reduced along x and not along y is a consequence of our gauge choice. A choice of $\boldsymbol{A} = B(-y, 0, 0)$, for example, would lead to a reduction of the Brillouin zone along y.

The energy eigenvalues obtained by solving Eq. (10.86) are given by

$$E = \pm 2t\sqrt{\cos^2 k_x a + \cos^2 k_y a}. \qquad (10.87)$$

We thus find that the energy spectrum now has two minima at $(k_x, k_y) = (0, 0)$ and $(k_x, k_y) = (0, \pi/a)$, as shown in Figure 10.3. In this context, it is useful to note that $(k_x, k_y) = (\pi/a, 0)$ does not fall within the reduced Brillouin zone and $(k_x, k_y) = (0, -\pi/a)$ is identified with $(0, \pi/a)$. Thus, the energy dispersion now constitutes $q = 2$ minima in the lower (negative) branch. It turns out that this feature is general; for a flux $\Phi/\Phi_0 = 2\pi p/q$ one has q minima of

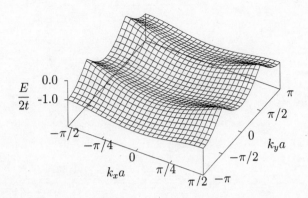

Figure 10.3 Plot of the lower energy branch (in units of t) obtained in Eq. (10.87) as a function of k_x and k_y (in units of $1/a$) within the first magnetic Brillouin zone given by $-\pi/2 \le k_x a \le \pi/2$ and $-\pi \le k_y a \le \pi$ for $\Phi/\Phi_0 = 1/2$. The two energy minima occur at $(k_x, k_y) = (0,0)$ and $(0,\pi)$, as discussed in the text after Eq. (10.87).

the dispersion within the first Brillouin zone. The positions of these minima within the Brillouin zone depend on our gauge choice; however, their number is independent of it and depends only on the strength of the magnetic field.

We end this section by noting that analysis carried out here for square lattice can be applied to any other lattices. Without getting into the details of such analysis, we note that the lattice geometry plays a crucial role in determining the details of the reduced Brillouin zone, the energy dispersion, and the number of its maxima.

☐ **Exercise 10.12** *Repeat the analysis leading to Eq. (10.85) starting from Eq. (10.78) choosing $\boldsymbol{A} = B(-y, 0, 0)$.*

☐ **Exercise 10.13** *Write down the Schrödinger equation, using a gauge of your choice, for a particle with the nearest neighbor hopping on a square lattice with a magnetic field such that $\Phi/\Phi_0 = 2\pi/3$. Find the number of energy branches.*

☐ **Exercise 10.14** *Consider a 2D triangular lattice and a particle hopping on it with the nearest neighbor hopping amplitude t. A magnetic field B is applied perpendicular to the lattice. Write down the Schrödinger equation and express it as a coupled eigenvalue equation in momentum space for $\Phi/\Phi_0 = 2\pi/3$.*

Part Three

Approximations

Chapter 11

Approximation methods

Many problems that we come across in quantum systems cannot be solved exactly. In fact, exactly solvable problems form only a small fraction of quantum problems that one encounters in realistic physical systems. In the absence of such exact solutions of the Schrödinger equation, one needs to resort to several methods providing approximate solutions. In this chapter we shall list some of these methods for time-independent Hamiltonians.

11.1 Variational method

The variational method is often used for a quick estimation of the ground state energy of a Hamiltonian whose exact eigenvalue and eigenstates are unknown. We first describe this method in general terms and then follow the discussion up with examples.

11.1.1 The method

To elucidate the technique, we first note that a trial wavefunction $|\psi_{\text{trial}}\rangle$ for any quantum Hamiltonian H always satisfies the inequality

$$E_{\text{trial}} = \frac{\langle \psi_{\text{trial}}|H|\psi_{\text{trial}}\rangle}{\langle \psi_{\text{trial}}|\psi_{\text{trial}}\rangle} \geq E_0, \tag{11.1}$$

where E_0 is the exact ground state energy of H. To see why this holds, first let us expand $|\psi_{\text{trial}}\rangle$ in the eigenbasis of H:

$$|\psi_{\text{trial}}\rangle = \sum_n c_n |n\rangle, \tag{11.2}$$

where $|n\rangle$ is n^{th} eigenstate of H with energy E_n. Since the eigenstates of H form a complete basis, this can always be done. The eigenstates $|n\rangle$, as well as the coefficients c_n that appear in Eq. (11.2), can be determined only if we can solve the Hamiltonian problem exactly. But, as we will see, this knowledge is not necessary for what follows.

Using Eq. (11.2), one can write

$$E_{\text{trial}} = \frac{\sum_n |c_n|^2 E_n}{\sum_n |c_n|^2} = E_0 + \frac{\sum_n |c_n|^2 (E_n - E_0)}{\sum_n |c_n|^2} \geqslant E_0, \qquad (11.3)$$

where we have used the fact that $(E_n - E_0) \geqslant 0$ for all n. We note that the equality holds when the trial wavefunction matches the exact one, because then $c_n = \delta_{n0}$.

The usefulness of the method lies in the fact that simple trial wavefunctions, which respect some basic symmetry properties of the actual wavefunction, yield a pretty good match for the exact ground state energy. To understand this point, we first note that if we choose a wavefunction that differs from the actual wavefunction to $\text{O}(\epsilon)$, i.e., if $c_n = \text{O}(\epsilon)$ for $n \neq 0$, Eq. (11.3) requires that $E_{\text{trial}} = E_0 + \text{O}(\epsilon^2)$. This result is general and holds irrespective of the distribution of c_n's. Thus, it is useful to choose a trial wavefunction that has the same symmetry as the ground state wavefunction of H. This helps in maximizing overlap between them. For example, if the wavefunction is known to vanish at a certain point, it is best to take a trial wavefunction that vanishes at that point. However, there is no unique recipe or general guideline for the choice of the trial wavefunction.

In practice it is useful to allow for variational parameters in the trial wavefunction such that the variation of these parameters leads one to obtain the closest approximation to the true ground state energy. The reason why such parameters are useful once again follows from the fact that $E_{\text{trial}} \geq E_0$. Consider a trial wavefunction $|\psi_{\text{trial}}(\{\lambda_i\})\rangle$, where λ_i for $i = 1, 2, \ldots, N$ are variational parameters. The corresponding trial ground state energy is given by

$$E_{\text{trial}}(\{\lambda_i\}) = \frac{\langle \psi_{\text{trial}}(\{\lambda_i\})|H|\psi_{\text{trial}}(\{\lambda_i\})\rangle}{\langle \psi_{\text{trial}}(\{\lambda_i\})|\psi_{\text{trial}}(\{\lambda_i\})\rangle}. \qquad (11.4)$$

The minimum value of such trial ground state energy is clearly given by parameters $\{\lambda_i\} = \{\lambda_i^0\}$ for which

$$\frac{\partial E_{\text{trial}}(\{\lambda_i\})}{\partial \lambda_i} = 0,$$

$$\det \left[\frac{\partial^2 E_{\text{trial}}(\{\lambda_i\})}{\partial \lambda_i \partial \lambda_j} \right] > 0. \qquad (11.5)$$

Since $E_{\text{trial}} \geq E_0$ for any $\{\lambda_i\}$, the lowest value of E_{trial} must be closest to E_0.

It turns out that the method of variational principle gives remarkably accurate results and is quite useful in computing estimates of ground state energy of complicated quantum systems where the exact wavefunction is not known. However, there are two issues with this method that the reader must be careful about. First, it only gives an upper bound to the ground state energy; it is not possible to know, for a generic complicated quantum system, how accurately it estimates the energy. Second, the method does not work for excited states except for the first excited state that is known to be orthogonal to the ground state. Thus, if one can find a trial wavefunction $|\psi_{\text{trial}}^{\text{ex}}\rangle$ such that $\langle \psi_{\text{trial}} | \psi_{\text{trial}}^{\text{ex}} \rangle = 0$, then one can estimate the first excited state energy by using the same techniques, along with the inequality

$$\frac{\langle \psi_{\text{trial}}^{\text{ex}} | H | \psi_{\text{trial}}^{\text{ex}} \rangle}{\langle \psi_{\text{trial}}^{\text{ex}} | \psi_{\text{trial}}^{\text{ex}} \rangle} \geq E_1. \tag{11.6}$$

□ **Exercise 11.1** *Prove Eq. (11.6).*

11.1.2 Examples

To illustrate the usefulness of the method, we consider a simple system whose ground state energy is known. We consider the Hamiltonian, consisting of a kinetic energy term T and a potential energy V, given by

$$H = -\frac{\hbar^2}{2m} \frac{d^2}{dx^2} - V_0 \delta(x) = T + V. \tag{11.7}$$

The ground state energy of H was derived exactly in §7.6, and the result is

$$E_0 = -\frac{mV_0^2}{2\hbar^2}. \tag{11.8}$$

To estimate this energy, let us choose a Gaussian normalized trial wavefunction given by

$$\psi_{\text{trial}}(x) = \left(\frac{2k^2}{\pi} \right)^{1/4} e^{-k^2 x^2}, \tag{11.9}$$

where k is the variational parameter. The wavefunction is normalized. The estimate of the ground state energy can be found as

$$\langle H \rangle = E_{\text{trial}}(k) = \langle T \rangle + \langle V \rangle,$$

$$\langle T \rangle = -\sqrt{\frac{2k^2}{\pi}} \frac{\hbar^2}{2m} \int_{-\infty}^{\infty} dx \, e^{-k^2 x^2} \frac{d^2}{dx^2} e^{-k^2 x^2} = \frac{\hbar^2 k^2}{2m},$$

$$\langle V \rangle = -\sqrt{\frac{2k^2}{\pi}} V_0 \int_{-\infty}^{\infty} dx \, e^{-2k^2 x^2} \delta(x) = -V_0 \sqrt{\frac{2k^2}{\pi}},$$

$$E_{\text{trial}}(k) = \frac{\hbar^2 k^2}{2m} - V_0 \sqrt{\frac{2k^2}{\pi}}. \tag{11.10}$$

Minimizing $E_{\text{trial}}(k)$ with respect to k yields $k_0 = (2m^2 V_0^2/(\pi \hbar^2))^{1/2}$. Substituting this value of k_0 in $E_{\text{trial}}(k)$, one gets

$$E_{\text{trial}}(k_0) = -\frac{mV_0^2}{\pi \hbar^2}, \tag{11.11}$$

which is indeed larger than the actual value of the ground state energy quoted in Eq. (11.8). Remarkably, even with this simple single parameter trial wavefunction, we get a reasonably close estimate of the ground state energy.

□ **Exercise 11.2** *Consider a particle in an infinite potential well*

$$V(x) = \begin{cases} 0 & \text{for } -a \leq x \leq a, \\ \infty & \text{otherwise.} \end{cases} \tag{11.12}$$

(a) *Using a trial wavefunction (unnormalized) $\psi(x) = |x|^p - a^p$, find the variational energy E_p of the particle as a function of the parameter p.*

(b) *Minimize E_p with respect to p and find the best estimate to the ground state energy of the particle. What is the value of optimal p?*

□ **Exercise 11.3** *Consider a harmonic oscillator Hamiltonian $H = \hat{p}^2/2m + m\omega_c^2 x^2/2$.*

(a) *Using a trial wavefunction (unnormalized) $\psi(x) = \exp[-|x|/a]$, find the variational energy E_a of the particle as a function of the parameter a.*

(b) *Minimize E_a with respect to a and find the best estimate to the ground state energy of the particle. What is the value of optimal a?*

□ **Exercise 11.4** *Consider the Hamiltonian of the helium atom given by*

$$H = \frac{\hat{\boldsymbol{p}}_1^2 + \hat{\boldsymbol{p}}_2^2}{2m} - 2\alpha \hbar c \left(\frac{1}{|\boldsymbol{r}_1|} + \frac{1}{|\boldsymbol{r}_2|} \right) + \frac{\alpha \hbar c}{|\boldsymbol{r}_1 - \boldsymbol{r}_2|}, \tag{11.13}$$

where $\hat{\boldsymbol{p}}_1$ and $\hat{\boldsymbol{p}}_2$ are momentum operators for the two electrons whose coordinates are $\boldsymbol{r}_1$ and $\boldsymbol{r}_2$, and α is fine structure constant. Taking cue from the fact that the ground state wavefunction of the hydrogen atom is given by $\psi(r) = (\pi a_B^3)^{-1/2} \exp[-|\boldsymbol{r}|/a_B]$, where $a_B = \hbar/m\kappa$ is the Bohr radius, use a trial wavefunction

$$\psi(\boldsymbol{r}_1, \boldsymbol{r}_2) = \sqrt{\frac{\eta^3}{\pi a_B^3}} e^{-\eta(|\boldsymbol{r}_1| + |\boldsymbol{r}_2|)/a_B}, \tag{11.14}$$

where η is the variational parameter. Find the variational ground state energy as a function of η and minimize it to find the estimate to the ground state energy of the helium atom.

☐ **Exercise 11.5** *Consider the 1D Hamiltonian $H = \hat{p}^2/(2m) + V(x)$, where $\hat{p}$ denotes the momentum operator and*

$$V(x) = \begin{cases} c_0 x & \text{for } x > 0, \\ \infty & \text{otherwise,} \end{cases} \tag{11.15}$$

$c_0 > 0$ *being a constant.*

(a) *Choose an appropriate trial wavefunction consistent with the symmetry of the problem and the requirement that $\psi(x)$ has to vanish at $x = 0$ and $x = \infty$.*

(b) *Now consider one such choice*

$$\psi(x) = N x e^{-\alpha x} \Theta(x), \tag{11.16}$$

where N is the normalization and $\Theta(x)$ denotes the Heaviside step function. Find the variational ground state energy E_{var} in terms of α and c_0 and hence find the value of α in terms of c_0 which minimizes E_{var}.

11.2 Non-degenerate perturbation theory

In this section, we develop the basic formalism for the perturbation theory for non-degenerate quantum states. This is also known as the Raleigh–Schrödinger perturbation theory. The first step towards developing this formalism involves writing the Hamiltonian of the quantum system as

$$H[\lambda] = H_0 + \lambda H_1, \tag{11.17}$$

where H_0 is exactly solvable and H_1 is the perturbation part. Here λ is a constant parameter which is used to keep track of the order of perturbation and is typically set to unity at the end of the calculation with the understanding that one wants to know the eigenstates and eigenvalues of $H = H[1] = H_0 + H_1$. The word *non-degenerate* in the title of this section means that we concentrate on the corrections, due to the presence of H_1, on a state which is a non-degenerate eigenstate of H_0. It is irrelevant whether there is any degeneracy in the other eigenstates of the Hamiltonian.

11.2.1 The method

To proceed further, we denote the eigenstates and eigenvalues of H_0 by $|m^{(0)}\rangle$ and $E_m^{(0)}$, i.e.,

$$H_0|m^{(0)}\rangle = E_m^{(0)}|m^{(0)}\rangle. \tag{11.18}$$

We then postulate that the eigenstates and eigenvalues of the Hamiltonian of Eq. (11.17) can be written as a power series in λ:

$$E_m = E_m^{(0)} + \lambda E_m^{(1)} + \lambda^2 E_m^{(2)} + \cdots, \tag{11.19a}$$

$$|m\rangle = |m^{(0)}\rangle + \lambda|m^{(1)}\rangle + \lambda^2|m^{(2)}\rangle + \cdots, \tag{11.19b}$$

where the ellipses indicate higher-order terms in λ. Next, one demands that $H[\lambda]|m\rangle = E_m|m\rangle$ and equates terms to each order in λ. Thus, one gets

$$(H_0 + \lambda H_1)\Big(|m^{(0)}\rangle + \lambda|m^{(1)}\rangle + \lambda^2|m^{(2)}\rangle + \cdots\Big)$$

$$= (E_m^{(0)} + \lambda E_m^{(1)} + \lambda^2 E_m^{(2)} + \cdots)\Big(|m^{(0)}\rangle + \lambda|m^{(1)}\rangle + \lambda^2|m^{(2)}\rangle + \ldots\Big). \tag{11.20}$$

The λ-independent terms in this equation give back Eq. (11.18). The first-order terms in λ give

$$H_0|m^{(1)}\rangle + H_1|m^{(0)}\rangle = E_m^{(0)}|m^{(1)}\rangle + E_m^{(1)}|m^{(0)}\rangle, \tag{11.21a}$$

the second-order terms give

$$H_0|m^{(2)}\rangle + H_1|m^{(1)}\rangle = E_m^{(1)}|m^{(1)}\rangle + E_m^{(2)}|m^{(0)}\rangle + E_m^{(0)}|m^{(2)}\rangle, \tag{11.21b}$$

and so on. Eq. (11.21) encapsulates the perturbative corrections to the energy and wavefunction of an arbitrary eigenstate $|m^{(0)}\rangle$. In particular, Eq. (11.21a) provides us with lowest-order perturbative results. To see an expression for the lowest-order correction to the energy eigenvalue, we take the inner product of Eq. (11.21a) with $\langle m^{(0)}|$. Since $\langle m^{(0)}|H_0 = \langle m^{(0)}|E_m^{(0)}$ and the eigenstates of H_0 are orthogonal to one another, one gets

$$E_m^{(1)} = \langle m^{(0)}|H_1|m^{(0)}\rangle. \tag{11.22}$$

We now want to find the change of the eigenvector in first order in λ. Since the eigenstates of H_0 form a complete basis, we can surely write

$$\big|m^{(1)}\big\rangle = \sum_n \big|n^{(0)}\big\rangle \big\langle n^{(0)} \big| m^{(1)}\big\rangle. \tag{11.23}$$

However, if we take the norm of both sides of Eq. (11.19b), since both $|m\rangle$ and $|m^{(0)}\rangle$ are assumed to be normalized to unity, the $\mathrm{O}(\lambda)$ terms tell us that

$$\big\langle m^{(0)} \big| m^{(1)}\big\rangle = 0. \tag{11.24}$$

This means that we can write

$$|m^{(1)}\rangle = \sum_{n \neq m} |n^{(0)}\rangle \langle n^{(0)} | m^{(1)}\rangle \tag{11.25}$$

since the term with $n = m$ does not contribute. To obtain the coefficients, we take the inner product of Eq. (11.21a) with $\langle n^{(0)}|$, which gives

$$\langle n^{(0)} | H_1 | m^{(0)}\rangle = \left(E_m^{(0)} - E_n^{(0)} \right) \langle n^{(0)} | m^{(1)}\rangle \tag{11.26}$$

for $n \neq m$. Putting this into Eq. (11.25), we can write

$$|m^{(1)}\rangle = \sum_{n \neq m} \frac{\langle n^{(0)} | H_1 | m^{(0)}\rangle}{E_m^{(0)} - E_n^{(0)}} |n^{(0)}\rangle. \tag{11.27}$$

Next, we obtain the second-order corrections of the energy eigenvalues and eigenvectors. To this end, we first take the inner product of Eq. (11.21b) with $\langle m^{(0)}|$. This gives

$$E_m^{(2)} = \langle m^{(0)} | H_1 | m^{(1)}\rangle. \tag{11.28}$$

Now, using Eqs. (11.24) and (11.27), we find

$$E_m^{(2)} = \sum_{n \neq m} \frac{|\langle n^{(0)} | H_1 | m^{(0)}\rangle|^2}{E_m^{(0)} - E_n^{(0)}}. \tag{11.29}$$

To obtain the second-order corrections to the eigenvectors, we again take the inner product of Eq. (11.21b), but this time with $\langle n^{(0)}| \neq \langle m^{(0)}|$ from the left. Using Eq. (11.22) along with Eq. (11.27), and after some straightforward algebra, we get

$$|m^{(2)}\rangle = \sum_{n,p \neq m} \frac{\langle n^{(0)} | H_1 | p^{(0)}\rangle \langle p^{(0)} | H_1 | m^{(0)}\rangle}{(E_m^{(0)} - E_n^{(0)})(E_m^{(0)} - E_p^{(0)})} |n^0\rangle$$
$$- \sum_{n \neq m} \frac{\langle m^{(0)} | H_1 | m^{(0)}\rangle \langle n^{(0)} | H_1 | m^{(0)}\rangle}{(E_m^{(0)} - E_n^{(0)})^2} |n^0\rangle$$
$$- \frac{1}{2} |m^{(0)}\rangle \sum_{n \neq m} \frac{|\langle m^{(0)} | H_1 | n^{(0)}\rangle|^2}{(E_m^{(0)} - E_n^{(0)})^2}. \tag{11.30}$$

□ **Exercise 11.6** *Supply the missing steps in Eq. (11.30).*

We now discuss a few properties related to the above expressions. First, from Eq. (11.22) we find that the first-order corrections to the energy of any eigenstate of H_0 is simply the expectation value of H_1 in that state. However, this is not the full correction, since the state itself changes due to such perturbation. This is reflected in the presence of higher-order corrections to the energy. Second, from the form of $E_m^{(2)}$, we find that this correction is always negative if $|m^{(0)}\rangle$ represents the ground state of H_0. Thus, the second-order perturbative correction to the ground state of a quantum system always leads to lowering of energy. Third, the change in the eigenstate $|m^{(0)}\rangle$ indicates the change in the state due to its perturbation-induced overlap with other eigenstates $|n^{(0)}\rangle$; the amount of this change depends on both the matrix elements $\langle n^{(0)}|H_1|m^{(0)}\rangle$ and the inverse of their energy difference $(E_m^{(0)} - E_n^{(0)})$. Thus, we find that the change in any state due to a perturbation is larger if there are more eigenstates of H_0 at nearby energies. Finally, we note that the efficacy of the perturbation theory requires $|\langle m^{(0)}|H_1|n^{(0)}\rangle| \ll |E_m^{(0)} - E_n^{(0)}|$ and thus systems with larger energy gaps are more amenable to this perturbative analysis. Note that the energy denominators in the various expressions never vanish, because we assumed the unperturbed state to be non-degenerate. The case of degeneracy will be discussed in §11.3.

11.2.2   Examples

As a first example of the application of the method outlined above, we consider the Hamiltonian of the *anharmonic oscillator*. This contains the Hamiltonian of the harmonic oscillator,

$$H_0 = \frac{\widehat{p}^2}{2m} + \frac{1}{2}m\omega_c^2 x^2, \tag{11.31}$$

and an additional term

$$H_1 = \frac{1}{4}bx^4. \tag{11.32}$$

The energy eigenvalues and eigenstates of H_0 are known: they were calculated in §7.7. We can now use them to find the corrections due to the anharmonic term. In the first order, we can apply Eq. (11.22) to obtain the energy corrections as

$$E_n^{(1)} = \frac{1}{4}b\left\langle n^{(0)}\left|x^4\right|n^{(0)}\right\rangle. \tag{11.33}$$

The matrix element appearing in this equation can be calculated either by using the wavefunctions given in Eq. (7.69, p. 170), or by using the operator method of §7.7.2. Here, we outline the latter method. Using Eqs. (7.71) and (7.72), we see that

$$\widehat{x} = (a + a^\dagger)\ell_0, \tag{11.34}$$

with ℓ_0 defined in Eq. (7.64, p. 170). In the matrix element for x^4, only the terms with the same number of a and $a^\dagger$ will contribute, and we will obtain

$$E_n^{(1)} = \frac{3}{4}b\ell_0^4(2n^2 + 2n + 1). \tag{11.35}$$

☐ **Exercise 11.7** *Verify Eq. (11.35) and find the correction to the eigenstates.*

To consider another example, we discuss the energy shift of the electron in the ground state of a hydrogen atom in the presence of an external electric field. The eigenstates and eigenenergies of the electron are denoted by $|n, l, m\rangle$ and $E_n^{(0)}$ respectively. We will calculate the energy shift of the electron in the ground state, with $n = 1$, due to an electric field $\mathcal{E}$. We take the electric field to be along the z-direction leading to $H_1 = e\mathcal{E}z$, where $-e$ is the charge of the electron. Then up to second order in perturbation theory, the correction to the energy of the electron due to the electric field is given by

$$\Delta E = \Delta E^{(1)} + \Delta E^{(2)}, \tag{11.36a}$$

where

$$\Delta E^{(1)} = e\mathcal{E}\langle 1, 0, 0|z|1, 0, 0\rangle, \tag{11.36b}$$

and

$$\Delta E^{(2)} = -\frac{1}{2}\alpha_0\mathcal{E}^2, \tag{11.36c}$$

with

$$\alpha_0 = 2e^2 \sum_{n,l,m \neq 1,0,0} \frac{|\langle n, l, m|z|1, 0, 0\rangle|^2}{E_n^{(0)} - E_0^{(0)}}. \tag{11.36d}$$

To evaluate these corrections, we first note that the unperturbed wavefunctions $|n, l, m\rangle$ can be written in position space as $\psi_{nlm}(r, \theta, \phi) = R_{nl}(r)Y_{lm}(\theta, \phi)$. Since $z = r\cos\theta = rY_{10}(\theta, \phi)$, the presence of $Y_{lm}(\theta, \phi)$ ensures that $\langle n, l, m|z|n, l', m'\rangle \sim \delta_{l l' \pm 1}\delta_{m m'}$. This leads to $\Delta E^{(1)} = 0$. Thus, the leading order corrections to the energy levels are quadratic in electric field and the coefficient of $\mathcal{E}^2$ in the expression for ΔE is called the *polarizability* of the atom. The exact evaluation of α_0 is not straightforward since Eq. (11.36) involves sum over both bound states with negative energies and continuum states with positive energies. However, it is possible to have an upper bound for α_0 in a relatively straightforward manner. To see this we first note that each term in the sum for expression of α_0 in Eq. (11.36d) is positive. Moreover, the denominators of all the terms in the sum satisfy the condition

$$E_n^{(0)} - E_1^{(0)} \geqslant E_2^{(0)} - E_1^{(0)} = \frac{3}{8} \frac{\hbar^2}{ma_B^2}, \qquad (11.37)$$

where m is the reduced mass, and a_B is the Bohr radius defined in Eq. (9.76, p. 234). Thus, one can obtain an upper bound for α_0 as

$$\alpha_0 < \frac{16me^2a_B^2}{3\hbar^2} \sum_{n,l,m} |\langle n, l, m|z|1, 0, 0\rangle|^2$$

$$= \frac{16me^2a_B^2}{3\hbar^2} \langle 1, 0, 0|z^2|1, 0, 0\rangle = \frac{16a_B^4}{3} \frac{me^2}{\hbar^2} = \frac{16K}{3} a_B^3, \qquad (11.38)$$

where in the last step we have used the standard result $\langle 1, 0, 0|z^2|1, 0, 0\rangle = a_B^2$ and $K = e^2/\hbar c\alpha$ is a constant which depends on the units used; it is 1 in Gaussian unit, $1/(4\pi\epsilon_0)$ in SI unit, and $1/(4\pi)$ in Heaviside–Lorentz unit as can be inferred from Eq. (9.64, p. 232). This upper bound estimate is close to the actual value of $\alpha_0 = 9Ka_B^3/2$. To obtain this exact value one needs to evaluate the sum exactly which is possible but extremely involved; we are not going to attempt this here. The energy shift in the presence of an external electric field is called the *Stark effect*. In particular, the example elaborated above is classified as *quadratic Stark effect*, since the energy shift is proportional to the square of the electric field. In contrast, one can also have a linear Stark effect, which will be discussed in §11.3.

In passing, we want to point out that the ground state of the hydrogen atom is not really non-degenerate. It has a 2-fold degeneracy due to spin. Yet we have used the formalism of the non-degenerate perturbation theory to calculate the energy shift. The reason is that the perturbation due to electric field produces a Hamiltonian that is spin-independent. So, all matrix elements of H_1 between states with different eigenvalues of S_z vanish. Spin-up and spin-down states can therefore be treated independently, treating each of them as non-degenerate for the purpose of this calculation.

☐ **Exercise 11.8** *In writing Eq. (11.38), we used the result*

$$\sum_{n,l,,m} \left| \langle n, l, m|z|1, 0, 0\rangle \right|^2 = \langle 1, 0, 0|z^2|1, 0, 0\rangle. \qquad (11.39)$$

Why is this relation true? Evaluate this expectation value using the wavefunction given in Chapter 9 to confirm that it is equal to a_B^2.

☐ **Exercise 11.9** *Consider the anharmonic oscillator Hamiltonian introduced in Eqs. (11.31) and (11.32).*

(a) *Treating the anharmonic term as perturbation, find the second-order energy correction to the nth oscillator state.*

(b) *Find the first-order correction to the wavefunction of the nth eigenstate.*

☐ **Exercise 11.10** *Consider a free particle inside an infinite potential well which extends from $x = -a$ to $x = a$. Next, apply a potential $H_1 = V_0 e^{-x^2/a^2}$.*

(a) *Find the first- and second-order corrections to the ground state energy of the particle.*

(b) *Find the first-order correction to the ground state wavefunction.*

☐ **Exercise 11.11** *Consider an infinite 1D potential well in the region $0 \leq x \leq a$. An additional potential of amplitude of $V_0 > 0$ is applied such that*

$$V(x) = \begin{cases} V_0 & \text{for } a/4 \leq x \leq 3a/4, \\ 0 & \text{otherwise.} \end{cases} \tag{11.40}$$

Compute the first- and second-order energy corrections due to this potential for the nth excited state. What is the first-order correction to the ground state wavefunction?

☐ **Exercise 11.12** *Consider an electron in the hydrogen atom in the $1s$ state with energy E_{1s}. The atom is perturbed by a 3D oscillator potential leading to $H_1 = K_0 r^2/2$. Find the first-order correction to the energy of the electrons.*

☐ **Exercise 11.13** *Consider a two-level Hamiltonian*

$$H = E_0 \sum_{a=0,1} (-1)^{a-1} |a\rangle\langle a| + E_1 \left(|1\rangle\langle 0| + |0\rangle\langle 1| \right), \tag{11.41}$$

*where $|1\rangle$ and $|0\rangle$ are the wavefunctions of the unperturbed levels with energy E_0 and $-E_0$ respectively, and we have $E_1 \ll E_0$. Compute the second-order energy corrections to the energies of both $|0\rangle$ and $|1\rangle$ and show that the energy gap widens irrespective of the sign of E_1. [**Note:** This is an example of level repulsion which is almost omnipresent in all quantum systems.]*

11.3 Degenerate perturbation theory

11.3.1 The method

The analysis of the previous section clearly breaks down in the presence of degeneracy. Thus, a new analysis is called for when the quantum state, whose correction is being sought, is degenerate in the absence of perturbation. To this end, we consider a quantum Hamiltonian H_0 which has an eigenvalue $E^{(0)}$

that is g_0-fold degenerate, i.e., there are g_0 linearly independent states which share the same eigenvalue. Let us denote these g_0 states by kets $|\alpha_i\rangle$, which means that we can write

$$H_0 |\alpha_i\rangle = E^{(0)} |\alpha_i\rangle \qquad (11.42)$$

for $i = 1, 2, \cdots, g_0$. These kets have been taken to be normalized, and orthogonal to one another. We note that any linear combination of $|\alpha_i\rangle$,

$$|\beta_i^{(0)}\rangle = \sum_j c_{ij} |\alpha_j\rangle, \qquad (11.43)$$

also satisfies

$$H_0 \left|\beta_i^{(0)}\right\rangle = E^{(0)} \left|\beta_i^{(0)}\right\rangle. \qquad (11.44)$$

The other eigenstates of H_0 outside the degenerate manifold are denoted by $|\gamma_a\rangle$ and obey $H_0 |\gamma_a\rangle = E^{(a)} |\gamma_a\rangle$. These eigenstates are orthogonal to all states in the degenerate manifold, i.e., $\langle \gamma_a | \alpha_i \rangle = 0$ for all i and a.

In the presence of H_1, the eigenvalue equation of the system can be written as

$$(H_0 + \lambda H_1)|\psi\rangle = \left(E^{(0)} + \lambda E^{(1)} + \cdots \right)|\psi\rangle, \qquad (11.45)$$

with

$$|\psi\rangle = |\beta_i^{(0)}\rangle + \lambda |\beta_i^{(1)}\rangle + \ldots \qquad (11.46)$$

where ellipses denote higher-order corrections. The parameter λ is used for convenience of keeping track of the order of perturbation as in the previous section, and shall be set to unity at the end of the calculation. We note that Eq. (11.46) suggests that when $\lambda \to 0$, i.e., in the absence of H_1, $|\psi\rangle \to |\beta^{(0)}\rangle$ which can in general be an arbitrary linear combination of $|\alpha_i\rangle$. Equating now the linear λ-terms of Eq. (11.45), we obtain

$$H_0 \left|\beta_i^{(1)}\right\rangle + H_1 \left|\beta_i^{(0)}\right\rangle = E^{(0)} \left|\beta_i^{(1)}\right\rangle + E^{(1)} \left|\beta_i^{(0)}\right\rangle. \qquad (11.47)$$

Taking the inner product with the bra $\langle \alpha_j |$ and using Eq. (11.44), we obtain

$$E^{(1)} c_{ij} = \sum_{j'} \langle \alpha_j | H_1 | \alpha_{j'} \rangle c_{ij'} = \sum_{j'} (H_1)_{jj'} c_{ij'}. \qquad (11.48)$$

Noting that Eq. (11.48) can be viewed as an eigenvalue equation for finding eigenvectors and eigenvalues of H_1 within the g_0-dimensional degenerate subspace, one finds that the first-order correction $E^{(1)}$ to the g_0 energy levels can be found as the solution to the equation

$$\det(H_1 - E^{(1)}\mathbb{1}) = 0. \tag{11.49}$$

In what follows, we shall assume that these levels are non-degenerate, *i.e.*, H_1 lifts the ground state degeneracy completely.

To obtain the first-order correction to $|\psi\rangle$ which has been denoted by $|\beta^{(1)}\rangle$ in Eq. (11.46), we now take the inner product of Eq. (11.45) with $\langle\gamma_a|$. This yields

$$\left(E^{(a)} - E^{(0)}\right)\left\langle\gamma_a\middle|\beta^{(1)}\right\rangle = -\langle\gamma_a|H_1|\beta^{(0)}\rangle \tag{11.50}$$

after setting $\lambda = 1$. This allows us to express the states $|\beta^{(1)}\rangle$ as

$$|\beta^{(1)}\rangle = \sum_a \frac{\langle\gamma_a|H_1|\beta^{(0)}\rangle}{E^{(0)} - E^{(a)}}\,|\gamma_a\rangle, \tag{11.51}$$

where the sum over a is only over states which are outside the degenerate ground state manifold. One cannot easily carry out this analysis to second order in perturbation theory following the same method outlined in §11.2. We only state here the expression for the second-order energy correction which is given by

$$E_0^{(2)} = \sum_a \frac{|\langle\gamma_a|H_1|\beta^{(0)}\rangle|^2}{E^{(0)} - E^{(a)}}. \tag{11.52}$$

Note that once again the sum includes only the states outside the degenerate subspace.

□ **Exercise 11.14** *Verify Eq. (11.52).*

11.3.2 Examples

Before closing this section, we discuss a representative example which would serve to illustrate the theory developed above. We choose the example of linear Stark effect in a hydrogen atom. The electric field-dependent part of the Hamiltonian is given by $H_1 = e\mathcal{E}z$, as used earlier in §11.2. This time, however, we focus on the energy shift of states with $n = 2$. First recall that for $n = 2$ there are four degenerate energy levels, viz., one $2s \equiv |2, 0, 0\rangle$, and three $2p \equiv |2, 1, \{0, \pm 1\}\rangle$. Since the electric field does not couple states with different m, one finds that the only non-zero matrix elements of H_1 are given by

$$\langle 2, 0, 0 | H_1 | 2, 1, 0\rangle = \langle 2, 1, 0 | H_1 | 2, 0, 0\rangle = 3 a_B e \mathcal{E}. \qquad (11.53)$$

The states with $m = \pm 1$ are therefore unaffected by the electric field, at least to the first order in perturbation. The states with $m = 0$ which are eigenstates of $H_0 + H_1$ are the combinations $|2, 0, 0\rangle \pm |2, 1, 0\rangle$, with eigenvalues $E^{(0)} \pm 3 a_B e \mathcal{E}$. The change of the eigenvalues is linear in the external electric field, which is why the effect is called the *linear Stark effect*. Note that this indicates the presence of a finite electric dipole moment which is permissible since these states are not eigenstates of the parity operator.

□ **Exercise 11.15** *Verify Eq. (11.53), using hydrogen atom wavefunctions given in Chapter 9.*

□ **Exercise 11.16** *Consider a hydrogen-like atom with spin–orbit coupling term given by $H_1 = K(\boldsymbol{L} \cdot \boldsymbol{S})$ where K is a constant and $\boldsymbol{L}$ and $\boldsymbol{S}$ are orbital and spin angular momentum operators. Assume the unperturbed wavefunctions to be $\psi_{nlm}(\boldsymbol{r}) = R_{nl}(r) Y_{lm}(\theta, \phi)$ in the coordinate representation.*

(a) *Show that the first-order shift to the energy levels is given by $\Delta E = K \hbar^2 \ell / 2$ if $j = \ell + 1/2$ and $\Delta E = -K \hbar^2 (\ell + 1)/2$ if $j = \ell - 1/2$ where j is the quantum number corresponding to the total angular momentum $\boldsymbol{J} = \boldsymbol{L} + \boldsymbol{S}$.*

(b) *Now consider a second perturbative Hamiltonian originates due to an applied magnetic field and is given by $H_2 = -K'(L_z + 2S_z)$. The total Hamiltonian of the system is $H_0 + H_1 + H_2$, where H_0 is the hydrogen atom Hamiltonian. Show that the first-order correction to an energy level E_n due to H_2 is given by $\Delta E = -K' m \hbar [1 \pm 1/(2\ell + 1)]$ for $K' \ll K \hbar$. Historically, such a splitting is termed as Zeeman effect.*

(c) *Show that in the other limit, when $K' \gg K \hbar$, the first-order energy shift due to both H_1 and H_2 is given by $\Delta E = -\hbar K'(m_l + 2 m_s) + K \hbar m_l m_s$ where m_l and m_s correspond to quantum numbers of L_z and S_z respectively. This phenomenon is called the Paschen–Back effect.*

□ **Exercise 11.17** *Consider three degenerate quantum states $|\alpha = 1, 2, 3\rangle$ with energy E_0. An interacting Hamiltonian which connects these states is given by*

$$H_1 = V_0 \Big(|1\rangle \langle 2| + |2\rangle \langle 3| \Big) + h.c., \qquad (11.54)$$

where h.c. denotes the Hermitian conjugate. Show that the new eigenvalues are $E_\pm = E_0 \pm \sqrt{2} V_0$ and E_0. Argue that the fact that one of the eigenenergies remains unchanged can be understood by looking into the matrix representation of H_1 in the degenerate subspace.

☐ **Exercise 11.18** *Consider a 2D harmonic oscillator with the Hamiltonian*

$$H_0 = \frac{1}{2m}(\hat{p}_x^2 + \hat{p}_y^2) + \frac{m\omega_c^2}{2}(x^2 + y^2). \tag{11.55}$$

The first excited state of such an oscillator forms a 2D degenerate subspace with each state having energy $E_1 = 2\hbar\omega_c$. Now consider a perturbation $H_1 = K_0 xy$ with $K_0 \ll m\omega_c^2$. Find the new energy values for the first excited states of the oscillator. What are the new eigenfunctions for these states?

11.4 WKB approximation

The WKB approximation is named after its inventors Gregor Wentzel, Hendrik Kramers, and Léon Brillouin and is a useful approximation method for problems which are either one-dimensional or which can be reduced to effective one-dimensional problems on account of their symmetries (such as 3D problems with radial potentials). The approximation mentioned here is at a slightly different footing in the sense that it is not necessarily perturbative in some small parameter. In the perturbation theory detailed in earlier sections, there was always a small parameter λ which served as an expansion parameter for the potentials. It was assumed that the quantities we are interested in calculating admit Taylor series expansion in this small parameter λ. This will not work in two kinds of cases. First, there may not be an obvious small parameter in the problem. Second, the quantity of our interest might vanish for $\lambda = 0$ but not be differentiable at $\lambda = 0$. An example is the function $f(\lambda) = \exp(-1/\lambda^2)$. The perturbation theory will not work for such cases. In such cases, the WKB approximation may become useful.

11.4.1 The method

To understand this method, let us first consider a particle moving in a constant potential V with energy $E > V$. The Schrödinger equation for such a particle is given by

$$-\frac{\hbar^2}{2m}\frac{d^2\psi(x)}{dx^2} + V\psi(x) = E\psi(x). \tag{11.56}$$

The solution of Eq. (11.56) is straightforward and yields

$$\psi(x) = Ne^{\pm ikx}, \qquad k = \sqrt{2m(E-V)}/\hbar. \tag{11.57}$$

Now imagine that the potential $V(x)$ is not a constant but is an arbitrary function of x. In general, in such cases it is impossible to obtain an exact solution for $\psi(x)$. However, if $V(x)$ varies slowly in space compared to the characteristic length scale $\lambda = 2\pi/k$ of the problem, then the potential remains practically constant over a distance λ. In such cases, the WKB approximation seeks a solution to the problem by postulating a trial wavefunction which is still oscillatory but whose amplitude and phase vary slowly with x. Such a wavefunction is typically written as

$$\psi(x) = N(x)e^{\pm i\alpha(x)}, \tag{11.58}$$

where $N(x)$ and $\alpha(x)$ are unknown functions of x. We note that the assumption which prompted such a choice is that $N(x)$ and $\alpha(x)$ vary slowly compared to λ; thus, they necessarily break down around points x_t where $E \simeq V(x_t)$ and $\lambda \to \infty$. These points are called *turning points* and we are going to discuss them separately later in the section.

To determine $N(x)$ and $\alpha(x)$ away from the turning points, we now substitute Eq. (11.58) into Eq. (11.56), using $V(x)$ instead of a constant V. After some straightforward algebra, one finds

$$\frac{d^2 N(x)}{dx^2} - N(x)\left(\frac{d\alpha(x)}{dx}\right)^2 + k^2(x)N(x)$$
$$= i\left(2\frac{dN(x)}{dx}\frac{d\alpha(x)}{dx} + N(x)\frac{d^2\alpha(x)}{dx^2}\right), \tag{11.59}$$

where

$$k(x) = \sqrt{2m(E - V(x))}/\hbar. \tag{11.60}$$

Since the left and right sides of Eq. (11.59) represent real and imaginary quantities respectively, they must vanish separately for the equation to hold. Vanishing of the imaginary part leads to the equation

$$2\frac{dN(x)}{dx}\frac{d\alpha(x)}{dx} + N(x)\frac{d^2\alpha(x)}{dx^2} = 0, \tag{11.61}$$

which means

$$\frac{d}{dx}\left(N^2(x)\frac{d\alpha(x)}{dx}\right) = 0, \tag{11.62}$$

or

$$N(x) = \frac{\mathcal{A}}{\sqrt{d\alpha(x)/dx}}, \tag{11.63}$$

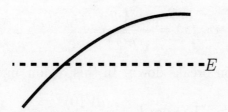

Figure 11.1 Schematic nature of the potential $V(x)$ analyzed in the text. The point where $V(x) = E$ is taken as the point with $x = 0$.

where $\mathcal{A}$ is a constant. On the other hand, the vanishing of the real part of Eq. (11.59) implies

$$\frac{d^2 N(x)}{dx^2} = N(x) \left[\left(\frac{d\alpha(x)}{dx} \right)^2 - k^2(x) \right]. \tag{11.64}$$

This differential equation cannot be solved exactly for general $V(x)$. However, when $N(x)$ is a slowly varying function of position and $d^2 N(x)/dx^2 \ll k^2(x)$, one can have an approximate solution $\alpha(x) \simeq \int_x dx' \, k(x')$. This leads to

$$\psi(x) = \frac{\mathcal{A}}{\sqrt{k(x)}} e^{\pm i\alpha(x)}, \qquad \alpha(x) = \int_x dx' \, k(x'). \tag{11.65}$$

Note that the form of the wavefunction is clearly untenable near the turning points where $k(x)$ vanishes, and therefore $\psi(x)$ blows up.

A similar analysis can be carried out for $E < V(x)$. All the steps leading to Eq. (11.65) go through with the change that $k(x) = i\kappa(x)$ is an imaginary quantity. This leads to, for $E < V(x)$, the WKB wavefunction

$$\psi(x) = \frac{\mathcal{A}}{\sqrt{\kappa(x)}} e^{\pm \int_x dx' \, \kappa(x')}. \tag{11.66}$$

Note that the solution with the positive exponent is only admissible if $V(x)$ is finite within a finite region of space.

Now we turn to the question of turning points. In our analysis, for simplicity, we shall use a single turning point at $x = 0$. Thus, for a potential

$V(x)$, we have $E = V(0)$. We shall further assume that for $x > 0$, $E < V(x)$ and for $x < 0$, $V(x) < E$, as seen in Figure 11.1. First, let us analyze the region $x > 0$. For $x \gg 0$, the WKB solution holds and one can write

$$\psi_{\text{far}}(x) = \frac{\mathcal{A}}{\sqrt{\kappa(x)}} e^{-\int^x dx' \kappa(x')}. \tag{11.67}$$

Near $x = 0$, this solution breaks down. In this region, one can write

$$V(x) \simeq E + x\frac{dV(0)}{dx}. \tag{11.68}$$

Thus, the Schrödinger equation in this regime can be written as

$$\frac{d^2\psi(x)}{dx^2} - k_0^3 x \psi(x) = 0, \tag{11.69a}$$

with

$$k_0 = \left(\frac{2m}{\hbar^2}\frac{dV(0)}{dx}\right)^{1/3}. \tag{11.69b}$$

This is the Airy differential equation that has been discussed in some detail in §C.4 of Appendix C. With the notation introduced there, the general solution can be written as

$$\psi_{\text{near}}(x) = c_1 \text{Ai}(k_0 x) + c_2 \text{Bi}(k_0 x), \tag{11.70}$$

where $\text{Ai}(z)$ and $\text{Bi}(z)$ are Airy functions of the first and the second kind.

Next, we patch the two wavefunctions in the overlap region (for $x > 0$) where both ψ_{far} and ψ_{near} hold. We note at the outset that the existence of such an overlap region is an assumption which is inherent in the analysis. In this region, the linear approximation to the potential has to hold; as a result, this region cannot be too far away from the turning point. Moreover, we also need the WKB wavefunction ψ_{far} to be accurate here; so one cannot be too close to the turning point. The existence of this region therefore clearly depends on the nature of the potential. Assuming that the region exists, we note the following. First, in this region $\alpha(x)$ can be approximated, using Eq. (11.68), as

$$\alpha(x) = \int^x dx' \sqrt{2m(E - V(x'))}/\hbar$$

$$\simeq \int^x dx' \, k_0^{3/2}\sqrt{x'} = \frac{2}{3}(k_0 x)^{3/2}, \tag{11.71}$$

so that the WKB wavefunction in the patching region can be written as

$$\psi_{\text{far}}(x) \simeq \frac{\mathcal{A}}{k_0^{3/4} x^{1/4}} e^{-\frac{2}{3}(k_0 x)^{3/2}}. \tag{11.72}$$

Second, we note that for $\psi_{\text{near}}(x)$, since the patching region is sufficiently far from the turning point at $x = 0$, one can use the large-x asymptotic forms of the Airy functions. Since $\text{Bi}(x) \sim \exp[x^{3/2}]$, it cannot be admitted; this leads to the condition $c_2 = 0$ in Eq. (11.70). Noting that $\text{Ai}(x) \sim x^{-1/4} \exp[-x^{3/2}]$ as shown in Eq. (E.32, p. 493), we find that $c_1 = \mathcal{A}/\sqrt{k_0}$. This leads to the WKB wavefunction for $x > 0$ to be

$$\psi_1(x) = \frac{\mathcal{A}}{\sqrt{|\kappa(x)|}} \exp\left(-\int^x dx' \, \kappa(x')\right). \tag{11.73}$$

Next we shall analyze the patching condition for the region $x < 0$ where $E > V(x)$. Here, in the patching region, one has

$$\alpha(x) = \int^{(-x)} dx' \, k(x') \simeq \frac{2}{3}(-k_0 x)^{3/2}, \tag{11.74}$$

which leads to the WKB wavefunction in the patching region for $x < 0$ to be

$$\psi'_{\text{far}}(x) \simeq \frac{1}{k_0^{3/4}(-x)^{1/4}} \left[\mathcal{A}_1 e^{\frac{2i}{3}(-k_0 x)^{3/2}} + \mathcal{A}_2 e^{\frac{2i}{3}(-k_0 x)^{3/2}} \right]. \tag{11.75}$$

The wavefunction near the turning point $x = 0$ in terms of the Airy functions reads $\psi_{\text{near}} = \mathcal{A} \, \text{Ai}[k_0 x]/k_0$. For large negative x, one has $\text{Ai}(x) \to \sin[2(-x)^{3/2}/3 + \pi/4]/(-x)^{1/4}$. Using this form and matching $\psi_{\text{near}}(x)$ with $\psi'_{\text{far}}(x)$ in the patching region for $x < 0$, one gets

$$\mathcal{A}_1 = \frac{k_0 \mathcal{A}}{2i\sqrt{\pi}} e^{i\pi/4}, \quad \mathcal{A}_2 = \frac{-k_0 \mathcal{A}}{2i\sqrt{\pi}} e^{-i\pi/4}. \tag{11.76}$$

Substituting Eq. (11.76) into Eq. (11.75), one gets the WKB wavefunction in the classical region as

$$\psi_2(x) = \frac{2\mathcal{A}}{\sqrt{k(x)}} \sin\left[\int_x^0 dx' \, k(x') + \frac{\pi}{4}\right]. \tag{11.77}$$

Thus, $\psi_1(x)$ and $\psi_2(x)$ yield the WKB wavefunction for the entire region (except near the turning point where it is given in terms of Airy function). The analysis can be easily extended to cases where the turning point occurs at $x \neq 0$ by a simple coordinate transformation. This joining of WKB wavefunctions across the turning points is sometimes referred to as the connection procedure.

11.4.2   Examples

We now present some representative examples to demonstrate the usefulness of the method. First, we consider a quantum system $H = p^2/2m + V(x)$ where the potential $V(x)$ is given by

$$V(x) = \begin{cases} \frac{1}{2}m\omega^2 x^2 & \text{for } x > 0, \\ \infty & \text{for } x \le 0. \end{cases} \tag{11.78}$$

This corresponds to a harmonic oscillator potential for $x > 0$ together with a hard wall at $x = 0$. Consequently, the wavefunction must satisfy $\psi(x = 0) = 0$. Note that there is no natural Taylor expansion of this potential in any small parameter.

To treat this problem, we note that for $x > 0$, one can write

$$k(x) = \sqrt{\frac{2m}{\hbar^2}\left(E - \frac{1}{2}m\omega^2 x^2\right)} = \frac{m\omega}{\hbar}\sqrt{x_t^2 - x^2}, \tag{11.79}$$

where

$$x_t = \sqrt{\frac{2E}{m\omega^2}} \tag{11.80}$$

is the position of the turning point. The WKB wavefunction for the problem is obtained by putting this expression for $k(x)$ into Eq. (11.77), and changing the upper limit of integration to x_t because that is the turning point in this problem, unlike in Eq. (11.77) where it was $x = 0$. Thus, the condition $\psi(x = 0) = 0$ requires that

$$\frac{m\omega}{\hbar}\int_0^{x_t} dx'\sqrt{x_t^2 - x'^2} = \left(n - \frac{1}{4}\right)\pi. \tag{11.81}$$

The integral on the left side can be easily carried out by substituting $x = x_t\sin\theta$, which gives

$$\frac{m\omega}{\hbar}\frac{\pi}{4}x_t^2 = \left(n - \frac{1}{4}\right)\pi. \tag{11.82}$$

Substituting the definition of x_t given in Eq. (11.80), we finally get the allowed energy levels to be

$$E_n = \left(n - \frac{1}{4}\right)2\hbar\omega, \tag{11.83}$$

which happens to coincide with the exact result for the half harmonic oscillator problem.

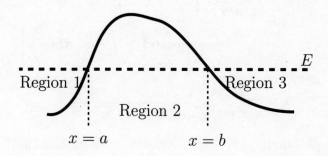

Figure 11.2 A potential barrier of unspecified shape. The points where $V(x)$ becomes equal to the energy of the incident particle have been marked.

☐ **Exercise 11.19** *Solve the half harmonic oscillator exactly, and verify that the energy eigenvalues are indeed given by Eq. (11.83).*

In the beginning of this section, we said that there are two kinds of situations in which the standard perturbation theory cannot be applied. The previous example is of one kind, where there is no obvious small parameter to guide the perturbation expansion. To identify the second kind of situation, we recall the problem of transmission through a potential barrier of height V_0 that was discussed with in §7.5. The expression given in Eq. (7.39, p. 165) goes to zero, as it should, if $V_0 \to \infty$, i.e., the small parameter $\xi = E_\star/V_0$ goes to zero. However, an attempt to find the transmission probability using the perturbation theory would surely fail since the expression for the probability contains the factor $e^{-2/\sqrt{\xi}}$, which clearly is not a differentiable function of ξ at $\xi = 0$.

Inspired by this example where an exact solution is available, we now outline how the WKB formalism can be applied to the problem of tunneling through an arbitrary potential barrier $V(x)$. For any specified energy E of the incident wave, there will now be three regions of interest, as shown schematically in Figure 11.2. Classical motion is possible in regions 1 and 3, but not in region 2. According to the discussion of the WKB formalism, the wavefunction in region 2 has the form

$$\psi_2(x) = \frac{1}{\sqrt{\kappa(x)}} \left[c_1 \exp\left(-\int_a^x dx'\, \kappa(x') \right) \right.$$
$$\left. + c_2 \exp\left(+\int_a^x dx'\, \kappa(x') \right) \right], \qquad (11.84)$$

where

$$\kappa(x') = \sqrt{2m(V(x') - E)/\hbar^2}. \qquad (11.85)$$

In a thick enough barrier, the exponentially growing term will cause trouble, so henceforth we assume that $c_2 = 0$. Then the values of the wavefunction at the two turning points are given by

$$\psi_2(a) = c_1/\sqrt{\kappa(a)}, \qquad \psi_2(b) = c_1 e^{-\Gamma}/\sqrt{\kappa(b)}, \qquad (11.86)$$

where

$$\Gamma = \int_a^b dx'\, \kappa(x'). \qquad (11.87)$$

From this one finds that the tunneling probability, to the leading order, is given by

$$T = \left| \frac{\psi_2(b)}{\psi_2(a)} \right|^2 \sim e^{-2\Gamma}. \qquad (11.88)$$

This analysis provides the leading order exponential behavior of the tunneling probability. However, a more careful analysis is required for thin barriers as well as for cases where it is important to know the prefactor of T. We shall not discuss these issues here.

To provide an explicit example where the integral shown in Eq. (11.87) can be performed, consider the following radial potential, shown in Figure 11.3:

$$V(r) = \begin{cases} -V_0 & \text{for } 0 < r < a, \\ +C/r & \text{for } r > a, \end{cases} \qquad (11.89)$$

where V_0 and C are both positive. This represents the Coulomb potential of a positively charged particle which is bound by an attractive potential at the core with radius a, and can be used to analyze alpha radioactive decay, for example. The inner turning point will definitely be at $r = a$. If the outer turning point is called $r = b$, it means

$$C/b = E, \qquad (11.90)$$

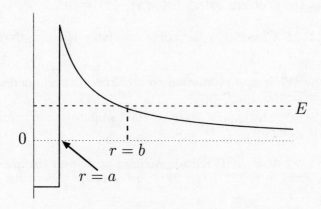

Figure 11.3 An inner potential well and an outer repulsive Coulomb potential, as denoted in Eq. (11.89). The energy of the particle crossing this potential has been denoted by E.

so that Eq. (11.87) can be written as

$$\Gamma = \frac{\sqrt{2mE}}{\hbar} \int_a^b dr \left(\frac{b}{r} - 1\right). \tag{11.91}$$

The integral can be easily evaluated by making the substitution $r = b\sin^2 \xi$, and the result is

$$\Gamma = \frac{\sqrt{2mE}}{\hbar} \left[b(\frac{\pi}{2} - \sin^{-1}\sqrt{\frac{a}{b}}) - \sqrt{a(b-a)} \right]. \tag{11.92}$$

If $a \ll b$, this expression reduces to

$$\Gamma = \sqrt{2mE} \, \frac{\pi b}{2\hbar}. \tag{11.93}$$

Once we replace b through Eq. (11.90), we see that $\Gamma \propto E^{-1/2}$, just as we had obtained in the exercise of §7.5. This dependence shows why this problem cannot be solved by the perturbation technique.

□ **Exercise 11.20** *Consider a particle outside a potential barrier between $0 \leq x \leq a$ whose height varies slowly with x with a mean value V_0. Consider a particle approaching the barrier from $x < 0$ with an energy $E \ll V_0$.*

  (a) *For $V(x) = V_0 + \delta V \sin(x/a)$ where $\delta V \ll V_0$, find an expression for the probability of the particle tunneling into the barrier.*

(b)   Repeat the problem with $V(x) = V_0 - \delta V x^2 / a^2$.

☐ **Exercise 11.21** *Consider a harmonic oscillator Hamiltonian $H = \hat{p}^2 / 2m + m\omega_c^2 x^2 / 2$.*

(a)   *Use the WKB approximation to find the allowed energies and compare them with the exact result.*

(b)   *For the nth stationary state of the oscillator, find the position of the turning points.*

☐ **Exercise 11.22** *Use the WKB approximation to find the allowed energies for the problem given in Ex. 11.10 (p. 291) and compare them with the exact result.*

Chapter 12

Time-dependent problems

In this chapter, we shall treat quantum mechanical problems where the system Hamiltonian either has explicit time dependence or has some parameter that is time-dependent. In either case, the Schrödinger equation is time-dependent and we are going to discuss techniques, either perturbative or exact, to solve such equations. In §12.1, we shall discuss the general formalism for treating such problems. Next, in §§ 12.2 and 12.3, we provide an analysis of the time-dependent Hamiltonian for a two-level system. This will be followed by §12.4, where perturbative solutions to time-dependent Schrödinger equation shall be outlined. Finally, we analyze some aspects of Hamiltonians with periodic time dependence in §12.5.

12.1   General formulation

Let us consider a time-dependent Hamiltonian $H(t)$ whose one or more parameters vary with time. The corresponding Schrödinger equation can be written as

$$i\hbar\frac{d}{dt}\left|\psi(t)\right\rangle = H(t)\left|\psi(t)\right\rangle. \tag{12.1}$$

The formal solution of such an equation can always be written in the form

$$\left|\psi(t)\right\rangle = U(t,t_0)\left|\psi(t_0)\right\rangle \tag{12.2}$$

for a suitably defined evolution operator U, which should be unitary if we take normalized states at all times. However, we cannot now say that $U(t,t_0) = \exp\left(-iH(t-t_0)/\hbar\right)$ as we did for time-independent Hamiltonians. The reason is that $H(t)$ need not commute with $H(t')$ if $t \neq t'$ for a time-dependent Hamiltonian. If we divide the time interval between t_0 and t into

n parts such that each part is an infinitesimal, we can write, ignoring higher powers of δt,

$$|\psi(t_k + \delta t)\rangle = \exp\left(-\frac{i}{\hbar}H(t_k)\delta t\right)|\psi(t_k)\rangle. \qquad (12.3)$$

In writing this, we have assumed that H does not change appreciably within the time interval t_k and $t_k + \delta t$; this serves, in practice, as a benchmark for how small δt needs to be. Stringing together relations like this for every part, we will obtain an equation of the form of Eq. (12.2), with

$$U(t, t_0) = U(t, t - \delta t)U(t - \delta t, t - 2\delta t)\cdots U(t_0 + \delta t, t_0), \qquad (12.4)$$

where each factor of U has the form given in Eq. (12.3). In the limit $\delta t \to 0$, it is written as

$$U(t, t_0) = T_t\left[\exp\left\{-\frac{i}{\hbar}\int_{t_0}^{t} dt'\, H(t')\right\}\right], \qquad (12.5)$$

where T_t is called the *time-ordering operator* which keeps the exponentials such as those occurring in Eq. (12.3) in proper temporal order, noting that they cannot be easily combined because of the non-commutativity.

We are going to discuss the properties of this solution in more detail in §12.4, in the context of developing a perturbative treatment for the time-dependent problem. Here we note that such a formal solution is not very easy to handle since it involves exponentiation of time-dependent non-commuting operators. So we shall proceed in a slightly different line for solving Eq. (12.1).

To this end, let us consider an arbitrary instant t_1 at which $H(t) \equiv H(t_1)$. Let us also denote the instantaneous eigenvalues and eigenstates of $H(t_1)$ as $E_n(t_1)$ and $|n(t_1)\rangle$ respectively:

$$H(t_1)|n(t_1)\rangle = E_n(t_1)|n(t_1)\rangle. \qquad (12.6)$$

At this stage, the choice of t_1 is arbitrary; it depends on the problem at hand. A smart choice of t_1, for example, could be dictated by the condition that $H(t_1)$ have a particularly simple form which enables us to obtain $E_n(t_1)$ and $|n(t_1)\rangle$ in a relatively simple manner.

Having made the choice of t_1, one can write

$$H(t) = H(t_1) + \delta H(t). \qquad (12.7)$$

Furthermore, one can expand the wavefunction in the instantaneous eigenbasis of $H(t_1)$:

$$|\psi(t)\rangle = \sum_n c_n(t)|n(t_1)\rangle. \qquad (12.8)$$

Substituting this into Eq. (12.1), we get

$$i\hbar\frac{d}{dt}\sum_n c_n|n(t_1)\rangle = \sum_n c_n(t)\Big[E_n(t_1) + \delta H(t)\Big]|n(t_1)\rangle. \qquad (12.9)$$

To make further progress, we take the inner product of Eq. (12.9) with $\langle m(t_1)|$. This leads to a set of coupled linear differential equations for $c_n(t)$ given by

$$i\hbar\frac{d}{dt}c_m - E_m(t_1)c_m(t) = \sum_n \Lambda_{mn}(t;t_1)c_n(t), \qquad (12.10)$$

where

$$\Lambda_{mn}(t;t_1) = \langle m(t_1)|\delta H(t)|n(t_1)\rangle. \qquad (12.11)$$

Note that the simplification we have achieved here is that we have reduced the operator equation of Eq. (12.1) to a coupled set of ordinary linear differential equations. The latter is almost always numerically easier to handle. Usually, for a generic $H(t)$, one needs to solve the set of equations implied in Eq. (12.10) numerically. However, there are a few cases where such equations allow for analytical solution. We are going to discuss one such case in the next section.

12.2 Time-dependent two-level system

The problem that we now discuss is that of the two-level system whose Hamiltonian has diagonal time-dependent terms and time-independent off-diagonal terms. For the case where the difference of the diagonal terms vary linearly with time, the time-dependent Schrödinger equation can be exactly solved. This problem was first solved independently by Landau, Zener, Stückelberg, and Majorana, so we can call it the LZSM solution.

To see how such a two-level problem may arise in quantum mechanics, let us consider a spin-$\frac{1}{2}$ particle in the presence of two crossed magnetic fields along x and z. Out of these, the field along z is chosen to depend linearly on time while that along x is taken to be a constant. The Hamiltonian of such a system is given by

$$H(t) = \mu B_0\Big((t/\tau)\sigma_z + \Delta\sigma_x\Big), \qquad (12.12)$$

where $-\mu$ is the magnetic moment of the particle and B_0 and Δ are constants. We note that one may always add a time-dependent term proportional to the identity matrix to this Hamiltonian; the solution will only change by a trivial phase factor. Such a Hamiltonian can most generally be written as

$$H'(t) = E_0(t)\mathbb{1} + H(t). \tag{12.13}$$

It is easy to check that any state $|\psi(t)\rangle$ that satisfies the Schrödinger equation for $H(t)$ is related to the solution $|\psi_2(t)\rangle$ of $H'(t)$ by

$$|\psi_2(t)\rangle = \exp\left[-\frac{i}{\hbar}\int^t dt'\, E_0(t')\right]|\psi_1(t)\rangle. \tag{12.14}$$

For the rest of this section, we shall therefore work with $H(t)$ which is traceless.

The Schorödinger equation can be written as

$$i\frac{d}{dt}|\psi(t)\rangle = \frac{\mu B_0}{\hbar}\left[(t/\tau)\sigma_z + \Delta\sigma_x\right]|\psi(t)\rangle. \tag{12.15}$$

We now use a dimensionless time variable defined as

$$\tilde{t} = \sqrt{\frac{\mu B_0}{\hbar\tau}}\, t. \tag{12.16}$$

It is easy to see that the Schrödinger equation, in this new variable, takes the form

$$i\hbar\frac{d}{d\tilde{t}}|\psi(\tilde{t})\rangle = \left(\tilde{t}\sigma_z + \tilde{\Delta}\sigma_x\right)|\psi(\tilde{t})\rangle, \tag{12.17}$$

where

$$\tilde{\Delta} = \sqrt{\frac{\mu B_0\tau}{\hbar}}\,\Delta. \tag{12.18}$$

If we denote the components of the state vector as

$$|\psi(\tilde{t})\rangle = \begin{pmatrix} c_1(\tilde{t}) \\ c_2(\tilde{t}) \end{pmatrix}, \tag{12.19}$$

then the equations for the components are

$$\left(i\frac{d}{d\tilde{t}} - \tilde{t}\right)c_1(\tilde{t}) = \tilde{\Delta}c_2(\tilde{t}), \tag{12.20a}$$

$$\left(i\frac{d}{d\tilde{t}} + \tilde{t}\right)c_2(\tilde{t}) = \tilde{\Delta}c_1(\tilde{t}). \tag{12.20b}$$

The next step constitutes decoupling the two equations. This is straightforward and leads to

$$\frac{d^2 c_2(\tilde{t})}{d\tilde{t}^2} + \left(\tilde{\Delta}^2 + \tilde{t}^2 - i\right) c_2(\tilde{t}) = 0. \tag{12.21}$$

Upon a variable transformation

$$z = \sqrt{2} e^{-i\pi/4} \tilde{t} = (1 - i)\tilde{t}, \tag{12.22}$$

we obtain

$$\frac{d^2 c_2(z)}{dz^2} + (\nu - \frac{1}{4}z^2 + \frac{1}{2}) c_2(z) = 0, \tag{12.23}$$

with

$$\nu = i\tilde{\Delta}^2/2. \tag{12.24}$$

This equation can be easily recognized to be the Weber equation introduced in Eq. (C.113, p. 471) of Appendix C. The solution of this equation requires specification of the boundary conditions, which we choose to be

$$c_2(\tilde{t} \to -\infty) = 0, \qquad |c_1(\tilde{t} \to -\infty)| = 1. \tag{12.25}$$

This means that we choose the system to be in the ground state of the instantaneous Hamiltonian at $\tilde{t} \to -\infty$. The solution we present below holds for the case when the dynamics starts at $t \to -\infty$ and ends at $t \to \infty$; a change in these conditions will lead to a different solution which we do not discuss here.

To obtain the solution for c_2, we note that the Weber function, which is a solution to Eq. (12.23), satisfies

$$D_{-\nu-1}(z) = 0 \quad \text{for } z = Re^{3\pi i/4} \quad \text{and } z = Re^{i\pi/4}, \tag{12.26}$$

when $R \to \infty$. Since $\tilde{t}$ is related to z by Eq. (12.22), one can write

$$c_2(t) = a_0 D_{-\nu-1}(-iz) = a_0 D_{-\nu-1}(-i\sqrt{2}e^{-i\pi/4}\tilde{t}), \tag{12.27}$$

so that $c_2(t \to -\infty) = 0$. Next, to obtain a_0, we need to impose the boundary condition which involves $c_1(t)$. For this we note that for $t \to -\infty$, c_2 has the asymptotic form

$$c_2(t \to -\infty) \simeq e^{-i(\nu+1)\pi/4} e^{-\tilde{t}^2/2} (\sqrt{2}\tilde{t})^{-\nu-1}. \tag{12.28}$$

Substituting this into Eq. (12.20b), we find that for $\tilde{t} \to -\infty$

$$c_1(t) \to \frac{a_0}{\tilde{\Delta}} \exp[-i\pi(\nu+1)/4](\sqrt{2}\tilde{t})^{-\nu} \exp[-i\tilde{t}^2/2]. \qquad (12.29)$$

Noting that ν is purely imaginary, the condition $|c_1(\tilde{t} \to -\infty)| = 1$ yields

$$|a_0| = \gamma^{1/2} \exp[-\pi\gamma^2/4], \quad \gamma = \tilde{\Delta}^2/2. \qquad (12.30)$$

To find the probability of the system to be in an excited state at $t \to \infty$, we need to find the value of $c_1(t)$ and $c_2(t)$ as $t \to \infty$. To this end, we note that for $t \to \infty$, the asymptotic form of $c_2(t)$ is

$$|c_2(\tilde{t} \to \infty)|^2 = \gamma e^{-\pi\gamma/2} |\lim_{\tilde{t} \to \infty} D_{-\nu-1}(\sqrt{2}e^{-3i\pi/4}\tilde{t})|^2. \qquad (12.31)$$

The asymptotic form of the Weber function for $\nu = i\tilde{\Delta}^2/2$ is

$$\lim_{\tilde{t} \to \infty} D_{-\nu-1}(\sqrt{2}e^{-3i\pi/4}\tilde{t}) = \frac{\sqrt{2\pi}}{\Gamma(i\gamma+1)} e^{i\tilde{t}^2/2}(\sqrt{2}\tilde{t})^{i\gamma}e^{-\pi\gamma/4}, \qquad (12.32)$$

where $\Gamma(x)$ denotes the gamma function. Thus, one can write

$$|c_2(\tilde{t} \to \infty)|^2 \simeq 2\pi\gamma e^{-\pi\gamma}/|\Gamma(i\gamma+1)|^2. \qquad (12.33)$$

Next we use a standard identity involving the gamma function:

$$|\Gamma(1+i\gamma)|^2 = \frac{\pi\gamma}{\sinh\gamma} = \frac{2\pi\gamma}{e^{\pi\gamma} - e^{-\pi\gamma}}. \qquad (12.34)$$

Substituting Eq. (12.34) into Eq. (12.33), we finally obtain

$$|c_2(\tilde{t} \to \infty)|^2 \simeq 1 - e^{-2\pi\gamma}. \qquad (12.35)$$

We note that since $|2\rangle$ denotes the instantaneous ground state of the system at $t \to \infty$, the probability of having the system in the excited state at late times is given by $P = |c_1(\tilde{t} \to \infty)|^2 = 1 - |c_2(\tilde{t} \to \infty)|^2$. Using Eq. (12.35), one gets

$$P = |c_1(t' \to \infty)|^2 = e^{-2\pi\gamma} = \exp[-\pi\Delta^2\mu B_0\tau/\hbar], \qquad (12.36)$$

where in the last expression we have used Eq. (12.18). Note that for $\tau \to \infty$, $P \to 0$, thus reproducing the adiabatic limit. Also, when $\tau \to 0$, $P \to 1$ which is the limit of sudden quench where the system does not have enough time to react to the drive and remains in the old ground state ($|1\rangle$ in the present case).

Eq. (12.36) has a simple interpretation and can be used to determine a qualitative criteria for the shift from one eigenstate to another in a driven system. Note that $1/\tau$ denotes the rate of change of instantaneous energy gap with respect to time: $\tau^{-1} \sim |d\Delta E/dt|$. Moreover, Δ denotes the lowest instantaneous energy gap of the two-level system. Thus, the criteria for formation of no excitation can be estimated as the square of the instantaneous energy gap $\epsilon(t)$ being large compared to the rate of its change:

$$\hbar \left| \frac{d\epsilon(t)}{dt} \right| \ll \epsilon^2(t). \qquad (12.37)$$

This ensures that $P \to 0$ and hence no excitation is formed. Extending this a little further one can use the equality $\hbar |d\epsilon/dt| \simeq \epsilon^2(t)$ as the criteria for minimal drive rate for excitation production. This is known as the *Landau adiabaticity criterion* and is quite useful for qualitatively estimating the excitation probability of a driven system.

☐ **Exercise 12.1** *Consider a particle in a ground state of a square well potential of width d and height V_0. At $t = 0$, the potential is instantly switched off. Find the probability of finding the particle in the nth excited state of the square well at time t_0. For what value of t_0 is this probability maximum?*

☐ **Exercise 12.2** *Consider a spin-$\frac{1}{2}$ particle whose Hamiltonian is given by a 2×2 matrix (in the σ_z basis) with elements $H_{11} = -H_{22} = \epsilon_0$ and $H_{12} = H_{21}^* = \Delta_0 e^{i\omega_D t}$. If the particle at $t = 0$ is in the $|\uparrow\rangle$ state, find the probability of its being in the $|\downarrow\rangle$ state at $t = T$. Find the value of T for which it is maximum.*

12.3 Level crossing

Another interesting variation can be discussed with time-dependent Hamiltonians. To illustrate the process, it is enough to talk about a two-level system, i.e., a system whose states belong to a 2D vector space. In this space, we will consider the time evolution of an initial state which can be denoted by

$$|\psi(0)\rangle = \begin{pmatrix} 1 \\ 0 \end{pmatrix} \qquad (12.38)$$

without any loss of generality. We take this as a basis vector, and the orthogonal vector as the other. Suppose in this basis, the Hamiltonian has the form

$$H(t) = E_\star \mathbb{1} + E_0 \begin{pmatrix} -\cos 2\theta + \epsilon(t) & \sin 2\theta \\ \sin 2\theta & \cos 2\theta \end{pmatrix}, \tag{12.39}$$

where all parameters except $\epsilon(t)$ are time-independent. The time-independent parts of this Hamiltonian are also quite general. The Hamiltonian has to be a Hermitian matrix. The off-diagonal elements can be complex, but their phases can be absorbed by suitably defining the phase of the basis vector orthogonal to the one shown in Eq. (12.38). The diagonal elements must be real, and can be adjusted by adding any multiple of the identity matrix, as discussed in §12.2.

We also note here that for $\epsilon(t) = 2\alpha t$, a transformation $E_\star \to E_\star - \epsilon(t)/2$ followed by $t \to t' = t - t_0$, where $\alpha t_0 = \cos 2\theta$, maps the Hamiltonian given by Eq. (12.39) to the Hamiltonian given in Eq. (12.12). Thus, the solution of the latter Hamiltonian discussed in the last section could be directly used for the former case. In what follows, we shall, however, concentrate on the situation where we want the dynamics to start for a finite t. To this end, we shall discuss the solution of the Hamiltonian given by Eq. (12.39) within an adiabatic approximation.

We can always define the time-dependent term in a way that $\epsilon(t) = 0$ at $t = 0$. At this time, then, the eigenvalues and eigenstates of the Hamiltonian are as follows:

Eigenvalue	:	Eigenvector
$-E_0$	:	$\lvert e_1 \rangle = \begin{pmatrix} \cos\theta \\ -\sin\theta \end{pmatrix},$
$+E_0$	:	$\lvert e_2 \rangle = \begin{pmatrix} \sin\theta \\ \cos\theta \end{pmatrix}.$

$$\tag{12.40}$$

This means that, in the absence of the time-dependent term, the initial state is given as the following superposition of the eigenstates:

$$\lvert \psi(0) \rangle = \cos\theta \, \lvert e_1 \rangle + \sin\theta \, \lvert e_2 \rangle. \tag{12.41}$$

In view of this equation, let us call θ the *mixing angle*, because it decides the probabilities of finding either of the eigenstates in the initial state.

☐ **Exercise 12.3** *Consider the case with $\epsilon = 0$, where the Hamiltonian is time-independent. Show that after a time t, the probability that the particle survives in the initial state is given by*

$$P_{\text{surv}}(t) = 1 - \frac{1}{2}\sin^2 2\theta \sin^2(E_0 t/\hbar). \tag{12.42}$$

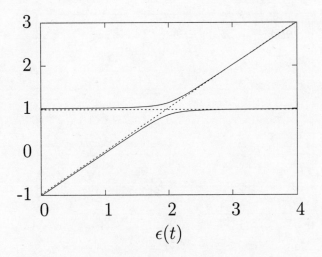

Figure 12.1 The solid lines show the variation of the eigenvalues with $\epsilon(t)$. The dashed straight lines are the diagonal elements of the Hamiltonian. We have used $\theta = 0.1$ for drawing this plot.

At any other time, in the presence of the time-dependent term, the eigenvalues of the instantaneous Hamiltonian are given by

$$E_{\pm}(t) = \frac{1}{2} E_0 \left(\epsilon(t) \pm \sqrt{\epsilon^2(t) + 4[1 - \epsilon(t) \cos \theta]} \right). \qquad (12.43)$$

The eigenstates are given by the same kind of formulas that are given in Eq. (12.40), but with θ replaced by $\theta(t)$, where

$$\tan 2\theta(t) = \frac{2 \sin 2\theta}{2 \cos 2\theta - \epsilon(t)}. \qquad (12.44)$$

If we have a situation where $\epsilon(t)$ increases with time, then this equation would imply that $\theta(t)$ increases with time, becoming equal to $\pi/4$ when $\epsilon(t) = 2 \cos 2\theta$. When $\epsilon(t)$ increases even further, $\tan 2\theta(t)$ becomes negative, i.e., $\theta(t)$ becomes larger than $\pi/4$, crawling up to the value $\pi/2$ for very large values of $\epsilon(t)$. The variation of $\theta(t)$ with $\epsilon(t)$ has been shown in Figure 12.2. In this figure, we have also shown the variation of $\sin^2 2\theta(t)$, which shows a nice resonance-like behavior.

How will the state of the system evolve if ϵ keeps increasing with time? Let us first discuss the answer heuristically. In the absence of time dependence in

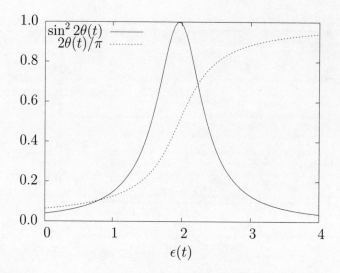

Figure 12.2 The dotted line shows how the mixing angle $\theta(t)$ varies with $\epsilon(t)$. The solid line shows the value of $\sin^2 2\theta(t)$.

the Hamiltonian, if an eigenstate is produced, it propagates as the eigenstate: only its phase changes with time. If we now include the time dependence but assume that the time dependence is very slow, we can guess that an eigenstate will still remain an eigenstate of the instantaneous Hamiltonian. This is called the *adiabatic approximation*, where the Hamiltonian changes so slowly that the state always finds enough time to adjust itself to the change.

Suppose now that the angle θ is very small, as has been used in making the plots of Figure 12.1. Initially, the particle is almost purely in the eigenstate $|e_1\rangle$. So, at the end of the journey, it also remains almost purely $|e_1(t)\rangle$. But if at this point $\epsilon(t)$ is large, we have $\theta(t) \approx \pi/2$, so the eigenstate becomes

$$|e_1(t)\rangle \approx \begin{pmatrix} 0 \\ 1 \end{pmatrix}. \tag{12.45}$$

Starting from the state given in Eq. (12.38), the particle would end up in a state that is orthogonal to it. This phenomenon is called *level crossing*. It is also called by the more explicit name *resonant level crossing*, because the conversion is accentuated by the resonant behavior shown in Figure 12.2, which means that the conversion probability, the complement of the survival probability given in Eq. (12.42), becomes large. Strictly speaking, the name is not very good because the levels do not really cross, as seen in Figure 12.1.

The heuristic argument given above can be improved, of course, in many ways. First, we can tackle the time dependence in a more formal way. Second, we can consider deviations from the adiabatic approximation, where there might be jumps from one eigenstate to another. Third, we need not consider small values of the mixing angle. Fourth, the particle needs to start its journey at a time when $\epsilon(t) = 0$. The initial as well as the final time may be arbitrary. Fifth, we can discuss a Hamiltonian where there is time dependence in other terms as well. We can also, given a specified time dependence, solve the Schrödinger equation numerically for the 2×2 Hamiltonian. We do not get into these details. But the essence of the surprise remains the same: the particle's state might change dramatically during the journey.

Level crossing can occur when, for example, a charged particle experiences a time-varying electric field. It has also been used in explaining the shortage of solar electron-neutrinos (ν_e's) detected on the earth, assuming that a level-crossing transition to another species of neutrinos takes place.

12.4 Time-dependent perturbation theory

In this section, we shall chart out the standard perturbative technique to obtain analytical handle on time-dependent problems for cases where $H(t)$ does not allow exact analytical solution of the Schrödinger equation. Most of the quantum mechanical systems with time-dependent Hamiltonians fall in this category.

12.4.1 Formalism

To describe this perturbation formalism, we consider the Hamiltonian to be of the form

$$H = H_0 + H_1(t), \tag{12.46}$$

where H_0 is the part of the Hamiltonian which can be exactly solved and $H_1(t)$ is the part to be treated perturbatively. We assume H_0 to be time-independent, a simplifying feature that is not strictly necessary as long as the solution to the Schrödinger equation with $H_0 \equiv H_0(t)$ is known exactly. Further, we shall assume that the corresponding eigenenergies E_m and eigenstates $|m\rangle$ satisfying

$$H_0 |m\rangle = E_m |m\rangle \tag{12.47}$$

are known.

The first step of the perturbation scheme is to write down the time-dependent state as

$$|\psi(t)\rangle = \sum_{\ell} c_{\ell}(t) \exp[-iE_{\ell}t/\hbar] |\ell\rangle, \tag{12.48}$$

with the initial condition $c_m(0) = \delta_{mm_0}$, which means that the system is in an eigenstate $|m_0\rangle$ of H_0 at $t = 0$. Note that we have kept an unusual phase factor $\exp[-iE_{\ell}t/\hbar]$ in the expansion of $|\psi(t)\rangle$. This should not be thought of as writing the coefficient in terms of its magnitude and phase. In fact, $c_m(t)$ can still be complex. The choice of separating out a phase becomes clear when we substitute the expression of $|\psi(t)\rangle$ in the Schrödinger equation $i\hbar\frac{d}{dt}|\psi(t)\rangle = (H_0 + H_1)|\psi(t)\rangle$ leading to

$$i\hbar\frac{d}{dt}c_m = \sum_{\ell} c_{\ell} e^{i(E_m - E_{\ell})t/\hbar} \langle m|H_1(t)|\ell\rangle. \tag{12.49}$$

Integrating, we obtain

$$c_m(t) = \frac{-i}{\hbar} \sum_{\ell} \int_0^t dt' \, c_{\ell}(t') e^{i(E_m - E_{\ell})t'/\hbar} \langle m|H_1(t)|\ell\rangle. \tag{12.50}$$

The phase factor in the expansion of $|\psi(t)\rangle$ can now be understood to be chosen such that $c_m(t)$ does not evolve for $H_1 = 0$. This allows us to write a perturbative expansion

$$c_m(t) = c_m^{(0)}(t) + c_m^{(1)}(t) + c_m^{(2)}(t) + \dots \tag{12.51}$$

with $c_m^{(0)}(t) = \delta_{mm_0}$. Substituting Eq. (12.51) into Eq. (12.50) and following the standard perturbative analysis outlined while discussing the time-independent perturbation theory, one obtains

$$c_m^{(1)}(t) = \left(\frac{-i}{\hbar}\right) \int_0^t dt' \, e^{i(E_m - E_{m_0})t'/\hbar} \langle m|H_1(t')|m_0\rangle,$$

$$c_m^{(2)}(t) = \left(\frac{-i}{\hbar}\right)^2 \sum_{\ell} \int_0^t dt_1 \int_0^{t_1} dt_2 \, e^{i(E_m - E_{\ell})t_1/\hbar} e^{i(E_{\ell} - E_{m_0})t_2/\hbar}$$
$$\times \langle m|H_1(t_1)|\ell\rangle \langle \ell|H_1(t_2)|m_0\rangle. \tag{12.52}$$

Thus, the probability of finding the state in the mth eigenstate at time t is given by

$$P_m(t) = |c_m(t)|^2 = \left|\sum_n c_m^{(n)}(t)\right|^2. \tag{12.53}$$

In what follows, we are going to analyze the behavior of $P_m(t)$ within the first- or second-order perturbation theory for different time dependence of $H_1(t)$.

Before delving on to specific examples, we note that one can reformulate the perturbation theory in terms of the evolution operator. We define the evolution operator in the interaction picture $U_I(t, t_0)$ as

$$|\psi(t)\rangle = U_I(t, t_0)|\psi(t_0)\rangle, \quad U_I(t_0, t_0) = \mathbb{1}, \tag{12.54}$$

where $\mathbb{1}$ is the identity operator. There is a slight change of notation from the definition given in §4.5: the initial time, which was 0 there, has been denoted by t_0 here. The H_0 part will be interpreted as the unperturbed Hamiltonian for the present problem. The integral equation for $U(t, t_0)$ will contain the other part, which will be denoted by H_I in what follows. With this notation, Eq. (4.52, p. 92) becomes

$$U_I(t, t_0) = \mathbb{1} - \left(\frac{-i}{\hbar}\right) \int_{t_0}^{t} dt' \, H_I(t') U_I(t', t_0). \tag{12.55}$$

Replacing the factor of U_I appearing on the right side by the integral again, and continuing like this, we obtain the perturbative expansion of the evolution operator:

$$U_I(t', t_0) = \sum_{n=0}^{\infty} \left(\frac{-i}{\hbar}\right)^n \int_{t_0}^{t} dt_1 \int_{t_0}^{t_1} dt_2 \ldots \int_{t_0}^{t_{n-1}} dt_n$$
$$\times H_I(t_1) H_I(t_2) \ldots H_I(t_n). \tag{12.56}$$

The series obtained in Eq. (12.56) is called the *Dyson series*. Comparing Eq. (12.48) and Eq. (12.54), we find that $U_I(t, t_0)$ is related to $c_m(t)$ through the relation

$$c_m(t) = \langle m|U_I(t, t_0)|m_0\rangle, \tag{12.57}$$

where $|m_0\rangle$ is the initial state. Indeed, it is easy to check from Eq. (12.56) that the evolution equation for $\langle m|U_I(t, t_0)|m_0\rangle$ is identical to Eq. (12.52). Thus, for standard single particle quantum mechanics, the two approaches are identical. However, the latter approach is more powerful while dealing with more complicated systems; this is particularly relevant for application of this method to field theory and/or many-particle quantum mechanics. A more detailed discussion of this point is beyond the scope of the present text. In the rest of this section, we shall use the wavefunction-based formalism charted out in Eq. (12.52).

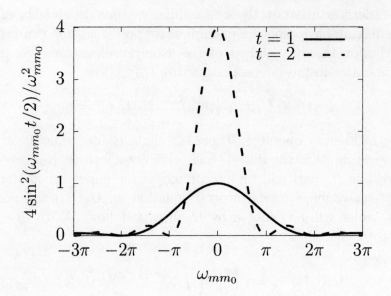

Figure 12.3 Transition probability as a function of the energy difference for two different times.

12.4.2   Transition rates

Next, we consider the case when $H_1(t)$ is independent of time: $H_1(t) \equiv H_1$. In view of our earlier assumption that H_0 is time-independent, this now means that the entire Hamiltonian is time-independent. However, we can ask time-dependent questions like how long it takes for a transition to occur, and our present formalism is appropriate for answering such questions.

In this case, the integrals involved in Eq. (12.52) can be evaluated in a straightforward manner. We first analyze the first-order term in Eq. (12.52), which yields

$$c_m^{(1)}(t) = e^{i\omega_{mm_0}t/2}\frac{\sin(\omega_{mm_0}t/2)}{\hbar\omega_{mm_0}/2}\left\langle m\left|H_1\right|m_0\right\rangle, \qquad (12.58)$$

where $\omega_{mm_0} = (E_m - E_{m_0})/\hbar$. If we restrict ourselves to the first order, then the transition probability to state $|m\rangle$ starting from a state $|m_0\rangle$ at time t is given by

$$P_m(t) = \frac{4|\langle m|H_1|m_0\rangle|^2 \sin^2(\omega_{mm_0}t/2)}{\hbar^2 \omega_{mm_0}^2}. \qquad (12.59)$$

Figure 12.3 shows the nature of variation of this probability with energy, assuming a constant matrix element for all states. It shows that the function is highly peaked near zero, almost negligible for values $|\omega_{mm_0}t| > 2\pi$. This means that the width of the peak goes like $1/t$. For discrete and well-separated states, it implies that transition with appreciable probability will occur between two states if they satisfy the relation $\omega_{mm_0} < 2\pi/t$. Thus, if we have a perturbing H_1 for only a time Δt, the energy range in which transitions will take place will be given by $\Delta E \Delta t \sim \hbar$. This relation may remind the reader of the uncertainty relation; however, it is to be noted that the situation here is fundamentally different from, for example, the uncertainty relation between x and p. In quantum mechanics, the latter uncertainty relation involves operators; in contrast, time and energy here are parameters, as discussed in §1.4 and §3.8.

The behavior of $P_m(t)$ for $E_m = E_{m_0}$ also requires some discussion. Let us consider the situation where there is a bunch of states denoted by $|\ell\rangle$ with $E_\ell = E_m$. This is almost always the case for systems where there is a continuum of final states. In this case one has

$$P(t) = \sum_{\ell; E_\ell \simeq E_m} \frac{4|\langle \ell|H_1|m_0\rangle|^2 \sin^2(\omega_{\ell m_0}t/2)}{\hbar^2 \omega_{\ell m_0}^2}. \tag{12.60}$$

Next, to evaluate the sum one defines a density of states $\rho(E)$ such that the number of quantum states between E and $E + dE$ is $\rho(E)dE$. This allows us to replace the sum over ℓ with the integral $\int dE_\ell\, \rho(E_\ell)$ and leads to

$$P(t) = \int dE_\ell\, \rho(E_\ell) \frac{4|\langle \ell|H_1|m_0\rangle|^2 \sin^2(\omega_{\ell m_0}t/2)}{\hbar^2 \omega_{\ell m_0}^2}. \tag{12.61}$$

In the long time limit, this expression can be further simplified, taking advantage of the almost unit probability of energy conserving transitions in this limit. Using the representation of the Dirac delta function

$$\lim_{a \to \infty} \frac{\sin^2(ax)}{ax^2} = \pi\delta(x), \tag{12.62}$$

one can evaluate the integral to obtain

$$P(t) = \frac{2\pi}{\hbar} \left| \langle \ell|H_1|m_0\rangle \right|_{\text{av}}^2 \rho(E_\ell)t, \tag{12.63}$$

where the subscripted 'av' denotes average over the matrix elements within the energy window of width dE_ℓ. This average can be understood to be the average transition probability for all states in the quasi-continuum within the energy window dE. These matrix elements are usually similar for states whose energies are very close; however, the summation procedure does not require this criteria. The rate of transition in this limit is given by

$$W_{m_0 \to \ell} = \frac{dP(t)}{dt} = \frac{2\pi}{\hbar} \left| \langle \ell | H_1 | m_0 \rangle \right|_{\mathrm{av}}^2 \rho(E_\ell). \tag{12.64}$$

This result is known as the *Fermi's golden rule*.

Next, we include the second-order terms. A straightforward calculation shows that in the long-time limit, the transition rate becomes

$$W'_{m_0 \to \ell} = \frac{2\pi}{\hbar} \rho(E_\ell) \left| \left(\langle \ell | H_1 | m_0 \rangle + \sum_j \frac{\langle \ell | H_1 | j \rangle \langle j | H_1 | m_0 \rangle}{E_j - E_{m_0}} \right)_{\mathrm{av}} \right|^2, \tag{12.65}$$

where j denotes intermediate states. The first term in W' shows the contribution from $c_m^{(1)}(t)$ and constitutes the probability of direct transition with energy conservation that is implied by the appearance of the Dirac delta function indicated through Eq. (12.62). The second term shows the contribution of $c_m^{(2)}(t)$. This contribution can be thought to be a result of two steps. In the first step, the system moves from state $|m_0\rangle$ to $|j\rangle$, while in the second it comes back from the state $|j\rangle$ to $|\ell\rangle$. In each of these two steps, energy need not be conserved since $E_j \neq E_\ell \simeq E_{m_0}$. The intermediate states constitute transitions to other states; such transitions are often called *virtual processes* and constitute a hallmark of quantum mechanics. Thus, at second and higher order in the perturbation theory, the transition rate receives contribution from virtual processes while the first-order term only shows direct contribution.

☐ **Exercise 12.4** *Prove Eq. (12.62). It is easy to see that the function vanishes for all $x \neq 0$ in the limit $a \to \infty$. Complete the proof by evaluating the integral of the function.*

☐ **Exercise 12.5** *Deduce Eq. (12.65).*

12.4.3   Harmonic perturbation

Finally, we consider examples of harmonic perturbation for which

$$H_1(t) = H_1 e^{i\omega_0 t} + H_1^\dagger e^{-i\omega_0 t}. \tag{12.66}$$

Here ω_0 is the external drive frequency. Note that H_1 need not be Hermitian in this case, since the sum of the two terms is Hermitian for any H_1. In what follows, we shall consider the first-order contribution. In the presence of a harmonic perturbation, the general expression of $c_m^{(1)}(t)$ given in Eq. (12.50) reads

$$c_m^{(1)}(t) = \frac{1}{\hbar} \left[\frac{\langle m|H_1|m_0\rangle \left(1 - e^{i(\omega_{mm_0}+\omega_0)t}\right)}{\omega_{mm_0} + \omega_0} \right.$$
$$\left. + \frac{\langle m|H_1^\dagger|m_0\rangle \left(1 - e^{i(\omega_{mm_0}-\omega_0)t}\right)}{\omega_{mm_0} - \omega_0} \right]. \qquad (12.67)$$

We note that in the long time limit where $t \gg |\omega_{mm_0} \pm \omega_0|^{-1}$, only the first or the second term contributes appreciably. The former contribution is large when $E_m - E_{m_0} \simeq -\hbar\omega_0$; this constitutes an example of induced emission where the system dumps energy $(E_{m_0} - E_m)$ to the external perturbation and comes down to a lower state. This process can take place only if $|m_0\rangle$ corresponds to an excited state. The latter process corresponds to stimulated absorption, when the system absorbs energy $E_m - E_{m_0}$ from the radiation and transits to a higher energy state. When the first term contributes appreciably the transition rate can be found, in an exactly similar way shown for the constant perturbation case, to be

$$W_{m_0 \to \ell}^{(1)\text{harmonic}} = \frac{2\pi}{\hbar} \left| \langle \ell|H_1|m_0\rangle \right|_{\text{av}}^2 \rho(E_\ell), \qquad (12.68)$$

where $E_\ell \simeq E_{m_0} - \hbar\omega$. Note that the system energy is not conserved as the system can dump energy to the time-dependent perturbing field. In contrast, when the second term of Eq. (12.67) dominates, one has

$$W_{\ell \to m_0}^{(2)\text{harmonic}} = \frac{2\pi}{\hbar} \left| \langle \ell|H_1|m_0\rangle \right|_{\text{av}}^2 \rho(E_\ell), \qquad (12.69)$$

where $E_\ell = E_{m_0} + \hbar\omega_0$, and we have used the identity

$$\langle \ell|H_1|m\rangle^* = \langle m|H_1^\dagger|\ell\rangle. \qquad (12.70)$$

We note that these two rates satisfy

$$W_{m_0 \to \ell}^{(1)\text{harmonic}} / \rho(E_\ell) = W_{m_0 \to \ell'}^{(2)\text{harmonic}} / \rho(E_{m_0}), \qquad (12.71)$$

which indicates a symmetry involving the transition rates and density of final states for stimulated emission and absorption processes in the presence of a harmonic perturbation. This is often referred to as the *principle of detailed balance*.

□ **Exercise 12.6** *A particle residing at an eigenstate $|m_0\rangle$ at $t = 0$ is subjected to a perturbation $H_1(t) = H' \tanh(\alpha t)$.*

(a) *Approximating the potential to be constant for $t > 1/\alpha$ and linearly varying for $t < 1/\alpha$, find the probability of transition to state $|m\rangle$ at $t \to \infty$ in terms of α and $[H']_{mm_0}$.*

(b) *Check your results for the limits $\alpha \to 0$ and $\alpha \to \infty$ and compare them with those obtained from constant perturbation ($H_1(t) = H'$).*

12.5   Periodic drive: Floquet analysis

A class of drive protocols that have garnered a lot of attention are periodic drives, for which $H(t) = H(t + nT)$, where $T = 2\pi/\omega_D$ is the drive period, ω_D is the drive frequency, and n is an integer. Such periodicity of the Hamiltonian allows one to analyze them using the Floquet theory which we now chart out.

To this end, we first note that the Schrödinger equation for such a system, given by

$$i\hbar \frac{d}{dt} |\psi(t)\rangle = H(t)|\psi(t)\rangle, \tag{12.72}$$

allows for solutions of the form

$$|\psi_\alpha(t)\rangle = e^{-i\epsilon_\alpha t/\hbar}|\Phi_\alpha(t)\rangle, \tag{12.73}$$

where

$$|\Phi_\alpha(t + nT)\rangle = |\Phi_\alpha(t)\rangle \tag{12.74}$$

for any integer n. This is similar to the *Bloch theorem* for periodic potentials where the wavefunctions can be written as $\psi(x) = \exp[ikx]u_k(x)$, where $u_k(x)$ is periodic in space. Both of these results follow from the *Floquet theorem*, which was discovered in the context of the theory of linear differential equations by the French mathematician Gaston Floquet in 1883. The theorem states that for a system of linear differential equations given by

$$\frac{dx(\tau)}{d\tau} = A(\tau)x(\tau), \tag{12.75}$$

where $x(\tau)$ is a column vector of length N and $A(\tau) = A(\tau + T)$ is an $N \times N$ square periodic matrix, the solution can be written in the form of

$$x(\tau) = e^{\mu\tau}y(\tau), \tag{12.76}$$

where $y(\tau)$ is periodic in time with period T. We do not delve into the proof of this theorem here. Instead, we note that in the context of quantum mechanics, the theorem is directly applicable with the identification $\mu \equiv ik$ and $\tau \equiv \boldsymbol{r}$ (so that $\mu\tau = i\boldsymbol{k} \cdot \boldsymbol{r}$) for the Bloch problem and $\mu \equiv -i\epsilon_\alpha/\hbar$ and $\tau \equiv t$ for the Floquet problem. This leads to the concept of quasi-momentum $\boldsymbol{k}$ for the Bloch problem and quasienergies ϵ_α for the Floquet problem. The word "quasi" here reflects the fact that there is no conserved momentum or energy in these problems since one does not have space or time translational symmetry. However, the periodic structure of the Hamiltonian allows one to define quantities that almost behave like momentum/energy in these cases. Substituting the solution of Eq. (12.73) into the Schrödinger equation, we see that Φ_α satisfies the equation

$$\left(H(t) - i\hbar\frac{d}{dt} \right) |\Phi_\alpha(t)\rangle = \epsilon_\alpha|\Phi_\alpha(t)\rangle. \tag{12.77}$$

Using the periodicity of $H(t)$, it is now easy to see that $|\Phi_{\alpha n}(t)\rangle = |\Phi_\alpha(t)\rangle \exp[-in\omega_D t]$ is also a solution with $\epsilon_\alpha \to \epsilon_\alpha + \hbar n\omega_D$. This allows one to define a *Floquet Brillouin zone* for the quasienergies $-\hbar\omega_D/2 \leq \epsilon_\alpha \leq \hbar\omega_D/2$; all solutions outside these zones can be found by translating the quasienergies by integer multiples of the drive frequency.

Another key quantity that allows us to understand properties of a driven system is the evolution operator $U(t,0)$. For periodically driven systems, U satisfies

$$U(t,0) = T_t\left[\exp\left(-\frac{i}{\hbar} \int_0^t dt' \, H(t') \right) \right]$$

$$= T_t\left[\exp\left(-\frac{i}{\hbar} \int_T^{t+T} dt' \, H(t') \right) \right] = U(t+T,T), \tag{12.78}$$

where we have used the time-periodicity of the Hamiltonian.

This indicates that using Floquet theorem one can write for $nT \leq t < (n+1)T$ $(n \in Z)$,

$$U(t,0) = P(t,0)e^{-in\Lambda(T)} = P(t,0)e^{-inH_F(T)T/\hbar}, \tag{12.79}$$

where $P(t,0)$ must be a periodic function of time. The phase $\Lambda(T)$ has been rewritten by introducing H_F, which has the dimension of energy and is called the *Floquet Hamiltonian* of the system. Its use is invoked to draw parallels with the standard time evolution operator, $\exp[-iHt/\hbar]$. Since $U(0,0) = 1$,

$P(0,0) = 1 = P(nT, 0)$, where the last equality holds for any integer n due to periodicity of $P(t,0)$. Using this periodicity property, one finds

$$U(T,0) = \exp[-iH_F T/\hbar] = T_t[e^{-i\int_0^T dt' H(t')/\hbar}], \qquad (12.80)$$

where we have written $H_F(T) \equiv H_F$ to avoid cluttering. We shall use this notation for the rest of this section; the readers are urged to remember that the Floquet Hamiltonian H_F is in general a function of the time period T. Eq. (12.80) indicates that if we have an eigenstate $|m\rangle$ of $U(T,0)$,

$$U(T,0)|m\rangle = \Lambda_m |m\rangle, \qquad (12.81)$$

then the eigenvalues of H_F will satisfy the relation

$$\Lambda_m = \exp\left(-\frac{i\epsilon_m T}{\hbar}\right). \qquad (12.82)$$

Note that this is in keeping with the general form of the eigenvalues of a unitary operator expressed through Eq. (2.151, p. 51).

Unfortunately, exact analytical computation of H_F is possible only in a few cases. For this reason, specific perturbation theories have been developed to obtain an analytic handle on this problem. Here we shall discuss one such theory, viz., the *Magnus expansion*. Such an expansion works in the high frequency regime where $\hbar\omega_D/E_0 \gg 1$ (E_0 is a typical energy scale of the Hamiltonian H). In this regime it is easy to see that $U(T,0)$ can be expanded in powers of T. Using Eq. (12.80), one can write in this regime

$$\begin{aligned}
U(T,0) &= \mathcal{T}_t[e^{-i\int_0^T dt' H(t')/\hbar}] \\
&= 1 + \sum_{m=1}^{\infty} \left(\frac{-i}{\hbar}\right)^m \int_0^T dt_1 \int_0^{t_1} dt_2 .. \int_0^{t_{m-1}} dt_m \, H(t_1)H(t_2)...H(t_m) \\
&= 1 + \sum_{m=1}^{\infty} \mathcal{R}_m, \qquad (12.83)
\end{aligned}$$

where at high frequencies each term in the expansion is smaller than its predecessors since they involve a higher number of time integrals. The simplest check for this statement can be done for $H(t) \equiv H$ where the mth term in the expansion is proportional to T^m.

Next we write $U(T,0)$ using the Floquet Hamiltonian and obtain

$$U(T,0) = \exp[-iH_F T/\hbar] = \exp[-iT\sum_{m=1}^{\infty} H_F^{(m)}/\hbar]$$

$$= 1 + \frac{-iT}{\hbar}H_F^{(1)} + \left[\frac{-iT}{\hbar}H_F^{(2)} + \frac{1}{2}\left(\frac{-iTH_F^{(1)}}{\hbar}\right)^2\right] + ..., (12.84)$$

where $H_F^{(m)}$ is the mth order Floquet Hamiltonian and the ellipsis in the last term of Eq. (12.84) indicates higher order terms in the expansion.

Comparing Eq. (12.83) and Eq. (12.84), we find the relations for $H_F^{(1)}$ and $H_F^{(2)}$ in terms of $H(t)$ as

$$\mathcal{R}_1 = \frac{-iT}{\hbar}H_F^{(1)}, \quad \mathcal{R}_2 = \left[\frac{iT}{\hbar}H_F^{(2)} + \frac{1}{2}\left(\frac{-iTH_F^{(1)}}{\hbar}\right)^2\right]. \quad (12.85)$$

Inverting these relations and after a few lines of straightforward algebra, one finds

$$H_F^{(1)} = \int_0^1 dz\, H(z),$$

$$H_F^{(2)} = \frac{T}{2i\hbar}\int_0^1 dx \int_0^x dy\, [H(x), H(y)], \quad (12.86)$$

where $z = t/T$, $x = t_1/T$, and $y = t_2/T$. One finds that at large frequency, this gives a perturbative series. The first term of this series is just the time-averaged Hamiltonian which brings out the simple structure of the Floquet Hamiltonian in the high frequency regime. As the frequency is reduced, the contributions of the higher-order terms become more important and the form of H_F becomes more complicated. The higher $(m > 2)$ order terms of these series can be computed in an analogous manner. We note here that the convergence of such an expansion is not easily guaranteed and is surprisingly tricky to ascertain. We shall not discuss this issue further here.

☐ **Exercise 12.7** *Using Eq. (12.85) derive the expression of $H_F^{(2)}$ given in Eq. (12.86).*

☐ **Exercise 12.8** *Relate $\mathcal{R}_3$ given in Eq. (12.83) to $H_F^{(3)}$ using Eq. (12.84). Obtain an expression for $H_F^{(3)}$ by inverting this relation and using Eq. (12.86).*

Finally, we provide one example where the Floquet Hamiltonian can be computed exactly. This constitutes a two-level system with a time-dependent magnetic field in the x-direction and a constant field along z. The Hamiltonian of such a system is given by

$$H(t) = -\mu B_z(\sigma_z + \beta(t)\sigma_x), \quad (12.87)$$

where μ is the magnetic moment, and $\beta(t)$ is a periodic square pulse, i.e.,

$$\beta(t) = \begin{cases} \beta_0 & \text{for } t \leqslant T/2, \\ -\beta_0 & \text{for } t > T/2. \end{cases} \tag{12.88}$$

In what follows, we shall scale all energies by $E_0 = \mu B_z$ and measure time period in units of $\hbar/E_0$. In these units, defining $T' = TE_0/\hbar$, one can write

$$U(T,0) = T_t\left[\exp\left(\frac{-i}{\hbar}\int_0^T dt\, H(t)\right)\right]$$
$$= e^{i(\sigma_z+\beta_0\sigma_x)T'/2}\, e^{i(\sigma_z-\beta_0\sigma_x)T'/2}. \tag{12.89}$$

We note that the time ordering involved in the definition of U becomes trivial for the square pulse protocol. This simplicity arises from the fact that $H(t)$ remains independent of time for $t \leqslant T/2$ leading to commutation of U at all times till $t \leqslant T/2$. Similarly for $t > T/2$, U commutes with itself at all times. Thus, the time ordering leads to a simple product of two terms as written down in Eq. (12.89). Eq. (12.89) can be readily simplified by using standard properties of Pauli matrices which leads to $U(T,0) = U_+U_-$, with

$$U_\pm = \begin{pmatrix} \cos\theta + i\cos\xi\sin\theta & \pm i\sin\xi\sin\theta \\ \pm i\sin\xi\sin\theta & \cos\theta - i\cos\xi\sin\theta \end{pmatrix}, \tag{12.90}$$

where $\theta = \sqrt{1+\beta_0^2}\,T'/2$ and $\tan\xi = \beta_0$. This leads to

$$U(T,0) = \begin{pmatrix} \cos\chi + in_z\sin\chi & n_y\sin\chi \\ -n_y\sin\chi & \cos\chi - in_z\sin\chi \end{pmatrix}, \tag{12.91}$$

with the definitions

$$\cos\chi = \frac{\beta_0^2 + \cos 2\theta}{1+\beta_0^2}, \tag{12.92a}$$

$$n_z = \frac{\sin 2\theta}{\sqrt{1+\beta_0^2}\sin\chi}, \tag{12.92b}$$

$$n_y = \frac{2\beta_0\sin^2\theta}{(1+\beta_0^2)\sin\chi}. \tag{12.92c}$$

In view of the definition given in Eq. (12.80), this means that

$$H_F = (\sigma_z n_z + \sigma_y n_y)\,\chi/T'. \tag{12.93}$$

In the high frequency limit, it can be easily verified that $\chi, \theta \to 0$. Thus, one can write

$$n_z \to \frac{2\theta \sqrt{1 + \beta_0^2}}{\chi (1 + \beta_0^2)} = \frac{T'}{\chi} \tag{12.94}$$

and $n_y \to 0$. Thus, in this limit, $H_F \to \sigma_z$, which is also the first-order Magnus result as can be verified from the expression of $H_F^{(1)}$ in Eq. (12.86).

☐ **Exercise 12.9** *Verify that $n_y^2 + n_z^2 = 1$, which guarantees unitarity of U in Eq. (12.91).*

☐ **Exercise 12.10** *Consider a time-dependent Hamiltonian $H(t) = E[\sigma_z \sin \omega t + \sigma_x]$ where E has the dimension of energy. Find the Floquet Hamiltonian of the system up to the second order in Magnus expansion.*

☐ **Exercise 12.11** *Consider the Hamiltonian $H(t) = E[A\sigma_z \cos \omega t + \sigma_x]$ where E has the dimension of energy and $A \gg 1$. In what follows, we shall use $H_0(t) = EA\sigma_z \cos(\omega t)$ and $H_1 = E\sigma_x$ and develop a perturbation theory (termed as the Floquet perturbation theory in the literature).*

(a) *Show that in the absence of H_1, the evolution operator is $U_0(t, 0) = \exp[-iE_0 A \sin(\omega_D t)\sigma_z/(\hbar \omega_D)]$. Using this and the definition of the Floquet Hamiltonian, show that $H_F^{(0)} = 0$.*

(b) *Obtain the first-order correction to U:*

$$U_1(T, 0) = \frac{-i}{\hbar} \int_0^T dt \, U_0^\dagger(t, 0) H_1 U_0(t, 0). \tag{12.95}$$

(c) *Find $H_F^{(1)}$. Show that it reduces to the Magnus result in the limit of high frequency.*

Chapter 13

Scattering theory

In this chapter, we shall discuss the formulation of scattering of a quantum particle from a potential. Throughout most of this chapter, we shall be interested in the state of the quantum particle for $r \gg a$, where a is the range of the potential. Our main focus shall be, following directions from classical scattering theory, on computing the differential scattering cross-section for such a particle. The formalism necessary for this purpose is discussed in §13.1 and §13.2. This is followed by a discussion of the perturbative schemes in §13.3 and partial wave analysis in §13.4. Next, we consider the scattering cross-section due to the long-range Coulomb potential in §13.5.

13.1   Lippmann–Schwinger formalism

Consider a quantum particle whose motion is governed by a Hamiltonian

$$H = H_0 + V \,, \tag{13.1}$$

where V is a short-range potential. In the absence of this potential, the particle is assumed to be in the eigenstates of H_0. For the rest of this section, we shall choose these eigenstates to be plane wave states, each characterized by a wave vector $\boldsymbol{k}$. The state vectors are denoted by $|\boldsymbol{k}\rangle$, i.e.,

$$H_0 \,|\boldsymbol{k}\rangle = E \,|\boldsymbol{k}\rangle \tag{13.2}$$

with

$$E = \frac{\hbar^2 \mathrm{k}^2}{2m}, \tag{13.3}$$

m being the mass of the particle and $\mathrm{k} = |\boldsymbol{k}|$.

We shall assume $|\boldsymbol{k}\rangle$ to be the initial state of the particle at $t = -\infty$. The wavefunction is given as

$$\psi_{\rm in}(\boldsymbol{r}) = \langle \boldsymbol{r} \,|\, \boldsymbol{k}\rangle = e^{i\boldsymbol{k}\cdot\boldsymbol{r}}, \tag{13.4}$$

which implies a plane wave propagating in the direction of $\boldsymbol{k}$. Whenever necessary, we will take this direction as the direction of the z-axis so that we can write the incoming wave as e^{ikz}. The incoming wave is then supposed to be coming from the $-z$-direction, i.e., moving in the $+z$-direction.

We shall focus on the state of the system far away from the range of the potential at $t = \infty$. The potential is centered around $\boldsymbol{r} = 0$ and has a finite range. Under these conditions, one can write down the time-independent Schrödinger equation as $H\,|\psi\rangle = E\,|\psi\rangle$ with a formal solution

$$|\psi\rangle = |\psi_{\rm in}\rangle + G_0 V|\psi\rangle, \tag{13.5}$$

where

$$G_0 = (E - H_0)^{-1} \tag{13.6}$$

is to be understood as the inverse of the operator $(E\mathbb{1} - H_0)$, and $\mathbb{1}$ is the identity operator. We shall discuss the significance of this operator in detail shortly. Eq. (13.5), with G_0 identified by Eq. (13.6), is called the *Lippmann–Schwinger equation*.

□ **Exercise 13.1** *Verify that Eq. (13.5) is indeed a solution to the stationary state Schrödinger equation by operating both sides by $E - H_0$.*

Instead of state vectors in the Hilbert space, we now focus on wavefunctions in the coordinate space and assume an incoming plane wave initial wavefunction as discussed. Taking the inner product of Eq. (13.5) with $\langle \boldsymbol{r}|$, we find

$$\psi(\boldsymbol{r}) = \langle \boldsymbol{r} \,|\, \psi\rangle = \psi_{\rm in}(\boldsymbol{r}) + \int d^3 r'\, G_0(\boldsymbol{r}, \boldsymbol{r}')V(\boldsymbol{r}')\psi(\boldsymbol{r}'), \tag{13.7}$$

where

$$G_0(\boldsymbol{r}, \boldsymbol{r}') = \langle \boldsymbol{r} \,\big|\, (E - H_0)^{-1} \,\big|\, \boldsymbol{r}'\rangle \tag{13.8}$$

is the inverse of the operator $E - H_0$ in the coordinate space. Note that G_0 depends on $\boldsymbol{r} - \boldsymbol{r}'$ only; this is a consequence of translational invariance of H_0.

Also, Eq. (13.6) can be written as $(E - H_0)G_0 = \mathbb{1}$, which in coordinate space reads as

$$(E - H_0)G_0(\boldsymbol{r} - \boldsymbol{r}') = \delta(\boldsymbol{r} - \boldsymbol{r}'). \tag{13.9}$$

Thus, G_0 can be identified as the Green function of the theory in the absence of the potential V. Strictly speaking, G_0 is the non-interacting Green function. The full Green function G satisfies $(E - H)G(\boldsymbol{r}, \boldsymbol{r}') = \delta(\boldsymbol{r} - \boldsymbol{r}')$ and is usually impossible to compute exactly. In the rest of this chapter, we shall refer to G_0 as the Green function keeping this distinction in mind.

To find an explicit expression for the Green function, we first define its Fourier transform through the relation

$$G_0(\boldsymbol{r} - \boldsymbol{r}') = \int \frac{d^3q}{(2\pi)^3} e^{i\boldsymbol{q}\cdot(\boldsymbol{r}-\boldsymbol{r}')} G_0(\boldsymbol{q}). \tag{13.10}$$

Putting it back into Eq. (13.9), we see that the Fourier transform satisfies the algebraic relation

$$\left(E - \frac{\hbar^2 \boldsymbol{q}^2}{2m}\right) G_0(\boldsymbol{q}) = 1. \tag{13.11}$$

Therefore,

$$G_0(\boldsymbol{r} - \boldsymbol{r}') = \int \frac{d^3q}{(2\pi)^3} e^{i\boldsymbol{q}\cdot(\boldsymbol{r}-\boldsymbol{r}')} \frac{1}{E - \hbar^2 \mathsf{q}^2/2m + i\eta}, \tag{13.12}$$

where $\mathsf{q} = |\boldsymbol{q}|$, and $\eta > 0$ is an infinitesimal quantity whose function is to consider the integral in Eq. (13.12) as a principal value integral to avoid the singularity at $E = \hbar^2 \mathsf{q}^2/2m$. The reason for the choice of $+i\eta$ in the denominator instead of $-i\eta$ is that it gives outgoing waves, as we desire to find, for $\boldsymbol{r} \to \infty$.

Eq. (13.11) tells us that $G_0(\boldsymbol{q})$ should be the inverse of the expression that multiplies it on the left side. It seems that, in writing the inverse Fourier transform in Eq. (13.12), we have used an extra term $+i\eta$. It should be noticed that it still gives the inverse. For any quantity x,

$$\lim_{\eta \to 0} \frac{1}{x + i\eta} = \frac{1}{x} - i\pi\delta(x). \tag{13.13}$$

The right side is still an inverse of x, because the extra term containing the Dirac delta vanishes when multiplied by x, as noted in Eq. (A.18, p. 427). It is this general form of the inverse that has been used in Eq. (13.12).

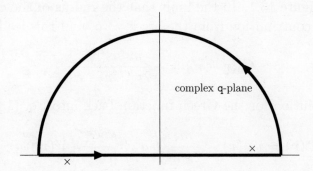

Figure 13.1 Integration contour for finding the Green function. The cross marks indicate the poles of the integrand for some arbitrary positive value of η.

It is not difficult to perform the integration that appears in Eq. (13.12). We can choose a spherical coordinate system with the polar angle θ being the angle between $\boldsymbol{q}$ and $\boldsymbol{r} - \boldsymbol{r}'$. We can start with the integration over the azimuthal angle, which is trivial, and obtain

$$G_0(\boldsymbol{r} - \boldsymbol{r}') = \frac{1}{4\pi^2} \int_0^\infty d\mathsf{q}\, \mathsf{q}^2 \frac{1}{E - \frac{\hbar^2 \mathsf{q}^2}{2m} + i\eta} \int_0^\pi d\theta\, \sin\theta e^{i\mathsf{q}|\boldsymbol{r}-\boldsymbol{r}'|\cos\theta}$$

$$= \frac{1}{4i\pi^2} \int_0^\infty d\mathsf{q} \frac{\mathsf{q}}{|\boldsymbol{r} - \boldsymbol{r}'|} \frac{[e^{i\mathsf{q}|\boldsymbol{r}-\boldsymbol{r}'|} - e^{-i\mathsf{q}|\boldsymbol{r}-\boldsymbol{r}'|}]}{E - \hbar^2 \mathsf{q}^2/2m + i\eta}. \tag{13.14}$$

In the second term within the square brackets, we make the variable transformation $\mathsf{q} \to -\mathsf{q}$, which allows us to write

$$G_0(\boldsymbol{r} - \boldsymbol{r}') = -\frac{2m}{\hbar^2} \frac{1}{4i\pi^2} \int_{-\infty}^\infty d\mathsf{q} \frac{\mathsf{q}}{|\boldsymbol{r} - \boldsymbol{r}'|} \frac{e^{i\mathsf{q}|\boldsymbol{r}-\boldsymbol{r}'|}}{\mathsf{q}^2 - \mathsf{k}^2 - i\eta'}, \tag{13.15}$$

using Eq. (13.3) for E.

The integration can be performed by going over to complex values of q, and choosing the contour shown in Figure 13.1. The residue theorem of complex analysis tells us that the integral of a function along a closed anticlockwise contour is given by

$$\oint_C dz\, f(z) = 2\pi i \sum (\text{residues at the poles enclosed}), \tag{13.16}$$

where the residue at a point z_0 is the coefficient of $1/(z - z_0)$ if the function $f(z)$ is expanded around $z = z_0$. In this case, there is only one pole enclosed, as seen from Figure 13.1. In the limit that the radius of the curved part goes to infinity, the contribution from the arc is zero, and this leads to

$$G_0(\boldsymbol{r} - \boldsymbol{r}') = -\frac{m}{2\pi\hbar^2}\frac{e^{ik|\boldsymbol{r}-\boldsymbol{r}'|}}{|\boldsymbol{r} - \boldsymbol{r}'|}. \tag{13.17}$$

Putting this solution for the Green function back into Eq. (13.7), we obtain

$$\psi(\boldsymbol{r}) = \psi_{\text{in}}(\boldsymbol{r}) - \frac{m}{2\pi\hbar^2}\int d^3r'\,\frac{e^{ik|\boldsymbol{r}-\boldsymbol{r}'|}}{|\boldsymbol{r} - \boldsymbol{r}'|}V(\boldsymbol{r}')\psi(\boldsymbol{r}'). \tag{13.18}$$

□ **Exercise 13.2** *Supply the missing steps that lead to Eq. (13.17), evaluating the residue and proving that the contribution from the curved part of Figure 13.1 vanishes.*

Since we are interested in the state of the scattered particle far away from the potential, we set $|\boldsymbol{r}| \gg |\boldsymbol{r}'|$ in Eq. (13.18). The exponent in the integrand can then be written as

$$k|\boldsymbol{r} - \boldsymbol{r}'| \simeq k\left(r - \frac{\boldsymbol{r}\cdot\boldsymbol{r}'}{r}\right) = kr - \boldsymbol{k}'\cdot\boldsymbol{r}', \tag{13.19}$$

with

$$\boldsymbol{k}' = k\boldsymbol{r}/r. \tag{13.20}$$

Thus, the direction of $\boldsymbol{k}'$ is the direction of $\boldsymbol{r}$, i.e, the direction of the scattered wave. Using this definition, we can write

$$\psi(\boldsymbol{r}) \simeq \psi_{\text{in}}(\boldsymbol{r}) + \frac{e^{ikr}}{r}f(\boldsymbol{k}, \boldsymbol{k}'), \tag{13.21}$$

with

$$f(\boldsymbol{k}, \boldsymbol{k}') = -\frac{m}{2\pi\hbar^2}\int d^3r'\,e^{i\boldsymbol{k}'\cdot\boldsymbol{r}'}V(\boldsymbol{r}')\psi(\boldsymbol{r}'). \tag{13.22}$$

This quantity is called the *scattering amplitude*. It has the dimension of length, and its $\boldsymbol{k}$-dependence comes from $\psi(\boldsymbol{r}')$.

Eq. (13.21) represents a superposition of an incoming plane wave with an outgoing spherical wave. The only assumption entering the equation is that we have taken $r \gg r'$. It should be noted that Eq. (13.21) is not a solution of the scattering problem by any means, because the right side of the equation contains $\psi(\boldsymbol{r})$ also, as part of the integrand defining the scattering amplitude. It is an integral equation which must be solved to obtain $\psi(\boldsymbol{r})$. This is usually difficult; therefore in most cases, one designs several approximation schemes to obtain analytic, albeit perturbative, solutions for $f(\boldsymbol{k}, \boldsymbol{k}')$.

☐ **Exercise 13.3** *Verify the following dimensions for the objects used in this section, and would be used in the rest of this chapter.*

$$\dim\left(\psi(\boldsymbol{r})\right) = 1,$$

$$\dim\left(\,|\boldsymbol{r}\rangle\,\right) = L^{-\frac{3}{2}},$$

$$\dim\left(\,|\psi\rangle\,\right) = L^{+\frac{3}{2}},$$

$$\dim\left(G_0(\boldsymbol{r},\boldsymbol{r}')\right) = M^{-1}L^{-5}T^2,$$

$$\dim\left(f(\boldsymbol{k},\boldsymbol{k}')\right) = L. \tag{13.23}$$

[**Note:** *The dimension of $\psi(\boldsymbol{r})$ is different from what we had used earlier for normalizable wavefunctions.*]

13.2   Scattering cross-section

The efficacy of a scattering process is usually denoted by the scattering cross-section. In this section, we find some relations between the cross-section and the scattering amplitude.

As already noted, Eq. (13.21) is a superposition of the incoming wave, $\psi_{\text{in}}(\boldsymbol{r})$, and the outgoing scattered wave, which can be separately written as

$$\psi_{\text{sc}}(\boldsymbol{r}) = \frac{e^{ikr}}{r} f(\boldsymbol{k},\boldsymbol{k}'). \tag{13.24}$$

The probability current density was defined in Eq. (3.86, p. 76). For the scattered wave, it is given by

$$\boldsymbol{J}_{\text{sc}} = \frac{\hbar}{m}\,\text{Im}\left(\psi_{\text{sc}}^*\boldsymbol{\nabla}\psi_{\text{sc}}\right) = \frac{\hbar}{m}\,\text{Im}\left(\psi_{\text{sc}}^*\frac{\partial}{\partial r}\psi_{\text{sc}}\right)\mathring{\boldsymbol{r}} + \cdots, \tag{13.25}$$

where the dots contain components perpendicular to the radial direction. Using the explicit form for the scattered wave, we obtain

$$\boldsymbol{J}_{\text{sc}} = \frac{\hbar\mathbf{k}}{mr^2}\left|f(\boldsymbol{k},\boldsymbol{k}')\right|^2\mathring{\boldsymbol{r}} + \cdots. \tag{13.26}$$

The total outgoing flux through a surface of constant r is therefore given by

$$\begin{aligned}
I_{\text{sc}} &= \int d\Omega\, r^2 \boldsymbol{J}_{\text{sc}}\cdot\mathring{\boldsymbol{r}}\\
&= \frac{\hbar\mathbf{k}}{m}\int d\Omega\,\left|f(\boldsymbol{k},\boldsymbol{k}')\right|^2,
\end{aligned} \tag{13.27}$$

where $d\Omega$ is an infinitesimal solid angle. On the other hand, the incident flux is

$$\boldsymbol{J}_{\text{in}} = \frac{\hbar}{m} \, \text{Im} \left(\psi_{\text{in}}^* \boldsymbol{\nabla} \psi_{\text{in}} \right), \tag{13.28}$$

which gives

$$|\boldsymbol{J}_{\text{in}}| = \frac{\hbar \mathbf{k}}{m}. \tag{13.29}$$

The scattering cross-section is then, by definition, the quantity

$$\sigma = \frac{I_{\text{sc}}}{|\boldsymbol{J}_{\text{in}}|} = \int d\Omega \left| f(\boldsymbol{k}, \boldsymbol{k}') \right|^2. \tag{13.30}$$

This is the connection between the scattering amplitude and the scattering cross-section. Equivalently, we can say that the differential scattering cross-section is given by

$$\frac{d\sigma}{d\Omega} = |f(\boldsymbol{k}, \boldsymbol{k}')|^2. \tag{13.31}$$

As the name indicates, the cross-section has the dimension of an area, i.e., square of length.

The angle between $\boldsymbol{k}$ and $\boldsymbol{k}'$ is called the *scattering angle*. For scattering from spherically symmetric potentials, the scattering amplitude can depend only on this angle θ, and is often written as $f(\theta)$.

Apart from Eq. (13.30), there is another important relation between the scattering cross-section and the scattering amplitude. To prove this relation, let us start with Eq. (13.22). Using the state vector notation, it can be written in the form

$$f(\boldsymbol{k}, \boldsymbol{k}') = -\frac{m}{2\pi\hbar^2} \left\langle \boldsymbol{k}' \left| V \right| \psi \right\rangle. \tag{13.32}$$

Consider now the forward scattering, i.e., scattering in which the scattered wave is along the incident wave. This means $\boldsymbol{k}' = \boldsymbol{k}$, and so

$$f(\boldsymbol{k}, \boldsymbol{k}) = -\frac{m}{2\pi\hbar^2} \left\langle \boldsymbol{k} \left| V \right| \psi \right\rangle. \tag{13.33}$$

Recalling that $|\psi_{\text{in}}\rangle = |\boldsymbol{k}\rangle$, from Eq. (13.5) we can write

$$\langle \boldsymbol{k}| = \langle \psi| - \langle \psi| (G_0 V)^\dagger = \langle \psi| - \langle \psi| V G_0^\dagger, \tag{13.34}$$

where in the last step we have used the fact that V is Hermitian. We can now write

$$f(\boldsymbol{k}, \boldsymbol{k}) = -\frac{m}{2\pi\hbar^2}\left(\langle\psi|V|\psi\rangle - \left\langle\psi\left|VG_0^\dagger V\right|\psi\right\rangle\right). \qquad (13.35)$$

The first term in the parenthesis is the expectation value of V in the state $|\psi\rangle$, which must be real since V is Hermitian. Therefore, if we consider the imaginary part of the forward scattering amplitude, we obtain

$$\operatorname{Im} f(\boldsymbol{k}, \boldsymbol{k}) = \frac{m}{2\pi\hbar^2}\operatorname{Im}\left\langle\psi\left|VG_0^\dagger V\right|\psi\right\rangle. \qquad (13.36)$$

Notice that we have written $G_0^\dagger$. We defined the operator G_0 in Eq. (13.6) as the inverse of a Hermitian operator. Since

$$(\mathscr{O}^\dagger)^{-1} = (\mathscr{O}^{-1})^\dagger \qquad (13.37)$$

for an operator $\mathscr{O}$, it follows that the inverse of a Hermitian operator is itself Hermitian, and so G_0 should be a Hermitian operator. If that were true, the matrix element appearing on the right side of Eq. (13.36) would have been real, making the expression vanish. The catch in this argument is that Eq. (13.37) is valid if the inverse really exists. Here we are talking of an operator which obviously has zero eigenvalues, and these are the ones that contribute to the imaginary part. This is also the reason that we had to put the term $i\eta$ in the denominator of Eq. (13.12).

To proceed, we introduce the T-matrix, defined by the relation

$$V|\psi\rangle = T|\psi_{\text{in}}\rangle = T|\boldsymbol{k}\rangle. \qquad (13.38)$$

The general expression for the scattering amplitude, Eq. (13.32), can be rewritten in terms of this newly introduced operator in the form

$$f(\boldsymbol{k}, \boldsymbol{k}') = -\frac{m}{2\pi\hbar^2}\langle\boldsymbol{k}'|T|\boldsymbol{k}\rangle, \qquad (13.39)$$

and the expression of Eq. (13.36) can be written as

$$\operatorname{Im} f(\boldsymbol{k}, \boldsymbol{k}) = \frac{m}{2\pi\hbar^2}\operatorname{Im}\left\langle\boldsymbol{k}\left|T^\dagger G_0^\dagger T\right|\boldsymbol{k}\right\rangle. \qquad (13.40)$$

Recall now from the discussion of §13.1 that

$$G_0 = (E - H_0)^{-1} = P[(E - H_0)^{-1}] - i\pi\delta(E - H_0), \qquad (13.41)$$

where P denotes the principal value. Therefore,

$$G_0^\dagger = P[(E - H_0)^{-1}] + i\pi\delta(E - H_0). \qquad (13.42)$$

The imaginary part of the forward scattering amplitude can then come only from the imaginary part of $G_0^\dagger$, and is given by

$$\text{Im} \, f(\boldsymbol{k}, \boldsymbol{k}) = \frac{m}{2\hbar^2} \left\langle \boldsymbol{k} \left| T^\dagger \delta(E - H_0) T \right| \boldsymbol{k} \right\rangle. \tag{13.43}$$

Let us now introduce a complete set of plane wave states in the middle, obtaining

$$\text{Im} \, f(\boldsymbol{k}, \boldsymbol{k}) = \frac{m}{2\hbar^2} \int \frac{d^3 k'}{(2\pi)^3} \left\langle \boldsymbol{k} \left| T^\dagger \delta(E - H_0) \right| \boldsymbol{k}' \right\rangle \left\langle \boldsymbol{k}' \left| T \right| \boldsymbol{k} \right\rangle$$

$$= \frac{m}{2\hbar^2} \int \frac{d^3 k'}{(2\pi)^3} \, \delta \left(\frac{\hbar^2 \mathbf{k}^2}{2m} - \frac{\hbar^2 \mathbf{k}'^2}{2m} \right) \left\langle \boldsymbol{k} \left| T^\dagger \right| \boldsymbol{k}' \right\rangle \left\langle \boldsymbol{k}' \left| T \right| \boldsymbol{k} \right\rangle. \tag{13.44}$$

Note that, by the definition of the adjoint operator, $\langle \boldsymbol{k} | T^\dagger | \boldsymbol{k}' \rangle = \langle \boldsymbol{k}' | T | \boldsymbol{k} \rangle^*$. Using an elementary property of Dirac delta functions shown in Eq. (A.20, p. 427), we can write

$$\delta \left(\frac{\hbar^2 \mathbf{k}^2}{2m} - \frac{\hbar^2 \mathbf{k}'^2}{2m} \right) = \frac{2m}{\hbar^2} \frac{\delta(\mathbf{k} - \mathbf{k}')}{2\mathbf{k}}. \tag{13.45}$$

Putting this into Eq. (13.44) and performing the integration on the magnitude of $\boldsymbol{k}'$, we obtain

$$\text{Im} \, f(\boldsymbol{k}, \boldsymbol{k}) = \frac{m}{2\hbar^2} \frac{m\mathbf{k}}{\hbar^2} \int \frac{d\Omega}{(2\pi)^3} \left| \left\langle \boldsymbol{k}' \left| T \right| \boldsymbol{k} \right\rangle \right|^2, \tag{13.46}$$

with $\mathbf{k}' = \mathbf{k}$. Using Eqs. (13.30) and (13.39), we finally obtain

$$\text{Im} \, f(\boldsymbol{k}, \boldsymbol{k}) = \frac{\mathbf{k}\sigma}{4\pi}, \tag{13.47}$$

where σ is the total cross-section. Equivalently, we can write

$$\sigma = \frac{4\pi}{\mathbf{k}} \, \text{Im}[f(\theta = 0)]. \tag{13.48}$$

This relation between the total cross-section and the forward scattering amplitude is called the *optical theorem*.

13.3   Born approximation

We have introduced the T-matrix in Eq. (13.38). Clearly, knowing the scattered wavefunction is equivalent to knowing the elements of the T-matrix, as is obvious from relations like Eq. (13.39). In this section, we shall outline a perturbative scheme for computing the T-matrix, and therefore $f(\boldsymbol{k}, \boldsymbol{k}')$.

For this, we operate both sides of Eq. (13.5) with V and use the definition of the T-matrix to obtain

$$T\left|\psi_{\text{in}}\right\rangle = V\left|\psi_{\text{in}}\right\rangle + VG_0 T\left|\psi_{\text{in}}\right\rangle. \tag{13.49}$$

Noting that $\left|\psi_{\text{in}}\right\rangle$ is an arbitrary initial state, one can interpret this equation as an operator identity:

$$T = V + VG_0 T. \tag{13.50}$$

Replacing the T on the right side by the entire expression, and continuing the process indefinitely, we obtain

$$T = V + VG_0 V + VG_0 VG_0 V + \cdots, \tag{13.51}$$

which can be written symbolically as

$$T = V(\mathbb{1} - G_0 V)^{-1}. \tag{13.52}$$

Eq. (13.51) represents a perturbative expansion of T in terms of V in the limit of weak V. In view of Eq. (13.39), we can write

$$f(\boldsymbol{k}, \boldsymbol{k}') = \sum_{n=0}^{\infty} f^{(n)}(\boldsymbol{k}', \boldsymbol{k}), \tag{13.53}$$

with

$$f^{(n)}(\boldsymbol{k}, \boldsymbol{k}') = -\frac{m}{2\pi\hbar^2}\langle \boldsymbol{k}'|V(G_0 V)^{n-1}|\boldsymbol{k}\rangle, \tag{13.54}$$

which means that the nth term in the expansion has $n-1$ factors of G_0 and n factors of V. In practice, the series for the scattering amplitude, given in Eq. (13.53), is truncated after n terms; this approximation is called the nth order *Born approximation*. For example, if we keep only $f^{(1)}$, the result will be said to be true in the first Born approximation, and so on.

The Born approximation allows for simple calculation of scattering amplitudes. For example, let us consider the well-known Yukawa potential

$$V(r) = \frac{e^{-\mu r}}{r}. \tag{13.55}$$

Within the first Born approximation, one gets $f(\boldsymbol{k'}, \boldsymbol{k}) \simeq f^{(1)}(\boldsymbol{k'}, \boldsymbol{k})$, where

$$f^{(1)}(\boldsymbol{k}, \boldsymbol{k'}) = -\frac{m}{2\pi\hbar^2}\langle \boldsymbol{k'}|V|\boldsymbol{k}\rangle = -\frac{m}{2\pi\hbar^2}\int d^3r \frac{e^{-\mu r}}{r}e^{i(\boldsymbol{k}-\boldsymbol{k'})\cdot\boldsymbol{r}}. \tag{13.56}$$

To evaluate this integral, we write $|\boldsymbol{k} - \boldsymbol{k'}| = \mathsf{q} = 2\mathsf{k}\sin\alpha/2$, where α denotes the angle between $\boldsymbol{k}$ and $\boldsymbol{k'}$. Using this we write

$$\begin{aligned}
f^{(1)}(\boldsymbol{k}, \boldsymbol{k'}) &= -\frac{m}{4\pi^2\hbar^2}\int rdr e^{-\mu r}\int d\theta \sin\theta e^{iqr\cos\theta} \\
&= -\frac{m}{2q\pi^2\hbar^2}\int dr e^{-\mu r}\sin qr \\
&= -\frac{m}{\pi^2\hbar^2}\frac{1}{\mu^2 + 4|\boldsymbol{k}|^2\sin^2(\alpha/2)}. \tag{13.57}
\end{aligned}$$

Thus, we find that the scattering amplitude is maximal in the forward scattering limit ($\alpha = 0$ or $\boldsymbol{k} \parallel \boldsymbol{k'}$) where $f(\boldsymbol{k}, \boldsymbol{k'}) \sim \mu^{-2}$ and minimal for backscattering ($\alpha = \pi$ or $\boldsymbol{k}$ is antiparallel to $\boldsymbol{k'}$) where $f(\boldsymbol{k}, \boldsymbol{k'}) \sim 1/(\mu^2+4|\boldsymbol{k}|^2)$. We also note that for $\mu = 0$ the forward scattering amplitude diverges. This divergence is an artifact of the fact that in this limit, the potential does not remain short range and our formalism is not expected to work in this limit. In fact, for $\mu = 0$, the potential is the Coulomb potential. Scattering from Coulomb potential will be discussed later in §13.5.

The first Born approximation, while yielding reasonable estimates of $f(\boldsymbol{k}, \boldsymbol{k'})$ for a weak scattering potential, has its limitations. For example, it provides qualitatively wrong answer for the imaginary part of $f(\boldsymbol{k}, \boldsymbol{k})$. To see this, we note that according to the first Born approximation, $f(\boldsymbol{k}, \boldsymbol{k}) = \langle\boldsymbol{k}|V|\boldsymbol{k}\rangle$, so $\mathrm{Im}[f(\boldsymbol{k}, \boldsymbol{k})] = 0$ for any Hermitian operator V. This contradicts the optical theorem, which was stated and proved in §13.2.

□ **Exercise 13.4** *Compute $f(\boldsymbol{k}, \boldsymbol{k'})$ for a potential $V = V_0 > 0$ for $r < a$ and $V = 0$ for $r > 0$ within first Born approximation. Show that f becomes independent of θ and ϕ for $E \simeq 0$.*

□ **Exercise 13.5** *Compute $f(\boldsymbol{k}, \boldsymbol{k'})$ within first Born approximation for a Gaussian potential $V(r) = V_0 e^{-r^2/a^2}$. Find σ in the limit of low energy $E \simeq 0$.*

13.4 Partial wave analysis

For potentials which preserve rotational symmetry, H is invariant under an arbitrary rotation in space. This means that $[\boldsymbol{L}, H] = 0$, where $\boldsymbol{L}$ is the angular momentum operator. Angular momentum channels with different values of the eigenvalue of $\boldsymbol{L}^2$ are therefore decoupled, and so it is advantageous to discus properties of such Hamiltonians for a fixed angular momentum. The same consideration holds for the scattering amplitude $f(\boldsymbol{k}, \boldsymbol{k}')$ which we discuss in this section. In this situation, the incoming plane wave allows for a θ dependence of the wavefunction; however, the wavefunction must be independent of the azimuthal angle ϕ. Stated differently, this means that the corresponding wavefunction must always have $m = 0$, where m is the azimuthal quantum number.

We first consider a free particle at $r \to \infty$:

$$\psi_{\boldsymbol{k}}(\boldsymbol{r}) = \exp[i\boldsymbol{k} \cdot \boldsymbol{r}] = \exp[ikr\cos\theta]. \tag{13.58}$$

Obviously this is the solution of the differential equation

$$(\nabla^2 + k^2)\psi_{\boldsymbol{k}}(\boldsymbol{r}) = 0. \tag{13.59}$$

Noting that the solution is independent of the azimuthal angle, we can expand it in terms of the Legendre polynomials:

$$\psi_{\boldsymbol{k}}(\boldsymbol{r}) = \sum_{\ell} (2\ell + 1) P_{\ell}(\cos\theta) R_{\ell}(r). \tag{13.60}$$

The differential equation for the radial part of this wavefunction was derived in §9.3 for arbitrary central potentials. For the present case of no potential, Eq. (9.29, p. 224) reduces to

$$\frac{d^2 R_{\ell}(r)}{dr^2} + \frac{2}{r}\frac{d R_{\ell}(r)}{dr} + \left[k^2 - \frac{\ell(\ell+1)}{r^2}\right] R_{\ell}(r) = 0. \tag{13.61}$$

This is the *spherical Bessel equation*, discussed in detail in Appendix C. The most general solution of this equation is

$$R_{\ell}(r) = a_{\ell} j_{\ell}(kr) + b_{\ell} \eta_{\ell}(kr), \tag{13.62}$$

where j_{ℓ} and η_{ℓ} denote spherical Bessel functions of the first and the second kinds respectively:

$$j_{\ell}(x) = (-x)^{\ell} \left(\frac{1}{x}\frac{d}{dx}\right)^{\ell} \frac{\sin x}{x}, \tag{13.63a}$$

$$\eta_{\ell}(x) = -(-x)^{\ell} \left(\frac{1}{x}\frac{d}{dx}\right)^{\ell} \frac{\cos x}{x}. \tag{13.63b}$$

The fact that the solution must be finite at origin sets $b_\ell = 0$ since η_ℓ diverges at the origin. Using the orthogonality relation of the Legendre polynomials given in Eq. (D.52, p. 481), we then obtain

$$R_\ell(r) = a_\ell j_\ell(kr) = \frac{1}{2} \int_{-1}^{+1} d(\cos\theta) \, P_\ell(\cos\theta) e^{ikr\cos\theta}. \qquad (13.64)$$

The coefficients a_ℓ will now depend on the normalization chosen for the spherical Bessel functions. In the normalization advocated in Eq. (13.63), it turns out that $a_\ell = i^\ell$ so that

$$\psi_{\boldsymbol{k}}(\boldsymbol{r}) = e^{ikr\cos\theta} = \sum_{\ell=0}^{\infty} j_\ell(kr) i^\ell P_\ell(\cos\theta)(2\ell+1). \qquad (13.65)$$

In Appendix E, it has been shown that the asymptotic form for the spherical Bessel functions is

$$j_\ell(x \to \infty) \sim \frac{\sin(x - \ell\pi/2)}{x}, \qquad (13.66)$$

where the '$\sim$' sign has been put to indicate that non-leading terms have been omitted in the expression. Using this one can write

$$\psi_{\boldsymbol{k}}(\boldsymbol{r} \to \infty) = \sum_{\ell=0}^{\infty} i^\ell (2\ell+1) P_\ell(\cos\theta) \frac{e^{i(kr-\ell\pi/2)} - e^{-i(kr-\ell\pi/2)}}{2ikr}. \qquad (13.67)$$

This means that for every ℓ there is an incoming and an outgoing wave at $r \to \infty$ with a relative *phase shift* $\ell\pi$.

□ **Exercise 13.6** *Take the expression of the Legendre polynomials for $\ell = 0, 1, 2$ and check explicitly that the integral of Eq. (13.64) is equal to $i^\ell j_\ell(kr)$, with j_ℓ defined in Eq. (13.63a).*

What happens when we turn on a localized potential? Since we are interested in the form of the wavefunction far away from the range of the potential, the net flux of incoming and outgoing waves across any surface at large r must be conserved. Thus such a localized potential may at most lead to ℓ-dependent shifts in the phase of the wavefunction. In the presence of such a potential, the wavefunction at $r \to \infty$ can be written as

$$\psi(r \to \infty) = \sum_{\ell=0}^{\infty} C_\ell \frac{e^{i(kr-\ell\pi/2+\delta_\ell)} - e^{-i(kr-\ell\pi/2+\delta_\ell)}}{r}, \qquad (13.68)$$

where δ_ℓ is the phase shift for each ℓ, and C_ℓ is an yet-undetermined coefficient. To determine C_ℓ, we note that the effect of the potential cannot change the incoming wave. Thus, comparing the second term in the right side of Eq. (13.68) with that of Eq. (13.67), we find

$$C_\ell = \frac{(2\ell + 1)}{2ik} P_\ell(\cos\theta) e^{i(\ell\pi/2 + \delta_\ell)}. \tag{13.69}$$

Substituting Eq. (13.69) into Eq. (13.68), we find

$$\psi_{\boldsymbol{k}}(r \to \infty) \sim \sum_{\ell=0}^{\infty} (2\ell + 1) P_\ell(\cos\theta) \frac{e^{i(kr + 2\delta_\ell)} - e^{-i(kr - \ell\pi)}}{2ikr}$$

$$= e^{ikz} + \sum_{\ell=0}^{\infty} \frac{e^{ikr}}{r} \left[\frac{e^{2i\delta_\ell} - 1}{2ik} \right] (2\ell + 1) P_\ell(\cos\theta), \tag{13.70}$$

where we have used Eq. (13.67) to extract the term e^{ikz}, noting that $i^\ell = e^{i\ell\pi/2}$.

Next we consider the form of the outgoing wave derived earlier in §13.1. The scattering amplitude was defined in Eq. (13.22). Since it is a function of θ only for rotationally invariant potentials, we can write

$$f(\boldsymbol{k}, \boldsymbol{k}') \equiv f(k, \theta) = \sum_\ell (2\ell + 1) P_\ell(\cos\theta) f_\ell(k). \tag{13.71}$$

Thus, the scattered wave will be given by

$$\psi_{\rm sc}(\boldsymbol{r} \to \infty) = \frac{e^{ikr}}{r} \sum_\ell (2\ell + 1) P_\ell(\cos\theta) f_\ell(k), \tag{13.72}$$

from Eq. (13.24). Comparing this expression with Eq. (13.68), we find

$$f_\ell(k) = \frac{e^{i\delta_\ell}}{k} \sin\delta_\ell. \tag{13.73}$$

The total scattering cross-section can be obtained by integrating $|f(k, \theta)|^2$ over all possible angles, as indicated in Eq. (13.31). The Legendre polynomials are usually normalized by the rule given in Eq. (D.52, p. 481). Using it we find

$$\sigma = \sum_\ell \sigma_\ell, \tag{13.74a}$$

where

$$\sigma_\ell = \frac{4\pi}{k^2} (2\ell + 1) \sin^2\delta_\ell. \tag{13.74b}$$

☐ **Exercise 13.7** *Starting from Eq. (13.31) and using Eqs. (13.71) and (13.73), perform the angular integrations to obtain Eq. (13.74).*

Eq. (13.71) and Eq. (13.74b) lead to two important relations. The first is called the *unitarity bound*. It states that in any angular momentum channel, the scattering cross-section is less than a maximum value given by

$$\sigma_\ell^{\max} = \frac{4\pi}{k^2}(2\ell + 1). \tag{13.75}$$

This result directly follows from Eq. (13.74b) since $\sin^2 \delta_\ell \leq 1$.

The other result is an alternative derivation of the optical theorem. From Eq. (13.71), we find that for $\theta = 0$, $P_\ell(\cos \theta) = 1$ and hence

$$f(\theta = 0) = \sum_\ell (2\ell + 1) f_\ell(k). \tag{13.76}$$

Therefore,

$$\operatorname{Im} f(\theta = 0) = \sum_\ell (2\ell + 1) \operatorname{Im} f_\ell(k) = \sum_\ell (2\ell + 1) \frac{\sin^2 \delta_\ell}{k}, \tag{13.77}$$

using the expression for f_ℓ given in Eq. (13.73). Comparing it with the expression of the total cross-section given through Eq. (13.74), we obtain the optical theorem, as given in Eq. (13.48).

We note that both these results hold irrespective of the nature of the phase shift δ_ℓ. Thus, they are completely general and independent of the details of the potential (as long as it is short ranged and rotationally symmetric).

Next, we would like to comment about the efficacy of the partial wave expansion. We note that if we needed to compute δ_ℓ for all ℓ, the method would be useless for computation. However, this is not the case; typically for a momentum $\hbar k$ (which is fixed by the incident energy $E = \hbar^2 k^2/2m$) and a range of the potential a_0, the maximum angular momentum channel which registers a non-trivial phase shift is $\hbar \ell_{\max} \simeq \hbar k a_0$ or $\ell_{\max} \simeq k a_0$. This ensures that as long we know δ_ℓ for $\ell \leq \ell_{\max}$, we may get an accurate estimate of the cross-section. This is particularly useful for low-energy incident particles and/or short-range scattering potentials where $\ell_{\max}$ remains small.

Let us therefore discuss the method of computation of the phase shifts δ_ℓ or equivalently of the quantities f_ℓ. To this end, we first note that we are talking of localized potentials here, so beyond a certain value of r, the potential term can be neglected. In this far regime the differential equation is the spherical

Bessel equation as shown in Eq. (13.61) and discussed in Appendix C, and so the solution of the Schrödinger equation can be written as

$$R_\ell(r) = A_\ell j_\ell(kr) + B_\ell \eta_\ell(kr), \tag{13.78}$$

where j_ℓ and η_ℓ are spherical Bessel functions of first and second kinds respectively. We now impose the condition that for $r \to \infty$, the solution of the Schrödinger equation must be in the form of a superposition of a plane wave and an outgoing spherical wave. For this, it is better to reorganize the terms of Eq. (13.78) in the form

$$R_\ell(r) = A'_\ell j_\ell(kr) + B'_\ell h_\ell^{(1)}(kr), \tag{13.79}$$

where

$$h_\ell^{(1)}(x) = j_\ell(x) + i\eta_\ell(x) \tag{13.80}$$

is called the *spherical Hankel function* of the first kind. The advantage of this combination is that its asymptotic form is given by

$$h_\ell^{(1)}(x \to \infty) \sim (-i)^{\ell+1} \frac{e^{ix}}{x}, \tag{13.81}$$

as can be easily verified by using Eq. (13.63). Therefore, the B'_ℓ terms comprise the outgoing spherical wave. The rest, i.e., the A'_ℓ terms, must then represent the plane wave. According to the discussion leading to Eq. (13.65), we must then obtain $A'_\ell = i^\ell$, and the wavefunction can be written as

$$\psi(r \to \infty) = \sum_\ell (2\ell + 1) P_\ell(\cos\theta) R_\ell(r)$$

$$= e^{ikz} + \frac{e^{ikr}}{kr} \sum_\ell (-i)^{\ell+1}(2\ell + 1) B'_\ell P_\ell(\cos\theta). \tag{13.82}$$

Comparing it with Eq. (13.21), we see that the scattering amplitude is given by

$$f(\boldsymbol{k}, \boldsymbol{k}') = \frac{1}{k} \sum_\ell (-i)^{\ell+1}(2\ell + 1) B'_\ell P_\ell(\cos\theta), \tag{13.83}$$

from which we can identify f_ℓ through Eq. (13.71), and then the phase shift δ_ℓ through Eq. (13.73).

The coefficients B'_ℓ depend on the specific potential that is responsible for the scattering, and can be obtained only by matching the far solution of

Eq. (13.82) with the wavefunction within the range of the potential. Here we present the steps for a specific choice of the potential.

The simplest example is a spherical infinite barrier (often called a hard sphere) such that $V(r) = \infty$ for $r < a$ and $= 0$ for $r > a$. Here the boundary condition is $\psi(r = a) = 0$. Since this has to be satisfied for all θ, we can equate each partial wave contribution to zero to obtain

$$B'_\ell = i^{2\ell+1} \frac{j_\ell(ka)}{h_\ell^{(1)}(ka)}. \tag{13.84}$$

One can now identify f_ℓ through Eq. (13.83) and use it to obtain

$$\sigma = \frac{4\pi}{k^2} \sum_{\ell=0}^{\infty} \left| \frac{j_\ell(ka)}{h_\ell^{(1)}(ka)} \right|^2 \simeq \frac{4\pi}{k^2} \sum_{\ell=0}^{\infty} (ka)^{2(2\ell+1)} \frac{2^\ell \ell!}{(2\ell)!} \frac{1}{2\ell + 1}, \tag{13.85}$$

where the expansion has been carried out for low-energy scattering for which $ka \ll 1$. Interestingly, if one retains only the first term ($\ell = 0$) of the expansion which is accurate if $E \ll \hbar^2/(2ma^2)$, one obtains $\sigma \simeq 4\pi a^2$ which is four times the geometric cross-section and is the total surface area of the sphere. This indicates the wave nature of the incident particle at low energies; whereas classical particles see only a cross-section of the sphere, the waves see its entire surface area.

☐ **Exercise 13.8** *Consider the expression of B'_ℓ obtained from Eq. (13.84). Use this to obtain f_ℓ using Eq. (13.83). Finally, use it to obtain the expression for σ given in Eq. (13.85).*

☐ **Exercise 13.9** *Consider a spherical delta function shell potential $V(r) = V_0 \delta(r/r_0 - 1)$. Show that the low-energy scattering cross-section is given by $\sigma = 4\pi r_0^2/(1 + \mu^2)$ where $\mu = \hbar^2/(2mV_0 r_0^2)$.*

☐ **Exercise 13.10** *Consider a delta function potential $V(r) = A\delta^3(\boldsymbol{r})$ in three dimensions.*

   (a) *What is the dimension of the constant A?*

   (b) *By dimensional arguments, show that the low-energy cross-section is proportional to $A^2 m^2/\hbar^4$, independent of the energy.*

   (c) *Which partial waves will be responsible for the scattering of a particle of mass m via this potential?*

   (d) *Find the proportionality constant that appears in the expression for the cross-section.*

   (e) *Argue how this result might be obtained as a limit of the result of Ex. 13.9.*

☐ **Exercise 13.11** *Consider a finite spherical well or barrier such that $V = V_0$ for $r < a$ and $V = 0$ outside. Find the phase shift in either case for $\ell = 0$ by matching the outside and inside wavefunctions and their derivatives at $r = a$. Show that for $V_0 < 0$ and for large $|V_0| \gg E$, one may have $\delta_{\ell=0} = \pi$ so that one has $\sigma_{\ell=0} = 0$. This is called the* Ramsauer–Townsend *effect.*

☐ **Exercise 13.12** *Consider the situation where the phase shift due to ℓ^{th} partial wave is given by*

$$\tan\delta_\ell = \frac{\Gamma/2}{E - E_0}. \tag{13.86}$$

Show that

$$\sigma_\ell = (4\pi/k^2)(2\ell + 1)\frac{(\Gamma/2)^2}{(E - E_0)^2 + (\Gamma/2)^2} \tag{13.87}$$

in this case. Find the value of E for which σ_ℓ saturates the unitarity bound. Show that the pole of σ_ℓ can be interpreted as a metastable particle with $E \simeq E_0$ and a lifetime $\hbar/\Gamma$. This phenomenon is called the Breit–Wigner *resonance.*

13.5 Coulomb scattering

So far, we talked about scattering from short-range potentials, for which the amplitude of the scattered wavefunction dies out sufficiently fast at large distances. This formalism fails for a Coulomb potential, as was indicated in connection with the calculation of scattering amplitude by a Yukawa potential, given in Eq. (13.57).

To set up the problem from scratch, let us consider an incoming plane wave,

$$\psi_{\text{in}}(\boldsymbol{r}) = e^{ikz}, \tag{13.88}$$

with energy

$$E = \frac{\hbar^2 k^2}{2m}. \tag{13.89}$$

The Schrödinger equation would be

$$\nabla^2\psi(\boldsymbol{r}) + \frac{2m}{\hbar^2}\left(E - \frac{C}{r}\right)\psi(\boldsymbol{r}) = 0, \tag{13.90}$$

where the constant C involves the product of the charges of the particle producing the Coulomb field as well as of the particle being scattered.

In order to find the desired solution, it is useful to work in the *parabolic cylindrical coordinates*, defined by

$$\xi = r - z, \qquad \chi = r + z, \qquad \phi. \tag{13.91}$$

In this system, the expression for the Laplacian is

$$\nabla^2 = \frac{4}{\xi + \chi} \left[\frac{\partial}{\partial \xi} \left(\xi \frac{\partial}{\partial \xi} \right) + \frac{\partial}{\partial \chi} \left(\chi \frac{\partial}{\partial \chi} \right) \right] + \frac{1}{\xi \chi} \frac{\partial^2}{\partial \phi^2}. \tag{13.92}$$

We have a spherically symmetric potential so that the scattered wave should be independent of ϕ. Therefore, the equation becomes

$$\frac{\partial}{\partial \xi} \left(\xi \frac{\partial \psi}{\partial \xi} \right) + \frac{\partial}{\partial \chi} \left(\chi \frac{\partial \psi}{\partial \chi} \right) + \left(\frac{1}{4} (\xi + \chi) k^2 - \frac{mC}{\hbar^2} \right) \psi = 0, \tag{13.93}$$

using Eq. (13.89).

Let us now write the wavefunction in the form

$$\psi(\boldsymbol{r}) = e^{ikz} F(\boldsymbol{r}). \tag{13.94}$$

The first factor is nothing but the incoming wave. All modifications due to scattering is encoded in the function F. However, there is something that can be guessed about this function. In the solution, we want a factor e^{ikr}, that can appear only if this part F contains a factor of $e^{ik(r-z)}$, i.e., of $e^{ik\xi}$. Thus, we suspect that F is a function of ξ only and write, more explicitly,

$$\psi(\boldsymbol{r}) = e^{\frac{1}{2} ik(\chi - \xi)} F(\xi). \tag{13.95}$$

Putting this form into Eq. (13.93), we obtain a differential equation for the function $F(\xi)$:

$$\xi F''(\xi) + (1 - ik\xi) F'(\xi) - \frac{mC}{\hbar^2} F = 0. \tag{13.96}$$

This equation does not contain χ. Comparing this equation with Eq. (C.108, p. 471) in Appendix C, we realize that the solutions are the *confluent hypergeometric functions*. The general solution can be written, by using the shorthand

$$\gamma = -mC/k\hbar^2, \tag{13.97}$$

as

$$F(\xi) = M(-i\gamma, 1; ik\xi). \tag{13.98}$$

Of course, the general solution of the second-order differential equation is a linear combination of the two independent solutions. The right combination will be determined by the behavior of the wavefunction at the origin. The remaining overall normalization, although irrelevant, can be fixed from the normalization of the incoming plane wave.

We now present, without proof, the form of this relevant combination of functions when $-ix$ becomes very large:

$$M(-i\gamma, 1; x) \longrightarrow a_\gamma \left[e^{i\gamma \ln(-ix)}(1 - \gamma^2/x + \cdots) \right.$$
$$\left. + \frac{\Gamma(1+i\gamma)}{\Gamma(-i\gamma)} \frac{e^x}{x} e^{-i\gamma \ln(-ix)} \right], \qquad (13.99)$$

where a_γ is an overall constant which is irrelevant for calculating the cross-section. In our case, the independent variable is $x = ik\xi$. Thus, the limit shown in this equation corresponds to large values of $k\xi$, or $k(r - z)$, i.e., $kr(1 - \cos\theta)$. For all values of θ except $\theta = 0$, this limit therefore corresponds to large kr, which is exactly what we want. In this limit, then, multiplying it with the factor e^{ikz} that was present in the expression for $\psi(\boldsymbol{r})$, we see that the wavefunction can be written as the sum of two terms:

$$\psi_1(\boldsymbol{r}) = a_\gamma e^{ikz+i\gamma \ln k(r-z)} \left(1 - \frac{\gamma^2}{ik(r-z)} + \cdots \right), \qquad (13.100a)$$

$$\psi_2(\boldsymbol{r}) = a_\gamma \frac{\Gamma(1+i\gamma)}{\Gamma(-i\gamma)} \frac{e^{ikr-i\gamma \ln k(r-z)}}{ik(r-z)}. \qquad (13.100b)$$

Note that ψ_1 has roughly the form of the plane wave. In the exponent, the term with $\ln k(r-z)$ can be neglected for large r. On the other hand, ψ_2 contains the factors necessary for qualifying as the scattered wave. Noting that

$$r - z = r(1 - \cos\theta) = 2r\sin^2(\theta/2), \qquad (13.101)$$

we can write

$$\psi_2(\boldsymbol{r}) = \psi_{\rm sc}(\boldsymbol{r}) = a_\gamma \frac{e^{ikr-i\gamma \ln(2kr)}}{r} f_c(\theta), \qquad (13.102)$$

where $f_c(\theta)$ is the scattering amplitude appropriate for Coulomb scattering, given by

$$f_c(\theta) = \frac{\Gamma(1+i\gamma)}{\Gamma(-i\gamma)} \frac{e^{-i\gamma \ln(\sin^2(\theta/2))}}{2ik\sin^2(\theta/2)}. \qquad (13.103)$$

Since the *gamma function* has the properties

$$\Gamma(\zeta + 1) = \zeta\Gamma(\zeta), \qquad \Gamma(\zeta^*) = [\Gamma(\zeta)]^* \tag{13.104}$$

for any complex ζ, we can write

$$\frac{\Gamma(1 + i\gamma)}{\Gamma(-i\gamma)} = i\gamma\frac{\Gamma(i\gamma)}{\Gamma(-i\gamma)} = i\gamma e^{2i\delta_0}, \tag{13.105}$$

where δ_0 is the phase of $\Gamma(i\gamma)$. So the scattering length is given by

$$f_c(\theta) = \frac{\gamma}{2k}e^{2i\delta_0}\left(\sin^2(\theta/2)\right)^{-i\gamma - 1}. \tag{13.106}$$

The probability current densities of the incident and scattered waves can be calculated in exactly the same way as was done in §13.2, giving the result

$$\frac{d\sigma}{d\Omega} = \left|f_c(\theta)\right|^2. \tag{13.107}$$

Putting in the expression for $f_c(\theta)$, we obtain

$$\frac{d\sigma}{d\Omega} = \frac{\gamma^2}{4k^2\sin^4(\theta/2)} = \frac{C^2}{16E^2\sin^4(\theta/2)}, \tag{13.108}$$

using the definition of γ from Eq. (13.97) and the expression for E given in Eq. (13.89). This is the Rutherford scattering formula, which Rutherford derived from purely classical considerations. It is just a coincidence that the quantum mechanical formula is exactly the same.

Looking back at the derivation, we see why the formalism developed in earlier sections of this chapter did not apply for Coulomb scattering. Eq. (13.102), for example, shows that there is an r-dependent phase shift. The scattered wave is definitely not of the form specified in Eq. (13.24): it falls slower than $1/r$ for large values of r. It is this property that classifies the Coulomb force as a long-range force.

Part Four

Advanced topics

Chapter 14

Identical particles

We mentioned in Chapter 5 that the wavefunction of a system consisting of several identical particles has some special symmetries, related to the permutation of the particles. In this chapter, we discuss the issue in detail, throwing light on some consequences of the symmetry.

14.1   Permutation symmetry

Consider first the Hamiltonian of N identical free particles of mass m:

$$H_0 = \sum_{a=1}^{N} \frac{\boldsymbol{p}_a^2}{2m},\tag{14.1}$$

where $\boldsymbol{p}_a$ is the momentum of the ath particle. Obviously, this Hamiltonian has the property that it remains unchanged if we change the labels a on the particles, i.e., if we perform any permutation of the particles. Now consider the Hamiltonian

$$H = \sum_{a=1}^{N} \frac{\boldsymbol{p}_a^2}{2m} + V(\boldsymbol{r}_1, \boldsymbol{r}_2, \cdots, \boldsymbol{r}_N),\tag{14.2}$$

where the potential is also invariant under any permutation of the particles. Hamiltonians involving identical particles must have such symmetry.

Any *permutation* of N objects can be seen as a product of a number of *transpositions*, i.e., permutations involving only two objects. For example, if we consider three objects, to be called 1, 2, and 3, and consider the permutation $\langle 321 \rangle$, meaning that the object 3 has come to the first place and object 1 at the third place, leaving object 2 undisturbed, it is easily seen that it can be thought of as the result of the following operations:

$$\langle 123 \rangle \to \langle 213 \rangle \to \langle 231 \rangle \to \langle 321 \rangle . \tag{14.3}$$

In the first step, only the first two objects have been interchanged, followed by the interchange of the last two in the second step, and so on. Therefore, it is sufficient to focus our attention to the transpositions while discussing permutations. For the purposes of illustration, we will always use P_{12}, which interchanges the first two objects.

If we consider a basis state of N particles in the coordinate representation, the transposition acts on it as follows:

$$P_{12} \left| \boldsymbol{r}_1, \boldsymbol{r}_2, \cdots, \boldsymbol{r}_N \right\rangle = \left| \boldsymbol{r}_2, \boldsymbol{r}_1, \cdots, \boldsymbol{r}_N \right\rangle . \tag{14.4}$$

Notice that if we apply the same operator again, we get back the same state as the original one:

$$\left(P_{12} \right)^2 \left| \boldsymbol{r}_1, \boldsymbol{r}_2, \cdots, \boldsymbol{r}_N \right\rangle = \left| \boldsymbol{r}_1, \boldsymbol{r}_2, \cdots, \boldsymbol{r}_N \right\rangle , \tag{14.5}$$

implying the operator relation

$$\left(P_{12} \right)^2 = \mathbb{1}. \tag{14.6}$$

Thus, the eigenvalues of the transposition operator can be only $+1$ and -1. This result is true for the transposition of any two objects. In other words, we should be able to write

$$P_{ab} \left| \boldsymbol{r}_1, \boldsymbol{r}_2, \cdots, \boldsymbol{r}_N \right\rangle = \pm \left| \boldsymbol{r}_1, \boldsymbol{r}_2, \cdots, \boldsymbol{r}_N \right\rangle \tag{14.7}$$

for any interchange.

It is easy to see what it means for the wavefunction in the coordinate representation. For a state vector $\left| \Psi \right\rangle$, we define the wavefunction by the relation

$$\left\langle \boldsymbol{r}_1, \boldsymbol{r}_2, \cdots, \boldsymbol{r}_N \mid \Psi \right\rangle = \psi(\boldsymbol{r}_1, \boldsymbol{r}_2, \cdots, \boldsymbol{r}_N), \tag{14.8}$$

where the time dependence of the state vector, as well as of the associated wavefunction, has not been explicitly mentioned. Let us now try to evaluate the matrix element $\left\langle \boldsymbol{r}_1, \boldsymbol{r}_2, \cdots, \boldsymbol{r}_N \right| P_{12} \left| \Psi \right\rangle$. On the one hand, using Eq. (14.4), we find that it should equal $\psi(\boldsymbol{r}_2, \boldsymbol{r}_1, \cdots, \boldsymbol{r}_N)$. On the other hand, using

Eq. (14.7), we find that it should equal $\pm\psi(\boldsymbol{r}_1, \boldsymbol{r}_2, \cdots, \boldsymbol{r}_N)$. Thus, we obtain the result:

$$\psi(\boldsymbol{r}_2, \boldsymbol{r}_1, \cdots, \boldsymbol{r}_N) = \pm\psi(\boldsymbol{r}_1, \boldsymbol{r}_2, \cdots, \boldsymbol{r}_N), \tag{14.9}$$

and similar ones for the exchange of any other pair of objects.

It should be commented that the basis states need not be just the position eigenstates. For example, if we consider identical particles with spin, the basis states would comprise direct products of position eigenstates and spin eigenstates. In that case, the transposition operator P_{12} will exchange not only the positions, but also the spin states of the two particles. Eq. (14.6) applies only when all parameters pertaining to particle 1 are exchanged with the corresponding parameters of particle 2.

It turns out that the appearance of the sign in Eq. (14.7) is not random. There are certain kinds of particles for which all multiparticle states have the positive sign in Eq. (14.7) for any interchange, and there are other kinds of particles for which the negative sign appears for any interchange. The first kind of particles are called *bosons* and the second kind are called *fermions*.

14.2 Bosons and fermions

There is an intimate connection between the spin of a particle and the symmetry properties of multiparticle wavefunctions. In Chapter 8, we discussed that the eigenvalue of $\boldsymbol{S}^2$ of a system must be of the form $s(s+1)\hbar^2$, where s can be either an integer or a half-integer. It turns out that the objects with integral spin are bosons, whereas the objects with half-integral spin are fermions. There is no rigorous proof of this connection between the spin and the permutation properties, at least within the realm of non-relativistic quantum mechanics. In relativistic quantum mechanics, some connection can be established, subject to some mild assumptions. We do not discuss this connection in this book.

For a collection of identical bosons then, the wavefunction has to be completely symmetric, i.e., symmetric with respect to the exchange of any two particles. For a collection of fermions, the wavefunction has to be completely antisymmetric.

The simplest way of constructing a totally symmetric wavefunction is to take a product wavefunction for all particles and add to it all permutations obtained by interchanging particles. For example, if we start with a product state of three particles, $\psi_{I_1}(1)\psi_{I_2}(2)\psi_{I_3}(3)$, where I_1, I_2, I_3 stand for different combinations of quantum numbers that characterize a state, then the symmetric combination, apart from a possible normalizing factor, will be

$$\psi_S = \psi_{I_1}(1)\psi_{I_2}(2)\psi_{I_3}(3) + \psi_{I_2}(1)\psi_{I_3}(2)\psi_{I_1}(3) + \psi_{I_3}(1)\psi_{I_1}(2)\psi_{I_2}(3)$$
$$+ \psi_{I_1}(1)\psi_{I_3}(2)\psi_{I_2}(3) + \psi_{I_2}(1)\psi_{I_1}(2)\psi_{I_3}(3) + \psi_{I_3}(1)\psi_{I_2}(2)\psi_{I_1}(3),$$

$$(14.10)$$

where all permutations have been included. Similarly, to make a completely antisymmetric combination, one should put minus signs for the terms which involve odd permutations of the objects. There is a simple way to write this combination:

$$\psi_A = \begin{Vmatrix} \psi_{I_1}(1) & \psi_{I_2}(1) & \psi_{I_3}(1) \\ \psi_{I_1}(2) & \psi_{I_2}(2) & \psi_{I_3}(2) \\ \psi_{I_1}(3) & \psi_{I_2}(3) & \psi_{I_3}(N) \end{Vmatrix}, \qquad (14.11)$$

where the double line on both sides imply the determinant of the matrix enclosed. This way of representing the wavefunction is called the *Slater determinant*. The normalization factor is omitted here as well. For both cases, the generalization to larger number of particles is obvious.

The Slater determinant shows very well that if we try to put two fermions in the same state, the wavefunction will vanish. This is the *exclusion principle* forwarded by Pauli, which played a key role in understanding the *periodic table* of elements. In Chapter 9, we showed that the state of an electron in an atom is characterized by the quantum numbers n, l, m_l, and m_s. For $n = 1$, we must have $l = 0$ and $m_l = 0$, so there can be at most two electrons with $n = 0$, one for each value of m_s. Hence, there are two elements in the first row of the periodic table, hydrogen with one sole electron, and helium with two electrons. For $n = 2$, there are rooms for two more electrons with $l = 0$, and six more with $l = 1$. This explains the elements in the second row, from lithium to neon. Subsequent rows can also be explained in the same way.

But how far does the restriction of the exclusion principle apply? Does it mean that if we have an electron somewhere with $n = 0$ and say $m_s = +\frac{1}{2}$, no other electron anywhere can have the same quantum numbers? Not really. To understand the restriction, let us consider two electrons, and construct the antisymmetrized wavefunction with them:

$$\psi(\boldsymbol{r}_1, \boldsymbol{r}_2) = \frac{1}{\sqrt{2}}\Big(\phi(\boldsymbol{r}_1)\chi(\boldsymbol{r}_2) - \phi(\boldsymbol{r}_2)\chi(\boldsymbol{r}_1)\Big). \qquad (14.12)$$

This time we have been careful about the normalization factor, because we will be calculating probabilities shortly. The wavefunctions of individual particles, $\phi(\boldsymbol{r})$ and $\chi(\boldsymbol{r})$, are normalized.

Let us now ask what is the probability of seeing an electron at a point $\boldsymbol{r}$, irrespective of the position of the other electron. This will be given by

$$P(\boldsymbol{r}) = \int d^3\boldsymbol{r}_2 \left|\psi(\boldsymbol{r}, \boldsymbol{r}_2)\right|^2 + \int d^3\boldsymbol{r}_1 \left|\psi(\boldsymbol{r}_1, \boldsymbol{r})\right|^2. \tag{14.13}$$

The first term on the right side is the expression for the electron 1 to be at $\boldsymbol{r}$, and the second term is for the electron 2 at $\boldsymbol{r}$. In each case, we have integrated out the position of the other electron. Using Eq. (14.12) now, we obtain

$$P(\boldsymbol{r}) = |\phi(\boldsymbol{r})|^2 \int d^3\boldsymbol{r}' \, |\chi(\boldsymbol{r}')|^2 + |\chi(\boldsymbol{r})|^2 \int d^3\boldsymbol{r}' \, |\phi(\boldsymbol{r}')|^2$$
$$- 2\operatorname{Re}\left(\phi(\boldsymbol{r})\chi^*(\boldsymbol{r}) \int d^3\boldsymbol{r}' \, \phi^*(\boldsymbol{r}')\chi(\boldsymbol{r}')\right). \tag{14.14}$$

Suppose now the two electrons are very far off, so there is no overlap between their wavefunctions. In mathematical terms, this means

$$\phi(\boldsymbol{r})\chi(\boldsymbol{r}) = 0 \qquad \text{for all } \boldsymbol{r}. \tag{14.15}$$

Then, since the individual wavefunctions are normalized, we obtain

$$P(\boldsymbol{r}) = |\phi(\boldsymbol{r})|^2 + |\chi(\boldsymbol{r})|^2. \tag{14.16}$$

If the electron of our interest is in the state ϕ, then this equation says that the probability of finding it at $\boldsymbol{r}$ is just $|\phi(\boldsymbol{r})|^2$, since at the points of interest $\chi(\boldsymbol{r})$ would vanish because of Eq. (14.15). Similarly, if the electron of our interest is in the state χ, we find that the probability is $|\chi(\boldsymbol{r})|^2$. This is exactly what we would have obtained if we had not considered the other electron at all. This means that the requirement of antisymmetrization is relevant only if the particles have overlapping wavefunctions. If the two particles are far enough so that there is no overlap, the antisymmetrization does not apply. The same statements can be made about symmetrization of bosons.

In many of the practical problems that we have solved in earlier chapters of this book, the wavefunctions die out exponentially at large distances. This means that the wavefunction is nowhere strictly zero, and so it seems that Eq. (14.15) cannot hold anywhere. However, Eq. (14.15) should be seen as an approximate estimate rather than a strict condition, related to the accuracy of measurements. In other words, if an overlap is such that the resulting contributions to probabilities cannot be measured in our experiments, then we still set those contributions to zero.

☐ **Exercise 14.1** *Suppose we have gaseous hydrogen at 300 K temperature at normal atmospheric pressure. Calculate the average distance between two atoms. If all atoms are in the ground state, use the ground state wavefunctions from Table 9.1 (p. 239) to obtain the overlap of the wavefunctions of two atoms midway between their average distance.*

☐ **Exercise 14.2** *The wavefunction of two spin-$\frac{1}{2}$ fermions is expressed in terms of their spatial coordinates and total spin S. If the spatial part is antisymmetric, what are the allowed values of S_z?*

☐ **Exercise 14.3** *Consider the wavefunction of three spin 1/2 fermions such that the spin part of the wavefunction is*

$$\phi_{1,2,3} = \frac{1}{\sqrt{3}}\left(|\downarrow\rangle_1|\uparrow\rangle_2|\downarrow\rangle_3 + |\downarrow\rangle_1|\downarrow\rangle_2|\uparrow\rangle_3 + |\uparrow\rangle_1|\downarrow\rangle_2|\downarrow\rangle_3.\right). \qquad (14.17)$$

What symmetry property of the spatial part of the wavefunction may be inferred from the information given for $\phi_{1,2,3}$. Write a spatial wavefunction that will be consistent with the exchange properties.

☐ **Exercise 14.4** *Consider a bound state of three fermions. Assume the spatial part of the wavefunction to be symmetric. The total wavefunction contains a spin part, and also a part involving isospin, whose mathematical properties are similar to that of spin. Each particle has spin $\frac{1}{2}$, and the eigenstates of S_z are denoted by $|\uparrow\rangle$ and $|\downarrow\rangle$. The isospin of the particles is also $\frac{1}{2}$, and the eigenstates of I_z are called $|p\rangle$ and $|n\rangle$. Thus, barring the spatial part, the state of each particle can be written as $|p_\uparrow\rangle$, $|p_\downarrow\rangle$, $|n_\uparrow\rangle$, and $|n_\downarrow\rangle$. Find the spin-isospin wavefunction of the three particles in a combination with $S = \frac{1}{2}$ and $I = \frac{1}{2}$, which must be antisymmetric in the exchange of any two particles.*

14.3 Scattering of identical particles

Something very special needs to be taken into account if we consider the scattering of two identical particles. The point has been explained through Figure 14.1.

The initial state of the scattering process has two particles. The final state also has two. Let us start with the case when the two particles are different, say A and B. The scattering is elastic, so the final state particles are also A and B. In the leftmost figure in Figure 14.1, let us say that the A particle comes in from the bottom left corner and the B particle from the top right corner. Then, if the A particle goes out through the top left of the figure and the B particle through the bottom right, the scattering process can be imagined as shown in Figure 14.1b. On the other hand, if the final A particle emerges at the bottom right corner, the process will look like that shown in Figure 14.1c.

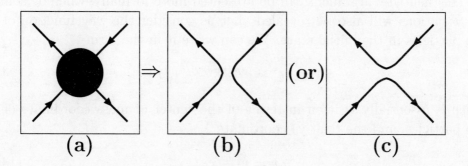

Figure 14.1 The left picture shows two incoming and two outgoing particles. The two pictures on the right side of the arrow sign show two different ways that the scattering might have happened. There is no way to distinguish between the two processes.

But now suppose that we do not have two kinds of particles: it is a scattering of identical particles in the initial state, resulting in the same particles in the final state. Then there will be no way to distinguish between Figure 14.1b and Figure 14.1c. For classical particles, one might contemplate taking a movie of the process to determine which of the initial balls went which way. For a quantum mechanical scattering, that is not allowed, because monitoring the scattering process would disturb the process, so much so that the monitored process and the unmonitored process must be treated as different processes altogether.

Let us again go back to the case of the scattering between two different particles. Suppose the scattering angle is θ for the situation in Figure 14.1b. Then the differential scattering cross-section for this process would be

$$\frac{d\sigma}{d\Omega} = |f(\theta)|^2, \tag{14.18}$$

as indicated by Eq. (13.31, p. 334). For the situation in Figure 14.1c then, the scattering angle would be $\pi - \theta$, and therefore the differential cross-section would be

$$\frac{d\sigma}{d\Omega} = |f(\pi - \theta)|^2. \tag{14.19}$$

If instead we have two identical particles, we cannot distinguish between the two situations, and therefore the scattering length will be given by $f(\theta) + f(\pi - \theta)$, and the differential cross-section would be

$$\frac{d\sigma}{d\Omega} = \left| f(\theta) + f(\pi - \theta) \right|^2. \tag{14.20}$$

This heuristic argument can be presented more formally, whereby some of the variations will also be revealed. Let us consider the wavefunction of the two particles in the initial state. We can write it in the form

$$\Psi(\boldsymbol{r}_1, \boldsymbol{r}_2) = e^{i\boldsymbol{p}\cdot(\boldsymbol{r}_1+\boldsymbol{r}_2)/\hbar}\psi(\boldsymbol{r}), \tag{14.21}$$

which is essentially written in terms of the center of mass coordinate of the two particles, and the relative coordinate

$$\boldsymbol{r} = \boldsymbol{r}_1 - \boldsymbol{r}_2. \tag{14.22}$$

Let us assume that the particles are spinless. They are thus bosons, and the wavefunctions must satisfy the relation

$$\Psi(\boldsymbol{r}_1, \boldsymbol{r}_2) = \Psi(\boldsymbol{r}_2, \boldsymbol{r}_1). \tag{14.23}$$

Using it on Eq. (14.21), we obtain the constraint

$$\psi(\boldsymbol{r}) = \psi(-\boldsymbol{r}). \tag{14.24}$$

If we use spherical polar coordinates, this equation can be written as

$$\psi(r, \theta, \phi) = \psi(r, \pi - \theta, \pi + \phi). \tag{14.25}$$

Remember that in the analysis of §13.1, we took the wavefunction to be in the form

$$\psi(\boldsymbol{r}) = e^{i\boldsymbol{k}\cdot\boldsymbol{r}} + \frac{e^{ikr}}{r}f(\theta), \tag{14.26}$$

as shown in Eq. (13.21, p. 332) in a slightly different notation. Certainly, we cannot use this expression for the part of the wavefunction depending on the relative coordinates of the two identical particles, because it does not satisfy Eq. (14.24). We should take

$$\psi(\boldsymbol{r}) = e^{i\boldsymbol{k}\cdot\boldsymbol{r}} + e^{-i\boldsymbol{k}\cdot\boldsymbol{r}} + \frac{e^{ikr}}{r}\Big(f(\theta) + f(\pi - \theta)\Big). \tag{14.27}$$

It is now clear how Eq. (14.20) follows.

This final result can change if the particles have spin. Now the wavefunction will also contain a spin part. The overall wavefunction, including both space and spin parts, should be symmetric under the exchange symmetry forp bosons and antisymmetric for fermions. So now the spatial part, by itself, might be symmetric or antisymmetric, depending on the symmetry of the spin part. If the spin part is such that the spatial part turns out to be symmetric, the arguments given above are still valid. However, if the spatial part is antisymmetric, then it should be written in the form

$$\psi(\boldsymbol{r}) = e^{i\boldsymbol{k}\cdot\boldsymbol{r}} - e^{-i\boldsymbol{k}\cdot\boldsymbol{r}} + \frac{e^{ikr}}{r}\Big(f(\theta) - f(\pi - \theta)\Big). \tag{14.28}$$

The differential cross-section would then be given by

$$\frac{d\sigma}{d\Omega} = \Big|f(\theta) - f(\pi - \theta)\Big|^2. \tag{14.29}$$

□ **Exercise 14.5** *Show, using partial wave analysis, that for scattering of two identical particles, only channels with even values of the angular momentum ℓ contribute to the scattering amplitude.*

Chapter 15

Relativistic corrections

Quantum mechanics originally developed for describing non-relativistic systems. It was a natural question to ask how the formalism can be extended to relativistic systems. Besides, it turned out that some of the properties of non-relativistic systems can be understood naturally in the light of the relativistic theory. One example is the spin of the electron, which has to be introduced in an ad-hoc manner in non-relativistic theory, but appears naturally in the relativistic theory, as we will explain in this chapter. The aim of this chapter is to indicate, rather than elaborate, the kind of questions that can be asked and answered using hints of relativity. A full-fledged relativistic theory takes us beyond quantum mechanics, as we argue in §15.1. We also clarify that when we talk of relativity, we only mean the special theory. The general theory of relativity is not discussed at all.

15.1   Conflict between relativity and quantum mechanics

The special theory of relativity can be interpreted as a theory of the geometry of the 4D space–time, composed of the three spatial dimensions and time. Thus, time and space are treated on equal footing in special relativity.

This is what causes a conflict with quantum mechanics. We have used time as a parameter with respect to which we study the evolution of systems. On the other hand, the spatial coordinates indicating the position are treated as operators, acting on the vector space of state vectors. There is no way that one can make a truly relativistic quantum theory if one maintains this status for time and space.

There are two ways out of the impasse. First, we might contemplate making time an operator as well, just as the position coordinates are. In §3.8, we have argued that this cannot be done. The second alternative is to make the spatial coordinates parameters, just like time is. In this case, one needs to study the evolution of objects in space and time, and the objects to be studied should therefore be some kind of functions of position and time. In physics, a function of space and time is called a *field*. Examples are electromagnetic and gravitation fields, which can depend on both position and time. In order to do *relativistic quantum mechanics*, i.e., do quantum mechanics in a way that is compatible with the special theory of relativity, one must therefore do some kind of field theory. The resulting structure is called *quantum field theory*, which will not be discussed in this book.

> It should be noted that the word *field* means something quite different in mathematics. It signifies a collection of elements on which addition and multiplication can be defined, obeying certain conditions. The real numbers, for example, constitute a field in the mathematical sense. So do the complex numbers. This usage should not be confused with the use of the word in physics.

There is an alternative that will be followed here. We will start with a relativistic Hamiltonian, identify the leading corrections that it inflicts on the non-relativistic Hamiltonian of the same system, and apply the formalism of non-relativistic quantum mechanics, as developed in the earlier chapters, using a Hamiltonian that includes those leading corrections. This will at least give us a flavor of things that might be expected from a truly relativistic theory. It is in this spirit that we undertake the entire discussion of this chapter.

15.2 Dirac Hamiltonian

The Schrödinger equation is given by

$$i\hbar\frac{d}{dt}\left|\psi\right\rangle = H\left|\psi\right\rangle. \tag{15.1}$$

This equation has first-order derivatives with respect to time. In order to have time and space on equal footing, the equation for the wavefunction in the coordinate representation must involve first-order derivatives with respect to the position coordinates of the wavefunction. In the operator approach, then, the Hamiltonian must be linear in the momentum operator. Moreover, the free particle Hamiltonian cannot depend on the position coordinates. Therefore, the most general possible form of the Hamiltonian is

$$H = c\boldsymbol{\alpha}\cdot\boldsymbol{p} + \beta mc^2, \tag{15.2}$$

where $\boldsymbol{\alpha}$ and β do not depend on position or momentum, and c is the speed of light in the vacuum which has been used to make $\boldsymbol{\alpha}$ and β dimensionless. Putting this Hamiltonian into Eq. (15.2) and taking inner product of both sides with a position eigenstate $|r\rangle$, we obtain

$$i\hbar\frac{\partial}{\partial t}\psi(\boldsymbol{r},t) = -i\hbar c\boldsymbol{\alpha}\cdot\boldsymbol{\nabla}\psi(\boldsymbol{r},t) + \beta mc^2\psi(\boldsymbol{r},t),\qquad(15.3)$$

or

$$i\hbar\left[\frac{1}{c}\frac{\partial}{\partial t} + \boldsymbol{\alpha}\cdot\boldsymbol{\nabla}\right]\psi(\boldsymbol{r},t) = \beta mc\psi(\boldsymbol{r},t).\qquad(15.4)$$

This equation, the Schrödinger equation with the linear Hamiltonian of Eq. (15.2), is called the *Dirac equation* because the Hamiltonian was conjectured by Paul Dirac.

However, the linear Hamiltonian must be consistent with the relativistic energy–momentum relation of a free particle:

$$E^2 = c^2\boldsymbol{p}^2 + m^2c^4.\qquad(15.5)$$

Let us therefore square the operator of Eq. (15.2). For the unknown guests $\boldsymbol{\alpha}$ and β, we will assume nothing except that they are independent of the momentum, and therefore commute with the momentum. Then we obtain

$$H^2 = \left(c\sum_i \alpha_i p_i + \beta mc^2\right)\left(c\sum_j \alpha_j p_j + \beta mc^2\right)$$
$$= c^2\sum_{i,j}\alpha_i\alpha_j p_i p_j + mc^3\sum_i(\alpha_i\beta + \beta\alpha_i) + m^2c^4\beta^2.\qquad(15.6)$$

The terms quadratic in the momentum operator can be rewritten as

$$\frac{1}{2}\sum_{i,j}\Big((\alpha_i\alpha_j + \alpha_j\alpha_i) + (\alpha_i\alpha_j - \alpha_j\alpha_i)\Big)p_i p_j.\qquad(15.7)$$

Since the components of the momentum operator commute, i.e., $p_i p_j = p_j p_i$, the term involving the difference of the α matrices sum up to zero. Then, comparing this term with the expression of Eq. (15.5), we find that we need the α matrices to satisfy the relation

$$\left\{\alpha_i,\alpha_j\right\} = 2\delta_{ij},\qquad(15.8a)$$

where we have the anticommutator on the left side. Similarly, comparing other terms of Eq. (15.6) with Eq. (15.5), we see that we need

$$\left\{\alpha_i, \beta\right\} = 0, \tag{15.8b}$$

$$\beta^2 = 1. \tag{15.8c}$$

Because the α's and the β anticommute among themselves, they have to be matrices. They are called *Dirac matrices*. We thus need four mutually anticommuting matrices. We know of three mutually anticommuting 2×2 matrices, viz., the Pauli matrices. But it is impossible to find another non-null 2×2 matrix that anticommutes with all three of them. We will indicate in Ex. 15.4 that these matrices are traceless. Since the square of each of the four matrices is the unit matrix, their eigenvalues must therefore be $+1$ and -1. To obtain a traceless matrix, the matrices must be even dimensional. Therefore, we need 4×4 matrices. Accordingly, the wavefunction appearing in Eq. (15.4) must be a column matrix with four components.

☐ **Exercise 15.1** *Show that it is impossible to find a non-null 2×2 matrix that anticommutes with all three Pauli matrices.*

☐ **Exercise 15.2** *There is an alternative notation for the Dirac matrices, which is particularly useful for doing the fully relativistic theory:*

$$\gamma_0 \equiv \beta, \qquad \gamma_i \equiv \beta\alpha_i \quad \text{for } i = 1, 2, 3. \tag{15.9}$$

Show that the anticommutation relations of Eq. (15.8) can be summarized into the relation

$$\left\{\gamma_\mu, \gamma_\nu\right\} = 2g_{\mu\nu}, \tag{15.10}$$

where the indices μ and ν run from 0 to 3, and

$$g_{\mu\nu} = \text{diag}(+1, -1, -1, -1). \tag{15.11}$$

☐ **Exercise 15.3** *Show that the matrix*

$$\gamma_5 = i\gamma_0\gamma_1\gamma_2\gamma_3 \tag{15.12}$$

satisfies the anticommutation relations

$$\left\{\gamma_\mu, \gamma_5\right\} = 0, \tag{15.13}$$

and the commutation relations

$$\left[\alpha_i, \gamma_5\right] = 0. \tag{15.14}$$

☐ **Exercise 15.4** *If a matrix A anticommutes with another matrix B and $B^2 = 1$, then show that A is traceless.* [**Hint:** $\mathrm{Tr}(AB^2) = \mathrm{Tr}(BAB)$ because of the cyclic property of traces.]

☐ **Exercise 15.5** *The α_i and β matrices must be Hermitian so that the Dirac Hamiltonian of Eq. (15.2) is Hermitian. Show that this implies that the Hermiticity property of the γ-matrices can be summarized by writing*

$$\gamma_\mu^\dagger = \gamma_0 \gamma_\mu \gamma_0, \tag{15.15}$$

i.e., γ_0 is Hermitian whereas the γ_i's are anti-Hermitian.

☐ **Exercise 15.6** *There is no unique way of specifying the elements of the four Dirac matrices. Show that if we have a set of Dirac matrices and another set defined by $\tilde{\alpha}_i = U\alpha_i U^\dagger$, $\tilde{\beta} = U\beta_i U^\dagger$ for any unitary matrix U, then the tilded avatars also satisfy Eq. (15.8).*

15.3   Spin

Angular momentum of a free particle should be conserved. As argued in Chapter 5, this means that the angular momentum operator should commute with the free particle Hamiltonian. Let us check whether that is the case with the orbital angular momentum when the Hamiltonian is the Dirac Hamiltonian given in Eq. (15.2).

$$\begin{aligned} \left[L_i, H\right] &= c \sum_{j,k} \sum_l \varepsilon_{ijk} \left[x_j p_k, \alpha_l p_l\right] \\ &= i\hbar c \sum_{j,k} \varepsilon_{ijk} \alpha_j p_k. \end{aligned} \tag{15.16}$$

The commutator is not zero! This means that there must be some other contribution to the angular momentum, viz., the spin, whose components satisfy the relation

$$\left[S_i, H\right] = -i\hbar c \sum_{j,k} \varepsilon_{ijk} \alpha_j p_k, \tag{15.17}$$

so that the total angular momentum

$$\boldsymbol{J} = \boldsymbol{L} + \boldsymbol{S} \tag{15.18}$$

is conserved.

The spin operator should be independent of momentum, and therefore should comprise the Dirac matrices only. It is easily seen that the combination

$$S_i = -\frac{1}{4}i\hbar \sum_{j,k} \varepsilon_{ijk}\alpha_j\alpha_k \tag{15.19}$$

satisfies Eq. (15.17). This is the spin operator. As indicated in the preamble to this chapter, the existence of spin comes out from the free Hamiltonian.

It says more. Take, for example, any one component of the spin operator and consider the square of it.

$$S_1^2 = -\frac{1}{4}\hbar^2(\alpha_2\alpha_3)^3. \tag{15.20}$$

Eq. (15.8) says that the square of each α matrix is the unit matrix, and that two different α matrices anticommute. Therefore, we obtain

$$S_1^2 = \frac{1}{4}\hbar^2\mathbb{1}. \tag{15.21}$$

The eigenvalues of this matrix are $\pm\frac{1}{2}\hbar$, which tells us that the Dirac Hamiltonian describes a spin-$\frac{1}{2}$ particle.

☐ **Exercise 15.7** *Show that for any set of operators A, B, C, D which are associative,*

$$\left[AB, CD\right] = A\{B,C\}D - AC\{B,D\} + \{A,C\}DB - C\{A,D\}B. \tag{15.22}$$

☐ **Exercise 15.8** *Using Eq. (15.22), show that the spin operators defined in Eq. (15.19) satisfy the commutation relations*

$$\left[S_i, S_j\right] = i\hbar \sum_k \varepsilon_{ijk}S_k, \tag{15.23}$$

as they should.

☐ **Exercise 15.9** *Use Eq. (15.19) and the Levi-Civita symbol identities given in Eq. (5.113, p. 132) to show that the spin operators satisfy the relation*

$$S_iS_j = \frac{1}{4}\hbar^2\delta_{ij} + \frac{1}{2}\hbar \sum_k \varepsilon_{ijk}S_k. \tag{15.24}$$

[**Note:** *This is reminiscent of the product of two Pauli matrices that follow from Eq. (8.67, p. 192). It is no coincidence. The matrices $\frac{1}{2}\hbar\boldsymbol{\sigma}$ are the generators of the spin-$\frac{1}{2}$ representation of the rotation algebra.*]

15.4   Magnetic moment

We now show that the Dirac equation, modified to include interactions with a magnetic field, specifies the magnetic moment of a Dirac particle.

The recipe for introducing magnetic interactions was discussed in Chapter 10. For a Dirac particle of charge q, this recipe yields the Hamiltonian

$$H = c\boldsymbol{\alpha} \cdot (\boldsymbol{p} - q\boldsymbol{A}) + \beta mc^2, \tag{15.25}$$

where $\boldsymbol{A}$ is the vector potential. The Dirac equation for the energy eigenvalue E is therefore

$$\left[c\boldsymbol{\alpha} \cdot (\boldsymbol{p} - q\boldsymbol{A}) + \beta mc^2 \right]\psi = E\psi. \tag{15.26}$$

To extract information out of this equation by comparing with the terms obtained in the non-relativistic version of the problem, we need to obtain terms quadratic in the momentum operator. Therefore, we consider the operator H^2. Since $H^2\psi = H(E\psi) = E^2\psi$, we can write

$$\left[c^2 \Big(\boldsymbol{\alpha} \cdot (\boldsymbol{p} - q\boldsymbol{A}) \Big)^2 + m^2 c^4 \right]\psi = E^2\psi, \tag{15.27}$$

using Eq. (15.8).

We now consider the term involving the α-matrices:

$$\Big(\boldsymbol{\alpha} \cdot (\boldsymbol{p} - q\boldsymbol{A}) \Big)^2 = \sum_{i,j} \alpha_i \alpha_j (p_i - qA_i)(p_j - qA_j). \tag{15.28}$$

Note that Eq. (15.19) can be inverted to write

$$\left[\alpha_j, \alpha_k \right] = \frac{4i}{\hbar} \sum_i \varepsilon_{ijk} S_i. \tag{15.29}$$

Combining it with Eq. (15.8a), we can write

$$\alpha_i \alpha_j = \delta_{ij} + \frac{2i}{\hbar} \sum_k \varepsilon_{ijk} S_k. \tag{15.30}$$

Putting this back into Eq. (15.28) and noting that the components of $\boldsymbol{p}$ commute among themselves and so do the components of $\boldsymbol{A}$, we obtain

$$\Big(\boldsymbol{\alpha} \cdot (\boldsymbol{p} - q\boldsymbol{A}) \Big)^2 = (\boldsymbol{p} - q\boldsymbol{A})^2 - \frac{2iq}{\hbar} \sum_{i,j,k} \varepsilon_{ijk} S_k (p_i A_j + A_i p_j)$$

$$= (\boldsymbol{p} - q\boldsymbol{A})^2 - \frac{2iq}{\hbar} \sum_{i,j,k} \varepsilon_{ijk} S_k (p_i A_j - A_j p_i), \tag{15.31}$$

using the antisymmetry of the Levi-Civita symbol in the last step to rename the dummy indices. In the coordinate representation, putting $p_j = -i\hbar\partial_j$, we find that, acting on any function f of the coordinates,

$$
\begin{aligned}
(p_i A_j - A_j p_i)f &= -i\hbar\Big(\partial_i(A_j f) - A_j \partial_i f\Big) \\
&= -i\hbar(\partial_i A_j)f,
\end{aligned}
\tag{15.32}
$$

where in the last step the parentheses imply that the derivative does not operate on things outside. Putting this back, we obtain

$$
\begin{aligned}
\Big(\boldsymbol{\alpha}\cdot(\boldsymbol{p} - q\boldsymbol{A})\Big)^2 &= (\boldsymbol{p} - q\boldsymbol{A})^2 - 2q\sum_{i,j,k}\varepsilon_{ijk}S_k(\partial_i A_j) \\
&= (\boldsymbol{p} - q\boldsymbol{A})^2 - 2q\boldsymbol{S}\cdot\boldsymbol{B},
\end{aligned}
\tag{15.33}
$$

where $\boldsymbol{B}$ is the magnetic field.

Putting this back into Eq. (15.27), we obtain

$$
c^2\Big[(\boldsymbol{p} - q\boldsymbol{A})^2 - 2q\boldsymbol{S}\cdot\boldsymbol{B}\Big]\psi = (E^2 - m^2 c^4)\psi.
\tag{15.34}
$$

Remember that we are dealing with a relativistic equation. The energy eigenvalue E must therefore be the relativistic energy, which includes the mass energy. The non-relativistic energy is given by

$$
E' = E - mc^2.
\tag{15.35}
$$

Thus,

$$
E^2 - m^2 c^4 \approx 2mc^2 E',
\tag{15.36}
$$

neglecting terms which involve smaller powers of c and are therefore insignificant.

Putting this back into Eq. (15.34), we obtain

$$
\left[\frac{1}{2m}(\boldsymbol{p} - q\boldsymbol{A})^2 - \frac{q}{m}\boldsymbol{S}\cdot\boldsymbol{B}\right]\psi = E'\psi.
\tag{15.37}
$$

Comparing this with the non-relativistic eigenvalue equation obtained from Eq. (10.3, p. 254), we see that there is an extra term, indicating the interaction with the magnetic field. It shows that the magnetic moment of a Dirac particle is related to its spin by

$$
\boldsymbol{\mu} = \frac{q}{m}\boldsymbol{S}.
\tag{15.38}
$$

If we compare it with Eq. (10.65, p. 269), we see that the Landé g-factor for a Dirac particle is 2.

Relativistic corrections to non-relativistic expressions come with suppression factors of inverse powers of c. The non-relativistic expression is recovered in the limit $c \to \infty$. One might wonder why the spin term in Eq. (15.37) does not show such suppression. The answer is that we have used the SI prescription in writing electromagnetic potentials starting from Eq. (15.25), which obscures many factors of c. In Gaussian units or Heaviside-Lorentz units, a factor of $1/c$ explicitly accompanies all occurrences of A and therefore of B, and it is easy to realize that the magnetic moment interaction term is a relativistic correction.

15.5 Dirac particle in a central potential

We now consider the Dirac particle in a potential V. This might even be an electromagnetic scalar potential of the Coulomb type. We do not even assume, until the very end, that the potential is central. However, we do assume that there is no vector potential in what we are going to consider now. The Dirac equation for the energy eigenvalue E is

$$\left[c\boldsymbol{\alpha} \cdot \boldsymbol{p} + \beta mc^2 + V \right] \psi = E\psi. \tag{15.39}$$

In the non-relativistic approximation, the dominant term in the Hamiltonian is the mass term, mc^2. If the other terms are neglected altogether, the solutions for ψ would be the eigenvectors of the matrix β. Since $\beta^2 = 1$, the eigenvalues are $+1$ and -1, both doubly degenerate.

When the momentum and the potential terms are added, we expect the deviation of the eigenvalues and eigenvectors. The solutions of Eq. (15.39) that we are looking for must then be of the form

$$\psi = \xi + \chi, \tag{15.40}$$

where ξ is some combination of the two eigenvectors of β with the positive eigenvalue, and χ is a combination of eigenvectors with the negative eigenvalue, i.e.,

$$\beta\xi = \xi, \qquad \beta\chi = -\chi. \tag{15.41}$$

Putting Eq. (15.40) into Eq. (15.39) and using Eq. (15.41), we obtain

$$\left[c\boldsymbol{\alpha} \cdot \boldsymbol{p} + mc^2 + V - E \right]\xi + \left[c\boldsymbol{\alpha} \cdot \boldsymbol{p} - mc^2 + V - E \right]\chi = 0. \tag{15.42}$$

Multiplying this equation from the left by the matrix β and using Eqs. (15.8b) and (15.41), we get

$$\left[-c\boldsymbol{\alpha} \cdot \boldsymbol{p} + mc^2 + V - E \right]\xi + \left[c\boldsymbol{\alpha} \cdot \boldsymbol{p} + mc^2 - V + E \right]\chi = 0. \tag{15.43}$$

Adding and subtracting Eqs. (15.42) and (15.43), we obtain the equations

$$\left[E - mc^2 - V\right]\xi = c\boldsymbol{\alpha} \cdot \boldsymbol{p}\chi, \tag{15.44a}$$

$$c\boldsymbol{\alpha} \cdot \boldsymbol{p}\xi = (E + mc^2 - V)\chi. \tag{15.44b}$$

From Eq. (15.44b), we deduce

$$\chi = (E + mc^2 - V)^{-1}c\boldsymbol{\alpha} \cdot \boldsymbol{p}\xi. \tag{15.45}$$

So far, we have not made any approximation. If we talk about the positive energy solutions and use the non-relativistic definition of energy given in Eq. (15.35), we note that the right side of Eq. (15.45) has the huge contribution of $2mc^2$ sitting in the denominator, so that χ is small. We therefore put Eq. (15.45) into Eq. (15.44a) and study the equation for the dominant contribution ξ:

$$E'\xi = V\xi + c\boldsymbol{\alpha} \cdot \boldsymbol{p}(2mc^2 + E' - V)^{-1}c\boldsymbol{\alpha} \cdot \boldsymbol{p}\xi. \tag{15.46}$$

Suppose we neglect $E' - V$ altogether in comparison with $2mc^2$. Recall that for arbitrary vectors $\boldsymbol{A}$ and $\boldsymbol{B}$, Eq. (15.30) can be written in the form

$$(\boldsymbol{\alpha} \cdot \boldsymbol{A})(\boldsymbol{\alpha} \cdot \boldsymbol{B}) = \boldsymbol{A} \cdot \boldsymbol{B} + \frac{2i}{\hbar}\boldsymbol{S} \cdot (\boldsymbol{A} \times \boldsymbol{B}). \tag{15.47}$$

In particular, it says $(\boldsymbol{\alpha} \cdot \boldsymbol{p})^2 = \boldsymbol{p}^2$, so we obtain

$$E'\xi = \left[\frac{\boldsymbol{p}^2}{2m} + V\right]\xi. \tag{15.48}$$

This is the completely non-relativistic formula, as expected. We now want to find the lowest-order corrections to this equation in the parameter $1/c$. For this, we put

$$(2mc^2 + E' - V)^{-1} = \frac{1}{2mc^2}\left(1 - \frac{E' - V}{2mc^2}\right). \tag{15.49}$$

Putting this back into Eq. (15.46) and using Eq. (15.47), we get

$$E'\xi = \left[V + \left(1 - \frac{E'}{2mc^2}\right)\frac{\boldsymbol{p}^2}{2m} + \frac{1}{4m^2c^2}\boldsymbol{\alpha} \cdot \boldsymbol{p}\, V \boldsymbol{\alpha} \cdot \boldsymbol{p}\right]\xi. \tag{15.50}$$

The last term needs special care. Since V is a function of coordinates, we have

$$[V, \boldsymbol{p}] = i\hbar(\boldsymbol{\nabla}V). \tag{15.51}$$

Using this, we write

$$\boldsymbol{\alpha}\cdot\boldsymbol{p}\,V\,\boldsymbol{\alpha}\cdot\boldsymbol{p} = \boldsymbol{\alpha}\cdot\Big(V\boldsymbol{p} - i\hbar(\boldsymbol{\nabla}V)\Big)\boldsymbol{\alpha}\cdot\boldsymbol{p}$$
$$= V\boldsymbol{p}^2 - i\hbar\boldsymbol{\nabla}V\cdot\boldsymbol{p} + 2\boldsymbol{S}\cdot(\boldsymbol{\nabla}V\times\boldsymbol{p}), \qquad (15.52)$$

using Eq. (15.47) once again. Putting this back into Eq. (15.50), we obtain the equation for ξ in the form

$$E'\xi = \left[V + \frac{\boldsymbol{p}^2}{2m} - \frac{E'-V}{2mc^2}\frac{\boldsymbol{p}^2}{2m} \right.$$
$$\left. - \frac{i\hbar}{4m^2c^2}\boldsymbol{\nabla}V\cdot\boldsymbol{p} + \frac{1}{2m^2c^2}\boldsymbol{S}\cdot(\boldsymbol{\nabla}V\times\boldsymbol{p}) \right]\xi. \qquad (15.53)$$

On the right side, the term containing $E'-V$ already has a suppression factor of $1/c^2$. Therefore, in it, we can use the approximation of Eq. (15.48) and write the final equation in the form

$$E'\xi = \left[V + \frac{\boldsymbol{p}^2}{2m} - \frac{\boldsymbol{p}^4}{8m^3c^2} \right.$$
$$\left. - \frac{i\hbar}{4m^2c^2}\boldsymbol{\nabla}V\cdot\boldsymbol{p} + \frac{1}{2m^2c^2}\boldsymbol{S}\cdot(\boldsymbol{\nabla}V\times\boldsymbol{p}) \right]\xi. \qquad (15.54)$$

This is the final equation, including all terms or order $1/c^2$, for the dominant component of the solution of the Dirac equation.

Let us try to interpret the correction terms of order $1/c^2$ that appear in this equation. The term involving $\boldsymbol{p}^4$ is independent of the external potential V, and occurs even for a free particle. It is the leading correction obtained by putting the relativistic energy expression of Eq. (15.5) into the definition of the non-relativistic energy in Eq. (15.35).

As for the terms that depend on the potential, we assume that the potential is central:

$$V = V(r), \qquad (15.55)$$

so that

$$\boldsymbol{\nabla} = \frac{V'}{r}\boldsymbol{r}, \qquad (15.56)$$

V' being the derivative with respect to r. Then the last term in the approximate Hamiltonian of Eq. (15.54) can be written as

$$V_{\text{s.o.}} = \frac{V'}{2m^2c^2r}\boldsymbol{S}\cdot\boldsymbol{L}, \qquad (15.57)$$

where L is the orbital angular momentum operator. The term therefore represents *spin–orbit interaction*, as indicated by the subscript letters on the left side. This kind of interaction was mentioned several times earlier. The remaining term in Eq. (15.54) is also a relativistic correction, but it does not have an easy intuitive explanation.

While discussing the hydrogen atom in Chapter 9, we commented that the spectrum did not depend on the orbital angular momentum eigenvalues. The non-relativistic Hamiltonian was found to commute with L^2, L_z, and S_z. If we add the relativistic corrections of Eq. (15.57), certainly the resulting form does not commute with L_z and S_z because of the presence of the spin–orbit interaction term. However, since the total angular momentum of the electron is given by Eq. (15.18), we can write

$$2\boldsymbol{L} \cdot \boldsymbol{S} = \boldsymbol{J}^2 - \boldsymbol{L}^2 - \boldsymbol{S}^2, \qquad (15.58)$$

and so the Hamiltonian still commutes with J^2 and J_z. Thus, after including relativistic corrections, the hydrogen atom wavefunctions should be given in the form $|njm_j\rangle$, where $j(j+1)\hbar^2$ and $m_j\hbar$ are the eigenvalues of $\boldsymbol{J}^2$ and J_z.

☐ **Exercise 15.10** *Show, by explicitly evaluating the commutators, that L_z and S_z do not commute with $\boldsymbol{L} \cdot \boldsymbol{S}$, but $J_z = L_z + S_z$ does.*

15.6 Further prospects

We said that this chapter was not going to be a comprehensive discussion of relativistic quantum mechanics. We only presented a few indications of how relativistic corrections justify or modify the non-relativistic results obtained in earlier chapters of this book.

In this chapter, we have focussed only on the theory of spin-$\frac{1}{2}$ particles. While this is of utmost importance when considering properties of ordinary matter since electrons fall into this class of particles, there are certainly particles with other values of spin which are of interest. For example, photons have spin equal to 1. We have not discussed the relativistic quantum theory of photons, or of any other particle of spin not equal to $\frac{1}{2}$.

We have kept our discussion limited to 3D systems. The Dirac Hamiltonian can be defined in any other number of spatial dimensions. The form will be the same as that in Eq. (15.2), except that the number of α-matrices will be equal to the number of spatial dimensions. The anticommutation relations of Eq. (15.8) will still follow. Moreover, if for odd number of spatial dimensions we write $D = 2k_0 - 1$ and for even dimensions $D = 2k_0$, the $\boldsymbol{\alpha}$ and β matrices will have 2^{k_0} rows and columns. This means that some of the physical consequences will be different. For example, in 2D, where $k_0 = 1$, there will be two 2×2 α-matrices. Together with the matrix β (which is also a 2×2 matrix), we will need three mutually anticommuting matrices, and the Pauli matrices can be used for this purpose. The wavefunctions will therefore have two components. Eq. (15.19) cannot be used as the spin operator in D dimensions if $D \neq 3$, because the Levi-Civita symbol of D dimensions would have D indices. For the same reason, the definition of the orbital angular momentum cannot be given as a cross product. However, the definitions

$$L_{ij} = x_i p_j - x_j p_i, \tag{15.59a}$$

$$S_{ij} = -\frac{1}{4} i\hbar [\alpha_i, \alpha_j] \tag{15.59b}$$

would work. In fact, for any dimension other than 3, the rotation operators should have two vector indices, corresponding to the plane in which the rotation is taking place.

☐ **Exercise 15.11** *Show that*

$$[L_{ij}, L_{kl}] = i\hbar \Big(\delta_{ik} L_{jl} - \delta_{jk} L_{il} + \delta_{jl} L_{ik} - \delta_{il} L_{jk} \Big). \tag{15.60}$$

[**Note:** *This is the form for the rotation algebra in arbitrary number of dimensions.*]

☐ **Exercise 15.12** *Show that the spin operators defined in Eq. (15.59b) obey the same algebra.*

Even in the context of spin-$\frac{1}{2}$ particles in 3D, we have brushed aside a few issues. It was intentional. We mentioned that the wavefunctions are 4-component spinors. Therefore, there would be four linearly independent solutions to the eigenvalue equations. Two of them can be interpreted as the two spin states of the spin-$\frac{1}{2}$ particle. We made no comment about the other two. It turns out that the other two solutions correspond to another particle that has the same mass and opposite charge compared to the particle we have investigated. It is called the *antiparticle*. A discussion of antiparticles would require a full treatment of the relativistic theory. As we said in §15.1, this would take us to relativistic quantum field theory, which is a subject by itself.

There is one way of obtaining two-component spinors even in 3D. If we talk about massless particles, there is no matrix β in the Hamiltonian. We then need only three mutually anticommuting α-matrices, and so the Pauli matrices are sufficient for the job. Therefore, two-component spinors can represent massless particles. The Dirac equation for massless particles is called the *Weyl equation*.

This does not mean that there is no antiparticle associated with a Weyl particle. There are still positive and negative energy solutions, and therefore antiparticles. To see why then two components are enough, we note that we can satisfy Eq. (15.8a) by taking $\boldsymbol{\alpha} = \eta\boldsymbol{\sigma}$, where $\eta = \pm 1$. The Weyl equation would turn out to be

$$\eta c \boldsymbol{\sigma} \cdot \widehat{\boldsymbol{p}}\psi = E\psi, \tag{15.61}$$

where now we explicitly put the operator sign on the momentum to facilitate the ensuing discussion. If we seek free particle solutions for which $\langle \boldsymbol{r}|\psi\rangle = \exp(-i\boldsymbol{p} \cdot \boldsymbol{r}/\hbar)\phi$, where ϕ is now a 2-component spinor that is independent of the coordinates, we will obtain $E = \pm c\mathrm{p}$. Putting the positive energy solution back into Eq. (15.61), we obtain

$$\boldsymbol{\sigma} \cdot \mathring{\boldsymbol{p}}\phi = \eta\phi. \tag{15.62}$$

Since $\boldsymbol{\sigma}$ represents the spin and $\mathring{\boldsymbol{p}}$ the unit vector along the momentum, the left side contains the spin component along the direction of momentum, called *helicity*. Eq. (15.62) then tells us that the positive energy solution can occur only with the eigenvalue η of the helicity. Similarly, the negative energy solution will have helicity equal to $-\eta$. So, only one spin projection state would be available to the particle as well as to the antiparticle, making the total number of independent solutions equal to two.

We do not discuss the Weyl equation further because massless particles are relativistic systems par excellence, and would require a truly relativistic treatment. However, it was recently found that low-energy excitations in certain material may obey an effective Weyl equation or Dirac equation even though these systems are non-relativistic. This issue will be briefly discussed in §17.4.

Chapter 16

Interpretation of quantum mechanics

Quantum mechanics urges us to look at anything in two complementary ways: as particle or as wave. The two descriptions are very different, and therefore intuitively some of the basic questions of quantum mechanics seem difficult to reconcile. In this chapter, we will outline some of the problems with this reconciliation.

16.1   Waves vs particles

Let us start with something as simple as a double-slit experiment, described in §1.5. In Figure 1.2 (p. 13), we have given a schematic representation of the apparatus, and also the resulting interference pattern obtained on the screen. For the convenience of description, let us say that we are talking of interference of light.

It is easy to understand the experiment if we think in terms of waves. Light waves are coming from the source. Some part of the wavefront escapes through one hole and some escapes through the other. On the other side, waves arriving through the two slits superpose and produce the interference pattern.

What if we now want to describe the same phenomenon using light particles, or photons? Should we now say, in order to stay as close to the wave description as possible, that the photons have gone through both slits?

It might naively seem that there is nothing wrong with that. After all, there are many photons in a beam of light. So, it is certainly conceivable that some of them have gone through one slit and some through the other.

But if that happened, each photon would have ended up in a region directly behind one slit or the other, with maybe a little scatter because of the finite

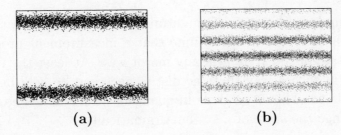

(a) (b)

Figure 16.1 The left panel shows the pattern that would have resulted if, in a double-slit experiment, each photon had gone through one slit or the other. The interference pattern for unmonitored photons is shown schematically on the right panel.

widths of the slits and also the finite size of the source. The pattern on the screen would have looked something like what we see in Figure 16.1a. But in reality, we observe the interference pattern which, on the plane of the screen, looks schematically like what has been shown in Figure 16.1b.

To escape the embarrassment, one will then have to say that a photon also must have gone through both slits, just like a wave can go through both. There is no way of saying which slit the photon has passed through, so we have to admit that the photon could have passed through both slits.

But wait! Why did we say that there is no way to say which slit the photon passes through? Can we not fit some kind of a detector behind one of the slits, which will record some signal if the photon passes through that slit? We can certainly put such a detector, but then the pattern on the screen will not resemble that of Figure 16.1b. In fact, it would then look like that of Figure 16.1a. In summary, the interference pattern is obtained only if we do not know which slit any particular photon has passed through, i.e., when we accept the fact that any photon passes through both slits, with a certain probability for both.

The wavefunction of the photons contains these possibilities of going through either slit. But once we make some kind of a measurement to ascertain whether the photon has passed through a particular slit, the measurement process changes the wavefunction. In technical terms, it is said that the wavefunction *collapses*. From the probabilities of going through both slits, the changed wavefunction chooses one. The chosen state is no more the original state: it tells you which slit the photon went through.

16.2   Collapse of a wavefunction

This is the major difference of quantum mechanics with classical physics. In classical physics, we always assume that a measurement process does not disturb the state of the system in any major way. At least the measurements can be made in such a way that the disturbance can be held below any level that one wants to. This is not true in quantum mechanics. The measurement processes change the state of a system dramatically.

It may be easy to attach a name such as *collapse* of wavefunctions to the phenomenon, but it is not easy to understand what exactly happens. Going back to the double-slit experiment, what we seem to be saying is that when we detect a photon passing through a particular slit, we cannot really say that it was supposed to go through that slit even if we had not made the detection. Without any kind of detection mechanism, the photon would have passed through both slits. It went through the upper slit, say, because we measured it. Our measurements forced it to go through the upper slit. Otherwise, it would have passed through both slits.

It does not really help if we say that our detection mechanism was placed *behind* the slits, and therefore it could not have affected which slit the photon went through. The wavefunction will collapse no matter where we put our detector, and we will lose the interference fringes.

To make matters worse, it is not clear how the photon recognizes that there is a detector behind one of the slits? It may not be difficult to realize that there is something. But, instead of a photon detector, if we had put a dictionary in its place, would it have caused the wavefunction to collapse? How would the photon know that the dictionary is not a detector at all? Or suppose we had put a common balance behind a slit. It is certainly a measuring instrument, but cannot measure anything about the path of a photon. What would happen to the wavefunction in this case?

The crucial question seems to be this: what constitutes a measurement process? We will come back to this question in various forms in the rest of this chapter.

16.3   EPR paradox

Albert Einstein, Boris Podolsky, and Nathan Rosen posed an interesting question about the collapse of wavefunction, through which they pointed out that quantum mechanics is, at best, incomplete. The essential idea behind their argument can be explained without using the specific example that they used. We are using a variant that was first used by David Bohm.

Suppose two spin-$\frac{1}{2}$ particles are produced in the decay of a spinless particle. This means that the state of the two particles is given by

$$|\psi\rangle = \frac{1}{\sqrt{2}}\Big(|\uparrow\rangle_1|\downarrow\rangle_2 - |\downarrow\rangle_1|\uparrow\rangle_2\Big), \tag{16.1}$$

where the notation $|\uparrow\rangle_1$, for example, means that particle 1 is in the state with the eigenvalue $+1$ of the operator σ_z. In the rest frame of the parent particle, the two particles are created with equal and opposite momenta.

If one asks, at this stage, what is the value of S_z for the first particle, that is a question that cannot be answered. The value can be positive or negative, both. The same can be said about the second particle. The values are entangled, in a sense that will be elaborated later in §16.4.

The two particles move away from the position of their original parent in this stage. But now suppose one makes a measurement of the spin of particle 1 and finds it in a state with σ_z value equal to $+1$. This means that because of the measurement process, the state of the first particle has collapsed to $|\uparrow\rangle$. The state of the two-particle system is no more that shown in Eq. (16.1). Since the original state was spinless, the second particle must now have collapsed to the state $|\downarrow\rangle$.

This may appear disturbing to the reader for various reasons. First, we get to know the state of the second particle without performing any measurement on it, which is contrary to what we had been talking about in the context of the double-slit experiment. Second, at the time of the measurement, the second particle is at some distance from the first one. So, when did the wavefunction collapse for the second particle? If it collapsed at the same time, then that implies an instantaneous propagation of information, something in contradiction with the special theory of relativity. And if it did not collapse at the same time, then it is equally puzzling to think about what caused the collapse, and how. However all these questions can be answered by noting that the second observer needs to know separately about the initial state of the particles to come to any conclusion about the state of the first particle after his/her measurement. That information could only have been communicated to him/her at the speed of light. Moreover, the information he/she gathers on the first particle from the measurement can also be communicated at best at the speed of light. Thus his/her conclusion does not constitute propagation of information with speed faster than light speed. Also, as we shall see later, it is an inherent quantum property of the initial state, viz., entanglement, which allows him/her to conclude about the state of the first particle. This property has no classical analogue and we shall discuss this in detail in the next section. We note here that Einstein, Podolsky, and Rosen (EPR), on the basis of this feature of quantum states, were tempted to believe that quantum mechanics is incomplete in some way since it does not specify the mechanism of the collapse of wavefunction.

16.4   Entanglement

The main problem with the EPR kind of argument is that the states of two particles may be entangled, i.e., the state of one particle might depend on the state of the other. More formally, we can say that although the Hilbert space of the particles 1 and 2 are of the form $H_1 \times H_2$, the state in Eq. (16.1) is not of the form $|\psi\rangle_1 \times |\phi\rangle_2$. In general, a state of two particles can be written in the form

$$\sum_{i,j} C_{ij} |\psi_i\rangle_1 \times |\phi_j\rangle_2. \tag{16.2}$$

It can be written as a direct product of states of the two particles if the coefficients C_{ij} are of the form

$$C_{ij} = A_i B_j \tag{16.3}$$

for each pair of values of the indices i and j. If that does not happen, then the state is entangled.

There is another way that an entangled state can be characterized. For that, we need to construct the density matrix in the state given in Eq. (16.1). Each particle can have two different states, so there are four basis states. Accordingly, the density matrix is a 4×4 matrix. In the basis $|\uparrow\rangle_1|\uparrow\rangle_2$, $|\uparrow\rangle_1|\downarrow\rangle_2$, $|\downarrow\rangle_1|\uparrow\rangle_2$, $|\downarrow\rangle_1|\downarrow\rangle_2$, in that order, the density matrix is

$$\rho = |\psi\rangle\langle\psi| = \begin{pmatrix} 0 & 0 & 0 & 0 \\ 0 & \frac{1}{2} & -\frac{1}{2} & 0 \\ 0 & -\frac{1}{2} & \frac{1}{2} & 0 \\ 0 & 0 & 0 & 0 \end{pmatrix}. \tag{16.4}$$

This density matrix pertains to a pure state, viz., the state given in Eq. (16.1), and indeed it is easy to show that

$$\rho^2 = \rho, \tag{16.5}$$

as is appropriate for a pure state.

After we have made a measurement of one of the particles, say particle 1, we are left with the state of only the other particle. This can be expressed by a *reduced density matrix* of particle 2, defined as

$$\rho_2 = \sum_i \langle \psi_i|_1 \Big(|\psi\rangle \langle \psi| \Big) |\psi_i\rangle_1. \tag{16.6}$$

Although naively it looks like this expression contains only inner products and should therefore be a number, it is in fact an operator since ρ is obtained from a sum of the direct product of states of two particles, whereas the inner products have been taken only with respect to the states of particle 1. The summation is over all states *of only one particle*; we therefore call it partial trace. The result of such a partial trace will still be a matrix, but now the basis states will be the states of only particle 2, i.e., it will be a 2×2 matrix.

Calculation of the reduced density matrix is straightforward. Consider the state of Eq. (16.1). Since

$$\langle \uparrow|_1 \Big(|\uparrow\rangle_1 |\downarrow\rangle_2 - |\downarrow\rangle_1 |\uparrow\rangle_2 \Big) = |\downarrow\rangle_2 \tag{16.7}$$

and

$$\langle \downarrow|_1 \Big(|\uparrow\rangle_1 |\downarrow\rangle_2 - |\downarrow\rangle_1 |\uparrow\rangle_2 \Big) = -|\uparrow\rangle_2, \tag{16.8}$$

we obtain

$$\rho_2 = \langle \uparrow|_1 \Big(|\psi\rangle \langle \psi| \Big) |\uparrow_1\rangle + \langle \downarrow|_1 \Big(|\psi\rangle \langle \psi| \Big) |\downarrow_1\rangle$$

$$= \frac{1}{2} \Big(|\downarrow\rangle_2 \langle \downarrow|_2 + |\uparrow\rangle_2 \langle \uparrow|_2 \Big). \tag{16.9}$$

Written in array form, this is

$$\rho_2 = \begin{pmatrix} \frac{1}{2} & 0 \\ 0 & \frac{1}{2} \end{pmatrix}. \tag{16.10}$$

Note that this matrix does not satisfy Eq. (16.5). It is a mixed state. Although we started with a pure state, after a measurement on particle 1 leading to a partial tracing out of the density matrix, the reduced state of the other particle is a mixed state. The appearance of such mixed reduced states when the density matrix corresponding to a pure quantum state is partially traced out is a crucial feature of quantum mechanics that leads to entanglement.

The trace over states of particle 1 constitutes a loss of information regarding its state since the resultant reduced density matrix does not have all information contained in ρ. Such a loss can be characterized by an entropy known as *entanglement entropy*. This entropy can be directly computed from the reduced density matrix ρ_2. There are several measures of such an entropy and we list two classes of them here.

The first class of such measures contains those which can be directly computed from the eigenvalues λ_j of the reduced density matrix. These are called von-Neumann and n^{th} order Renyi entropies $(n > 1)$ and are defined, in terms of the reduced density matrix ρ_2 defined in Eq. (16.6), as

$$S_{\text{vn}} = -\text{Tr}[\rho_2 \ln \rho_2] = -\sum_j \lambda_j \ln \lambda_j, \qquad (16.11\text{a})$$

$$S_{\text{ren}}^{(n)} = \frac{1}{1-n} \ln[\text{Tr}\,\rho_2^n] = \frac{1}{1-n} \ln \sum_j \lambda_j^n. \qquad (16.11\text{b})$$

There are two properties of these entanglement measures that are worth noting. First, since for any density matrix, $\text{Tr}\,\rho = 1$, all its eigenvalues must satisfy the condition $\lambda_j \leq 1$. This ensures that

$$S_{\text{vn}}, S_{\text{ren}}^{(n)} \geqslant 0. \qquad (16.12)$$

Their zero values occur for pure states where $\lambda_j = 0$ or 1 for all j. Second, such entropies are always zero if one starts from a product initial state. To see this, consider a density matrix for such states which can be written as

$$\rho_s = \rho_1 \otimes \rho_2. \qquad (16.13)$$

Now imagine we carry out a trace with respect to ρ_1. It is clear that such a trace does not change the eigenvalues of ρ_2. Thus, for such configurations, the reduced density matrix also resembles a pure state with eigenvalues $\lambda_j = 1, 0$. This ensures that $S_{\text{vn}}, S_{\text{ren}}^{(n)} = 0$ for such states.

The fact that density matrices corresponding to separable states can be written in a product form allows one to design another class of measure for entanglement called *negativity*. To define this, we need to invoke the concept of a partial transpose of a density matrix. For separable density matrices given by ρ_s, such a transpose corresponds to taking a transpose of ρ_1 leading to

$$\rho_s' = \rho_1^T \otimes \rho_2. \qquad (16.14)$$

The key feature is to note that ρ_s' is an allowed density matrix for a quantum system in the sense that all its eigenvalues will remain positive. This follows from the fact that both ρ_1 and ρ_2 are matrices with positive eigenvalues and unit trace. Thus, for separable density matrices, the sign of the eigenvalues do not change under partial transpose.

For generic density matrices that cannot be written as a direct product, the partial transpose may lead to negative eigenvalues. To see this, let us first go back to the separable density matrix ρ_s and define

$$(\rho_s)_{i,\alpha;j,\beta} = (\rho_1)_{ij}(\rho_2)_{\alpha\beta}. \tag{16.15}$$

Under partial transpose, the matrix element then goes to

$$(\rho_s')_{i,\alpha;j,\beta} = (\rho_s)_{j,\alpha;i,\beta}. \tag{16.16}$$

The definition of partial transpose can now be generalized for an arbitrary density matrix ρ by demanding that it leads to $\rho \to \rho'$ such that $\rho_{i,\alpha;j,\beta}' = \rho_{j,\alpha;i,\beta}$. The negative eigenvalues of the matrix ρ' is then taken to be the signature of the presence of non-separability and hence entanglement for ρ; this measure is called negativity.

The measure of negativity can be computed for the density matrix corresponding to the spin singlet given by Eq. (16.4) as follows. For this matrix $\rho_{1,0;1,0} = \rho_{0,1;0,1} = 1/2$ and $\rho_{1,0;0,1} = \rho_{0,1;1,0} = -1/2$. The partial transpose of this matrix with respect to particle 1 is given by

$$\rho' = \begin{pmatrix} 0 & 0 & 0 & -\frac{1}{2} \\ 0 & \frac{1}{2} & 0 & 0 \\ 0 & 0 & \frac{1}{2} & 0 \\ -\frac{1}{2} & 0 & 0 & 0 \end{pmatrix}. \tag{16.17}$$

The eigenvalues of ρ' are $1/2, 1/2, 1/2, -1/2$ showing the presence of one negative eigenvalue bearing the signature of non-separability.

☐ **Exercise 16.1** *Consider a state of two spin-$\frac{1}{2}$ particles of the form suggested in Eqs. (16.2) and (16.3). Construct the reduced density matrix for particle 2 and verify that it pertains to a pure state.*

☐ **Exercise 16.2** *An impure singlet or a* Werner state *corresponds to a 4×4 density matrix of two spins whose elements are given by*

$$(\rho_w)_{i,\alpha;j,\beta} = x\rho_{i,\alpha;j,\beta} + \frac{1}{4}(1-x)\delta_{ij}\delta_{\alpha\beta}, \tag{16.18}$$

where $0 \leqslant x \leqslant 1$ denotes the singlet fraction and ρ is given by Eq. (16.4).

(a) *Write ρ_w as a 4×4 matrix.*

(b) *Find ρ_w', i.e., the partial transpose of ρ_w with respect to spin 1.*

(c) *Show that ρ_w' has eigenvalues $(1+x)/4$ (three fold degenerate) and $(1-3x)/4$. Hence, show that the system is entangled for $x > 1/3$. Note that the results reduce to that of pure singlets for $x = 1$ as it should.*

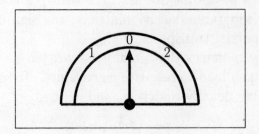

Figure 16.2 Schematic diagram of a detector.

16.5   When does a measurement end?

The main problem seems to be associated with the question: when does a measurement end? Without a clear-cut answer, we will face paradoxes everywhere. Let us discuss the nature of these paradoxes.

16.5.1   A generic description of paradoxes

Let us go back to the double-slit experiment. Suppose we have placed a detector of some sort behind the slits which would tell us which slit a particular photon passes through. We do not need a description of the detector. However, in very general terms, we can say that the detector will have to have three different states, no matter how the states are manifested. One of these states must be the *inactive state*, which is how the detector should be found when no photon passes through the slits, or the detector is turned off. And then, if the photon passes through slit 1, the detector must appear in a different state. If the photon passes through slit 2, the detector must end up in yet another state. In Figure 16.2, we have indicated these states schematically by 0, 1, and 2, with something like the hand of a clock to indicate the state that the detector is in at a given moment.

First, consider the situation where the detector is not there at all. One photon is released. As said earlier, in this case quantum theory will tell us that the photon will pass through both slits.

Now the detector is brought in, and again one photon is released. While going through the slits, a collapse of the wavefunction takes place, so the detector hand moves either to 1 or to 2.

But wait! How do we know which place the detector hand has moved to? If it is impossible to tell, without performing a measurement, which slit the photon has gone through, by the same token we should say that it is impossible to tell where the detector hand has ended up without making a measurement on it. So, that will require another measuring apparatus, with again some generic pointers or hands on it. And then we will need another apparatus to measure the final position of this second apparatus. And the argument repeats. How do we get out of this argument? How will the measurement end?

One might think that we are making a problem out of nothing. In the first step, when the pointer moved to one of the two positions, we could just look at the pointer to check whether it now points towards 1 or towards 2, and that ought to be it. There is no second step, or any of the subsequent steps.

The problem in this exit route lies in the use of the words "look at." First, remember that our description of the measuring instrument was purely schematic. The actual instrument might not have any pointer at all, at least not anything that a human eye can see. For example, suppose the instrument has some chemical arrangement which produces a particular type of gas if a photon passes through slit 1, and a different gas if the photon passes through slit 2. In that case, we definitely need a second phase to determine which kind of gas has been produced.

Second, who looks at the pointer? A human being? Does that mean that a process can be called a measurement only if it is visible to the human eye? We will come back to this question in a little while.

16.5.2 Schrödinger cat paradox

A very famous paradox regarding the termination of the measurement process was proposed by Schrödinger. This can be stated as follows.

Consider a cat in a sealed room with no communication with the environment outside. There is also a phial of some deadly poison, but the phial is sealed. There is a hammer just next to the phial. The hammer is connected to a switch outside the room. If the switch is turned on, some electrical signal would flow and the hammer would strike the phial, breaking it. The poison will then kill the cat.

What activates the switch? It is connected to a detector set up to detect the path of a photon through a double-slit experiment in the next room. If we release a photon and the detector finds that the photon passes through slit 1, say, then the switch is activated and the cat dies. If the photon passes through slit 2, the switch is not activated, and the cat stays alive.

Now a scientist named S performed an experiment by sending just one single photon through the double-slit apparatus. The switch was either activated or not, depending on which slit the photon passed through. Scientist S did not go check the fate of the cat in the room.

How was the cat at this stage? Was it alive or was it dead? If we want to follow the kind of logic that we have been using for photons, we should say that the cat was still both, because no measurement had been performed on it. Its wavefunction had not collapsed; it is in an equal linear superposition of states $|\text{alive}\rangle$ and $|\text{dead}\rangle$ so that $|\psi_{\text{cat}}\rangle = (|\text{alive}\rangle + |\text{dead}\rangle)/\sqrt{2}$.

That is a bit uncomfortable, and that is the paradox. What is more surprising in this line of argument is what happened next morning when S went to take a look at the cat. Let us say that S found the cat dead. When did that happen? It was neither alive nor dead, or both alive and dead, until S looked. Did the wavefunction collapse when S finally looked at it?

We are back to the same question: When does the wavefunction collapse, and when does the measurement end? What is it that brings us out of the seemingly infinite loop of argument presented in §16.5.1?

16.6   Searching for answers

We have raised the question. There is no answer that is universally accepted. We try to outline a few avenues in which the answer has been sought.

16.6.1   When does collapse occur?

One answer was hinted at earlier. The collapse happens when scientist S of §16.5.2 looks at the cat in the morning. One should carefully note here that measurement simply means action of any detector, which is not a part of the closed quantum system comprising the cat and the poison vial, on the state of the cat. It does not mean that a human being (viz., scientist S) has to actually look; an automatic camera, for example, would have the same effect.

This interpretation still raises the uncomfortable question about the state of the cat anytime before the measurement. What does it mean to say that it was both alive and dead? But this question can be easily avoided by saying that indeed that was the state. How do we know what kind of a state that is? Well, we cannot, because any measurement to probe the state of the cat inside the room would destroy the state, ushering the collapse.

16.6.2   A matter of size?

It might be thought that the size of the object matters. Quantum mechanics works for small objects, but when it comes to a macroscopic object such as a cat, we should use only classical concepts.

This would be a big blow, because it would mean that quantum mechanics does not provide the real underlying theory of everything. There are different laws for large objects. Even apart from the philosophical uneasiness that results, there is the question of where to draw the line between small and large, and how. And no line will be fully satisfactory, because we know of macroscopic systems also which show quantum behavior, like superconductors and superfluids. In fact, the electrical and thermal properties of all metals, semiconductors, and insulators require quantum mechanics for their explanation.

16.6.3   Many universes?

Hugh Everett III proposed another way out in 1957. According to him, when a wavefunction collapses into one of the many possibilities that existed before, the remaining possibilities are not really lost. They are realized in some other universe, with which we do not interact at all.

That is not very satisfactory. Science depends on testable experiments. If a theory talks about other universes with which we cannot communicate, we do not even know how to test the theory.

16.6.4   Hidden variables?

We should seriously consider another possibility: that the probabilistic nature of the quantum states arise due to lack of information, and the systems are really deterministic if those extra pieces of information are available. This idea, originally due to Louis de Broglie and further developed by David Bohm, was a serious contender to the probabilistic interpretation of quantum mechanics for some time.

Let us explain. When we toss a coin, we do not know whether it will land with the head up or the tail up. We do not conclude that the coin's dynamics is probabilistic, i.e., it will never be possible to predict the outcome with certainty. We cannot predict the outcome because we do not pay attention to the details of the tossing process. If we had known exactly where the knock was given to it, and the force imparted by the knock, maybe also the air resistance and some other parameters, we could have predicted with absolute certainty the outcome of any coin toss. Could it be the same for a quantum system? Can we say that there are *hidden variables* of a quantum system, and if we knew the values of all of them, we could have predicted, for example, which slit a particular photon would go through in a double-slit experiment?

Remember what we had said earlier regarding measurement. According to quantum mechanics, a quantity does not have a pre-assigned value irrespective of the measurement. The value is assigned by the collapse that happens because of the measurement. If there are hidden variables and the theory is deterministic at the core, then we should not make this statement. We should say, instead, that the values are predetermined, only that we do not know them because we do not have the information of a number of hidden variables. This would be a huge change in the nature of the underlying reality, and would solve many of the paradoxes concerning quantum mechanics, some of which we have described earlier.

How can we test this hypothesis? If the extra variables are indeed hidden, does that not mean that they are inaccessible, like what we had encountered in the explanation with *many universes*? Well, not really. For this solution, fortunately, great progress has been achieved, and it has been proved that the hidden variable hypothesis cannot be correct. This progress has been achieved through Bell inequalities, which we discuss next.

16.7   Bell inequalities

The phrase *Bell inequalities* does not refer to a fixed set of inequalities. It was shown by John Bell that there are inequalities among physical measurements which must be obeyed by any local deterministic theory, but can be violated by quantum mechanics. Any inequality of this sort is called a Bell inequality, many of which have been discovered not by Bell but by others. If measurements are done in the disputed region, one can decide whether the underlying theory is deterministic. We will give an example of such relations.

Let us first see what kind of inequalities can be obtained from observables in a deterministic theory. Suppose we are considering three different observables, each of which can take only two values, which we represent by a and b. Thus, for the three observables taken together, there can be eight possible outcomes, *aaa*, *aab*, etc., up to *bbb*. Suppose we make a huge number of measurements for the three variables and make a list of the results obtained in each. In the list, there will be a certain number of rows with *aaa*, certain number of *aab*, etc. Suppose now we count the number of rows with *aa* for the first two variables, without paying any attention to the value of the third variable. Let us denote this combination as *aao*. In other words,

$$N(aao) = N(aaa) + N(aab). \tag{16.19}$$

Let us similarly count the number of rows with the result *boa*, i.e., with *b* for the first variable and *a* for the third, no matter what the second one is. Obviously,

$$N(boa) = N(baa) + N(bba). \tag{16.20}$$

Thus,

$$N(aao) + N(boa) = N(aaa) + N(aab) + N(baa) + N(bba)$$
$$= \Big(N(aaa) + N(baa)\Big) + N(aab) + N(bba). \tag{16.21}$$

The object shown in parentheses is nothing but $N(oaa)$. Since the remaining numbers are non-negative, we conclude that

$$N(aao) + N(boa) - N(oaa) \geqslant 0. \tag{16.22}$$

This is an example of the kind of inequalities that we have been talking about. We will now see how for a suitable choice of the three variables, this inequality can be violated by quantum mechanics.

For the three different observables mentioned in the argument above, let us choose the components of spin of a spin-$\frac{1}{2}$ particle in three different directions. Each measurement can yield the value $+\frac{1}{2}\hbar$ or $-\frac{1}{2}\hbar$, which takes the place of *a* and *b* in Eq. (16.22), and the inequality must hold in a local deterministic theory if these values are pre-determined.

To see what quantum mechanics says about the three numbers that appear in Eq. (16.22), let us say that all three directions of measurement lie in the *x*-*z* plane, so the operator for the spin component along that direction can be written as

$$S_\theta = \frac{1}{2}\hbar(\sigma_z \cos\theta + \sigma_x \sin\theta)$$
$$= \frac{1}{2}\hbar \begin{pmatrix} \cos\theta & \sin\theta \\ \sin\theta & -\cos\theta \end{pmatrix} \tag{16.23}$$

for some value of θ. The eigenstates are

$$|\theta\uparrow\rangle = \begin{pmatrix} \cos\frac{\theta}{2} \\ \sin\frac{\theta}{2} \end{pmatrix}, \qquad |\theta\downarrow\rangle = \begin{pmatrix} \sin\frac{\theta}{2} \\ -\cos\frac{\theta}{2} \end{pmatrix} \qquad (16.24)$$

corresponding to the positive and the negative eigenvalues respectively.

In order to find the value of the quantities that appear in Eq. (16.22), we need to make measurements of the spin component along two different directions. How do we do that? We can use the idea that was put forward in connection with the EPR paradox. We will take two particles, originally in a state with combined spin equal to zero, so that

$$|\psi_{\text{in}}\rangle = \frac{1}{\sqrt{2}}\Big(|\uparrow\rangle\,|\downarrow\rangle - |\downarrow\rangle\,|\uparrow\rangle \Big), \qquad (16.25)$$

where, in each term on the right side, the first state refers to particle 1 and the second to particle 2. The states $|\uparrow\rangle$ and $|\downarrow\rangle$ can be taken as the states given in Eq. (16.24) for any fixed value of θ: it does not matter what value is taken. We will use the value $\theta = 0$ in what follows.

Now, the two particles are traveling in two directions. We measure the spin component of particle 1 along a direction α, and at the same time measure the spin component of particle 2 in the direction β. If for the particle 2, we obtain a positive value in the measurement, that would mean that the component of particle 1 in the same direction must be negative, and vice versa. Let us use the notation $a(\alpha\uparrow, \beta\uparrow)$ to denote the amplitude of finding both spin components to be positive for particle 1. This means that we are measuring particle 1 having positive spin in the α direction, and particle 2 having negative spin in the β direction. The quantum mechanical amplitude for obtaining this, starting from the initial state of Eq. (16.25), would be

$$\begin{aligned}
a(\alpha\uparrow, \beta\uparrow) &= \langle\psi_{\text{in}}\,|\,(\alpha\uparrow)(\beta\downarrow)\rangle \\
&= \frac{1}{\sqrt{2}}\Big((\cos\frac{\alpha}{2})(-\cos\frac{\beta}{2}) - (\sin\frac{\alpha}{2})(\sin\frac{\beta}{2}) \Big) \\
&= -\frac{1}{\sqrt{2}}\cos\frac{\beta-\alpha}{2}.
\end{aligned} \qquad (16.26)$$

So, the probability of obtaining such values in measurement is given by

$$P(\alpha\uparrow, \beta\uparrow) = \frac{1}{2}\cos^2\frac{\beta-\alpha}{2}. \qquad (16.27)$$

Similarly, we can calculate

$$P(\alpha\uparrow, \beta\downarrow) = \frac{1}{2}\sin^2\frac{\beta-\alpha}{2}. \qquad (16.28)$$

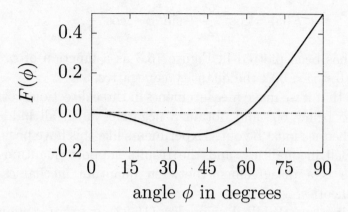

Figure 16.3 An example of Bell correlations. Local hidden variable theories predict the correlation to be non-negative at all angles.

Suppose now we think of three different directions, denoted by the angles α, β, and γ in the x-z plane. We make three sets of experiments. In the first one, we measure spin components in the directions α and β, without paying any attention to the direction γ. In the second one, we measure spin components only in directions 1 and 3. In the third, we measure spin components in only the directions β and γ. For each measurement, the result with a positive eigenvalue is to be considered similar to a and a negative value similar to b of Eq. (16.22). Then, the quantity on the left side of Eq. (16.22) would be proportional to

$$f(\alpha, \beta, \gamma) = \cos^2 \frac{\beta - \alpha}{2} + \sin^2 \frac{\gamma - \alpha}{2} - \cos^2 \frac{\beta - \gamma}{2}, \qquad (16.29)$$

where the constant of proportionality involves the total number of measurements made.

If Eq. (16.22) has to be satisfied, this quantity should be positive for all possible values of the three angles. We need not go to a multidimensional plot to show that this is not the case. We can just take some exemplary values to show that the quantity can be negative. Let us take

$$\beta - \gamma = \phi, \qquad \gamma - \alpha = 2\phi. \qquad (16.30)$$

Then the quantity appearing in Eq. (16.29) is given by

$$F(\phi) = \cos^2 \frac{3\phi}{2} + \sin^2 \phi - \cos^2 \frac{\phi}{2}. \tag{16.31}$$

This function has been plotted in Figure 16.3 as a function of ϕ. Clearly, it shows that for $0 < \phi < 60°$, the quantity is negative.

This means that if we make measurements in three directions corresponding to a value of ϕ below 60° and obtain a negative value, all hidden variable deterministic theories must be out. Experiments like this have been performed with various Bell inequalities, and the results support quantum mechanics wherever there is a disagreement between quantum mechanics and local deterministic theories.

What does it mean? While deriving Eq. (16.22), we were assuming that the spin components in different directions have well-defined values. Experiments show that that does not seem to be the case. The values are not pre-assigned. The quantum system restricts itself to a state corresponding to a particular eigenvalue of the observable at the time of measurement, because of the collapse of the wavefunction.

Chapter 17

Miscellaneous topics

In this chapter, we shall discuss a few problems that are not usually addressed for a first course in quantum mechanics. However, these topics are becoming increasingly important for modern-day research in several areas of theoretical physics. We have tried to make the range of topics discussed as broad as possible; however, they are by no means exhaustive.

17.1   Coherent states

In this section, we shall discuss coherent states. Such states can be constructed for a wide variety of quantum mechanical systems. We shall, however, restrict our discussion to coherent states of harmonic oscillators, because the concept and analytic treatment of coherent states are particularly simple for harmonic oscillators.

Simple harmonic oscillators were discussed in Chapter 7. The concept of coherent states of such oscillators comes from noting that the annihilation operator a annihilates the ground state: $a|0\rangle = 0$. It can be said that the ground state of the oscillator is an eigenstate of the annihilation operator with eigenvalue equal to zero. This raises a natural question: are there states $|\alpha\rangle$ that are eigenstates of a with eigenvalue α such that the equation

$$a|\alpha\rangle = \alpha|\alpha\rangle \tag{17.1}$$

is satisfied for some non-zero value of α? Note that α may be complex since a is not Hermitian.

> In §2.7.3, we commented that non-commuting operators can have some common eigenvectors. Here we see an example: the ground state $|0\rangle$ is an eigenstate of a as well as of the Hamiltonian H, although these two operators do not commute, as seen in Eq. (7.75, p. 172).

To find what these states are, we first note that one can always express $|\alpha\rangle$ in the number basis as

$$|\alpha\rangle = \sum_{n=0}^{\infty} c_n |n\rangle = \sum_{n=0}^{\infty} c_n \frac{(a^\dagger)^n}{\sqrt{n!}} |0\rangle. \qquad (17.2)$$

Thus, the states $|\alpha\rangle$ can be charted out if we can find the coefficients c_n, given by $c_n = \langle n|\alpha\rangle$. Using this along with Eqs. (17.1) and (17.2), one can write

$$c_n = \frac{1}{\sqrt{n!}} \langle 0|a^n|\alpha\rangle = \frac{\alpha^n}{\sqrt{n!}} c_0, \qquad (17.3)$$

where we have used the fact that $c_0 = \langle 0|\alpha\rangle$. This allows us to write

$$|\alpha\rangle = c_0 \sum_{n=0}^{\infty} \frac{\alpha^n}{\sqrt{n!}} |n\rangle = c_0 \sum_{n=0}^{\infty} \frac{(\alpha a^\dagger)^n}{n!} |0\rangle = c_0 e^{\alpha a^\dagger} |0\rangle. \qquad (17.4)$$

From the expression of $|\alpha\rangle$, it is easy to check, using commutation relation between a and $a^\dagger$, that Eq. (17.1) is indeed satisfied. The coefficient c_0 can be determined by demanding $\langle \alpha|\alpha\rangle = 1$. This leads to

$$c_0 = e^{-\frac{1}{2}|\alpha|^2}, \qquad (17.5)$$

apart from a possible, but inconsequential, global phase in the expression of c_0. Thus, the expression for the coherent state $|\alpha\rangle$ is

$$|\alpha\rangle = e^{-\frac{1}{2}|\alpha|^2 + \alpha a^\dagger} |0\rangle. \qquad (17.6)$$

Eq. (17.6) allows us to define an operator

$$\mathcal{D}(\alpha) = \exp[\alpha a^\dagger - \alpha^* a], \qquad (17.7)$$

which acts as the generator of coherent states:

$$|\alpha\rangle = D(\alpha)|0\rangle. \qquad (17.8)$$

This property follows directly from the commutation relation of the creation and annihilation operators.

☐ **Exercise 17.1** *Verify Eq. (17.8).* [**Hint:** *Use Baker–Campbell–Hausdorff formula, Eq. (5.84, p. 127), to evaluate* $e^{\alpha a^\dagger} e^{-\alpha^* a}$.]

In contrast to the standard set of eigenstates of Hermitian operators which we have encountered so far, the coherent states are not orthogonal. This is only expected, since a is not a normal operator owing to the commutation relation of Eq. (7.73, p. 172). Considering two such states $|\alpha\rangle$ and $|\beta\rangle$, we find

$$\langle\alpha|\beta\rangle = d_{\alpha\beta} = e^{-(|\alpha|^2+|\beta|^2)/2} \sum_{n=0}^{\infty} \frac{(\alpha^*\beta)^n}{n!} = e^{-(|\alpha|^2+|\beta|^2-2\alpha^*\beta)/2},$$

$$|d_{\alpha\beta}|^2 = e^{-|\alpha-\beta|^2} \neq 0. \tag{17.9}$$

Thus the set of these states forms what is known as an overcomplete basis. The quantity $|\alpha-\beta|$ therefore gives us a notion of the *distance* between any two coherent states in the Hilbert space.

Next, we discuss the time evolution of these coherent states under the action of the harmonic oscillator Hamiltonian

$$H = (a^\dagger a + \frac{1}{2})\hbar\omega. \tag{17.10}$$

The evolution operator $U(t,0)$ corresponding to this Hamiltonian is given by $U(t,0) = \exp[-iHt/\hbar]$. Its action on a coherent state $|\alpha\rangle$ yields the state at time t, given by

$$\begin{aligned}
|\alpha(t)\rangle &= U(t,0)|\alpha\rangle = e^{-\frac{1}{2}|\alpha|^2-i\omega t/2} \sum_{n=0}^{\infty} \frac{\alpha^n}{\sqrt{n!}} e^{-in\omega t}|n\rangle \\
&= e^{-\frac{1}{2}|\alpha|^2-i\omega t/2} \sum_{n=0}^{\infty} \frac{(\alpha\exp[-i\omega t])^n}{\sqrt{n!}}|n\rangle \\
&= e^{-\frac{1}{2}|\alpha|^2-i\omega t/2} \sum_{n=0}^{\infty} \frac{(\alpha\exp[-i\omega t]a^\dagger)^n}{n!}|0\rangle \\
&= e^{-i\omega t/2}|\alpha e^{-i\omega t}\rangle.
\end{aligned} \tag{17.11}$$

Apart from an overall phases that is independent of α, this is just a coherent state with $\alpha_t = \alpha e^{-i\omega t}$. This shows that the coherent states maintain their shape during dynamics. These states are not eigenstates of the simple harmonic oscillator Hamiltonian; they are therefore expected to evolve under its action. However, the construction of these states ensures that their evolution does not destroy their shapes. Thus, they form a coherent Gaussian wavepacket, i.e., a superposition of energy eigenstates that retain its Gaussian shape during dynamics.

It is interesting to note that these states provide a connection between the quantum and the classical oscillators. To see this, we first note from Eq. (17.11) that $id\alpha_t/dt = \omega\alpha_t$. Since α_t is complex, we can decompose this equation in terms of its real (α_R) and imaginary (α_I) parts as

$$\frac{d}{dt}\alpha_R = \omega\alpha_I, \quad \frac{d}{dt}\alpha_I = -\omega\alpha_R. \qquad (17.12)$$

Since $\alpha_R = \langle\alpha|(a + a^\dagger)/2|\alpha\rangle = \sqrt{m\omega/(2\hbar)}\langle\hat{x}\rangle$ and $\alpha_I = \sqrt{2m\hbar\omega}\langle\hat{p}\rangle$, we find that these equations are equivalent to

$$\frac{d}{dt}\langle\hat{x}\rangle = \frac{\langle\hat{p}\rangle}{m}, \quad \frac{d}{dt}\langle\hat{p}\rangle = -m\omega^2\langle\hat{x}\rangle. \qquad (17.13)$$

These are indeed the equations that follow from Ehrenfest theorem applied to the motion of the oscillator. They coincide with the classical equations of motion of a single harmonic oscillator with $\langle\hat{x}\rangle \to x$ and $\langle\hat{p}\rangle \to p$.

The uncertainty of position and momentum in a coherent state is of interest. It is easily seen that in such a state,

$$\langle x\rangle \equiv \langle\alpha|x|\alpha\rangle = \sqrt{\frac{\hbar}{2m\omega}}\langle\alpha|a + a^\dagger|\alpha\rangle$$

$$= \sqrt{\frac{\hbar}{2m\omega}}(\alpha + \alpha^*), \qquad (17.14)$$

where we have used Eq. (17.1) and its conjugate, $\langle\alpha|a^\dagger = \alpha^*\langle\alpha|$, in the last step. Similarly, one can find $\langle x^2\rangle$ and hence verify that the uncertainty in position is

$$\Delta x \equiv \sqrt{\langle x^2\rangle - \langle x\rangle^2} = \sqrt{\frac{\hbar}{2m\omega}}. \qquad (17.15)$$

By a similar procedure, one obtains

$$\Delta p = \sqrt{\frac{m\omega\hbar}{2}}, \qquad (17.16)$$

so that

$$\Delta x\Delta p = \frac{1}{2}\hbar. \qquad (17.17)$$

This is the minimum value of this product that is allowed by the uncertainty principle, as shown in §3.2. Note that this uncertainty product is independent of time, unlike for a free particle for which the product increases with time, as shown in Chapter 6.

□ **Exercise 17.2** *Verify Eqs. (17.15) and (17.16).*

It turns out that the concept of coherent states can be further developed to find representation of *squeezed states*. For harmonic oscillators, the coherent states discussed are states with equal distribution uncertainties corresponding to $\widehat{x}$ and $\widehat{p}$: $\Delta x \sqrt{m\omega} = \Delta p/\sqrt{m\omega} = \sqrt{\hbar/2}$. The squeezed states, in contrast, are constructed so that $\Delta x \sqrt{m\omega} \neq \Delta p/\sqrt{m\omega}$ although $\Delta x \Delta p = \hbar/2$. These states can, in general, be expressed as

$$|\alpha, s\rangle = \mathcal{D}(\alpha)\mathcal{F}(s)|0\rangle, \quad \mathcal{F}(s) = e^{(s^* a^2 - s(a^\dagger)^2)/2}, \qquad (17.18)$$

where s is an additional complex parameter that measures the degree of squeezing. It can be shown that for these states

$$\Delta x = \sqrt{\frac{\hbar}{2m\omega}}e^{-s}, \quad \Delta p = \sqrt{\frac{m\omega\hbar}{2}}e^{s}, \qquad (17.19)$$

which leads to the possibility of tuning uncertainty in position or momentum space as a function of the parameter s. These states are extremely important for several quantum optics applications. Furthermore, the concept of coherent states plays an important role in the study of spin systems where these states can be constructed following a similar but slightly different procedure. Similar states can also be constructed for the hydrogen atom problem. A detailed discussion of these states and their properties is beyond the scope of this book.

□ **Exercise 17.3** *Show that for any set of coherent states $|\alpha\rangle$*

$$\frac{1}{\pi}\int d^2\alpha|\alpha\rangle\langle\alpha| = 1. \qquad (17.20)$$

[**Note:** *This is called the completeness property of the coherent states.*]

□ **Exercise 17.4** *Show that the operator $\mathcal{D}(\alpha)$ satisfies*

$$\mathcal{D}^\dagger(\alpha)a\mathcal{D}(\alpha) = a + \alpha. \qquad (17.21)$$

[**Note:** *This is why $\mathcal{D}(\alpha)$ is often called the* displacement operator.]

□ **Exercise 17.5** *Show that there are no normalizable eigenstates that satisfies $a^\dagger|\alpha\rangle = \alpha|\alpha\rangle$, i.e., there is no normalizable eigenstate of $a^\dagger$.*

□ **Exercise 17.6** *Show that Δx and Δp for the squeezed states $|\alpha, s\rangle$ are given by Eq. (17.19).*

□ **Exercise 17.7** *Consider the operator*

$$B = p - ibx, \qquad (17.22)$$

where p and x are the momentum and position operators, and b is a real constant.

(a) *What is the dimension of b?*

(b) *Evaluate the commutator $[B, B^\dagger]$.*

(c) *A state $|\psi\rangle$ satisfies the condition $B|\psi\rangle = 0$. Find the uncertainties Δx and Δp for this state.*

(d) *Find the wavefunction $\langle x|\psi\rangle$ in the coordinate space. No need to normalize it. Hence argue why b should be positive for a bound state.*

17.2   Pancharatnam–Berry phase

In this section, we shall discuss a special kind of phase, called the Pancharatnam–Berry phase, which provides an insight into non-trivial effects of parameter space geometry in quantum mechanics. This phase was discovered, in the context of optics, by Shivaramakrishnan Pancharatnam in 1956 and later given a more general interpretation independently by Michael Berry in 1983. The non-trivial effects arising out of relative phase difference between wavefunctions in the presence of a magnetic field leads to the Aharanov–Bohm effect. This has been already discussed in Chapter 10. Here we shall concentrate on an intrinsic geometrical phase that does not require the presence of an external field.

To this end, let us consider a quantum mechanical system whose Hamiltonian depends on a set of parameters λ_i: $H \equiv H[\lambda_1, \lambda_2, \cdots] \equiv H[\{\lambda_i\}]$. We assume that the system is initially in some eigenstate of this Hamiltonian. We shall also suppose that this eigenstate $|n_0\rangle$, which may or may not be the ground state, is separated from its nearest eigenstates by an energy gap Δ. This will allow us to define a notion of *adiabaticity* as we shall soon see. Finally, in what follows, we shall assume all λ_i's to be dimensionless.

Next, let us consider the variation of one or more parameters λ_i as a function of time: $\lambda_i \to \lambda_i(t)$. The Schrödinger equation for the wavefunction starting from the state $|n_0\rangle$ at $t = 0$ is then given by

$$i\hbar\frac{d}{dt}|\psi(t)\rangle = H[\{\lambda_i(t)\}]|\psi(t)\rangle, \qquad (17.23)$$

where $|\psi(t)\rangle$ means the state at time t. We now express the wavefunction in terms of the instantaneous eigenfunctions $|n(t)\rangle$ of the Hamiltonian which satisfies $H[\{\lambda_i\}] \, |n(t)\rangle = E_n(t) \, |n(t)\rangle$ at any instant t. Thus, one writes

$$|\psi(t)\rangle = \sum_n c_n(t)|n(t)\rangle. \tag{17.24}$$

Note that the kets $|n(t)\rangle$ change with time in this representation. Such a time-dependent basis is often called the *moving basis* in the literature. Substituting Eq. (17.24) into Eq. (17.23), we find

$$i\hbar \frac{d}{dt} \sum_n c_n(t)|n(t)\rangle = \sum_n E_n(t)c_n(t)|n(t)\rangle, \tag{17.25}$$

with the boundary condition

$$c_n(0) = \delta_{nn_0}, \tag{17.26}$$

as stated earlier. From this, one obtains the equation for the coefficient $c_n(t)$ as

$$\left(i\hbar \frac{d}{dt} - E_n(t)\right) c_n(t) = -i\hbar \sum_m c_m \langle n(t)|\partial_t m(t)\rangle. \tag{17.27}$$

Note that the term in the right side of Eq. (17.27) arises because of our choice of the moving basis.

There is no approximation so far. However, Eq. (17.27) is still difficult to solve and at this point we take recourse to adiabaticity. Imagine that the system starts from a state $|n\rangle = |n_0\rangle$ and the parameters $\lambda_i(t)$ are varied adiabatically, i.e., very slowly. In particular, we assume that

$$\hbar \frac{\partial E_{n_0}(\lambda_1, \lambda_2...)}{\partial \lambda_i} \frac{d\lambda_i(t)}{dt} \ll \Delta^2, \tag{17.28}$$

where E_{n_0} is the energy of the initial eigenstate. This is similar to the adiabaticity condition used in Eq. (12.37, p. 311) for a two-level system with a time-dependent Hamiltonian. Then clearly at any point in time, in absence of excitation formation, $|c_{n_0}| \simeq 1 \gg |c_{m \neq n_0}| \simeq 0$. Thus, in the right side of Eq. (17.27), we can neglect all terms except the one for which $m = n_0$. In this case, Eq. (17.27) admits a simple solution

$$c_n(t) = \delta_{nn_0} \exp\left(-i \int_0^t dt_1 E_n(t_1)/\hbar\right) \exp\left(i\gamma_n^{\mathrm{PB}}\right), \tag{17.29}$$

with

$$\gamma_n^{\mathrm{PB}} = i \int_0^t dt_1 \langle n(t_1)|\partial_{t_1} n(t_1)\rangle$$

$$= i \sum_i \int_{\lambda_i^{\mathrm{ini}}}^{\lambda_i^{\mathrm{fin}}} d\lambda_i \langle n(\{\lambda_i\})|\partial_{\lambda_i} n(\{\lambda_i\})\rangle$$

$$= i \int_{\{\lambda_i^{\mathrm{ini}}\}}^{\{\lambda_i^{\mathrm{fin}}\}} d\boldsymbol{\lambda} \cdot \langle n(\{\lambda_i\})|\boldsymbol{\partial}_\lambda n(\{\lambda_i\})\rangle, \qquad (17.30)$$

where we have traded the integral over time to those over parameters λ_i and $\boldsymbol{\partial}_\lambda = \{\partial_{\lambda_1}, \partial_{\lambda_2}, \cdots, \partial_{\lambda_i}\}$. We note here that γ_n^{PB} is real due to the time derivative within the integral.

Several comments are in order here. First, this result indicates that the wavefunction, apart from the standard kinematic phase, possesses an additional phase which, in general, depends on its trajectory in the parameter space. This is the Pancharatnam–Berry phase. Second, although the derivation of this phase used time-evolution of the wavefunction, the final form given in Eq. (17.30) shows that the phase depends only on the geometry of the parameter space; the integral is taken over a curve generated by variation of the parameters λ_i. Thus, this phase is purely geometric in nature. Third, for arbitrary initial and final values of the parameters λ_i, the integral is over an open curve in parameter space and hence, in general, path dependent. However, it becomes path independent for cyclic trajectories $\lambda_i^{\mathrm{ini}} = \lambda_i^{\mathrm{fin}}$. This allows us to associate several universal (path-independent) features with this phase which we chart out below.

For closed trajectories in parameter space, the Pancharatnam–Berry phase given in Eq. (17.30) can be written as a line integral over a closed curve:

$$\gamma_n = i \oint d\boldsymbol{\lambda} \cdot \boldsymbol{A}_n(\boldsymbol{\lambda}), \qquad (17.31\mathrm{a})$$

where

$$\boldsymbol{A}_n(\boldsymbol{\lambda}) = \langle n(\{\lambda_i\})|\boldsymbol{\partial}_\lambda n(\{\lambda_i\})\rangle. \qquad (17.31\mathrm{b})$$

Eq. (17.31) makes clear the analogy of the geometric phase with the phase acquired by a wavefunction in the presence of an electromagnetic vector potential, as shown in Eq. (10.56, p. 265). The difference here is that the parameter space plays the role of physical space. Due to this analogy, $\boldsymbol{A}_n(\boldsymbol{\lambda})$ is called the *Berry potential*. For a two-dimensional parameter space with parameters λ_1 and λ_2, one can also define a field strength $\mathcal{F}_n$ corresponding to this potential:

$$\gamma_n = i \int d^2\lambda \, \mathcal{F}_n,$$
$$\mathcal{F}_n = \langle \partial_{\lambda_1} n | \partial_{\lambda_2} n \rangle - \langle \partial_{\lambda_2} n | \partial_{\lambda_1} n \rangle. \qquad (17.32)$$

$\mathcal{F}_n$ is often called the Berry curvature for the state $|n\rangle$.

We note that since the line integral of any smooth vector field over a closed curve can only be non-zero if it encloses a singular point, a non-zero value of γ_n for cyclic evolution usually indicates the presence of a singular point of the Berry curvature in the parameter space. A detailed analysis of such singularities leads to a classification scheme for electron wavefunctions in solids and is the key concept for understanding a new class of insulators called *topological insulators*.

We end this section with an illustration of the Berry phase for a simple case. Consider a two-dimension quantum Hamiltonian in the k_x-k_y plane given by $H = A_0 \boldsymbol{\sigma} \cdot \boldsymbol{k}$, where $A_0|\boldsymbol{k}|$ has the dimension of energy and $\boldsymbol{k} = (k_x, k_y)$. The reader is urged to show that diagonalization of this Hamiltonian leads to eigenenergies $E_\pm = \pm A_0|\boldsymbol{k}|$ with wavefunctions

$$\psi_\pm(\boldsymbol{k}) = \frac{1}{\sqrt{2}} \begin{pmatrix} 1 \\ \pm e^{i\theta_k} \end{pmatrix}, \quad \theta_k = \tan^{-1}(k_y/k_x). \qquad (17.33)$$

Note that $\theta_{\boldsymbol{k}}$, and hence the wavefunction, is ill-defined at the origin where $k_x = k_y = 0$. Next let us consider an adiabatic circular trajectory in the k_x-k_y space, where k_x and k_y are varied with $|\boldsymbol{k}| = k_0 \neq 0$. The corresponding Pancharatnam–Berry potential is given by

$$\mathcal{A}_{j+}(\boldsymbol{k}) = \psi_+^*(\boldsymbol{k}) \partial_{k_j} \psi_+(\boldsymbol{k}) = \frac{i}{2} \partial_{k_j} \theta(\boldsymbol{k}). \qquad (17.34)$$

☐ **Exercise 17.8** *When the trajectory encloses the origin, show that $\oint d\boldsymbol{k} \cdot \mathcal{A}_+(\boldsymbol{k}) = \pi$ leading to a finite phase.* [**Note:** *In contrast, for any trajectory which does not enclose the origin, it is zero.*]

17.3 Supersymmetric methods

The use of supersymmetry in field theory pertains to the presence of partner particles whose masses (energies) and couplings are identical if the symmetry is not broken. If such a symmetry is broken, one of the partners may acquire higher mass (energy).

In quantum mechanics, the analogy to supersymmetry pertains to two different Hamiltonians H_1 and H_2 whose spectra are identical except for a possible zero-energy state which may be present in the spectrum of either H_1 or H_2 (but not both). The presence of the zero-energy state is not mandatory. As we shall see, the presence or absence of this state is tied to the condition of supersymmetry breaking.

To see how all of these come about, let us consider an operator

$$\widehat{A} = \alpha \left(\widehat{p} + \frac{i\hbar}{\ell} \Phi(x) \right), \tag{17.35}$$

where $\widehat{p} = -i\hbar\partial_x$ is the usual momentum operator and $\Phi(x)$ is called the superpotential. There are two constants in the definition: ℓ has the dimension of length and the dimension of the real constant α is to be determined by Eq. (17.36) below. We note that $\widehat{A}$ is not Hermitian because of the factor of i; indeed it is easy to see that $\widehat{A}^\dagger = \alpha\left(\widehat{p} - (i\hbar/\ell)\Phi(x)\right)$. We now use $\widehat{A}$ and $\widehat{A}^\dagger$ to construct two Hamiltonians H_1 and H_2 given by

$$H_1 = \widehat{A}\,\widehat{A}^\dagger, \quad H_2 = \widehat{A}^\dagger\widehat{A}. \tag{17.36}$$

These are both Hermitian. Let us now assume that H_1 has known eigenvalues and energies:

$$H_1|n_1\rangle = \widehat{A}\widehat{A}^\dagger|n_1\rangle = \epsilon_n^{(1)}|n_1\rangle. \tag{17.37}$$

To find the eigenstates and eigenenergies of H_2, we now use

$$\widehat{A}^\dagger H_1|n_1\rangle = \epsilon_n^{(1)}\widehat{A}^\dagger|n_1\rangle \tag{17.38}$$

and use the definition of H_1 to write it as

$$\widehat{A}^\dagger\widehat{A}\widehat{A}^\dagger|n_1\rangle = \epsilon_n^{(1)}\widehat{A}^\dagger|n_1\rangle. \tag{17.39}$$

But this equation can also be written as

$$H_2\widehat{A}^\dagger|n_1\rangle = \epsilon_n^{(1)}A^\dagger|n_1\rangle. \tag{17.40}$$

Note that the eigenvalues of both operators shown in Eq. (17.36) are non-negative. Thus, Eq. (17.40) leads to two possibilities, depending on whether $\epsilon_n^{(1)}$ is zero or non-zero. First, if $\epsilon_n^{(1)} > 0$, H_2 must have a spectrum identical to that of H_1: $\epsilon_n^{(2)} = \epsilon_n^{(1)}$. The corresponding eigenstates are given by $|n_2\rangle = c_{n_1}A^\dagger|n_1\rangle$. We could have as well started from the eigenspectra of H_2 and carried out the same analysis. For any eigenstate $|m_2\rangle$ of H_2 with eigenenergy ϵ_m, H_1 will have an eigenstate with same energy and eigenfunction $|m_1\rangle = c_{m_1}A|m_2\rangle$.

□ **Exercise 17.9** *Show that the eigenvalues of H_1 and H_2 cannot be negative.*

□ **Exercise 17.10** *Show that the normalization coefficients are given by $c_{n_1} = 1/[\epsilon_n^{(1)}]^{1/2}$.*

The second possibility that follows is that either H_1 or H_2 might have a zero-energy state with eigenfunction $|0\rangle$ which satisfies $A^\dagger|0\rangle_1 = 0$ (if H_1 has a zero-energy state) or $A|0\rangle_2 = 0$ (if H_2 has a zero-energy state). We note two things regarding these possibilities. Writing A and $A^\dagger$ in real space and demanding $A^\dagger|0\rangle_1 = 0$ or $A|0\rangle_2 = 0$, we find that these zero-energy states, if they exist, must have a wavefuntion

$$\psi_{01}(x) = c_{01}\langle x|0\rangle_1 = c_{01}\exp\left(-\frac{1}{\ell}\int^x dx'\,\Phi(x')\right),$$

$$\psi_{02}(x) = c_{02}\exp\left(\frac{1}{\ell}\int^x dx'\,\Phi(x')\right). \tag{17.41}$$

Depending on the value of the integral, one of the exponents must be a rising function of x, and therefore will not be normalizable. The other one will be acceptable. Thus, either H_1 or H_2, but not both, can have zero-energy states. Note that it is possible that neither H_1 nor H_2 has zero-energy solutions. In the absence of zero-energy states for either H_1 or H_2, it turns out that $\epsilon_n^{(1)} = \epsilon_n^{(2)}$ for all eigenstates of H_1 and H_2. These Hamiltonians are termed as supersymmetric partners of each other. If there is a zero-energy state present in the spectrum of H_1, one has $\epsilon_n^{(2)} = \epsilon_{n+1}^{(1)}$. On the other hand, the presence of a zero-energy eigenstate of H_2 implies $\epsilon_n^{(2)} = \epsilon_{n-1}^{(1)}$. In either case, we say that the supersymmetry is broken.

The condition of supersymmetry breaking or the existence of zero-energy state depends on the property of the superpotential $\Phi(x)$. It turns out that if $\Phi(x)$ crosses zero an odd number of times for $-\infty < x < +\infty$, there will be zero-energy states in the spectrum of either H_1 or H_2. This statement follows from the so-called index theorem; in this book, we are not going to discuss this statement in any further details.

We now discuss a specific example where $\Phi(x) = \tanh(x/\ell)$. In this case, the two Hamiltonians H_1 and H_2 are given by

$$H_1 = \alpha^2\widehat{p}^2 + \left(\frac{\alpha\hbar}{\ell}\right)^2\left(\tanh^2(x/\ell) - \text{sech}^2(x/\ell)\right),$$

$$H_2 = \alpha^2\widehat{p}^2 + \left(\frac{\alpha\hbar}{\ell}\right)^2. \tag{17.42}$$

It is easy to see that H_2 has wavefunctions and eigenvalues given by

$$\psi_k^{(2)}(x) = e^{ikx}, \qquad \epsilon_k^{(2)} = \alpha^2\hbar^2\left(k^2 + (1/\ell)^2\right). \tag{17.43}$$

The formalism developed tells us that this will also be the spectrum of H_1 for all states with non-zero energy. The eigenfunctions of H_1 will be

$$\psi_k^{(1)}(x) = e^{ikx}\frac{k + (i/\ell)\tanh(x/\ell)}{\sqrt{k^2 + (1/\ell)^2}}. \tag{17.44}$$

It is also to be noted that the superpotential $\Phi(x) = \tanh(x/\ell)$ crosses the origin once at $x = 0$; consequently H_1 must have a zero-energy state.

□ **Exercise 17.11** *Derive the wavefunction corresponding to the zero eigenvalue of H_1.*

There is an alternative but equivalent formulation of supersymmetric quantum mechanics that we now want to outline. For this, instead of starting with differential operators A and $A^\dagger$, we define matrix operators Q and $Q^\dagger$ which satisfy the anticommutation relations

$$\{Q, Q\} = \{Q^\dagger, Q^\dagger\} = 0. \tag{17.45}$$

In terms of these, one defines a Hamiltonian H given by

$$H = \{Q, Q^\dagger\} \equiv QQ^\dagger + Q^\dagger Q. \tag{17.46}$$

Note that Eq. (17.45) implies that

$$Q^2 = (Q^\dagger)^2 = 0, \tag{17.47}$$

the null operator, which leads to the commutation relations

$$[Q, H] = [Q^\dagger, H] = 0. \tag{17.48}$$

□ **Exercise 17.12** *Show that the expectation value of this Hamiltonian is non-negative in any state.*

□ **Exercise 17.13** *Verify Eq. (17.48).*

The expectation value of $Q^\dagger Q$ is non-negative in any state, because

$$\langle\psi|Q^\dagger Q|\psi\rangle = \|Q|\psi\rangle\|^2. \tag{17.49}$$

The same thing can be said about the expectation value of $QQ^\dagger$. Therefore, the eigenvalues of H must be either positive or zero. Let us consider an eigenstate of H:

$$H \left| n \right> = E_n \left| n \right>. \tag{17.50}$$

Since both Q and $Q^\dagger$ commute with H, so does the combination $Q^\dagger Q$. Therefore,

$$HQ^\dagger Q \left| n \right> = Q^\dagger Q H \left| n \right> = E_n Q^\dagger Q \left| n \right>. \tag{17.51}$$

However, using Eq. (17.47), we see that

$$HQ^\dagger Q = (Q^\dagger Q)^2. \tag{17.52}$$

So, Eq. (17.51) reduces to the form

$$(Q^\dagger Q)^2 \left| n \right> = E_n Q^\dagger Q \left| n \right>. \tag{17.53a}$$

Exactly similarly, we can show that

$$(QQ^\dagger)^2 \left| n \right> = E_n QQ^\dagger \left| n \right>. \tag{17.53b}$$

It has already been pointed out, through Eq. (17.49), that the eigenvalues of $Q^\dagger Q$ are non-negative. According to the result of Ex. 2.32 (p. 36), all its eigenvectors with zero eigenvalue, augmented by the null vector, form a vector space, which is a subspace of the entire space of states. Let us call this $\mathbb{V}_0$. It is also easy to see that all other eigenvectors plus the null vector is a subspace, which we call $\mathbb{V}_+$. Using the same kind of consideration on $QQ^\dagger$, we can define a subspace $\mathbb{V}'_0$ containing all eigenvectors of $QQ^\dagger$ with zero eigenvalue, and another subspace $\mathbb{V}'_+$ that inclues all other eigenvectors. Any eigenstate of H will belong either to $\mathbb{V}_0$ or to $\mathbb{V}_+$, and at the same time it should also belong to $\mathbb{V}'_0$ or $\mathbb{V}'_+$.

Let us now suppose there is an eigenstate of H that belongs to both $\mathbb{V}_+$ and $\mathbb{V}'_+$. Note that since $\mathbb{V}_+$ contains no zero eigenvalue, $Q^\dagger Q$ is an invertible operator in $\mathbb{V}_+$, and so is $QQ^\dagger$ in V'_+. For states that belong to both $\mathbb{V}_+$ and $\mathbb{V}'_+$, we can therefore define inverses of the operators $Q^\dagger Q$ and $QQ^\dagger$. Applying these inverses on Eq. (17.53), we find

$$Q^\dagger Q \left| n \right> = E_n \left| n \right>, \tag{17.54a}$$
$$QQ^\dagger \left| n \right> = E_n \left| n \right>. \tag{17.54b}$$

If both these equations are satisfied, we obtain

$$H\left|n\right\rangle = Q^\dagger Q\left|n\right\rangle + QQ^\dagger\left|n\right\rangle = 2E_n\left|n\right\rangle, \tag{17.55}$$

which contradicts the definition of $\left|n\right\rangle$ given in Eq. (17.50). This means that there cannot be any state that belongs to both $\mathbb{V}_+$ and $\mathbb{V}'_+$.

What happens if an energy eigenstate $\left|0\right\rangle$ belongs to both $\mathbb{V}_0$ and $\mathbb{V}'_0$? It means that $Q^\dagger Q\left|0\right\rangle = 0$, $QQ^\dagger\left|0\right\rangle = 0$, so that $H\left|0\right\rangle = 0$, i.e., this is a state with energy eigenvalue equal to zero.

For any energy eigenstates with positive eigenvalue, we are therefore led to the conclusion that

$$\text{either}\quad \left\langle n\right|Q^\dagger Q\left|n\right\rangle = 0 \quad\text{or}\quad \left\langle n\right|QQ^\dagger\left|n\right\rangle = 0, \tag{17.56}$$

which means

$$\text{either}\quad Q\left|n\right\rangle = \left|\Omega\right\rangle \quad\text{or}\quad Q^\dagger\left|n\right\rangle = \left|\Omega\right\rangle. \tag{17.57}$$

This statement has an easy intuitive significance. For any $E_n \neq 0$, consider the two operators

$$c_n = Q/\sqrt{E_n}, \qquad c_n^\dagger = Q^\dagger/\sqrt{E_n}. \tag{17.58}$$

On the states in the subspace of this energy, they satisfy the relations

$$\{c_n, c_n\} = \{c_n^\dagger, c_n^\dagger\} = 0, \tag{17.59a}$$

$$\{c_n, c_n^\dagger\} = \mathbb{1}. \tag{17.59b}$$

These are reminiscent of the relations between the creation and annihilation operators encountered in §7.7 for the simple harmonic oscillator, with the difference that all commutators have been replaced by anticommutators. Thus, if we interpret c_n as an annihilation operator and $c_n^\dagger$ as a creation operator, we see that Eq. (17.59a) implies that two particles cannot be produced in a state or annihilated from a state. The *particles* that these operators create and annihilate are therefore like fermions, obeying the Pauli exclusion principle. For each energy $E_n > 0$, there can be therefore two possible states, one with zero fermion, which can be denoted by $\left|n_0\right\rangle$, and the other with one fermion, to be denoted by $\left|n_1\right\rangle$. They satisfy the relations

$$c_n\left|n_0\right\rangle = \left|\Omega\right\rangle, \qquad c_n^\dagger\left|n_1\right\rangle = \left|\Omega\right\rangle, \tag{17.60}$$

or equivalently,

$$Q\left|n_0\right\rangle = \left|\Omega\right\rangle, \qquad Q^\dagger\left|n_1\right\rangle = \left|\Omega\right\rangle, \tag{17.61}$$

in compliance with Eq. (17.57).

Thus, all positive energy eigenvalues of H are two-fold degenerate. These two degenerate states can be mapped to eigenstates of H_1 and H_2 respectively in the earlier formulation. The point is that, unlike having two different Hamiltonians as shown in Eq. (17.36), here we have only one Hamiltonian which is the sum of the two kinds, but when it acts on any energy eigenstate, only one of the two terms is relevant, because the other term annihilates the state, as indicated in Eq. (17.57).

There is a relation between the two kinds of states for a given energy. Consider states of the type $|n_1\rangle$. For them, Eq. (17.50) simply reads

$$Q^\dagger Q |n_1\rangle = E_n |n_1\rangle . \tag{17.62}$$

Assuming the states are normalized to unity, we can take an inner product with $\langle n_1|$ to obtain

$$E_n = \langle n_1| Q^\dagger Q |n_1\rangle = \| Q |n_1\rangle \|^2 . \tag{17.63}$$

Since $Q |n_1\rangle$ must be annihilated by Q because of Eq. (17.47), it must be a state of type $|n_0\rangle$. Thus we obtain

$$Q |n_1\rangle = \sqrt{E_n} |n_0\rangle . \tag{17.64a}$$

Exactly similarly, we can deduce the relation

$$Q^\dagger |n_0\rangle = \sqrt{E_n} |n_1\rangle . \tag{17.64b}$$

This two-fold degeneracy is the signature of supersymmetry.

The two kinds of states can be distinguished by defining the operator

$$F = [Q, Q^\dagger] . \tag{17.65}$$

It is easy to see that if it acts on $|n_1\rangle$, we will obtain

$$F |n_1\rangle = (QQ^\dagger - Q^\dagger Q) |n_1\rangle = -Q^\dagger Q |n_1\rangle = -E_n |n_1\rangle , \tag{17.66}$$

whereas on the state $|n_0\rangle$, we will have

$$F |n_0\rangle = (QQ^\dagger - Q^\dagger Q) |n_0\rangle = QQ^\dagger |n_0\rangle = E_n |n_0\rangle . \tag{17.67}$$

This means that both kinds of states are eigenstates of F, but the eigenvalue is negative for one kind of states and positive for the other kind.

We can write down an explicit 2×2 matrix representation for Q and $Q^\dagger$ within the two-dimensional subspace spanned by $|n_0\rangle$ and $|n_1\rangle$, both of which have energy E_n. Writing the states

$$|n_0\rangle = \begin{pmatrix} 1 \\ 0 \end{pmatrix}, \qquad |n_1\rangle = \begin{pmatrix} 0 \\ 1 \end{pmatrix}, \tag{17.68}$$

we can represent, in terms of the Pauli matrices in the combinations $\sigma_\pm = \frac{1}{2}(\sigma_x \pm i\sigma_y)$,

$$Q = \sqrt{E_n}\sigma_-, \quad Q^\dagger = \sqrt{E_n}\sigma_+ \tag{17.69}$$

within the 2×2 subspace of states with energy E_n. Note that this representation satisfies Eqs. (17.45) and (17.64) and yields $H = E_n\mathbb{1}$ (as expected within the 2D subspace spanned by $|n_0\rangle$ and $|n_1\rangle$). We also note that the operator P in this subspace defined by

$$P = \frac{1}{2E_n}[Q, Q^\dagger] = \sigma_z \tag{17.70}$$

satisfies the relations

$$P|n_0\rangle = |n_0\rangle, \qquad P|n_1\rangle = -|n_1\rangle. \tag{17.71}$$

For reasons related to quantum field theoretic treatment of supersymmetry which we will not discuss in detail, P is often called the fermion parity operator. A state with $P = 1$ is termed as bosonic while that with $P = -1$ is termed as fermionic.

Finally we discuss the computation of zero-energy in this formulation. This is exactly the same as in the earlier formulation involving A and $A^\dagger$. To see this, we note, from Eq. (17.46), that the existence of a zero-energy state requires $Q|0\rangle = Q^\dagger|0\rangle = 0$. The existence of such a state requires a normalizable solution of these equations and is, in general, not guaranteed. If such a state exists, supersymmetry is said to be broken, else it is said to be preserved.

□ **Exercise 17.14** *Consider a supersymmetric system for which $Q = (\hat{p}+i\alpha x)\sigma_-$ and $Q^\dagger = (\hat{p} - i\alpha x)\sigma_+$. Find H. Find all $E_n > 0$ and check for the existence of a zero-energy state.*

17.4 Lower-dimensional systems with linear Hamiltonians

In this section, we shall discuss quantum mechanics of a class of Hamiltonians with linear dispersion, which have become vastly important in a broad class of solid state systems which are commonly dubbed as topological materials. The reasons for their occurrence in these materials are varied. Sometimes they stem from the presence of a large interaction between orbital and spin degrees of freedom (spin–orbit interactions) of electrons in a solid; in other systems, they are used as effective low-energy description of particles near band-crossing points. In this section, we shall study their properties without getting into the details of the reason of their existence in realistic systems.

These Hamiltonians, in two spatial dimensions (which is what we are going to concentrate on) are typically given by

$$H = \hbar c \boldsymbol{\sigma} \cdot \boldsymbol{k}, \tag{17.72}$$

where $\hbar \boldsymbol{k} = (\hbar k_x, \hbar k_y)$ is the momentum vector, $\boldsymbol{\sigma} = (\sigma_x, \sigma_y)$ are the off-diagonal Pauli matrices representing the spin of the electrons, and c is a parameter representing some velocity scale which depends on the microscopic details. We note that H represents the massless Dirac Hamiltonian in two spatial dimensions where the Dirac matrices may be represented by Pauli matrices. Analogous Hamiltonian in 3D lead to the realization of Weyl fermions, which we shall not discuss.

First, we consider the eigenvalues and eigenvectors of H by solving $H\psi = E\psi$. For each $\boldsymbol{k}$, H represents a 2×2 matrix given by

$$H = \hbar c \begin{pmatrix} 0 & k_x - ik_y \\ k_x + ik_y & 0 \end{pmatrix}. \tag{17.73}$$

Thus, E can be found by finding eigenvalues of the matrix:

$$E_\pm = \pm \hbar c |\boldsymbol{k}|. \tag{17.74}$$

The corresponding eigenvectors are given by

$$\psi_\pm(\boldsymbol{k}) = \frac{1}{\sqrt{2}} \begin{pmatrix} 1 \\ \pm e^{i\theta_{\boldsymbol{k}}} \end{pmatrix}, \quad \theta_{\boldsymbol{k}} = \tan^{-1}(k_y/k_x). \tag{17.75}$$

Note that the two branches intersect at $|\boldsymbol{k}| = 0$ and the angle $\theta_{\boldsymbol{k}}$ is ill-defined at this point. This singularity leads to the presence of a non-trivial Berry phase for these systems as elucidated in §17.2.

It is interesting to note that, unlike the usual non-relativistic Hamiltonian, the Hamiltonian given in Eq. (17.72) does not commute with the components of the spin operator. However, the component of spin in the direction of momentum, $\boldsymbol{\sigma} \cdot \boldsymbol{\mathring{p}}$, commutes with the Hamiltonian. This combination is called *helicity*, which was also discussed in §15.6 in the context of the Weyl equation. As mentioned there, a positive energy eigenstate has helicity eigenvalue $+1$, whereas a negative energy eigenstate has helicity eigenvalue -1.

Another interesting consequence is the behavior of such a system in the presence of a constant magnetic field. We know that for any 2D system, a magnetic field parallel to the plane, say along the x-axis, simply adds a term $H_1 = -c_0 B \sigma_x$ where c_0 is proportional to the magnetic moment of the particle, and there is no orbital term due to such a field. In the present case, the Hamiltonian thus becomes

$$H(\boldsymbol{k}) = H_0(\boldsymbol{k}) + H_1 = \hbar c(\sigma_x(k_x - \alpha B) + \sigma_y k_y), \qquad (17.76)$$

where $\alpha = c_0/(\hbar c)$. Thus the presence of the magnetic field results in a constant shift of the momentum; it has no effect on their energy spectrum since such a shift is equivalent to adding a constant vector potential. This situation is in contrast to particles with the usual non-relativistic form of kinetic energy, for which the energy spectrum splits due to the presence of such a term.

But now suppose that the magnetic field is non-uniform. For example, let us assume that it varies linearly in the y direction: $B = B_0 y/a_0$, where a_0 has the dimension of length. Since, as noted, the term αB behaves like a vector potential, the particle will now feel a magnetic field which is the curl of it, i.e.,

$$\mathcal{B} = -\partial_y(\alpha B)\hat{\boldsymbol{z}} = -\alpha B_0/a_0 \hat{\boldsymbol{z}}. \qquad (17.77)$$

Note that $\mathcal{B}$ is not the field one applied to the particles, and in that sense it is fictitious; however, its effect leads to experimentally measurable consequences such as curving of the semiclassical trajectory of the electrons and leading to cyclotron motion. Thus, the effect of a parallel magnetic field on these electrons is qualitatively different from their Schrödinger counterparts.

Finally we address the transmission of Dirac particles through a potential barrier of height V_0 and width a as shown in Figure 7.5 (p. 163). We assume that the barrier is spin-independent, so it can be represented by a multiple of the unit matrix. Therefore, in region 2, the Hamiltonian is now given by

$$H'(\boldsymbol{k}) = \begin{pmatrix} V_0 & \hbar c(k_x - ik_y) \\ \hbar c(k_x + ik_y) & V_0 \end{pmatrix} \qquad (17.78)$$

whereas in regions 1 and 3, the Hamiltonian is still given by Eq. (17.73). Let us now consider a particle with energy E and transverse momentum k_y incident on the barrier from the left in region I with wavefunction ψ_+. The reflected wavefunction is given by ψ_-. The expressions of $\psi_\pm$ can be easily found from Eq. (17.75) and are given by

$$\psi_\pm(x, y, E) = \frac{e^{i(k_y y \pm k_x x)}}{\sqrt{2}} \begin{pmatrix} 1 \\ \pm e^{i\theta_k} \end{pmatrix}, \tag{17.79}$$

with

$$k_x = \sqrt{(E/\hbar c)^2 - k_y^2}. \tag{17.80}$$

The wavefunction in region 1 is thus given by

$$\psi_1 = \psi_+ + r\psi_- \tag{17.81}$$

where r is the reflection amplitude. Similarly in region 3, the transmitted wavefunction is given by

$$\psi_3 = t\psi_+. \tag{17.82}$$

In region 2, due to the presence of the potential, the equation describing these particles is governed by the Hamiltonian of Eq. (17.78). We assume that V_0 is spin-independent and hence it enters H' as a multiple of the identity matrix. The wavefunctions for right and left moving particles with energy E and transverse momentum k_y in region 2 are therefore given by

$$\psi'_\pm(x, y, E) = \frac{e^{i(k_y y \pm k'_x x)}}{\sqrt{2}} \begin{pmatrix} 1 \\ \pm e^{\pm i\theta_{k'}} \end{pmatrix}, \tag{17.83}$$

with

$$k'_x = \sqrt{\left(\frac{E - V_0}{\hbar c}\right)^2 - k_y^2} \tag{17.84}$$

and $\theta_{k'} = \tan^{-1}(k_y/k'_x)$. Thus in region 2, the wavefunction is given by

$$\psi_2 = p\psi'_+ + q\psi'_-, \tag{17.85}$$

where p and q are the amplitudes of right and left moving particles respectively.

Next, we would like to solve for r and t by matching boundary conditions on the wavefunction which comes from demanding continuity of probability current across the boundary. For a free particle with Hamiltonian $p^2/2m$, the probability current is given by $\boldsymbol{J} = \text{Im}(\psi^\dagger \boldsymbol{\nabla} \psi)$, as shown in Eq. (3.86, p. 76).

Thus, the current continuity across a finite barrier demands continuity of the wavefunction and its derivative. However, in the present case, due to the linear dispersion and matrix structure of the Dirac Hamiltonian, one can show that $J_x = \psi^\dagger \sigma_x \psi$; thus, it is enough to demand continuity of the wavefunction across the boundaries at $x = 0$ and $x = a$ (see Figure 7.5 (p. 163)). Using Eqs. (17.79) and (17.83) and demanding $\psi_1(x = 0) = \psi_2(x = 0)$ and $\psi_2(x = a) = \psi_3(x = a)$, we get

$$1 + r = p + q, \tag{17.86a}$$

$$e^{i\theta_k} - re^{-i\theta_k} = pe^{i\theta'_k} - qe^{-i\theta'_k}, \tag{17.86b}$$

$$pe^{ik'_x a} + qe^{ik'_x a} = te^{ik_x a}, \tag{17.86c}$$

$$pe^{i(k'_x a + \theta'_k)} - qe^{-i(k'_x a + \theta'_k)} = te^{i(k_x a + \theta_k)}. \tag{17.86d}$$

These equations may be easily solved to obtain expressions of r and t. Here we concentrate on the limit where $V_0 \gg E$ and $a \to 0$ such that the dimensionless barrier strength $\chi = V_0 a/(\hbar c)$ is held fixed. This is called the thin barrier limit. In this limit, $k'_x a \to \chi$ and $\theta'_k \to \pi$ and the solution to Eq. (17.86) yields for the transmission probability

$$\mathcal{T} = 1 - |r|^2 = \frac{\cos^2 \theta_k}{1 - \cos^2 \chi \sin^2 \theta_k}. \tag{17.87}$$

We find therefore that for Hamiltonians with linear dispersion, the transmission probability, in the thin barrier limit, is an oscillatory function of the dimensionless barrier strength.

☐ **Exercise 17.15** *Derive Eq. (17.87), starting from the boundary conditions in Eq. (17.86).*

Eq. (17.87) also points out two unconventional properties of the transmission probability. The first one is that for $\chi = n\pi$, for any integer n, we find perfect transmission. This phenomenon is called the transmission resonance and it has no analog for electrons obeying the non-relativistic Schrödinger equation. The second is that, for normal incidence, i.e., when $\theta_k = 0$, we find $\mathcal{T} = 1$. This paradoxical property was discovered for three-dimensional Dirac electrons by Oscar Klein right after Dirac discovered his equations and it goes by the name of *Klein paradox*. One simple way to understand this result is to appeal to the helicity property of these electrons discussed earlier in this section. For normal incidence, the electron has its spin along x and a reflection would require flipping its spin. However, the potential is a scalar in spin space and therefore scattering from it cannot change the spin orientation of a particle. Thus the barrier has no choice but to let the electron through. The reader is urged to think why the electron can be reflected from the barrier at other angles of incidence.

17.5 The path integral formulation revisited

In this section, we shall expand on the discussion of §6.5 and provide a more detailed derivation of path integrals starting from a Hamiltonian H of a quantum system. We shall deal with a system in 1D space and assume that such a Hamiltonian can be written as

$$H = \frac{\widehat{p}^2}{2m} + V(\widehat{x}), \tag{17.88}$$

where $\widehat{p}$ and $\widehat{x}$ are position and momentum operators. The potential $V(\widehat{x})$ shall be assumed to be a smooth non-singular function. We shall be interested in computing the probability amplitude $K(x_f, t_f; x_i, t_i)$ which is called the propagator and was defined in Eq. (6.32, p. 142). For a generic Hamiltonian H, it is easy to see that K can be written as

$$K(x_f, t_f; x_i, t_i) = \langle x_f | U(t_f, t_i) | x_i \rangle$$
$$|\psi(t)\rangle = U(t, t') |\psi(t')\rangle, \tag{17.89}$$

where

$$U(t, t') = \exp[-iH(t - t')/\hbar] \tag{17.90}$$

denotes the evolution operator for a quantum system governed by its Hamiltonian H and $|\psi(t)\rangle$ is the state of the system at time t. However, the computation of $|\psi(t_f)\rangle$ is not straightforward for a generic Hamiltonian H whose form is given by Eq. (17.88). The difficulty stems from non-commutativity of the two terms in H. Due to this non-commutativity one has

$$\exp\left[-i\frac{H}{\hbar}t\right] \neq \exp\left[-i\frac{\widehat{p}^2}{2m\hbar}t\right] \exp\left[-i\frac{V(\widehat{x})}{\hbar}t\right], \tag{17.91}$$

and this makes evaluation of U difficult.

The central idea of carrying out evaluation of U is that the non-commutativity of the two terms in H becomes unimportant if the time interval of evolution becomes small compared to all other times scales in the problem.

This is easily seen from the Baker–Campbell–Haussdorf formula, given in Eq. (5.84, p. 127). It shows that if two operators A and B are both proportional to some number ϵ, the terms in the exponent that make $e^{A+B} \neq e^A e^B$ due to the non-commutativity of A and B, all contain ϵ^n with $n \geqslant 2$. Thus, for small ϵ, the effect of non-commutativity can be neglected in the term $A + B$ in the exponent which depends linearly on ϵ. This allows us to write, for infinitesimal time of evolution Δt,

$$\exp\left[-i\frac{H}{\hbar}\Delta t\right] \simeq \exp\left[-i\frac{\widehat{p}^2}{2m\hbar}\Delta t\right] \exp\left[-i\frac{V(\widehat{x})}{\hbar}\Delta t\right] + \mathrm{O}\left((\Delta t)^2\right), \quad (17.92)$$

where Δt is understood to be the smallest timescale in the problem. The choice of Δt depends on the problem at hand when it comes to numerical evaluation of U. Its existence is also contingent on the non-singular nature of $V(\widehat{x})$; for singular $V(\widehat{x})$, the number of time slices required diverges as $\Delta t \to 0$. This shall be discussed later in more details.

Assuming that Δt exists, we now face the challenge of evaluation of U when the time of evolution $t_f - t_i$ is not small. To this end, one notes that since H is time-independent (so that $H(t_1)$ commutes with $H(t_2)$ for all t_1 and t_2), it is possible to write

$$U(t_f, t_i) = U(t_f, t_f - \Delta t)U(t_f - \Delta t, t_f - 2\Delta t)$$
$$\times \cdots U(t_i + 2\Delta t, t_i + \Delta t)U(t_i + \Delta t, t_i). \quad (17.93)$$

Thus, we have sliced the total time of evolution into $N = (t_f - t_i)/\Delta t$ slices. This technique is commonly referred to as *time slicing*. Thus, one can write

$$U(t_f, t_i) = \prod_{j=1}^{N} U(t_j, t_{j-1}), \quad t_j = t_i + j\Delta t. \quad (17.94)$$

To evaluate the amplitude $K(x_f, t_f; x_i, t_i)$ we need to compute the expectation value of U. Using Eqs. (17.89) and (17.94), we write

$$K(x_f, t_f; x_i, t_i) \equiv K = \langle x_f | \prod_{j=1}^{N} U(t_j, t_i) | x_i \rangle. \quad (17.95)$$

To make further progress we insert a complete set of states $|x_j\rangle$ between each slice using the identity $\int dx_j |x_j\rangle\langle x_j| = \mathbb{1}$ for any j. This yields

$$K = \int \left(\prod_{j=1}^{N-1} dx_j\right) \langle x_f | U(t_f, t_f - \Delta t) | x_{N-1}\rangle$$

$$\times \langle x_{N-1}|U(t_f - \Delta t, t_f - 2\Delta t)|x_{N-2}\rangle$$
$$\times \cdots \langle x_j|U(t_j, t_{j-1})|x_{j-1}\rangle \cdots \langle x_1|U(t_i + \Delta t, t_i)|x_i\rangle$$
$$= \prod_{j=1}^{N-1} \int dx_j \, U_{j,j-1}. \tag{17.96}$$

We now need to evaluate each of these matrix elements $U_{j,j-1}$. To this end, note that we have chosen the complete set of states to be the eigenbasis of $\widehat{x}$. Thus,

$$\exp[-iV(\widehat{x})\Delta t/\hbar]|x_j\rangle = \exp[-iV(x_j)\Delta t/\hbar]|x_j\rangle. \tag{17.97}$$

Using this we find

$$U_{j,j-1} = \int dx_j e^{-iV(x_j)\Delta t/\hbar} \langle x_j|e^{-i\widehat{p}^2 \Delta t/(2m\hbar)}|x_{j-1}\rangle. \tag{17.98}$$

At this point, it is natural to ask what prompts us to choose $V(\widehat{x})$ to act on the bra $\langle x_j|$, instead of the ket $|x_{j-1}\rangle$ while we evaluate the matrix element. These two choices lead to different results since $V(x_{j-1}) \neq V(x_j)$. In fact, it would also have been possible to use a symmetrized combination in the right side of Eq. (17.98) in which we would end up with a factor of $\exp[-i(V(x_j) + V(x_{j-1}))/(2\hbar)]$. However, for infinitesimal Δt, the difference in these choices contributes at higher order in Δt and therefore can be safely neglected provided $V(\widehat{x})$ is non-singular; for singular $V(\widehat{x})$, this procedure does not work for any finite Δt.

The next step is to evaluate the remaining part of $U_{j,j-1}$. To this end, we use the eigenbasis of $\widehat{p}$ and note that $\langle p|x\rangle = \exp[-ipx/\hbar]$. Thus, inserting two complete sets of basis states using $\int (dp/2\pi\hbar)|p\rangle\langle p| = \mathbb{1}$, we obtain

$$U_{j,j-1} = \int \frac{dx_j dp dp'}{(2\pi\hbar)^2} e^{-iV(x_j)\Delta t/\hbar} \langle x_j|p\rangle \langle p|e^{-i\widehat{p}^2 \Delta t/(2m\hbar)}|p'\rangle \langle p'|x_{j-1}\rangle. \tag{17.99}$$

Since the momentum eigenstates are orthogonal to each other, we get

$$\langle p|e^{-i\widehat{p}^2 \Delta t/(2m\hbar)}|p'\rangle = \delta(p - p')e^{-ip^2 \Delta t/(2m\hbar)}. \tag{17.100}$$

Notice that on the right side, we now have the momentum eigenvalue, not the momentum operator. Putting this back, we obtain

$$U_{j,j-1} = \int \frac{dx_j dp}{2\pi\hbar} \exp\left[-\frac{i}{\hbar}\left\{-p(x_j - x_{j-1}) + \left(\frac{p^2}{2m} + V(x_j)\right)\Delta t\right\}\right]. \tag{17.101}$$

One can now perform the integral over p since it is a Gaussian integral. A few lines of algebra yields

$$U_{j,j-1} = \sqrt{\frac{m}{2\pi i \Delta t}} \int dx_j \exp\left[\frac{i}{\hbar}\left\{\frac{m(x_j - x_{j-1})^2}{2\Delta t} - V(x_j)\Delta t\right\}\right].$$

(17.102)

Substituting Eq. (17.102) in Eq. (17.96), we find

$$U(t_f, t_i) = \int \left(\prod_{j=1}^{N} dx_j \frac{m}{2\pi i \hbar \Delta t}\right) \exp\left[\frac{i}{\hbar}\left\{\frac{m(x_j - x_{j-1})^2}{2\Delta t} - V(x_j)\Delta t\right\}\right].$$

(17.103)

Note that we have expressed the evolution operator in terms of x and t; $\widehat{x}$ and $\widehat{p}$ have been traded off in favor of a large number of additional integrals over x_j that need to be carried out. Finally, we need to take the limit $\Delta t \to 0$ and $N \to \infty$, keeping $N\Delta t = (t_f - t_i)$ finite. We define the shorthand notations

$$\lim_{\substack{N \to \infty \\ \Delta t \to 0}} \int \left(\prod_{j=1}^{N} dx_j \sqrt{\frac{m}{2\pi i \hbar \Delta t}}\right) = \int [\mathcal{D}x],$$

$$\lim_{\substack{N \to \infty \\ \Delta t \to 0}} \sum_{j=1}^{N} \Delta t = \int_{t_i}^{t_f} dt.$$

(17.104)

Using these in Eq. (17.103), we get

$$U(t_f, t_i) = \int [\mathcal{D}x]\, e^{i\mathbb{S}/\hbar},$$

(17.105)

where $\mathbb{S}$ is the action,

$$\mathbb{S} = \int_{t_i}^{t_f} dt\, \mathcal{L},$$

(17.106)

$\mathcal{L}$ being the Lagrangian:

$$\mathcal{L} = \frac{m}{2}\left(\frac{dx}{dt}\right)^2 - V(x).$$

(17.107)

Thus, we find that the evolution operator of a quantum system with a Hamiltonian H can be expressed in terms of its action. Note that the action of any quantum system has the dimension of $\hbar$ which must be the case for Eq. (17.105) to make sense.

Finally, using Eq. (17.105), one finds

$$K = \int_{x_i}^{x_f} [\mathcal{D}x]\, e^{i\mathcal{S}/\hbar}. \qquad (17.108)$$

This expression of K presents an alternative interpretation of probability amplitude. For a quantum system to transit from a state $|x_i\rangle$ to a state $|x_f\rangle$, there exists an infinite number of possible trajectories $x(t)$. Each of these trajectories must have the same end points x_i and x_f; however, each of them contributes to the amplitude with a weight $\exp[i\mathcal{S}(x)/\hbar]$. The weighted sum of all these probability amplitudes yields the total probability amplitude for the system to transit from the state $|x_i\rangle$ to the state $|x_f\rangle$ under the action of its Hamiltonian H.

The maximum contribution to the probability amplitude comes from the path for which $\delta\mathcal{S} = 0$; this condition defines the classical trajectory of the particle. This trajectory is given by x_{cl} which obeys the classical equation of motion

$$m\frac{d^2 x_{\mathrm{cl}}}{dt^2} = -\left.\frac{dV(x)}{dx}\right|_{x = x_{\mathrm{cl}}} \qquad (17.109)$$

with the boundary conditions $x_{\mathrm{cl}}(t_f) = x_f$ and $x_{\mathrm{cl}}(t_i) = x_i$. For a system for which $\mathcal{S} \gg \hbar$, any path apart from the classical one has negligible contribution as can be seen by expanding $\mathcal{S}$ about its stationary path. This limit is often colloquially called as $\hbar \to 0$ limit and the ratio $\mathcal{S}/\hbar$ is taken as a measure of classical nature of a system.

We also note that in this formulation, x and t are treated at an equal footing. This is in contrast to the Hamiltonian formulation where $\hat{x}$ (or $\hat{p}$) is an operator while t is a variable. Thus, the path integral formulation turns out to be useful for describing relativistic systems where space and time need to be treated on an equal footing. Finally, we note that we have glossed over the issue of the existence of the measure $[\mathcal{D}x]$ as defined in Eq. (17.104); to the best of our knowledge, there is no simple proof to show that the limits in the definitions of Eq. (17.104) exist.

☐ **Exercise 17.16** *Consider a harmonic oscillator Hamiltonian given by $H = \hat{p}^2/2m + m\omega_c^2 x^2/2$. For this Hamiltonian write down the path integral for K.*

☐ **Exercise 17.17** *In the last problem, the first step for computing K is to evaluate the contribution of the classical trajectory. Show that this trajectory* is given by $d^2 x_{cl}/dt^2 + \omega_c^2 x_{cl}^2 = 0$ and find its solution such that $x_{cl}(t_f) = x_f$ and $x_{cl}(t_i) = x_i$. Hence find the classical action S_{cl} in terms of x_f and x_i.

☐ **Exercise 17.18** *Finally, for the harmonic oscillator problem, show that the expression for K can be written as $K = \exp[iS_{cl}/\hbar]K_q$ and evaluate K_q by carrying out the relevant Gaussian path integral. What are the boundary conditions for this path integral?*

17.6   Non-Hermitian quantum mechanics

Throughout this book, we have studied quantum mechanical Hamiltonians which are Hermitian operators. This restriction ensured that their eigenvalues are real; moreover, this condition is also important for assignment of definite probability amplitudes to their eigenstates. However, there exists another formulation, known as bi-orthogonal quantum mechanics, that allows us to deal with non-Hermitian quantum Hamiltonians.

There are two kinds of systems for which this analysis is useful. First, with non-Hermitian Hamiltonians, one can discuss the quantum mechanical behavior of unstable systems. The eigenvalues of such a Hamiltonian will be complex. If a typical eigenvalue is $E - \frac{i}{2}\Gamma$ with real E and Γ, then the corresponding energy eigenstate at time t will satisfy the condition

$$|\psi(t)|^2 = e^{-i\Gamma t/\hbar}|\psi(0)|^2. \qquad (17.110)$$

Definitely, this describes the exponential decrease of probability of finding the unstable system at time t.

The other kind of problems are motivated by the fact that, although Hermitian matrices ensure real eigenvalues, one does not require a Hermitian matrix to obtain all real eigenvalues. It can be shown, and we will show, that if a Hamiltonian commutes with PT (where P denotes parity and T time reversal), all its eigenvalues will be real even if it is not Hermitian. This class of non-Hermitian Hamiltonians are termed as PT-symmetric Hamiltonians.

17.6.1   General formulation

First, let us treat the general class of non-Hermitian Hamiltonians. We will denote these Hamiltonians by the symbol S in order to distinguish them from the usual Hermitian Hamiltonians. Note that we can always write S in the form

$$S = H_0 - iH_1, \qquad (17.111)$$

where $H_0 = \frac{1}{2}(S + S^\dagger)$ and $H_1 = \frac{1}{2i}(S^\dagger - S)$ are both Hermitian matrices. In general, S is not a normal matrix in the sense defined in §2.8, because it does not commute with $S^\dagger$. The eigenstates of S and $S^\dagger$ will not be the same. The eigenvalues need not be the same as well. Let us define the eigenstates and eigenvalues of these two operators by the relations

$$S|\psi_s\rangle = \epsilon_s |\psi_s\rangle , \qquad S^\dagger |\phi_s\rangle = \epsilon'_s |\phi_s\rangle . \tag{17.112}$$

The adjoint of these equations, through the definition of Eq. (2.80, p. 38), can be written as

$$\langle\psi_s| S^\dagger = \epsilon^*_s \langle\psi_s| , \qquad \langle\phi_s| S = \epsilon'^*_s \langle\phi_s| . \tag{17.113}$$

One summarizes Eq. (17.112) by saying that $|\psi_s\rangle$ and $|\phi_s\rangle$ are the *right eigenstates* of S and $S^\dagger$ respectively, with eigenvalues ϵ_s and ϵ'_s. Similarly, Eq. (17.113) means that $\langle\psi_s|$ and $\langle\phi_s|$ are the *left eigenstates* of $S^\dagger$ and S respectively, with eigenvalues ϵ^*_s and ϵ'^*_s. To keep the discussion simple, we will assume that all these eigenvectors are non-degenerate.

Because the matrix S is not normal, the eigenstates are not orthogonal: $\langle\psi_{s_1}|\psi_{s_2}\rangle \neq \delta_{s_1 s_2}$ and $\langle\phi_{s_1}|\phi_{s_2}\rangle \neq \delta_{s_1 s_2}$. This can be easily seen by noting that

$$\langle\psi_{s_1}|\psi_{s_2}\rangle = \frac{2i}{\epsilon^*_{s_1} - \epsilon_{s_2}} \langle\psi_{s_1}|H_1|\psi_{s_2}\rangle, \tag{17.114a}$$

$$\langle\phi_{s_1}|\phi_{s_2}\rangle = \frac{2i}{\epsilon'^*_{s_1} - \epsilon'_{s_2}} \langle\phi_{s_1}|H_1|\phi_{s_2}\rangle, \tag{17.114b}$$

which follow by taking matrix elements of both sides of the identity $2iH_1 = S^\dagger - S$.

This non-orthogonality of eigenstates raises questions regarding the possibility of assignment of probability amplitude to wavefunctions as is customary in standard quantum mechanics. This difficulty can fortunately be circumvented by noting that

$$\langle\phi_{s_1}|S|\psi_{s_2}\rangle = \epsilon_{s_2}\langle\phi_{s_1}|\psi_{s_2}\rangle = \epsilon'^*_{s_1}\langle\phi_{s_1}|\psi_{s_2}\rangle. \tag{17.115}$$

To allow for consistency in the last two equalities of Eq. (17.115) for arbitrary s_1 and s_2, we must therefore have

$$\epsilon'^*_s = \epsilon_s \quad \text{for each } s, \tag{17.116a}$$

$$\langle\phi_{s_1}|\psi_{s_2}\rangle = 0 \quad \text{for } s_1 \neq s_2. \tag{17.116b}$$

This is an orthogonality relation: the left eigenstates are orthogonal to the right eigenstates. Using this relation, we can now assign probabilities to states. To this end, for every state $|\alpha\rangle$ we define a *dual state* $|\alpha'\rangle$ by the rule

$$|\alpha\rangle = \sum_s c_s |\psi_s\rangle, \quad |\alpha'\rangle = \sum_s c_s |\phi_s\rangle. \tag{17.117}$$

This allows one to write, using orthogonality relation given by Eq. (17.116),

$$\langle \alpha' | \alpha \rangle = \sum_s |c_s|^2. \tag{17.118}$$

We can then normalize each state by choosing $\sum_s |c_s|^2 = 1$. Using this, we find that the probability of overlap of a generic state $|\alpha\rangle$ with any eigenstate $|\psi_s\rangle$ involves both the state $|\alpha\rangle$ and its dual $|\alpha'\rangle$ and can be expressed as

$$p_s = \langle \phi_s | \alpha \rangle \langle \alpha' | \psi_s \rangle = |c_s|^2. \tag{17.119}$$

The expectation value of any operator O between a state $|\alpha\rangle$ and its dual $|\alpha'\rangle$ can be computed in this formalism using the rule

$$\langle O \rangle = \langle \alpha' | O | \alpha \rangle = \sum_{s_1, s_2} c_{s_1}^* c_{s_2} O_{s_1 s_2}, \tag{17.120}$$

where

$$O_{s_1, s_2} = \langle \phi_{s_1} | O | \psi_{s_2} \rangle. \tag{17.121}$$

Thus, the expectation value of a generic operator and the assignment of probability to a given state requires a state and its dual defined by Eq. (17.117); this distinguishes the bi-orthogonal formulation from standard Hermitian quantum mechanics.

□ **Exercise 17.19** *Verify that for $H_1 = 0$, when S is Hermitian, this formulation reduces to the standard one involving Hermitian Hamiltonians.*

□ **Exercise 17.20** *Prove Eq. (17.116a) by using directly the definition of eigenvalues. In other words, show that any eigenvalue of $S^\dagger$ must be the complex conjugate of an eigenvalue of S.*

□ **Exercise 17.21** *Consider the operator*

$$P_s = \frac{|\psi_s\rangle \langle \phi_s|}{\langle \phi_s | \psi_s \rangle}. \tag{17.122}$$

An operator like this, with the property $P_s^2 = P_s$, is called a projection operator. *Show that for any state $|\alpha\rangle$ and its dual $|\alpha'\rangle$,*

$$\sum_s \langle \alpha' | P_s | \alpha \rangle = \langle \alpha' | \alpha \rangle, \tag{17.123}$$

and hence argue that $\sum_s P_s = \mathbb{1}$.

17.6.2 A two-level system as example

As an instructive example of the formalism where unstable particles are involved, we consider a system consisting of particles called neutral *kaons*, K^0, and their antiparticles, $\bar{K}^0$. These are the states which are produced in high energy collisions between strongly interacting particles like the proton and the neutron. However, these are not the eigenstates of the full Hamiltonian. In the basis of K^0 and $\bar{K}^0$, the Hamiltonian is of the form

$$S = \begin{pmatrix} a & bp^2 \\ bq^2 & a \end{pmatrix}. \tag{17.124}$$

The parameters are all complex, except b which can be taken to be real without any loss of generality, in a way that ensures

$$|p|^2 + |q|^2 = 1. \tag{17.125}$$

The Hamiltonian is not Hermitian, indicative of the fact that the kaons are unstable particles. The eigenvalues, $a \pm bpq$, are complex. Let us write them by separating the real and imaginary parts:

$$a + bpq = m_L - i\gamma_L/2,$$
$$a - bpq = m_S - i\gamma_S/2. \tag{17.126}$$

The corresponding right eigenstates are

$$|K_L\rangle = p\,|K^0\rangle + q\,|\bar{K}^0\rangle,$$
$$|K_S\rangle = p\,|K^0\rangle - q\,|\bar{K}^0\rangle. \tag{17.127}$$

The meaning of the subscripts 'L' and 'S' on the right sides of Eqs. (17.126) and (17.127) will be explained soon.

□ **Exercise 17.22** *What are the left eigenstates of the matrix given in Eq. (17.124)?*

□ **Exercise 17.23** *What is the condition under which the Hamiltonian of Eq. (17.124) will be a normal matrix? Verify that the eigenvectors will be orthogonal if this condition is satisfied.*

The state $|K_L\rangle$ will evolve with time as

$$|K_L(t)\rangle = e^{-i(m_L - i\gamma_L/2)t/\hbar}\,|K_L(0)\rangle, \tag{17.128}$$

so that

$$\left\|K_L(t)\right\|^2 = e^{-\gamma_L t/\hbar}\left\|K_L(0)\right\|^2. \tag{17.129}$$

It means that $\hbar/\gamma_L$ is the lifetime of K_L. Similarly, $\hbar/\gamma_S$ is the lifetime of K_S. As it turns out, $\gamma_L \ll \gamma_S$, so K_L has a much larger lifetime. Hence the names K_L and K_S, with subscripts indicating *long-lived* and *short-lived*.

A very interesting illustration of quantum mechanical superposition is seen if one studies the time-evolution of the state $|K^0\rangle$. Obviously,

$$|K^0\rangle = \frac{1}{2p}\Big(|K_L\rangle + |K_S\rangle\Big). \tag{17.130}$$

Thus, at time t, we have

$$|K^0(t)\rangle = \frac{1}{2p}\Big(|K_L(t)\rangle + |K_S(t)\rangle\Big), \tag{17.131}$$

where $|K_L(t)\rangle$ is given in Eq. (17.128), and $|K_S(t)\rangle$ by a similar equation. One can ask what will be the probability of finding $|K^0\rangle$ in this state at time t. The answer is

$$\left|\langle K^0 | K^0(t)\rangle\right|^2 = \frac{1}{4}\left|e^{-i(m_L - i\gamma_L/2)t/\hbar} + e^{-i(m_S - i\gamma_S/2)t/\hbar}\right|^2$$

$$= \frac{1}{4}e^{-\gamma_L t/\hbar}\left[1 + e^{-\Delta\gamma t/\hbar} + 2e^{-\frac{1}{2}\Delta\gamma t/\hbar}\cos\left(\Delta m\, t/\hbar\right)\right], \tag{17.132}$$

where $\Delta m = |m_S - m_L|$ and $\Delta\gamma = \gamma_S - \gamma_L$. Because of the cosine term in the expression, the probability shows an oscillatory behavior with time, superposed on an exponential decay. This phenomenon is called *strangeness oscillation*, because there is a property called *strangeness* which is -1 for K^0 and $+1$ for $\bar{k}^0$.

17.6.3 PT-symmetric Hamiltonians

Finally, we discuss non-Hermitian Hamiltonians which commute with PT:

$$[PT, S] = 0. \tag{17.133}$$

Let us consider a simultaneous eigenstate of PT and the Hamiltonian S:

$$PT |\psi\rangle = \eta |\psi\rangle, \tag{17.134a}$$

$$S |\psi\rangle = \epsilon |\psi\rangle. \tag{17.134b}$$

Now note that

$$SPT |\psi\rangle = S\eta |\psi\rangle = \eta\epsilon |\psi\rangle, \tag{17.135}$$

whereas

$$PTS \, |\psi\rangle = PT\epsilon \, |\psi\rangle = \epsilon^*\eta \, |\psi\rangle , \qquad (17.136)$$

since T is an antilinear operator, as discussed in Chapter 5. Eq. (17.133) then tells us that

$$(\epsilon^* - \epsilon)\eta \, |\psi\rangle = 0. \qquad (17.137)$$

Since η cannot be zero because of the fact that $(PT)^2 = \mathbb{1}$, we conclude that the energy eigenvalues must be real. We note that the above analysis shows that an eigenvalue of S is real *if* the corresponding eigenstate is also an eigenstate of PT; however, this need not be the case with all eigenvalues of S as demonstrated below.

This reality of eigenvalues can be demonstrated by a simple 2D example. Let us consider a 2×2 Hamiltonian linear in momentum:

$$S = c_0 \tau_x (p_x - ip_y) + \tau_z m, \qquad (17.138)$$

where $c_0 > 0$ and $m > 0$ are real-valued constants and p_x and p_y denote momenta along x and y respectively. The matrices τ_x and τ_z are the same as the Pauli matrices first introduced in Eq. (4.26, p. 87), represented by a different letter here so that it is not thought of as the spin operator. The Hamiltonian acts on state vectors with two components.

The time-reversal operator for S is simply given by complex conjugation, $\mathcal{T} = C_0$, together with the change in the signs of momenta p_x and p_y and without any factor $i\tau_y$ as was mentioned in §8.5. As was noted in connection with Eq. (5.26, p. 111), the parity operator in 2D changes $x \to -x$, leaving y unchanged. Thus, $p_x \to -p_x$ whereas p_y is unchanged. Applying the time-reversal operator on the Hamiltonian of Eq. (17.138), we see that the first term changes $c_0\tau_x(p_x - ip_y) \to c_0\tau_x(-p_x - ip_y)$ while the second term remains invariant. Applying parity to S, we find that $c_0\tau_x(p_x - ip_y) \to c_0\tau_x(-p_x - ip_y)$ while the second term of S again remains invariant. Thus S changes under both parity and time reversal; however, it remains invariant under successive application of P and T and is therefore PT symmetric.

The eigenvalues and (unnormalized) eigenvectors of these Hamiltonian are given by

$$E_{\pm}(p_x, p_y) = \pm\sqrt{m^2 + c_0^2(p_x^2 - p_y^2) - 2ic_0^2 p_x p_y},$$

$$|\psi_{\pm}(p_x, p_y)\rangle = \begin{pmatrix} 1 \\ \dfrac{E_{\pm}(p_x, p_y) - m}{c_0(p_x - ip_y)} \end{pmatrix}. \qquad (17.139)$$

We note that for $c_0|p_y| < m$ and $p_x = 0$, $PT|\psi_\pm\rangle = |\psi_\pm\rangle$; this is also the condition for real eigenvalues. For $m < c_0|p_y|$ and $p_x = 0$, the states break PT symmetry (since $|\psi_\pm(0, p_y)\rangle \to |\psi_\mp(0, p_y)\rangle$ under the action of PT) and the spectrum becomes imaginary with the imaginary eigenvalues occuring in pairs. At $c_0 p_y = m$ and $p_x = 0$, the two eigenstates coalesce to give a zero-energy mode which is an eigenstate of τ_y. Such points in the spectrum of non-Hermitian Hamiltonian separating PT preserved and PT broken phases are called *exceptional points*. Such points have no analogue in Hermitian quantum systems.

□ **Exercise 17.24** *Operate both sides of Eq. (17.134a) and use $(PT)^2 = \mathbb{1}$ to show that the eigenvalues η satisfy the condition $|\eta|^2 = 1$.*

□ **Exercise 17.25** *Consider a spin-$\frac{1}{2}$ particle in a magnetipc field $B = (B_x, B_y, B_z) = (c_0, 0, i)$. Construct the non-Hermitian Hamiltonian for the system and find the value of c_0 at the exceptional points in its spectrum.*

Part Five

Appendices

Appendix A

Delta function

The Kronecker delta, for a set of discrete indices $i, j, \cdots$, is defined as

$$\delta_{ij} = \begin{cases} 1 & \text{if } i = j, \\ 0 & \text{if } i \neq j. \end{cases} \tag{A.1}$$

It follows trivially from the definition that

$$\sum_i A_i \delta_{ij} = A_j, \tag{A.2}$$

since only the term with $i = j$ will give non-trivial contribution to the sum. A special case, where $A_i = 1$ for all i, can be written as

$$\sum_i \delta_{ij} = 1. \tag{A.3}$$

The Dirac delta function is a generalization of the idea of Kronecker delta for continuous variables. In order to pave the way, note that we can define the Kronecker delta as a function of only one variable in the following way:

$$\delta_{i-j} = \begin{cases} 1 & \text{if } i - j = 0, \\ 0 & \text{otherwise.} \end{cases} \tag{A.4}$$

Eqs. (A.2) and (A.3) can also be written with the same notation. If now we consider i and j to be continuous variables x and x', we can arrive at the generalization by replacing the summations by integrations in Eqs. (A.2) and (A.3), i.e., we can define a function $\delta(x)$ with the properties

$$\int_{-\infty}^{+\infty} dx\, f(x)\delta(x - x') = f(x'), \tag{A.5}$$

$$\int_{-\infty}^{+\infty} dx\, \delta(x - x') = 1. \tag{A.6}$$

Just as Eq. (A.3) is a special case of Eq. (A.2), the property of Eq. (A.6) is a special case of Eq. (A.5).

It would be difficult to write the generalization of Eq. (A.1) or of Eq. (A.4) for the continuous case. Eq. (A.5) certainly suggests that

$$\delta(x - x') = 0 \qquad \text{if } x \neq x', \tag{A.7}$$

so that the functional values of $f(x)$ for all $x \neq x'$ are irrelevant for the result of the integration. However, the *value* of the function $\delta(x - x')$ at $x = x'$ is not well-defined: it must be infinite so that the integral in Eq. (A.6) gives a non-zero answer despite the fact that all contribution to the integral comes only from one point, viz., $x = x'$.

A good way of visualizing the delta function is through various limiting procedures. For example, consider a rectangular distribution function defined by:

$$\phi(x, a) = \begin{cases} \frac{1}{a} & \text{for } -\frac{a}{2} < x < +\frac{a}{2}, \\ 0 & \text{otherwise.} \end{cases} \tag{A.8}$$

Obviously

$$\int_{-\infty}^{+\infty} dx \, \phi(x, a) = 1 \tag{A.9}$$

for any value of a. In the limit $a \to 0$, $\phi(x, a)$ will be equal to zero for $x \neq 0$. Thus,

$$\lim_{a \to 0} \phi(x, a) = \delta(x). \tag{A.10}$$

In fact, any singly peaked probability distribution reduces to the *delta function* in the limit that the standard deviation of the probability distribution is taken to zero. For example, starting from the normal or *Gaussian distribution*, we can write

$$\lim_{\sigma \to 0} \frac{1}{\sigma\sqrt{2\pi}} \exp\left(-\frac{x^2}{2\sigma^2}\right) = \delta(x). \tag{A.11}$$

□ **Exercise A.1** *Argue that the delta function can be expressed as the limiting form of the* Cauchy *distribution function as well:*

$$\lim_{\alpha \to 0} \frac{1}{\pi} \frac{\alpha}{x^2 + \alpha^2} = \delta(x). \tag{A.12}$$

Because of the problem of defining $\delta(x)$ for $x = 0$, mathematicians refuse to call $\delta(x)$ a *function*. Rather, due to its close similarity with probability distributions, they prefer to call it a *delta distribution*. We will ignore this subtle semantic point.

The delta function can also be defined as an integral. Note that Eq. (A.5) implies

$$\int_{-\infty}^{+\infty} dx \; e^{-2\pi i k x} \delta(x) = 1, \tag{A.13}$$

i.e., the Fourier transform of the delta function is 1. Taking the inverse Fourier transform, we can write

$$\int_{-\infty}^{+\infty} dk \; e^{2\pi i k x} = \delta(x). \tag{A.14}$$

When applied to physics problems, it is more convenient to use the exponent in the form $ipx/\hbar$, so Eq. (A.14) can be rewritten as

$$\frac{1}{2\pi\hbar} \int_{-\infty}^{+\infty} dp \; e^{ipx/\hbar} = \delta(x). \tag{A.15}$$

Obviously, we can exchange the roles of x and p in the formula and write

$$\frac{1}{2\pi\hbar} \int_{-\infty}^{+\infty} dx \; e^{ipx/\hbar} = \delta(p). \tag{A.16}$$

□ **Exercise A.2** *Prove Eq. (A.15) by the method of contour integration.*

The following properties of the delta function should be obvious from its definition:

$$\delta(x) = \delta(-x), \tag{A.17}$$
$$x\delta(x) = 0, \tag{A.18}$$
$$f(x)\delta(x - a) = f(a)\delta(x - a), \tag{A.19}$$
$$\delta(ax) = \frac{1}{|a|} \; \delta(x). \tag{A.20}$$

Further, we can formally define the derivative of the delta function:

$$\delta'(x) \equiv \frac{d}{dx} \; \delta(x). \tag{A.21}$$

Then, applying integration by parts, we can write

$$\int dx\ f(x) x \delta'(x) = -\int dx\ \delta(x) \frac{d}{dx}\Big(x f(x) \Big)$$

$$= -\int dx\ \delta(x)\Big(f(x) + x f'(x) \Big). \qquad (A.22)$$

The second term vanishes because of Eq. (A.18), and so the result can be written symbolically as

$$x \delta'(x) = -\delta(x). \qquad (A.23)$$

Finally, we want to see the value of the delta function when its argument is a function of the independent variable x, i.e., expressions of the form $\delta(F(x))$. Obviously, the function is zero at all points where $F(x)$ has a non-zero value. Suppose $F(x)$ vanishes only at one point, $x = x_0$. Then, near this point, we can write

$$F(x) = F(x_0) + (x - x_0) F'(x_0) + \cdots = (x - x_0) F'(x_0) + \cdots . \qquad (A.24)$$

Thus,

$$\delta\Big(F(x) \Big) = \delta\Big((x - x_0) F'(x_0) \Big) = \frac{1}{|F'(x_0)|}\ \delta(x - x_0), \qquad (A.25)$$

using Eq. (A.20). The generalization to functions which have multiple zeros is obvious, at least so long as the zeros are well separated. If the zeros are at the points $x = x_i$, then

$$\delta\Big(F(x) \Big) = \sum_i \frac{1}{|F'(x_i)|}\ \delta(x - x_i). \qquad (A.26)$$

A related function is useful as well. It is called the *Heaviside step function* and denoted by $\Theta(x)$. It is defined as

$$\Theta(x) = \begin{cases} 0 & \text{if } x < 0, \\ 1 & \text{if } x > 0. \end{cases} \qquad (A.27)$$

The value of the function at $x = 0$ is not of much importance: it is chosen as $\frac{1}{2}$ by some and as 0 by others. The relation with the Dirac delta function can be expressed in the following way:

$$\Theta(x) = \int_{-\infty}^{x} dy\ \delta(y). \qquad (A.28)$$

Appendix B

Solution of second-order differential equations

Since a great many of the examples of quantum systems described in this book involve solutions of second-order differential equations, in this appendix we lay down the basic techniques for finding these solutions. We do not care to be completely general. We only discuss the kind of equations that we encounter for the purpose of this book. The stationary state form of the Schrödinger equation has a wavefunction of the form

$$\Psi(\boldsymbol{r}, t) = e^{-iEt/\hbar}\psi(\boldsymbol{r}), \tag{B.1}$$

where E is the energy eigenvalue. The coordinate dependent part of the wavefunction is a solution of the equation

$$H\psi(\boldsymbol{r}) = E\psi(\boldsymbol{r}). \tag{B.2}$$

In the coordinate representation, H is a differential operator. For non-relativistic particles, since the kinetic energy is quadratic in momentum, the differential equation is second-order. However, the specialty of this equation is that it is a homogeneous equation, i.e., all terms are linear in $\psi(\boldsymbol{r})$, so given any two solutions, arbitrary linear combinations of them would also constitute a solution. We will discuss only homogeneous equations here.

B.1   General considerations

We will discuss differential equations involving only one independent variable x. For 1D systems, this is how the equation presents itself in practical problems. For higher-dimensional problems, one can often reduce the equation

429

to this form, a process that has been described in the text in many places, notably in Chapter 9.

The most general homogeneous equation of second-order has the form

$$a(x)\frac{d^2y}{dx^2} + b(x)\frac{dy}{dx} + c(x)y = 0. \tag{B.3}$$

If $a(x) = 0$, this is not a second-order differential at all, and that case need not be discussed. For $a(x) \neq 0$, we can divide by $a(x)$ and write the equation in the form

$$\frac{d^2y}{dx^2} + f(x)\frac{dy}{dx} + g(x)y = 0. \tag{B.4}$$

Most of the time, we will be using this simplified form for the differential equation.

One important aspect of the solution is that there can be only two solutions that are linearly independent. To show this, let us say that Eq. (B.3) has three different solutions $y_1(x)$, $y_2(x)$, and $y_3(x)$. Each of them would satisfy Eq. (B.3), a condition that we can write succinctly as

$$\begin{pmatrix} y_1'' & y_1' & y_1 \\ y_2'' & y_2' & y_2 \\ y_3'' & y_3' & y_3 \end{pmatrix} \begin{pmatrix} a(x) \\ b(x) \\ c(x) \end{pmatrix} = 0. \tag{B.5}$$

This is a set of three homogeneous equations. Since the functions $a(x)$, $b(x)$, and $c(x)$ cannot all be zero, the determinant of the square matrix must vanish. This square matrix is called the *Wronskian matrix*.

Let us now try to find the condition that the three solutions are linearly dependent, i.e., there exist constants α_1, α_2, and α_3, all of which are not zero, and which satisfy the condition

$$\alpha_1 y_1(x) + \alpha_2 y_2(x) + \alpha_3 y_3(x) = 0 \tag{B.6}$$

for all x. Taking the first and second derivatives of this equation, we would obtain

$$\alpha_1 y_1'(x) + \alpha_2 y_2'(x) + \alpha_3 y_3'(x) = 0,$$
$$\alpha_1 y_1''(x) + \alpha_2 y_2''(x) + \alpha_3 y_3''(x) = 0. \tag{B.7}$$

These three equations can be written as

$$\begin{pmatrix} y_1'' & y_2'' & y_3'' \\ y_1' & y_2' & y_3' \\ y_1 & y_2 & y_3 \end{pmatrix} \begin{pmatrix} \alpha_1 \\ \alpha_2 \\ \alpha_3 \end{pmatrix} = 0. \tag{B.8}$$

The condition for the existence of non-zero solutions for α_1, α_2, and α_3 is given by the vanishing of the determinant of the square matrix that appears in this equation. But this matrix is just the transpose of the Wronskian matrix that appeared in Eq. (B.5). Its determinant is equal to the determinant of the Wronskian matrix. Thus, we see that if we find three solutions to the differential equation of Eq. (B.3), they must be linearly dependent. Only two of them can be linearly independent. In other words, if we find any two solutions $y_1(x)$ and $y_2(x)$ that are linearly independent, the general solution of the differential equation can be written in the form

$$y(x) = c_1 y_1(x) + c_2 y_2(x), \tag{B.9}$$

where c_1 and c_2 are arbitrary coefficients. This is the comment made in the preamble to this appendix, that any linear superposition of two independent solutions is also a solution of a linear homogeneous differential equation.

B.2 Equations with constant coefficients

If the coefficients $f(x)$ and $g(x)$ are constants, let us denote them simply by f and g. In this case, the solution is obtained by putting an ansatz of the form

$$y(x) = Ae^{kx}, \tag{B.10}$$

where A and k are constants, i.e., they do not depend on x. Putting this guess into the differential equation of Eq. (B.4), we obtain that the constant k must satisfy the equation

$$k^2 + fk + g = 0. \tag{B.11}$$

This is a quadratic equation in k. Let the roots of this equation be k_1 and k_2. Then $e^{k_1 x}$ and $e^{k_2 x}$ are solutions to the differential equation. Clearly, the two solutions with exponents $k_1 x$ and $k_2 x$ are linearly independent if $k_1 \neq k_2$. The general solution is

$$y(x) = c_1 e^{k_1 x} + c_2 e^{k_2 x}, \tag{B.12}$$

where c_1 and c_2 are arbitrary constants, as encountered in Eq. (B.9).

If $g = \frac{1}{4}f^2$, the two roots of Eq. (B.11) are equal, say equal to k. In this case, the general solution is not of the form of Eq. (B.12). In fact, we need to find another solution that is linearly independent from e^{kx}.

To this end, let us start from the general case, when k_1 and k_2 are different, and we note that the two independent solutions can be written in various other ways, by making linear combinations of $e^{k_1 x}$ and $e^{k_2 x}$. Thus, for example, we can write the general solutions as

$$y(x) = c_1'\left(e^{k_1 x} + e^{k_2 x}\right) + c_2'\left(\frac{e^{k_1 x} - e^{k_2 x}}{k_1 - k_2}\right). \tag{B.13}$$

We can now consider the limit $k_1 \to k_2$, whereby the first solution becomes e^{kx} and the second one becomes

$$\frac{d}{dk}\, e^{kx} = xe^{kx}. \tag{B.14}$$

Indeed, putting this function back into the differential equation, we find that it is a solution if $g = \frac{1}{4}f^2$. So, the general solution for this case is given by

$$y(x) = \left(c_1 + c_2 x\right)e^{kx}. \tag{B.15}$$

B.3   Checking the asymptotic behavior

For differential equations in which $f(x)$ and $g(x)$ are not constants, we cannot make an easy ansatz for the solution. However, we might be able to check the nature of the solution for very large or very small values of x. Once that is known, it might be possible to make some guesses about the solution for general x.

Finding the leading behavior of the solution for large or small x is usually easier than solving the entire equation, because in this case we can consider only the asymptotic behavior of the functions $f(x)$ and $g(x)$, which will, in general, be somewhat simpler than the full functions. However, without a knowledge of the behavior of the derivatives of y, we still cannot tell which terms would dominate the left side of Eq. (B.4).

To get around this problem, we consider factorizing the solution as follows:

$$y(x) = u(x)v(x). \tag{B.16}$$

If we put it back into Eq. (B.4), we obtain

$$u''v + v''u + 2u'v' + (u'v + v'u)f + uvg = 0, \tag{B.17}$$

where we have used the primes to denote the derivatives, and omitted the explicit reference to the arguments of the functions, because all functions have the same argument x.

So far, the factors u and v are completely arbitrary. Let us now choose v such that

$$2v' + fv = 0. \tag{B.18}$$

The motivation for this choice is twofold. First, the equation for v is a first-order differential equation, which can be easily solved. In fact, the solution is

$$v(x) = \exp\left(-\frac{1}{2}\int dx\, f(x)\right). \tag{B.19}$$

The second motivation for the choice of Eq. (B.18) is that this ensures that there is no term involving the first derivative of u in Eq. (B.17). Indeed, Eq. (B.18) gives

$$v'' = -\frac{1}{2}\left(f'v + v'f\right) = -\frac{1}{2}f'v + \frac{1}{4}f^2v, \tag{B.20}$$

which can be put back into Eq. (B.17) to obtain

$$u'' - \left(\frac{1}{2}f' + \frac{1}{4}f^2 - g\right)u = 0. \tag{B.21}$$

We have therefore managed to show that, given any differential equation that contains both second and first derivatives, the solution can be expressed as a product of two functions, one of which satisfies a first-order differential equation, and the other a second-order equation without any first derivative. This form given in Eq. (B.21) is suitable for finding limiting behavior of the solution for large or small values of x.

□ **Exercise B.1** *Eq. (B.19) involves an indefinite integral, and therefore the solution must involve an indeterminable constant. Why is this arbitrariness irrelevant?*

As an example of the merit of this technique, we show how it works for a differential equation with constant coefficients. We take the differential equation of a damped harmonic oscillator,

$$\frac{d^2y}{dt^2} + 2b\frac{dy}{dt} + \omega^2 y = 0, \tag{B.22}$$

where b and ω are constants, and we call the independent variable t in view of the physical situation addressed. In the notation used in Eq. (B.4), we have here

$$f = 2b, \qquad g = \omega^2. \tag{B.23}$$

So, Eq. (B.19) gives us

$$v(t) = e^{-bt}. \tag{B.24}$$

The differential equation for the other factor in the solution, Eq. (B.21), now reads

$$\frac{d^2 u}{dt^2} + (\omega^2 - b^2)u = 0. \tag{B.25}$$

For $b^2 < \omega^2$, the solution of this equation is in terms of a sine and a cosine function. Multiplying by $v(t)$ obtained earlier, we find the solution of Eq. (B.22) as

$$y(t) = e^{-bt}\Big(c_1 \cos \Omega t + c_2 \sin \Omega t\Big), \tag{B.26}$$

where

$$\Omega = \sqrt{\omega^2 - b^2}. \tag{B.27}$$

B.4 Series solution at ordinary points

B.4.1 General formulation

We have discussed the solution of Eq. (B.4) when the coefficients of the derivatives are constants. Now we venture into differential equations with non-constant coefficients. In this section, we want to consider the case where both $f(x)$ and $g(x)$ are analytic at the point $x = 0$. In this case, the point $x = 0$ is called an *ordinary point* of the differential equation. There is a method, called *Frobenius' method*, through which each linearly independent solution can be written in the form of a power series:

$$y(x) = \sum_{n=0}^{\infty} a_n x^n. \tag{B.28}$$

We want to comment that there is nothing sacrosanct about the point $x = 0$. We could try a series expansion in powers of $(x - x_0)$ for any x_0, provided x_0 is an ordinary point of the differential equation, i.e., both $f(x)$ and $g(x)$ are analytic at $x = x_0$. However, there is no loss of generality in discussing the series expansion around $x = 0$, because we can always make a transformation to a new variable $x' = x - x_0$, whereby expansion of functions of x around x_0 becomes equivalent to expansion of functions of x' around $x' = 0$.

Finding the solution now means finding the coefficients of the powers of x appearing in Eq. (B.28). These coefficients can be obtained by substituting the form of Eq. (B.28) into the differential equation of Eq. (B.4). Since $f(x)$ and $g(x)$ are analytic at $x = 0$, we can use the Taylor series expansion of these functions:

$$f(x) = \sum_{n=0}^{\infty} F_n x^n, \qquad g(x) = \sum_{n=0}^{\infty} G_n x^n, \tag{B.29}$$

where the F_n's and the G_n's are constants. Moreover, Eq. (B.28) implies

$$\frac{dy}{dx} = \sum_{n=0}^{\infty} n a_n x^{n-1} = \sum_{n=0}^{\infty} (n+1) a_{n+1} x^n,$$

$$\frac{d^2 y}{dx^2} = \sum_{n=0}^{\infty} n(n-1) a_n x^{n-2} = \sum_{n=0}^{\infty} (n+1)(n+2) a_{n+2} x^n. \tag{B.30}$$

Substitution into Eq. (B.4) gives

$$\sum_{n=0}^{\infty} (n+1)(n+2) a_{n+2} x^n + \sum_{p=0}^{\infty} \sum_{m=0}^{\infty} (m+1) a_{m+1} F_p x^{m+p}$$

$$+ \sum_{p=0}^{\infty} \sum_{m=0}^{\infty} a_m G_p x^{m+p} = 0. \tag{B.31}$$

In the last two terms which contain double summations, we replace the dummy index p by $n = m + p$. For any given n, the maximum value of the index m is n, so the sum now reads

$$\sum_{n=0}^{\infty} \left[(n+1)(n+2) a_{n+2} + \sum_{m=0}^{n} \left\{ (m+1) a_{m+1} F_{n-m} + a_m G_{n-m} \right\} \right] x^n = 0. \tag{B.32}$$

Since this resulting algebraic equation has to be valid at each point in an interval, the coefficients of each power of x on the left-hand side must be zero. Thus, the coefficients a_n satisfy the equations

$$(n+1)(n+2) a_{n+2} + \sum_{m=0}^{n} \left\{ (m+1) a_{m+1} F_{n-m} + a_m G_{n-m} \right\} = 0. \tag{B.33}$$

This equation is called the *recursion relation* among the coefficients. It shows that, given a_0 and a_1, we can find all a_n's iteratively. However, the coefficients a_0 and a_1 cannot be determined. They will serve as two arbitrary constants that are expected to appear in the solution of a second-order differential equation.

B.4.2   The harmonic oscillator equation

As a trivial example of the application of recursion equations for determining the a_n's, let us consider the equation

$$\frac{d^2y}{dx^2} + y = 0, \tag{B.34}$$

which means $f(x) = 0$ and $g(x) = 1$ in the general form of Eq. (B.4). Thus, $F_n = 0$ for all n, whereas $G_0 = 1$ and all other $G_n = 0$. Eq. (B.33) then becomes

$$(n+1)(n+2)a_{n+2} + a_n = 0, \tag{B.35}$$

which gives us a recursion formula between the coefficients:

$$a_{n+2} = -\frac{a_n}{(n+1)(n+2)}. \tag{B.36}$$

Taking a_0 and a_1 as two undetermined parameters as indicated earlier, we can determine all other a_n's. The solution is

$$y(x) = a_0 \cdot \left(1 - \frac{x^2}{2!} + \frac{x^4}{4!} + \cdots \right) + a_1 \cdot \left(x - \frac{x^3}{3!} + \frac{x^5}{5!} + \cdots \right)$$
$$= a_0 \cos x + a_1 \sin x. \tag{B.37}$$

B.5   Obtaining one solution from another

If one solution to a differential equation, $y_1(x)$, is known by the Frobenius' method, one can use it to find another linearly independent solution. This may look like a useless exercise when two series solutions are available. But we will see later that there are cases when only one power series solution exists, in which case it can be used to find the second solution by the method we are going to present now.

To see how this can be done, let us write the second solution as

$$y_2(x) = y_1(x)\xi(x), \tag{B.38}$$

where the task is now to determine $\xi(x)$. We now put this form into Eq. (B.4). The terms containing no derivative of $\xi(x)$ will vanish because $y_1(x)$ satisfies Eq. (B.4). Thus, we obtain the following constraint on the function $\xi(x)$:

$$y_1\xi'' + 2y_1'\xi' + fy_1\xi' = 0, \tag{B.39}$$

where the primes indicate derivatives with respect to x. If we write

$$\zeta(x) = \xi'(x), \tag{B.40}$$

then we see that Eq. (B.39) is just a first-order differential equation for $\zeta(x)$:

$$y_1 \zeta' = -2y_1' \zeta - f y_1 \zeta, \tag{B.41}$$

which can be rewritten as

$$\frac{\zeta'}{\zeta} = -2\frac{y_1'}{y_1} - f. \tag{B.42}$$

Integrating both sides, we obtain

$$\ln \zeta = -2 \ln y_1 - \int dx\ f(x), \tag{B.43}$$

apart from an arbitrary additive constant that results from the integration. Therefore,

$$\zeta(x) = \frac{1}{y_1^2} \exp\left(-\int dx\ f(x) \right), \tag{B.44}$$

apart from an overall multiplicative factor. Integrating this expression once again, one obtains $\xi(x)$, and therefore the second independent solution to the differential equation:

$$y_2(x) = y_1(x) \int dx\ \frac{1}{y_1^2} \exp\left(-\int dx\ f(x) \right). \tag{B.45}$$

The derivation shows clearly that it is another solution to the same differential equation. But is this solution linearly independent? If it is not, it means that $\xi(x)$, defined in Eq. (B.38), is a constant, i.e., $\zeta = 0$. Certainly that is a solution of the homogeneous equation Eq. (B.41), and this solution implies that a constant multiplied by $y_1(x)$ is also a solution. No surprise there. Any non-zero solution of Eq. (B.41) gives an $y_2(x)$ that is linearly independent of $y_1(x)$.

☐ **Exercise B.2** *Consider a second-order linear differential equation with constant coefficients, as was discussed in §B.2. Show that if one takes $y_1(x) = e^{k_1 x}$, Eq. (B.45) gives $y_2(x) = e^{k_2 x}$, apart from a constant factor.* [**Hint:** *Remember that Eq. (B.11) implies that $k_1 + k_2 = -f$.*]

☐ **Exercise B.3** *For the harmonic oscillator equation of Eq. (B.34), verify that if we take $y_1(x) = \sin x$, the prescription of Eq. (B.45) gives $y_2(x) = \cos x$ apart from a possible multiplicative constant, and vice versa.*

B.6 Solution at regular singular points

We have seen how the Frobenius' method gives power series solutions at ordinary points of a differential equation. If a point is not ordinary, it is called a singular point. In this section, we show that for a restricted class of singular points, at least one solution can be obtained by Frobenius' method.

B.6.1 Frobenius' method for power series solutions

The point $x = 0$ is called a regular singular point of the differential equation given in Eq. (B.4) if $xf(x)$ and $x^2g(x)$ are analytic at $x = 0$. A theorem by Frobenius asserts that, at such points, at least one solution of the differential equation can be obtained that is in the form of a power series in x, i.e., of the form

$$y(x) = \sum_{n=0}^{\infty} a_n x^{n+\kappa}. \tag{B.46}$$

Thus, κ is the smallest power of x that appears in the power series expression. The theorem also says that the power series sum will converge at least in the interval containing $x = 0$ where $f(x)$ or $g(x)$ are both differentiable.

For the case of regular singular points, it is convenient to use the notation

$$f(x) = \sum_{n=0}^{\infty} f_n x^{n-1} = \frac{f_0}{x} + f_1 + f_2 x + f_3 x^2 + \cdots, \tag{B.47a}$$

$$g(x) = \sum_{n=0}^{\infty} g_n x^{n-2} = \frac{g_0}{x^2} + \frac{g_1}{x} + g_2 + g_3 x + \cdots, \tag{B.47b}$$

rather than the definitions of the coefficients given in Eq. (B.29). Putting these forms of $f(x)$ and $g(x)$ into the differential equation and equating the coefficients of equal powers of x, we obtain the equations

$$(n+\kappa)(n+\kappa-1)a_n + \sum_{m=0}^{n} \Big((m+\kappa)f_{n-m} + g_{n-m}\Big)a_m = 0. \tag{B.48}$$

Note that there is an a_n-term within the sum. Extracting it out, we can write this equation as

$$\Big((n+\kappa)(n+\kappa-1+f_0) + g_0\Big)a_n + S_n = 0, \tag{B.49}$$

where S_n contains a sum over the coefficients with powers smaller than n:

$$S_n = \sum_{m=0}^{n-1} \Big((m + \kappa) f_{n-m} + g_{n-m} \Big) a_m. \tag{B.50}$$

Eq. (B.49) is the recursion relation.

Of special attention is the equation for $n = 0$. Since by definition κ is the smallest power, it implies that $a_0 \neq 0$. The sum in Eq. (B.50) contains no term if $n = 0$, and therefore $S_0 = 0$. In that case, the $n = 0$ equation obtained from Eq. (B.49) tells us that

$$\kappa^2 + (f_0 - 1)\kappa + g_0 = 0. \tag{B.51}$$

This equation does not say anything about the coefficients a_n. Rather, it determines the possible values of κ. It is called the *indicial equation*.

It is a quadratic equation. Let us call the two roots κ_+ and κ_-, choosing the names in a way that ensures

$$\kappa_+ \geqslant \kappa_-. \tag{B.52}$$

We can start with any one root. The coefficient a_0 must be non-zero in Eq. (B.46) because by definition κ is the smallest power in the series. Its value has to be arbitrary (though non-zero) because the solution has a multiplicative arbitrariness, i.e., if $y(x)$ is a solution then so is $cy(x)$ for any constant c. The lesson is that a_0 cannot be determined. But we can use the recursion relation to find the coefficients a_n, for $n > 0$, in terms of a_0. This gives us the series solution.

The recursion relation can be written in a way that is much simpler to handle. For this, first eliminate the κ^2 term by using the indicial equation, obtaining

$$n(n + 2\kappa + f_0 - 1)a_n = -S_n. \tag{B.53}$$

Next, note that since κ_+ and κ_- are roots of Eq. (B.51), we have

$$\kappa_+ + \kappa_- = 1 - f_0. \tag{B.54}$$

Using this relation, we can write the recursion relation in the form

$$n \Big(n + 2\kappa - \kappa_+ - \kappa_- \Big) a_n = -S_n. \tag{B.55}$$

It means that for the series solution with $\kappa = \kappa_+$, the factor multiplying a_n is $n(n + \nu)$ and for the solution with $\kappa = \kappa_-$, the factor is $n(n - \nu)$, where

$$\nu = \kappa_+ - \kappa_-, \tag{B.56}$$

the difference between the roots of the indicial equation. The recursion relation can then be written as

$$n(n + \nu)a_n = -S_n \qquad \text{for the larger root } \kappa_+, \tag{B.57a}$$
$$n(n - \nu)a_n = -S_n \qquad \text{for the smaller root } \kappa_-. \tag{B.57b}$$

Note that S_n also, in general, depends on the value of the root.

If a solution can be obtained this way for each root of the indicial equation and they are linearly independent, then our task is complete, since one expects to obtain two such solutions for a second-order differential equation. We will work out both kinds of examples, where this naive expectation works and where it does not. Frobenius' theorem, stated above, guarantees one series solution. We will then use the method of §B.5 to find the other independent solution.

B.6.2   A simple example

Let us take the differential equation

$$\frac{d^2y}{dx^2} + \frac{5}{2x}\frac{dy}{dx} + \frac{1-x}{2x^2}\,y = 0. \tag{B.58}$$

This means that we have

$$f_0 = \frac{5}{2}, \qquad g_0 = \frac{1}{2}, \qquad g_1 = -\frac{1}{2}, \tag{B.59}$$

whereas all other f_n's and g_n's are zero. The solutions of the indicial equation are:

$$\kappa_+ = -\frac{1}{2}, \qquad \kappa_- = -1. \tag{B.60}$$

Thus, for this differential equation,

$$\nu = \frac{1}{2} \tag{B.61}$$

and

$$S_n = g_1 a_{n-1} = -\frac{1}{2}a_{n-1} \tag{B.62}$$

irrespective of the value of the indicial root.

For the larger root, $\kappa_1 = -\frac{1}{2}$, the recursion relation of Eq. (B.57a) reads

$$a_n = \frac{a_{n-1}}{n(2n+1)}. \tag{B.63}$$

Taking the coefficient a_0 arbitrarily to be equal to 1, we obtain

$$a_1 = \frac{1}{3}, \qquad a_2 = \frac{a_1}{10} = \frac{1}{30}, \tag{B.64}$$

and so on, so we can write this solution of the differential equation as

$$y_1(x) = x^{-\frac{1}{2}}\left(1 + \frac{x}{3} + \frac{x^2}{30} + \frac{x^3}{630} + \cdots\right). \tag{B.65}$$

For the other solution, we plug $\kappa = -1$ in the recursion relation, which gives

$$a_n = \frac{a_{n-1}}{n(2n-1)}. \tag{B.66}$$

The coefficients of all powers can be evaluated now if we assume a_0 to be equal to 1 as before, and we obtain the second solution:

$$y_2(x) = x^{-1}\left(1 + x + \frac{x^2}{6} + \frac{x^3}{90} + \cdots\right). \tag{B.67}$$

The most general solution of the differential equation of Eq. (B.58) is therefore of the form given in Eq. (B.9), where y_1 and y_2 are given by Eqs. (B.65) and (B.67).

☐ **Exercise B.4** *Apply the method on the differential equation*

$$\frac{d^2y}{dx^2} + \frac{3}{2x}\frac{dy}{dx} - \frac{1-x}{x^2}y = 0 \tag{B.68}$$

to show that the general solution is of the form of Eq. (B.9), with

$$y_1(x) = x^{\frac{1}{2}}\left(1 - \frac{1}{5}x + \frac{1}{70}x^2 + \cdots\right),$$
$$y_2(x) = x^{-1}\left(1 + x + \frac{1}{2}x^2 + \cdots\right). \tag{B.69}$$

We now show that either of the solutions of Eq. (B.58) could have been derived from the other by the application of Eq. (B.45). To check, let us first note that, for this differential equation,

$$\exp\left(-\int dx\, f(x)\right) = \exp\left(-\frac{5}{2}\ln x\right) = x^{-\frac{5}{2}}. \tag{B.70}$$

Therefore, taking $y_1(x)$ from Eq. (B.65) and applying Eq. (B.45), we obtain

$$
y_2(x) = x^{-\frac{1}{2}} \left(1 + \frac{x}{3} + \frac{x^2}{30} + \frac{x^3}{630} + \cdots \right)
$$
$$
\times \int dx \, x \left(1 + \frac{x}{3} + \frac{x^2}{30} + \frac{x^3}{630} + \cdots \right)^{-2} x^{-\frac{5}{2}}. \qquad \text{(B.71)}
$$

Using the power series expansion of the term in the integrand, it is easy to see that the solution is exactly the same as that given in Eq. (B.67).

☐ **Exercise B.5** *For the same differential equation, start now with $y_2(x)$ and apply the same procedure to check that you find the solution given in Eq. (B.65).*

☐ **Exercise B.6** *Consider the differential equation*

$$
\frac{d^2y}{dx^2} - \frac{1}{x}\frac{dy}{dx} + \frac{y}{x^2} = 0. \qquad \text{(B.72)}
$$

Show that the indicial equation has only one solution, $\kappa = 1$. Show that the series solution corresponding to this value of κ is simply

$$
y_1(x) = x, \qquad \text{(B.73)}
$$

apart from a multiplicative constant. Then use Eq. (B.45) to show that the other independent solution is

$$
y_2(x) = x \ln x, \qquad \text{(B.74)}
$$

once again ignoring an overall multiplicative constant.

☐ **Exercise B.7** *Consider the differential equation*

$$
\frac{d^2y}{dx^2} + \frac{l+m+1}{x}\frac{dy}{dx} + \frac{lm}{x^2}y = 0. \qquad \text{(B.75)}
$$

Show that for $l \neq m$, the solutions are

$$
y_1(x) = x^{-l}, \qquad y_2(x) = x^{-m}. \qquad \text{(B.76)}
$$

What would be the two independent solutions if $l = m$?

B.6.3 Possibility of problems with series solutions

If two linearly independent series solutions are obtained of a certain differential equation, each solution is derivable from the other in the manner shown in §B.5. However, there are situations where two independent series solutions are not available. Our task now will be to identify the cases where that might happen, and find the second solution if it does.

Note that we only need to consider the values $\nu \geqslant 0$, because of the definition given in Eq. (B.56). When $\nu = 0$, i.e., the roots of the indicial equation are degenerate, we obviously cannot get two independent Frobenius solutions. There is only one possible recursion relation in this case, and therefore one series solution.

Among the positive values of ν, there is no problem with the recursion relation if ν is not an integer. If ν is an integer, there is still no problem if we try a series solution with the larger indicial root κ_+. This is the power solution guaranteed by Frobenius' theorem.

For the smaller root κ_-, however, the left side of Eq. (B.57b) vanishes for $n = \nu$, which can pose problems. In fact, two possibilities may arise.

Case 1: If $S_\nu \neq 0$, Eq. (B.57b) presents an inconsistency for $n = \nu$. This implies that a second power series solution will not be obtained. The second independent solution then must be obtained from Eq. (B.45).

Case 2: If, in Eq. (B.57b), S_ν happens to be zero, then a_ν becomes arbitrary for the solution with κ_- as the lowest power. Coefficients of higher powers of x will then be obtained in terms of a_ν, and this part will be identical to the solution obtained with κ_+. Thus, basically the general solution will be of the form of Eq. (B.9), where $y_1(x)$ starts with the power κ_+ and can have any number of terms, whereas $y_2(x)$ will contain terms starting from the power κ_- and going up to $\kappa_+ - 1$.

To summarize, a second solution is not available if $\nu = 0$, and also may not be available if ν is a positive integer. To obtain the second solution in such cases, and also to show where the second solution is in fact available, let us start from the series solution of Eq. (B.46) with $\kappa = \kappa_+$ and put it into Eq. (B.45). This gives

$$y_2(x) = y_1(x) \int \frac{dx}{x^{2\kappa_+}} \left(\sum_{m=0}^{\infty} a_m x^m \right)^{-2} \exp\left(-f_0 \ln x - \sum_{m=1}^{\infty} \frac{f_m x^m}{m} \right).$$

$$\text{(B.77)}$$

Note that

$$\exp\left(-f_0 \ln x \right) = x^{-f_0}, \qquad \text{(B.78)}$$

so that in the integrand, we have a factor $x^{-(2\kappa_+ +f_0)}$. Using Eq. (B.54) to eliminate f_0, we can rewrite the expression for the second solution in the form

$$y_2(x) = y_1(x) \int \frac{dx}{x^{\nu+1}} \sum_{m=0}^{\infty} b_m x^m, \qquad (B.79)$$

with suitably defined b_m's, since all other factors in the integrand can be written as a power series in x, starting with the constant term.

The integration yields a power series in x if ν is not an integer. If ν is an integer, zero or positive, then it should be borne in mind that the term with $m = \nu$ in the summation integrates to $b_\nu \ln x$. For the sake of simplicity of notation, let us denote b_ν by B from now on. Separating this log term out, we can write the second solution as

$$y_2(x) = By_1(x)\ln x + y_1(x) \int \frac{dx}{x^{\nu+1}} \left(\sum_{m=0}^{\infty}{}' b_m x^m \right), \qquad (B.80)$$

where now the sum is primed, reminding us that the term $m = \nu$ is omitted from the sum. The log term shows that the solution is not analytic at $x = 0$, which is the reason it cannot be found as a power series. If, however, it happens that $B = 0$, i.e., the b_ν term is absent in the integrand of Eq. (B.79), then $y_2(x)$ is also a power series solution. This is the case mentioned earlier where the right side of Eq. (B.49) vanishes.

□ **Exercise B.8** *Show that the condition $B = 0$ is the same as the condition that the right side of Eq. (B.49) vanishes.*

Each term in the sum of Eq. (B.80) will now integrate to become a power of x. The integral, multiplied by $y_1(x)$, is thus also a power series in x. What is the minimum power? In $y_1(x)$, the minimum power is κ_+. In the integral, the term corresponding to $m = 0$ integrates to $b_0 x^{-\nu}$. Thus, the minimum power of x in the power series part of the solution is $\kappa_+ - \nu$, which is equal to κ_- because of the definition of ν. Therefore, we can write the second solution as

$$y_2(x) = By_1(x)\ln x + x^{\kappa_-} \sum_{m=0}^{\infty} d_m x^m, \qquad (B.81)$$

where the coefficients d_m will involve the quantities b_m as well as the coefficients in $y_1(x)$. Once we find these coefficients, we know the second solution. If $B = 0$, we have a power series solution, and we will have to set d_0 arbitrarily and find the coefficients of the higher-order terms. If $B \neq 0$, we can always set

$B = 1$ by utilizing the multiplicative arbitrariness. Once this freedom is used, the coefficients d_m are completely fixed, and can be determined by putting the solution into the original differential equation. The terms containing $\ln x$ will cancel since y_1 itself is a solution of the differential equation. In the other terms, we can equate powers of x, as before, and thus obtain the d_m's. Rather than giving a general formula for d_m, we will illustrate the procedure with some examples given below.

B.6.4   A variety of examples

Example 1:   Consider the differential equation of Eq. (B.72). Here,

$$f_0 = -1, \qquad g_0 = 1, \tag{B.82}$$

whereas all other f_n's and g_n's vanish. The indicial equation is

$$(\kappa - 1)^2 = 0. \tag{B.83}$$

The two roots are identical, viz., $\kappa = 1$. To obtain the series solution, one starts with $a_0 = 1$ and notes that the recursion relation gives all other a_n's to be zero. Hence, we obtain the solution given in Eq. (B.73). To obtain the other solution, instead of using Eq. (B.45) directly, let us start with the form dictated by Eq. (B.81):

$$y_2(x) = x \ln x + \sum_{m=0}^{\infty} d_m x^{m+1}. \tag{B.84}$$

Once we put it into the differential equation, we obtain

$$\sum_{m=1}^{\infty} m^2 d_m x^{m-1} = 0. \tag{B.85}$$

Since this will have to be obeyed for arbitrary x, we conclude that $d_m = 0$ for $m \geqslant 1$. Note that the term with d_0 does not appear in the sum, so d_0 can be non-zero and we can therefore write the general solution of the differential equation as

$$y(x) = c_1 x + c_2(x \ln x + d_0 x). \tag{B.86}$$

But it is useless to write this d_0 term here, since it is anyway included in $c_1 x$ with arbitrary c_1. Thus, the general solution can be written as

$$y(x) = c_1 x + c_2 x \ln x. \tag{B.87}$$

This result agrees with what we had written in the statement of Ex. B.6 (p. 442).

Example 2: Next we take up the differential equation

$$\frac{d^2y}{dx^2} - \frac{1}{x}\frac{dy}{dx} - \frac{3(1-x)}{x^2}\,y = 0. \tag{B.88}$$

Now the indicial equation is

$$\kappa^2 - 2\kappa - 3 = 0, \tag{B.89}$$

whose roots are

$$\kappa_+ = 3, \qquad \kappa_- = -1. \tag{B.90}$$

The recursion relation for this differential equation is

$$n(n \pm 4)a_n = -S_n, \tag{B.91}$$

where

$$S_n = g_1 a_{n-1} = 3a_{n-1}. \tag{B.92}$$

Thus, we obtain

$$a_n = -\frac{3a_{n-1}}{n(n \pm 4)}. \tag{B.93}$$

For the larger root $\kappa = 3$, taking $a_0 = 1$ arbitrarily, we obtain

$$a_1 = -\frac{3}{5}, \qquad a_2 = -\frac{1}{4}a_1 = \frac{3}{20}, \tag{B.94}$$

and so on. So the series solution with this indicial root is

$$y_1(x) = x^3 - \frac{3}{5}x^4 + \frac{3}{20}x^5 + \cdots. \tag{B.95}$$

If we try to find a series solution with the other root, $\kappa_- = -1$, the equation for the coefficient a_4 becomes inconsistent. This means that the series solution does not exist. The solution must then be of the form given in Eq. (B.80). To obtain the coefficients d_m, we set $B = 1$ and plug the solution into the differential equation to obtain

$$\frac{2y_1'}{x} - \frac{2y_1}{x^2} + \sum_{m=0}^{\infty} \Big((m-1)(m-2)x^{m-3} - (m-1)x^{m-3}$$

$$-3x^{m-3} + 3x^{m-2} \Big) d_m = 0. \tag{B.96}$$

Simplifying the terms, and writing the coefficients of the y_1 solution as a_m for the sake of brevity, we get

$$\sum_{m=0}^{\infty} \left(2(m+2)a_m x^{m+1} + m(m-4)d_m x^{m-3} + 3d_m x^{m-2} \right) = 0. \quad \text{(B.97)}$$

We can now determine the d_m's by equating different powers of x to zero in this equation. The coefficient of x^{-3} vanishes automatically. For the other powers of x, we obtain the following equations:

$$\begin{aligned}
\text{Coeff. of } x^{-2} &\Rightarrow -3d_1 + 3d_0 = 0, \\
\text{Coeff. of } x^{-1} &\Rightarrow -4d_2 + 3d_1 = 0, \\
\text{Coeff. of } x^0 &\Rightarrow -3d_3 + 3d_2 = 0, \\
\text{Coeff. of } x^1 &\Rightarrow 4 + 3d_3 = 0, \quad \text{(B.98)}
\end{aligned}$$

and so on. Solving these equations, one gets

$$d_0 = d_1 = \frac{16}{9}, \qquad d_2 = d_3 = -\frac{4}{3}. \quad \text{(B.99)}$$

Thus, the second solution of the differential equation is given by

$$y_2(x) = y_1(x) \ln x + \frac{1}{x} \left(\frac{16}{9} + \frac{16}{9}x - \frac{4}{3}x^2 - \frac{4}{3}x^3 + \cdots \right). \quad \text{(B.100)}$$

Example 3: We next provide an example where the indicial roots differ by an integer, and yet there are two power series solutions. The differential equation is

$$x^2 \frac{d^2 y}{dx^2} + (x^2 + 3x)\frac{dy}{dx} - 3y = 0. \quad \text{(B.101)}$$

Dividing throughout by x^2, we find, in the notation of Eq. (B.47), that all f_m's and g_m's are zero except the following ones:

$$f_0 = 3, \qquad f_1 = 1, \qquad g_0 = 3. \quad \text{(B.102)}$$

The indicial equation is

$$\kappa^2 + 2\kappa - 3 = 0, \quad \text{(B.103)}$$

with the solutions

$$\kappa_+ = 1, \qquad \kappa_- = -3. \quad \text{(B.104)}$$

In the evaluation of the sum S_n, only f_1 contributes, so

$$S_n = (n - 1 + \kappa)a_{n-1}. \tag{B.105}$$

The recursion relation is therefore

$$n(n \pm 4)a_n = -(n - 1 + \kappa)a_{n-1}. \tag{B.106}$$

For the larger root $\kappa_+ = 1$, this recursion relation becomes

$$a_n = -\frac{1}{n + 4}a_{n-1}. \tag{B.107}$$

Taking $a_0 = 1$ arbitrarily, we can evaluate all coefficients of the power series and obtain one solution:

$$y_1(x) = x\left(1 - \frac{x}{5} + \frac{x^2}{6 \cdot 5} - \frac{x^3}{7 \cdot 6 \cdot 5} + \cdots\right). \tag{B.108}$$

Now let us look at the smaller indicial root, $\kappa_- = -3$. The left side of the recursion relation, Eq. (B.106), now becomes zero for $n = 4$. However, note that S_n also vanishes for this indicial root when $n = 4$. In fact, the factor $n - 4$ cancels from both sides of the recursion relation and we can write

$$a_n = -\frac{1}{n}, \tag{B.109}$$

so that the sum of $a_n x^n$ becomes equal to e^{-x}. Remembering the factor of x^κ that goes with it, we see that the solution is

$$y_2(x) = \frac{1}{x^3}e^{-x}. \tag{B.110}$$

Earlier, we said that in such cases, the second solution $y_2(x)$ can be written with only a finite number of terms, the highest power being $x^{\kappa_+ - 1}$. But in Eq. (B.110), we see an infinite series. There is no conflict between the two statements. In fact, we can just as well write

$$y_2(x) = \frac{1}{x^3} - \frac{1}{x^2} + \frac{1}{2!x} - \frac{1}{3!}, \tag{B.111}$$

which represents the first four terms obtained by expanding the exponential. The sum of the remaining infinite number of terms is $y_1(x)/4!$, which is anyway contained in the general solution of the form of Eq. (B.9).

Example 4: We now consider the differential equation

$$x(1-x)\frac{d^2y}{dx^2} + (2-3x)\frac{dy}{dx} - y = 0. \tag{B.112}$$

Dividing throughout by $x(1-x)$ and using the power series expansion for $(1-x)^{-1}$ that is valid near $x = 0$, we can write this equation in the form

$$\frac{d^2y}{dx^2} + (\frac{2}{x} - 3)\Big(\sum_{r=0}^{\infty} x^r\Big)\frac{dy}{dx} - \Big(\frac{1}{x} + \sum_{r=0}^{\infty} x^r\Big)y = 0. \tag{B.113}$$

This means that, in the notation of Eq. (B.47), here we have

$$f_0 = 2, \qquad f_m = -1 \text{ for } m > 0,$$
$$g_0 = 0, \qquad g_m = -1 \text{ for } m > 0. \tag{B.114}$$

The indicial roots are

$$\kappa_+ = 0, \qquad \kappa_- = -1. \tag{B.115}$$

The recursion relation for the coefficient a_n of the power series is

$$n(n \pm 1)a_n = -S_n. \tag{B.116}$$

The evaluation of the coefficients is straightforward for the larger root, as always, and a power series solution $y_1(x)$ can be obtained.

☐ **Exercise B.9** *Show that, for the differential equation being discussed, the power series solution for the larger indicial root has the form*

$$y_1(x) = 1 + \frac{x}{2} + \frac{x^2}{3} + \frac{x^3}{4} + \cdots = -\frac{\ln(1-x)}{x}. \tag{B.117}$$

As for the smaller indicial root, we see that there can be a problem with the evaluation of a_1. For this root, Eq. (B.50) gives

$$S_1 = -f_1a_0 + g_1a_0 = 0. \tag{B.118}$$

Hence there is no inconsistency in the recursion relation for $n = 1$. In fact, Eq. (B.116) shows that a_1 can be arbitrarily chosen. We choose $a_1 = 0$ and thereby obtain the solution as

$$y_2(x) = \frac{1}{x}. \tag{B.119}$$

So the general solution of this differential equation is of the form

$$y(x) = \frac{1}{x}\Big(c_1 \ln(1-x) + c_2\Big), \tag{B.120}$$

with arbitrary constants c_1 and c_2.

☐ **Exercise B.10** *Verify that if we had taken a_1 to be some non-zero constant for the smaller indicial root, we would have obtained, in $y_2(x)$, terms with higher orders of x which would have amounted to a constant times $y_1(x)$, which would be irrelevant in the expression for the general solution.*

B.6.5   Revisiting ordinary points

We discussed ordinary points of differential equations earlier in §B.4. Ordinary points can also be seen as regular singular points, with the restrictions

$$f_0 = 0, \qquad g_0 = 0, \qquad g_1 = 0, \tag{B.121}$$

in the notation of Eq. (B.47). We can therefore use our newly obtained wisdom to see why two independent series solutions are guaranteed at ordinary points of a differential equation.

Looking at Eq. (B.51), we see that Eq. (B.121) implies that the indicial equation for ordinary points is

$$\kappa^2 - \kappa = 0, \tag{B.122}$$

which means the roots are given by

$$\kappa_+ = 1, \qquad \kappa_- = 0. \tag{B.123}$$

Indeed, in the solutions we discussed in §B.4, we saw that one solution starts with a constant term, corresponding to $\kappa = 0$, and the other starts with a linear term in x, corresponding to $\kappa = 1$.

Series solution with $\kappa = \kappa_+$ is always guaranteed. For the other solution, the recursion relation of Eq. (B.49) reads

$$n(n-1)a_n = S_n, \tag{B.124}$$

since the difference of indicial roots is equal to 1 at ordinary points. A problem can occur with the evaluation of a_1 for this root if S_1 is non-zero. However, looking at the definition of Eq. (B.50), we find that the sum there consists only of the term with $m = 0$, and with $\kappa = 0$, we have

$$S_1 = g_1 a_0. \tag{B.125}$$

We already said that $g_1 = 0$ at an ordinary point. Hence, there is no obstruction to the evaluation of a_1. In fact, a_1 can be chosen arbitrarily, and we always chose the a_1 corresponding to the root $\kappa = 0$ to be zero.

☐ **Exercise B.11** *Solve the following differential equations. In each case, find two linearly independent solutions.*

$$\frac{d^2y}{dx^2} + x^2y = 0, \qquad\qquad \text{(B.126a)}$$

$$x^2\frac{d^2y}{dx^2} - n\frac{dy}{dx} + ny = 0. \qquad\qquad \text{(B.126b)}$$

Appendix C

Special functions

We have laid down the basic techniques for solving second-order homogeneous differential equations. In this appendix, we apply these techniques to some differential equations of special interest, which appeared at various parts of the text. We perform series solutions to identify the functions that emerge as solutions of these equations. Note that for some of these differential equations, the point $x = 0$ is an ordinary point so that two independent series solutions are possible. For some others, the point $x = 0$ is a regular singular point. We do not point out which differential equation belongs to which category, because it is easy to make that decision from the look of any equation.

C.1   Hermite equation

The Hermite differential equation is

$$\frac{d^2y}{dx^2} - 2x\frac{dy}{dx} + 2\alpha y = 0, \tag{C.1}$$

where α is a constant. In the general notation introduced in Eq. (B.29), here we have

$$F_1 = -2, \qquad G_0 = 2\alpha, \tag{C.2}$$

all other F_n's and G_n's being zero. Eq. (B.33) now becomes

$$(n + 1)(n + 2)a_{n+2} - 2na_n + 2\alpha a_n = 0, \tag{C.3}$$

or equivalently

$$a_{n+2} = -\frac{2(\alpha - n)}{(n + 1)(n + 2)}\, a_n. \tag{C.4}$$

The general solution is therefore of the form

$$y(x) = a_0 \cdot \left(1 - \frac{2\alpha}{2!}x^2 + \frac{2^2\alpha(\alpha - 2)}{4!}x^4 - \frac{2^3\alpha(\alpha - 2)(\alpha - 4)}{6!}x^6 + \cdots \right)$$

$$+ a_1 \cdot \left(x - \frac{2(\alpha - 1)}{3!}x^3 + \frac{2^2(\alpha - 1)(\alpha - 3)}{5!}x^5 - \cdots \right). \qquad \text{(C.5)}$$

If α is a non-negative integer, one of the two series terminates after a finite number of terms. The solutions are therefore polynomials, and they are called *Hermite polynomials*, and denoted by $H_n(x)$ where n is the integral value of α. For any $H_n(x)$, the relative magnitude of different co-coefficients is given in Eq. (C.5). If we take the convention of normalizing the polynomials by the conditions

$$a_0 = \frac{(-1)^r n!}{r!} \qquad \text{for } n = 2r, \qquad \text{(C.6a)}$$

$$a_1 = \frac{(-1)^r 2 \times n!}{r!} \qquad \text{for } n = 2r + 1, \qquad \text{(C.6b)}$$

we obtain the following expressions for the first few polynomials:

$$H_0(x) = 1,$$
$$H_1(x) = 2x,$$
$$H_2(x) = 4x^2 - 2,$$
$$H_3(x) = 8x^3 - 12x,$$
$$H_4(x) = 16x^4 - 48x^2 + 12,$$
$$H_5(x) = 32x^5 - 160x^3 + 120x, \qquad \text{(C.7)}$$

and so on. In a more compact form, one can write

$$H_n(x) = \sum_{s=0}^{\lfloor n/2 \rfloor} \frac{(-1)^s n!}{s!(n - 2s)!} (2x)^{n-2s}, \qquad \text{(C.8)}$$

where $\lfloor n/2 \rfloor$ means the largest integer that is not bigger than $n/2$. Clearly,

$$H_n(-x) = (-1)^n H_n(x). \qquad \text{(C.9)}$$

Of course, it has to be remembered that for each value of α, there are two linearly independent solutions, given by the two series that multiply a_0 and a_1. We will have more discussion on this issue later in the form of Ex. C.4 (p. 457).

C.2 Legendre equation

The Legendre equation is of the form

$$(1 - x^2)\frac{d^2y}{dx^2} - 2x\frac{dy}{dx} + l(l+1)y = 0. \tag{C.10}$$

The constant multiplying y in the last term on the left-hand side has been taken, curiously, as $l(l+1)$. This is no loss of generality. We could have taken it as α as well. The reason for taking the constant in this form will be obvious shortly.

Note that if we divide the entire equation by $(1-x^2)$ to put it in the form given in Eq. (B.4), then both $f(x)$ and $g(x)$ are singular at $x = \pm 1$. However, this poses no problem for us, because both these functions are analytic at $x = 0$, and we are interested in finding power series solutions at $x = 0$. Whatever solutions we find, they will be valid until the singular points are reached, i.e., in the interval $-1 < x < +1$. The issue of the singular points $x = \pm 1$ will be discussed later.

Since

$$\frac{1}{1-x^2} = \sum_{r=0}^{\infty} x^{2r}, \tag{C.11}$$

we find that the non-vanishing F_n's and G_n's are given by

$$F_{2r+1} = -2, \qquad G_{2r} = l(l+1) \tag{C.12}$$

for all integers $r \geqslant 0$. Eq. (B.33) now takes the form

$$(n+1)(n+2)a_{n+2} = 2\sum_{\substack{m \leqslant n \\ (n-m \text{ odd})}} (m+1)a_{m+1} - l(l+1)\sum_{\substack{m \leqslant n \\ (n-m \text{ even})}} a_m. \tag{C.13}$$

Since this is valid for any n, we can replace $n+2$ by n in this equation, which gives

$$n(n-1)a_n = 2\sum_{\substack{m \leqslant n-2 \\ (n-m \text{ odd})}} (m+1)a_{m+1} - l(l+1)\sum_{\substack{m \leqslant n-2 \\ (n-m \text{ even})}} a_m. \tag{C.14}$$

We now subtract this equation from the earlier one involving a_{n+2}. On the right-hand side, only $m = n-1$ will contribute to the difference involving odd values of $n-m$, and $m = n$ to the other. Thus, we get

$$(n+1)(n+2)a_{n+2} - n(n-1)a_n = 2na_n - l(l+1)a_n, \tag{C.15}$$

or

$$a_{n+2} = \frac{n(n+1) - l(l+1)}{(n+1)(n+2)} \, a_n. \qquad (C.16)$$

The solution is therefore of the form

$$y(x) = a_0 \cdot \left(1 - \frac{l(l+1)}{2!} x^2 - \frac{l(l+1)[6 - l(l+1)]}{4!} x^4 + \cdots \right)$$
$$+ \, a_1 \cdot \left(x + \frac{2 - l(l+1)}{3!} x^3 + \frac{[12 - l(l+1)][2 - l(l+1)]}{5!} x^5 + \cdots \right).$$
$$(C.17)$$

Notice that the solutions are valid only in the region $|x| < 1$, as commented before.

□ **Exercise C.1** *Show that the recursion relation of Eq. (C.16) follows more simply by trying a series solution on the original equation, Eq. (C.10), without dividing out by $1 - x^2$.*

The mystery for taking the constant in the form $l(l+1)$ is clear from the recursion relation, Eq. (C.16). The point is that if l happens to be a non-negative integer, one of the series solutions terminates. This part of the solution is then a polynomial. These polynomials, for different values of l, are called *Legendre polynomials*, and are conventionally denoted by $P_l(x)$. The first few of these polynomials are as follows:

$$P_0(x) = 1,$$
$$P_1(x) = x,$$
$$P_2(x) = \frac{1}{2}(3x^2 - 1),$$
$$P_3(x) = \frac{1}{2}(5x^3 - 3x),$$
$$P_4(x) = \frac{1}{8}(35x^4 - 30x^2 + 3),$$
$$P_5(x) = \frac{1}{8}(63x^5 - 70x^3 + 15x), \qquad (C.18)$$

and so on. Of course, the normalization is fixed only by convention. We have chosen the normalization by imposing the condition

$$P_l(1) = 1. \qquad (C.19)$$

The resulting form of the polynomials is consistent with the *Rodriguez formula*,

$$P_l(x) = \frac{1}{2^l l!} \frac{d^l}{dx^l}\left[(x^2 - 1)^l\right]. \qquad (C.20)$$

One expects two linearly independent solutions of a second-order differential equation. Seeing so many solutions in Eq. (C.18), one should not be confused and start thinking that many solutions have been obtained. The different functions shown in Eq. (C.18) are not solutions to the same differential equation. For each value of l, Eq. (C.10) represents a different equation. Thus, for each value of l, we expect two linearly independent solutions. Eq. (C.18) shows one such solution for some selected values of l. There are the other solutions, of course, which are present in Eq. (C.17), even when l is an integer. They are called the Legendre functions of the second kind and are usually denoted by $Q_l(x)$.

For example, if l is an even integer, the polynomial solution is obtained from the function that multiplies a_0 in Eq. (C.17), because this is the part where the series terminates. Then $Q_l(x)$ is the function that multiplies a_1. Similarly, for odd l, the part that multiplies a_0 is $Q_l(x)$. We can list a few functions for small integral values of l:

$$Q_0(x) = x + \frac{1}{3}x^3 + \frac{1}{5}x^5 + \cdots,$$

$$Q_1(x) = 1 - x^2 - \frac{1}{3}x^4 + \cdots,$$

$$Q_2(x) = x - \frac{2}{3}x^3 - \frac{1}{5}x^5. \qquad (C.21)$$

These Q_l solutions, called *Legendre functions of the second kind*, can be derived from the P_l's by the method outlined in §B.5. For this equation,

$$f(x) = -\frac{2x}{1 - x^2}, \qquad (C.22)$$

so that Eq. (B.44) gives

$$\zeta(x) = \frac{1}{P^2(x)}\frac{1}{1 - x^2}, \qquad (C.23)$$

apart from an overall multiplicative constant, where $P(x)$ denotes the polynomial solution obtained in Eq. (C.18). Therefore, the other independent solution is given by

$$Q(x) = P(x)\int dx\, \frac{1}{P^2(x)}\frac{1}{1 - x^2}. \qquad (C.24)$$

For each $P_l(x)$, one obtains a corresponding $Q_l(x)$. We now evaluate a few of these functions for low values of the constant l.

For $l = 0$, we got $P_0(x) = 1$. Therefore,

$$Q_0(x) = \int dx \, \frac{1}{1 - x^2} = \frac{1}{2} \ln \left(\frac{1 + x}{1 - x} \right). \tag{C.25}$$

Next, we use $P_1(x) = x$ to obtain

$$Q_1(x) = x \int dx \, \frac{1}{x^2} \frac{1}{1 - x^2}$$
$$= x \int dx \left(\frac{1}{1 - x^2} + \frac{1}{x^2} \right) = \frac{x}{2} \ln \left(\frac{1 + x}{1 - x} \right) - 1. \tag{C.26}$$

These are the *Legendre functions of the second kind*. Note that these solutions diverge at $x = \pm 1$, which are singular points of the differential equation. More specifically, if we write these solutions by using the independent variable $\tilde{x} = 1 + x$ or $\tilde{x} = 1 - x$, the expressions will contain a $\ln \tilde{x}$ term, which cannot be written as a power series in $\tilde{x}$. In §B.6, we have seen that it is a feature that is expected at singular points of certain differential equations.

□ **Exercise C.2** *Verify a few more:*

$$Q_2(x) = \frac{3x^2 - 1}{4} \ln \left(\frac{1 + x}{1 - x} \right) - \frac{3x}{2},$$

$$Q_3(x) = \frac{5x^3 - 3x}{4} \ln \left(\frac{1 + x}{1 - x} \right) - \frac{5}{2} x^2 + \frac{2}{3}. \tag{C.27}$$

□ **Exercise C.3** *Expand the logarithms in power series about x to show that the Legendre functions of the second kind obtained here are the same as those found in Eq. (C.21).*

□ **Exercise C.4** *For the Hermite differential equation given in Eq. (C.1), we found only one set of solutions for various values of the parameter α. There is, of course, another set. Find these other solutions by the method of §B.5 and verify that for each α, it coincides with the other series, which does not terminate in Eq. (C.5),*

C.3 Associated Legendre equation

There is an important and interesting differential equation that is closely linked with the Legendre equation. It contains one more parameter which we will denote by m. The equation is:

$$(1 - x^2) \frac{d^2 y}{dx^2} - 2x \frac{dy}{dx} + \left(l(l + 1) - \frac{m^2}{1 - x^2} \right) y = 0. \tag{C.28}$$

The solution of this equation is denoted by the symbol $P_l^m(x)$, where the superscripted m should be thought of as an index, and not a power.

Obviously, if $m = 0$, Eq. (C.28) reduces to the Legendre differential equation, so we can write

$$P_l^0(x) = P_l(x). \tag{C.29}$$

For positive integral values of m, it is easy to see that the following expression is a solution to Eq. (C.28):

$$P_l^m(x) = (-1)^m \, (1 - x^2)^{m/2} \, \frac{d^m}{dx^m} \Big(P_l(x) \Big). \tag{C.30}$$

Since $P_l(x)$ is a polynomial of degree l, the derivatives clearly vanish for $m > l$. Thus, non-zero solutions are obtained if

$$|m| \leqslant l. \tag{C.31}$$

For negative values of m, the same solution is allowed, since the differential equation contains only m^2.

The functions defined in Eq. (C.30) are called *associated Legendre functions*. Sometimes, one also uses the name *associated Legendre polynomials*, although the functions are not polynomials for odd values of m. Yet the name may be tolerable because if we make a change of variable by writing $x = \cos\theta$, the functions $P_l^m(\cos\theta)$, though not polynomials in $\cos\theta$ for any value of m, will contain a finite number of terms if written in terms of powers of $\cos\theta$ and $\sin\theta$. For example, the associated Legendre polynomials for a few low values of l and non-zero values of m are:

$$\begin{aligned}
P_1^1(\cos\theta) &= -\sin\theta, \\
P_2^1(\cos\theta) &= -3\cos\theta\sin\theta, \\
P_2^2(\cos\theta) &= 3\sin^2\theta, \\
P_3^1(\cos\theta) &= -\frac{3}{2}(5\cos^2\theta - 1)\sin\theta, \\
P_3^2(\cos\theta) &= 15\cos\theta\sin^2\theta, \\
P_3^3(\cos\theta) &= -15\sin^3\theta.
\end{aligned} \tag{C.32}$$

C.4 Airy equation

The Airy equation is:

$$\frac{d^2y}{dx^2} - xy = 0. \tag{C.33}$$

The relative magnitude and sign between the coefficients is not important. It is easily seen that if $A(x)$ is a solution of Eq. (C.33), then the solution of $y'' - b^3 xy = 0$ would be $A(x/b)$.

This is the first equation we encounter for which $F_0 = G_0 = 0$. This creates a special situation when we try to apply Eq. (B.33) for $n = 0$, because there is no term in the summation that appears there. Therefore, for $n = 0$, Eq. (B.33) simply gives

$$a_2 = 0 \tag{C.34}$$

in the series solution. To obtain the other coefficients, we put $G_1 = -1$ while all other G_n's, and all F_n's, are zero. This gives the recursion relation

$$(n+1)(n+2)a_{n+2} - a_{n-1} = 0, \tag{C.35}$$

which means

$$a_{n+3} = \frac{a_n}{(n+2)(n+3)}. \tag{C.36}$$

Therefore, the general series solution is of the form

$$y(x) = a_0 y_1(x) + a_1 y_2(x), \tag{C.37}$$

where

$$y_1(x) = 1 + \frac{x^3}{2 \cdot 3} + \frac{x^6}{2 \cdot 3 \cdot 5 \cdot 6} + \cdots, \tag{C.38a}$$

$$y_2(x) = x + \frac{x^4}{3 \cdot 4} + \frac{x^7}{3 \cdot 4 \cdot 6 \cdot 7} + \cdots. \tag{C.38b}$$

These forms are helpful if one wants the functions for small values of the argument. For very large values, on the other hand, it is useful to define some linear combinations:

$$\mathrm{Ai}(x) = \frac{3^{-2/3}}{\Gamma(\frac{2}{3})} y_1(x) - \frac{3^{-4/3}}{\Gamma(\frac{4}{3})} y_2(x),$$

$$\mathrm{Bi}(x) = \frac{3^{-1/6}}{\Gamma(\frac{2}{3})} y_1(x) + \frac{3^{-5/6}}{\Gamma(\frac{4}{3})} y_2(x). \tag{C.39}$$

These combinations are called *Airy functions* of first and second kind respectively. The leading behavior of these functions for large values of the argument will be discussed in §E.2.

☐ **Exercise C.5** *Show that the coefficients in the series solution can be succinctly written in the form*

$$a_{3n} = \frac{\Gamma(\frac{2}{3})}{9^n n! \Gamma(n + \frac{2}{3})} \, a_0,$$

$$a_{3n+1} = \frac{\Gamma(\frac{4}{3})}{9^n n! \Gamma(n + \frac{4}{3})} \, a_1,$$

$$a_{3n+2} = 0. \tag{C.40}$$

☐ **Exercise C.6** *Starting from Ai(x), derive the first two terms of the Airy function of the second kind, Bi(x), by using the technique described in §B.5.*

C.5   Bessel equation

The Bessel equation is

$$x^2 \frac{d^2 y}{dx^2} + x \frac{dy}{dx} + (x^2 - \alpha^2) y = 0. \tag{C.41}$$

Dividing throughout by x^2, we see that for this equation

$$f(x) = \frac{1}{x}, \qquad g(x) = 1 - \frac{\alpha^2}{x^2}. \tag{C.42}$$

Thus, the only non-zero coefficients f_n and g_n are given by

$$f_0 = 1, \qquad g_0 = -\alpha^2, \qquad g_2 = 1. \tag{C.43}$$

The indicial equation in this case reads

$$\kappa^2 - \alpha^2 = 0. \tag{C.44}$$

Since α^2 appears in the differential equation, we can take α to be non-negative without any loss of generality. The difference of the two roots is 2α, so the recurrence relation is

$$n(n \pm 2\alpha) a_n + a_{n-2} = 0, \tag{C.45}$$

where the plus sign pertains to the larger root, i.e., α, and the minus sign to the root $-\alpha$. For the larger root, the recursion relation gives

$$a_2 = -\frac{1}{2(2\alpha + 2)} a_0,$$

$$a_4 = -\frac{1}{4(2\alpha + 4)} a_2 = \frac{1}{2 \cdot 4(2\alpha + 2)(2\alpha + 4)} a_0, \qquad \text{(C.46)}$$

and so on. Putting these coefficients in, we can write the solutions known as *Bessel functions* of the first kind:

$$J_\alpha(x) = \sum_{r=0}^{\infty} \frac{(-1)^r}{r!\,\Gamma(\alpha + r + 1)} \left(\frac{x}{2}\right)^{\alpha + 2r}, \qquad \text{(C.47)}$$

where use has been made of the property $\alpha\Gamma(\alpha) = \Gamma(\alpha + 1)$, and the normalization has been fixed by making the conventional choice

$$a_0 = \frac{1}{2^\alpha \Gamma(\alpha)}. \qquad \text{(C.48)}$$

Quite often, these are the functions that are referred to as *Bessel functions*, without the specifier *of the first kind*. Note that they satisfy the relation

$$J_\alpha(-x) = (-1)^\alpha J_\alpha(x). \qquad \text{(C.49)}$$

The solution with the other indicial root is $J_{-\alpha}(x)$. If α is not an integer or half-integer, the two roots differ by a non-integer, and are therefore linearly independent. The general solution for the Bessel equation in this case is

$$y(x) = c_1 J_\alpha(x) + c_2 J_{-\alpha}(x). \qquad \text{(C.50)}$$

In fact, this form of the solution is also valid when α is a half-integer. The reason can be seen from the recursion relation. For the smaller root, the recursion relation has a factor that vanishes when $n = 2\alpha$. This is the place that might pose problems for the series solution. But, for half-integral α, this value of n is an odd integer, and the recursion relation, having jumps of 2, does not allow a non-zero coefficient for the corresponding power of x. Thus, there is no problem.

The only problem that remains is for integral values of α, which we will denote by ν from now on. In fact, the expression in Eq. (C.47) shows what the problem is. Note that

$$J_{-\nu}(x) = \left(\frac{x}{2}\right)^{-\nu} \sum_{r=0}^{\infty} \frac{(-1)^r}{\Gamma(r+1)\,\Gamma(r - \nu + 1)} \left(\frac{x}{2}\right)^{2r}, \qquad \text{(C.51)}$$

where we have written $r!$ in terms of the Gamma function for the sake of future convenience. The factor $\Gamma(r - \nu + 1)$ blows up for all integral values of r when $r \leqslant \nu - 1$. These values of r then do not contribute to the sum, and we can as well start the sum from $r = \nu$:

$$J_{-\nu}(x) = \left(\frac{x}{2}\right)^{-\nu} \sum_{r=\nu}^{\infty} \frac{(-1)^r}{\Gamma(r + 1)\,\Gamma(r - \nu + 1)} \left(\frac{x}{2}\right)^{2r}. \tag{C.52}$$

We now introduce $s = r - \nu$ and rewrite the expression as a sum over s:

$$J_{-\nu}(x) = \left(\frac{x}{2}\right)^{-\nu} \sum_{s=0}^{\infty} \frac{(-1)^{s+\nu}}{\Gamma(s + \nu + 1)\,\Gamma(s + 1)} \left(\frac{x}{2}\right)^{2s+2\nu}$$

$$= (-1)^\nu J_\nu(x). \tag{C.53}$$

We see then that $J_{-\nu}(x)$ is not linearly independent of $J_\nu(x)$. Therefore, we need to find a different solution that would be linearly independent of $J_\nu(x)$.

To this end, one defines the Neumann functions by the relation

$$N_\alpha(x) = \frac{J_\alpha(x)\cos\alpha\pi - J_{-\alpha}(x)}{\sin\alpha\pi}. \tag{C.54}$$

It is easy to see that when $\alpha = \nu$, an integer, then both numerator and denominator vanish, so we can apply l'Hospital rule to write, for an integer ν,

$$N_\nu(x) = \frac{1}{\pi}\frac{\partial}{\partial\alpha}\Big(J_\alpha(x) - (-1)^\nu J_{-\alpha}(x)\Big)\Big|_{\alpha=\nu}. \tag{C.55}$$

It is easily seen that for very small values of x, the leading term for $J_\nu(x)$ is proportional to x^ν for $\nu \geqslant 0$. Also, from Eq. (C.55), we see that in the same limit, the leading term for the Neumann function $N_\nu(x)$ goes like $x^{-\nu}$. Thus, at $x = 0$, the Neumann functions are not defined. If a physical problem demands solutions in terms of Bessel functions which are well defined at $x = 0$, we must disregard the Neumann functions, and take the solutions as the J_ν functions.

If we want a solution which is well behaved at infinity, we can use neither the Bessel functions nor the Neumann functions. However, we can define the linear combinations

$$H_n^{(1)}(x) = J_n(x) + iN_n(x),$$
$$H_n^{(2)}(x) = J_n(x) - iN_n(x), \tag{C.56}$$

which are called Hankel functions of the first and second kind. One of these two will have acceptable behavior at infinity. We will give more details about such combinations as we discuss the spherical Bessel functions next.

C.6   Spherical Bessel equation

The spherical Bessel equation is

$$x^2\frac{d^2y}{dx^2} + 2x\frac{dy}{dx} + \Big(x^2 - n(n+1)\Big)y = 0. \tag{C.57}$$

If we define

$$z = y\sqrt{x}, \tag{C.58}$$

it is straightforward to see that z satisfies the equation

$$x^2\frac{d^2z}{dx^2} + x\frac{dz}{dx} + \Big(x^2 - \big(n+\tfrac{1}{2}\big)^2\Big)z = 0, \tag{C.59}$$

which is the Bessel equation with n replaced by $n + \tfrac{1}{2}$. Thus, the solutions of the spherical Bessel equation are $1/\sqrt{x}$ times the Bessel functions. Usually they are denoted as

$$j_n(x) = \sqrt{\frac{\pi}{2x}}\, J_{n+\frac{1}{2}}(x). \tag{C.60}$$

The factor of $\sqrt{\pi/2}$ is purely conventional.

□ **Exercise C.7** *Using the series expansion of the Bessel functions given in Eq. (C.47), show that*

$$j_0(x) = \frac{\sin x}{x},$$
$$j_1(x) = \frac{\sin x}{x^2} - \frac{\cos x}{x},$$
$$j_2(x) = \frac{3-x^2}{x^3}\sin x - \frac{3}{x^2}\cos x. \tag{C.61}$$

□ **Exercise C.8** *Show that the functions*

$$S_n(x) = xj_n(x) = \sqrt{\frac{\pi x}{2}}\, J_{n+\frac{1}{2}}(x) \tag{C.62}$$

are solutions of the differential equation

$$x^2\frac{d^2y}{dx^2} + \Big(x^2 - n(n+1)\Big)y = 0. \tag{C.63}$$

The spherical Neumann functions, $\eta_n(x)$, can be obtained from the functions $j_n(x)$ by using the method of §B.5. It can also be defined through a relation resembling Eq. (C.60), which is

$$\eta_n(x) = (-1)^{n+1}\sqrt{\frac{\pi}{2x}}\, J_{-n-\frac{1}{2}}(x).$$
(C.64)

As in Eq. (C.60), the numerical factors are purely conventional.

☐ **Exercise C.9** *Show that the first few spherical Neumann functions are as follows:*

$$\eta_0(x) = -\frac{\cos x}{x},$$
$$\eta_1(x) = -\frac{\cos x}{x^2} - \frac{\sin x}{x},$$
$$\eta_2(x) = -\frac{3-x^2}{x^3}\cos x - \frac{3}{x^2}\sin x.$$
(C.65)

Solutions like those shown in Eqs. (C.61) and (C.65) can be obtained iteratively by defining the functions $\xi_n(x)$ through the relation

$$y_n(x) = x^n \xi_n(x).$$
(C.66)

If we put this into Eq. (C.57), we see that the ξ_n's satisfy the differential equation

$$x\frac{d^2\xi_n}{dx^2} + 2(n+1)\frac{d\xi_n}{dx} + x\xi_n = 0.$$
(C.67)

Taking a further derivative, one obtains

$$x\frac{d^3\xi_n}{dx^3} + (2n+3)\frac{d^2\xi_n}{dx^2} + x\frac{d\xi_n}{dx} + \xi_n = 0.$$
(C.68)

Substituting ξ_n from Eq. (C.67), we obtain

$$x\frac{d^3\xi_n}{dx^3} + 2(n+1)\frac{d^2\xi_n}{dx^2} + \left(x - \frac{2(n+1)}{x}\right)\frac{d\xi_n}{dx} = 0.$$
(C.69)

We now define

$$\chi_n(x) = \frac{1}{x}\frac{d\xi_n}{dx}.$$
(C.70)

Then Eq. (C.69) can be written in the form

$$x\frac{d^2(x\chi_n)}{dx^2} + 2(n+1)\frac{d(x\chi_n)}{dx} + (x^2 - 2(n+1))\chi_n = 0,$$
(C.71)

which is the same as

$$x\frac{d^2\chi_n}{dx^2} + 2(n+2)\frac{d\chi_n}{dx} + x\chi_n = 0. \tag{C.72}$$

Comparing it with Eq. (C.67), we see that

$$\chi_n(x) = \frac{1}{x}\frac{d\xi_n}{dx} = \alpha\xi_{n+1}(x), \tag{C.73}$$

where α is an overall multiplicative factor which can be fixed only through some convention.

Now the solutions can be iteratively found. We note that Eq. (C.67) can be solved easily for $n = -1$:

$$\xi_{-1}(x) = \sin x \quad \text{or} \quad \cos x. \tag{C.74}$$

Spherical Bessel functions for all integral values of n, for $n \geqslant 0$, can be found then by using Eqs. (C.66) and (C.73). The solutions have been shown in Eq. (13.63, p. 339) with a certain convention about the multiplicative constant factors and are not repeated here.

Clearly it is seen that the spherical Bessel functions are regular at $x = 0$, whereas the spherical Neumann functions diverge. Hence, if we need a solution that is regular at the origin, we can use only the spherical Bessel functions, as was done in §9.4.

For the infinite spherical well discussed in §9.4.1, the wavefunction vanishes outside. However, that is not the case for the finite potential well discussed in §9.4.2. So, for that problem, we needed functions that fall sufficiently fast for large values of its argument. Neither $j_n(x)$ nor $\eta_n(x)$ is suitable for that purpose. However, from the expressions given in Eqs. (C.61) and (C.65), we see that if we define the *spherical Hankel functions* as the linear combinations

$$h_n^{(1)}(x) = j_n(x) + i\eta_n(x),$$
$$h_n^{(2)}(x) = j_n(x) - i\eta_n(x), \tag{C.75}$$

then

$$h_0^{(1)}(x) = -\frac{ie^{ix}}{x},$$
$$h_1^{(1)}(x) = -\frac{ie^{ix}}{x^2} - \frac{e^{ix}}{x},$$
$$h_2^{(1)}(x) = -i\frac{(3-x^2)e^{ix}}{x^3} - \frac{3e^{ix}}{x^2}, \tag{C.76}$$

whereas the functions of the second kind are complex conjugates of these for real x. Thus, if x is a positive imaginary number, these functions will fall like $e^{-|x|}$ as the absolute value of the number goes to infinity. This fact was used in §9.4 for the finite potential well.

C.7 Modified Bessel equation

The modified Bessel equation is

$$x^2 \frac{d^2 y}{dx^2} + x \frac{dy}{dx} - (x^2 + \alpha^2)y = 0. \tag{C.77}$$

Note that the only difference with the Bessel equation is the sign of the term $x^2 y$. Thus, if we define

$$x' = ix \tag{C.78}$$

and rewrite Eq. (C.77) by using x' as the independent variable instead of x, we would obtain exactly the Bessel equation. So, the solution is obtained by replacing x by ix in any solution of the Bessel equation. Of course, an overall normalizing factor is optional, and one usually defines the *modified Bessel functions* of the first kind as

$$I_\alpha(x) = \frac{1}{i^\alpha} \, J_\alpha(ix). \tag{C.79}$$

The motivation for introducing the factor $1/i^\alpha$ is easily understood by looking at Eq. (C.47), which clearly gives

$$I_\alpha(x) = \sum_{r=0}^{\infty} \frac{1}{r!\,\Gamma(\alpha + r + 1)} \left(\frac{x}{2}\right)^{\alpha + 2r}. \tag{C.80}$$

There are also modified Bessel functions of the second kind, which are defined as

$$K_\alpha(x) = \frac{\pi}{2} \frac{I_{-\alpha}(x) - I_\alpha(x)}{\sin \alpha\pi}. \tag{C.81}$$

□ **Exercise C.10** *Show that for an integer ν,*

$$I_\nu(x) = I_{-\nu}(x). \tag{C.82}$$

C.8 Laguerre equation

The Laguerre equation is

$$x\frac{d^2y}{dx^2} + (1-x)\frac{dy}{dx} + \alpha y = 0, \tag{C.83}$$

where α is a constant. We see that the non-vanishing coefficients in the functions $f(x)$ and $g(x)$ are as follows:

$$f_0 = 1, \qquad f_1 = -1, \qquad g_1 = \alpha. \tag{C.84}$$

The indicial equation is

$$\kappa^2 = 0, \tag{C.85}$$

i.e., κ has only one possible value. Putting in this value of κ as well as the coefficients f_n and g_n, we find that the recursion equation is

$$a_n = \frac{n-1-\alpha}{n^2}\, a_{n-1}. \tag{C.86}$$

The solution is therefore an arbitrary constant multiplying the so-called *Laguerre function*, given by

$$L_\alpha(x) = 1 - \alpha x + \frac{\alpha(\alpha-1)}{(2!)^2}x^2 - \frac{\alpha(\alpha-1)(\alpha-2)}{(3!)^2}x^3 + \cdots. \tag{C.87}$$

Obviously, the series terminates if α is a non-negative integer. The resulting polynomials are called *Laguerre polynomials*. Taking $a_0 = 1$, we can write down the expression for these polynomials as follows:

$$L_n(x) = \sum_{s=0}^{n}(-1)^s\binom{n}{s}\frac{x^s}{s!}. \tag{C.88}$$

Explicit forms for the first few polynomials are given below:

$$L_0(x) = 1,$$
$$L_1(x) = -x + 1,$$
$$L_2(x) = \frac{1}{2}x^2 - 2x + 1,$$
$$L_3(x) = -\frac{1}{6}x^3 + \frac{3}{2}x^2 - 3x + 1. \tag{C.89}$$

Note that

$$L_n(0) = 1 \tag{C.90}$$

because of our choice of $a_0 = 1$.

C.9   Generalized Laguerre equation

A more generalized form of the Laguerre equation is

$$x\frac{d^2y}{dx^2} + (l+1-x)\frac{dy}{dx} + \alpha y = 0, \tag{C.91}$$

with two constants l and α.

Compared to the Laguerre equation, the only difference in the functions $f(x)$ and $g(x)$ defined in Eq. (B.4) is that the coefficient f_0 is different:

$$f_0 = l+1, \tag{C.92}$$

where all other f_n's and g_n's have the same values that they have for the Laguerre equation. Because of the different value of f_0, the indicial equation is now different:

$$\kappa(\kappa + l) = 0. \tag{C.93}$$

We will be interested only in the case where l is a non-negative integer. Hence, the two solutions for the indicial equation differ by an integer. The corresponding solutions of the differential equation are therefore not independent. We can consider only the solution with $\kappa = 0$, for which the recursion equation is

$$a_n = \frac{n-1-\alpha}{n(n+l)}a_{n-1}. \tag{C.94}$$

The series terminates if α is any non-negative integer. The resulting polynomial solutions are called *associated Laguerre polynomials* and will be denoted by $L_m^l(x)$, where m is an integer denoting the value of α. Note that, unlike the case of associated Lagendre functions, there is nothing that tells us about any limit on l for a fixed value of m, or vice versa. The two parameters are independent, as can be seen in §9.6 where they appear in the text.

We need not write down the solutions explicitly for $l = 0$ because they are the Laguerre polynomials described above. For $l > 0$, we can fix the normalization by taking

$$a_0 = \frac{(l+1)(l+2)\cdots(l+m)}{m!} = \frac{(l+m)!}{l!m!}. \tag{C.95}$$

Then, we obtain the following series for the associated Laguerre polynomials with

$$L_0^l(x) = 1,$$
$$L_1^l(x) = -x + l + 1,$$

$$L_2^l(x) = \frac{1}{2}\Big[x^2 - 2(l+2)x + (l+1)(l+2)\Big],$$

$$L_3^l(x) = \frac{1}{6}\Big[-x^3 + 3(l+3)x^2 - 3(l+2)(l+3)x$$

$$+(l+1)(l+2)(l+3)\Big]. \tag{C.96}$$

In writing these forms, we have normalized the polynomials in such a way that, for $l = 0$, they reduce to the Laguerre polynomials given in Eq. (C.89).

C.10   Hypergeometric equation

This equation has the form

$$x(1-x)\frac{d^2y}{dx^2} + \Big(\gamma - (1+\alpha+\beta)x\Big)\frac{dy}{dx} - \alpha\beta y = 0, \tag{C.97}$$

where α, β, and γ are constants. Dividing both sides by $x(1-x)$ and expanding around $x = 0$, we can write the equation as

$$\frac{d^2y}{dx^2} + \left(\sum_{r=0}^{\infty} x^r\right)\left[\left(\frac{\gamma}{x} - (1+\alpha+\beta)\right)\frac{dy}{dx} - \frac{\alpha\beta}{x}y\right] = 0. \tag{C.98}$$

We find that the non-zero coefficients of the functions $f(x)$ and $g(x)$ are

$$f_0 = \gamma, \qquad f_r = \gamma - (1+\alpha+\beta) \quad \text{for } r > 0,$$
$$g_0 = 0, \qquad g_r = -\alpha\beta \quad\quad\quad \text{for } r > 0. \tag{C.99}$$

Therefore, the indicial equation reads

$$\kappa(\kappa - 1) + \kappa\gamma = 0, \tag{C.100}$$

so κ can be either 0 or $1 - \gamma$. In what follows, we assume that γ is not an integer, so the difference between the two indicial roots is not an integer either.

The recursion relation of Eq. (B.49) comes out to be

$$(n + \kappa)(n + \kappa - 1 + \gamma)a_n = -S_n \tag{C.101}$$

for the present case, where

$$S_n = \sum_{m=0}^{n-1}\Big((m+\kappa)(1+\alpha+\beta-\gamma) + \alpha\beta\Big)a_m. \tag{C.102}$$

Since this has to be true for any n, we can also replace n by $n+1$ in Eq. (C.101), which gives the equation

$$(n + \kappa + 1)(n + \kappa + \gamma)a_{n+1} = -S_{n+1}. \qquad (C.103)$$

Note that in the expression for S_n, the coefficient of a_m within the sum is independent of n. Thus, in the sum S_{n+1}, the terms up to $m = n - 1$ are already given by the left-hand side of Eq. (C.102). Putting that in and adding the extra term that is present in Eq. (C.103), i.e., the term with $m = n$, we obtain

$$a_{n+1} = \frac{(n + \kappa + \alpha)(n + \kappa + \beta)}{(n + \kappa + 1)(n + \kappa + \gamma)} \, a_n. \qquad (C.104)$$

This is the recursion relation.

The series solution with $\kappa = 0$ can thus be written as

$$F(\alpha, \beta, \gamma; x) = 1 + \frac{\alpha\beta}{1!\gamma} \, x + \frac{\alpha(\alpha + 1)\beta(\beta + 1)}{2! \cdot \gamma(\gamma + 1)} \, x^2 + \cdots$$

$$= 1 + \sum_{r=1}^{\infty} \frac{(\alpha)_r(\beta)_r}{r!(\gamma)_r} \, x^r, \qquad (C.105)$$

taking the constant a_0 to be equal to 1, and using the shorthand notation

$$(\alpha)_r \equiv \alpha(\alpha + 1) \cdots (\alpha + r - 1) = \frac{\Gamma(\alpha + r)}{\Gamma(\alpha)}, \qquad (C.106)$$

sometimes called the rising factorial or the *Pochhammer symbol*. This defines the *hypergeometric series*. The name is derived from the fact that for $\alpha = 1$ and $\beta = \gamma$, the series reduces to a geometric series.

The other solution, with $\kappa = 1 - \gamma$, can be written as

$$x^{1-\gamma}F(\alpha + 1 - \gamma, \beta + 1 - \gamma, 2 - \gamma; x) \qquad (C.107)$$

by using the same notation as that used in writing Eq. (C.105). Because we have used the Taylor series for $1/(1 - x)$ at $x = 0$ for writing Eq. (C.99), the hypergeometric series converges only for $|x| < 1$.

Quite often, instead of denoting the hypergeometric function simply as $F(\alpha, \beta, \gamma; x)$, one uses the notation $_2F_1(\alpha, \beta, \gamma; x)$, where the numbers used as subscripts denote the number of parameters whose rising factorials go to the numerator and the denominator of the series expansion shown in the last step of Eq. (C.105). With this notation, the confluent hypergeometric function, defined in Eq. (C.109) below, would be denoted by $_1F_1(b, c; x)$. We do not use this notation in this book, since the parameters are anyway given after the name of the function, and no confusion can arise.

C.11   Confluent hypergeometric equation

The differential equation

$$x\frac{d^2y}{dx^2} + (c-x)\frac{dy}{dx} - by = 0, \tag{C.108}$$

where a and c are constants, is called the *confluent hypergeometric equation*, and its solutions called confluent hypergeometric functions. The series solution is of the form

$$M(b,c;x) = 1 + \sum_{r=1}^{\infty} \frac{(b)_r}{r!(c)_r}\, x^r, \tag{C.109}$$

where the rising factorial notation, $(b)_r$, was defined in Eq. (C.106). These functions are called the *confluent hypergeometric functions* of the first kind. The other solution, i.e., the same functions of the second kind, has been indicated in Ex. C.13 below.

Before ending this section, we note that both the hypergeometric and the confluent hypergeometric functions are related to several other special functions. We provide some of these relations as exercises and leave their proofs to the reader.

□ **Exercise C.11** *Derive the series solution for the confluent hypergeometric equation using the method used earlier for other equations.*

□ **Exercise C.12** *Show that the confluent hypergeometric function can be viewed as a limit of the hypergeometric function:*

$$M(b,c;x) = \lim_{a\to\infty} F(a,b,c;x/a). \tag{C.110}$$

□ **Exercise C.13** *Show that the other solution for Eq. (C.108) is given by*

$$\tilde{M}(b,c;x) = x^{-b}F(b, 1+b-c; -1/x). \tag{C.111}$$

□ **Exercise C.14** *Show that the Bessel functions of the first kind can be viewed as a limit of the confluent hypergeometric function:*

$$J_\nu(x) = \frac{(x/2)^\nu}{\Gamma(\nu+1)} \lim_{b\to\infty} M\left(b, \nu+1; -\frac{x^2}{4b}\right). \tag{C.112}$$

□ **Exercise C.15** *Define the functions*

$$D_\nu(x) = 2^{\nu/2}e^{-x^2/4}M\left(\frac{1}{2}\nu, \frac{1}{2}, x^2\right). \tag{C.113}$$

Show that they satisfy the differential equation

$$\frac{d^2 D_\nu}{dx^2} + \left(\nu + \frac{1}{2} - \frac{1}{4}x^2\right) D_\nu = 0. \tag{C.114}$$

[**Note:** *These functions are called* parabolic cylinder functions, *or alternatively* Weber *functions.*]

☐ **Exercise C.16** *Show that the confluent hypergeometric function $M(-n, \ell + 1, x)$ for positive integer n is related to the associated Laugurre polynomial $L_n^\ell(x)$ by the relation*

$$L_n^\ell(x) = \binom{n+\ell}{n} M(-n, \ell + 1, x). \tag{C.115}$$

☐ **Exercise C.17** *Show that the Legendre polynomial $P_m(x)$ can be written in terms hypergeometric functions as*

$$P_m(x) = F(-m, m + 1, 1, (1 - x)/2). \tag{C.116}$$

Appendix D

Further properties of special functions

We have introduced many functions in Appendix C. Here, we discuss some of their properties. Defined as a solution of a certain differential equation, special functions are defined only up to an arbitrary multiplicative constant. However, in this appendix, we assume some particular normalization for each of these functions. Some of these preferred normalization conditions have already been mentioned in Appendix C. Others will be mentioned as and when necessary.

D.1 Generating functions

Different polynomials introduced earlier in the appendix can be seen as coefficients in the power series expansion of some algebraic functions. To illustrate the point, we show here how to derive the generating functions of some sets of orthogonal polynomials.

a) Hermite polynomials

Let us consider the sum of all Hermite polynomials $H_n(x)$, weighted by the factor $t^n/n!$. Using the expression for the Hermite polynomials given in Eq. (C.8), we can write

$$\sum_{n=0}^{\infty} \frac{t^n}{n!} H_n(x) = \sum_{n=0}^{\infty} \sum_{s=0}^{\lfloor n/2 \rfloor} \frac{(-1)^s}{s!(n-2s)!} (2x)^{n-2s} t^n, \tag{D.1}$$

where the symbol $\lfloor N \rfloor$ means the largest integer less than or equal to N. Let us make a change in the variables by introducing

$$r = n - 2s. \tag{D.2}$$

473

We can now eliminate n from the right side of Eq. (D.1). The upper limit of the summation on s now implies that r cannot be negative. Thus, we obtain

$$\sum_{n=0}^{\infty}\frac{t^n}{n!}H_n(x) = \sum_{r=0}^{\infty}\sum_{s=0}^{\infty}\frac{(-1)^s}{s!r!}(2x)^r t^{r+2s}$$

$$= \sum_{r=0}^{\infty}\frac{(2xt)^r}{r!}\sum_{s=0}^{\infty}\frac{(-1)^s t^{2s}}{s!}. \tag{D.3}$$

Both sums are trivial, and we obtain

$$\sum_{n=0}^{\infty}\frac{t^n}{n!}H_n(x) = e^{2xt-t^2}. \tag{D.4}$$

This means that if we expand e^{2xt-t^2} in a power series in t, the coefficient of $t^n/n!$ will give us the Hermite polynomial $H_n(x)$. The function e^{2xt-t^2} is called the *generating function* for Hermite polynomials.

b)   Laguerre polynomials

For Laguerre polynomials, similarly, we use Eq. (C.88) to write

$$\sum_{n=0}^{\infty}L_n(x)t^n = \sum_{n=0}^{\infty}\sum_{s=0}^{n}(-1)^s\binom{n}{s}\frac{x^s t^n}{s!}. \tag{D.5}$$

The double sum ensures that $n \geqslant s$. Thus, if we make the change of variable to

$$r = n - s, \tag{D.6}$$

the sum on r can go from 0 to ∞, and we obtain

$$\sum_{n=0}^{\infty}L_n(x)t^n = \sum_{s=0}^{\infty}(-1)^s\frac{x^s}{s!}\sum_{r=0}^{\infty}\binom{r+s}{s}t^{r+s} \tag{D.7}$$

We can first perform the sum over r, obtaining

$$\sum_{n=0}^{\infty}L_n(x)t^n = \sum_{s=0}^{\infty}(-1)^s\frac{x^s t^s}{s!}(1-t)^{-s-1}. \tag{D.8}$$

The remaining sum gives an exponential series so that we obtain

$$\sum_{n=0}^{\infty}L_n(x)t^n = \frac{1}{1-t}\exp\left(-\frac{xt}{1-t}\right), \tag{D.9}$$

which establishes the right-hand side of this equation as the generating function for Laguerre polynomials.

c) Bessel functions

We now look at Bessel functions of the first kind. Using the power series representation given in Eq. (C.47) and noting that for integral n and r we can write $\Gamma(n + r + 1) = (n + r)!$, we obtain

$$\sum_{n=-\infty}^{\infty} J_n(x)t^n = \sum_{n=-\infty}^{\infty} \sum_{r=0}^{\infty} \frac{(-1)^r}{r!\,(n+r)!} \left(\frac{x}{2}\right)^{n+2r} t^n. \tag{D.10}$$

Note that the sum over n really starts from $n = 1 - r$. For $n < 1 - r$, the factor $(n+r)!$, which should be seen as a gamma function, diverges, and so the summand becomes zero. We now introduce a different integer-valued index

$$s = n + 2r. \tag{D.11}$$

Then we eliminate n from the double sum using s and r:

$$\sum_{n=-\infty}^{\infty} J_n(x)t^n = \sum_{s=0}^{\infty} \sum_{r=0}^{\infty} \frac{(-1)^r}{r!\,(s-r)!} \left(\frac{x}{2}\right)^s t^{s-2r}. \tag{D.12}$$

Here also, the sum over s actually starts from $s = r + 1$, but we have taken advantage of the fact that the summand vanishes for smaller values of s. The sum over r can be performed by noting that

$$\left(t - \frac{1}{t}\right)^s = \sum_{r=0}^{s} \frac{(-1)^r s!}{r!\,(s-r)!} t^{s-2r}. \tag{D.13}$$

In fact, it does not matter if we extend the sum over r all the way to infinity in this equation, because the terms for $r > s$ will have the factorial of a negative number in the denominator, which should be interpreted as a gamma function, and which is infinite. Therefore, putting this into Eq. (D.12), we obtain

$$\sum_{n=-\infty}^{\infty} J_n(x)t^n = \sum_{s=0}^{\infty} \frac{1}{s!} \left(\frac{x}{2}\right)^s \left(t - \frac{1}{t}\right)^s$$

$$= \exp\left(\frac{x}{2}\left(t - \frac{1}{t}\right)\right). \tag{D.14}$$

The final exponential function is therefore the generating function of Bessel functions of the first kind.

☐ **Exercise D.1** *Show that*

$$e^{ix\sin\theta} = \sum_{n=-\infty}^{\infty} e^{in\theta} J_n(x),$$

$$e^{ix\cos\theta} = \sum_{n=-\infty}^{\infty} i^n e^{in\theta} J_n(x). \tag{D.15}$$

d) Spherical Bessel functions

For spherical Bessel functions of the first kind, using Eqs. (C.60) and (C.47), we can write

$$j_n(x) = \frac{\sqrt{\pi}}{2} \sum_{r=0}^{\infty} \frac{(-1)^r}{r!\,\Gamma(n+r+\frac{3}{2})} \left(\frac{x}{2}\right)^{n+2r}. \tag{D.16}$$

Therefore,

$$\sum_{n=0}^{\infty} \frac{t^n}{n!} j_{n-1}(x) = \frac{\sqrt{\pi}}{2} \sum_{n=0}^{\infty} \sum_{r=0}^{\infty} \frac{(-1)^r t^n}{n!\,r!\,\Gamma(n+r+\frac{1}{2})} \left(\frac{x}{2}\right)^{n-1+2r}. \tag{D.17}$$

For any integer N, one has

$$\Gamma(N+\frac{1}{2}) = \frac{(2N)!}{2^{2N}\,N!}\,\sqrt{\pi}, \tag{D.18}$$

which can be easily proved by using the properties $\Gamma(z+1) = z\Gamma(z)$ and $\Gamma(\frac{1}{2}) = \sqrt{\pi}$. Using this property, we obtain

$$\sum_{n=0}^{\infty} \frac{t^n}{n!} j_{n-1}(x) = \frac{1}{2} \sum_{n=0}^{\infty} \sum_{r=0}^{\infty} \frac{(-1)^r t^n 2^{2(n+r)}(n+r)!}{n!\,r!\,(2n+2r)!} \left(\frac{x}{2}\right)^{n-1+2r}. \tag{D.19}$$

We now introduce a new integer variable

$$s = n + r \tag{D.20}$$

and eliminate r to write

$$\sum_{n=0}^{\infty} \frac{t^n}{n!} j_{n-1}(x) = \frac{1}{2} \sum_{n=0}^{\infty} \sum_{s=n}^{\infty} \frac{(-1)^{s-n} t^n 2^{2s} s!}{n!\,(s-n)!\,(2s)!} \left(\frac{x}{2}\right)^{2s-n-1}. \tag{D.21}$$

Since $(s-n)!$ diverges for $n > s$, we can equivalently write the double sum in the form

$$\sum_{n=0}^{\infty} \frac{t^n}{n!} j_{n-1}(x) = \frac{1}{2} \sum_{s=0}^{\infty} \sum_{n=0}^{s} \frac{(-1)^{s-n} t^n 2^{2s}}{(2s)!} \binom{s}{n} \left(\frac{x}{2}\right)^{2s-n-1}$$

$$= \sum_{s=0}^{\infty} \frac{1}{(2s)!} \sum_{n=0}^{s} \binom{s}{n} (-1)^{s-n} (2xt)^n x^{2s-2n-1}. \quad \text{(D.22)}$$

The sum over n is now easily done, because it is a binomial series. It gives

$$\sum_{n=0}^{\infty} \frac{t^n}{n!} j_{n-1}(x) = \frac{1}{x} \sum_{s=0}^{\infty} \frac{(2xt - x^2)^s}{(2s)!}. \quad \text{(D.23)}$$

The remaining sum is also easy, yielding

$$\sum_{n=0}^{\infty} \frac{t^n}{n!} j_{n-1}(x) = \frac{1}{x} \cos \sqrt{x^2 - 2xt}. \quad \text{(D.24)}$$

☐ **Exercise D.2** *Show similarly that for spherical Bessel functions of the second kind, one can use the generating function*

$$\sum_{n=0}^{\infty} \frac{t^n}{n!} \eta_{n-1}(x) = \frac{1}{x} \sin \sqrt{x^2 - 2xt}. \quad \text{(D.25)}$$

☐ **Exercise D.3** *For the Legendre polynomials, show that the generating function is given by the relation*

$$\frac{1}{\sqrt{1 - 2xt + t^2}} = \sum_{n=0}^{\infty} P_n(x) t^n. \quad \text{(D.26)}$$

D.2 Recurrence relations

Recurrence relations are useful because, in a sequence of functions defined through an integer n, they help find the function for a certain value of n in terms of the function for lower values of n, which means that the functions can be constructed iteratively. We give a few examples. Of course, these are not exhaustive by any means. There are many more, including ones that involve derivatives.

a) Legendre functions

We start from the generating function given in Eq. (D.26). Taking the derivatives of both sides with respect to t, we get

$$\frac{x-t}{(1-2xt+t^2)^{3/2}} = \sum_{n=0}^{\infty} n P_n(x) t^{n-1}. \tag{D.27}$$

This can be rewritten as

$$(x-t)\sum_{n=0}^{\infty} P_n t^n = (1-2xt+t^2)\sum_{n=0}^{\infty} n P_n(x) t^{n-1}, \tag{D.28}$$

or

$$\sum_{n=0}^{\infty} \left(x P_n t^n - P_n t^{n+1} - n P_n t^{n-1} + 2xn P_n t^n - n P_n t^{n+1} \right) = 0. \tag{D.29}$$

Since this relation has to be satisfied for any value of t, the coefficient of each power of t must equal zero. Collecting the powers of t^n, we obtain

$$x P_n - P_{n-1} - (n+1) P_{n+1} + 2xn P_n - (n-1) P_{n-1} = 0, \tag{D.30}$$

or

$$(n+1) P_{n+1} = (2n+1) x P_n - n P_{n-1}. \tag{D.31}$$

It can be easily seen that this recurrence relation defines the Legendre polynomials. To prove that, let us start from the function $x P_n(x)$. Since P_n is an nth degree polynomial, $x P_n$ must be a polynomial of degree $n+1$. Therefore, it can be expressed as a superposition of all Legendre polynomials up to P_{n+1}, i.e., by an expression of the form

$$x P_n(x) = \sum_{r=0}^{n+1} b_r P_r(x). \tag{D.32}$$

The orthogonality relation of the Legendre polynomials, to be proved in Eq. (D.52), then implies that

$$b_r = \frac{1}{2r+1} \int_{-1}^{+1} dx\, x P_n(x) P_r(x). \tag{D.33}$$

It is easy to see that $b_n = 0$, since the integrand is then an odd function of x. It is also easy to see that $b_r = 0$ if $r < n-1$, because then $x P_r$ is an

$(r+1)^{\text{th}}$ degree polynomial, which can be expanded by Legendre polynomials of degrees 0 to $r+1$, which are all orthogonal to P_n. Thus, we can write

$$xP_n(x) = \alpha P_{n-1}(x) + \beta P_{n+1}(x). \tag{D.34}$$

Using Rodriguez formula and integrating by parts, the coefficients α and β can be determined, and the result is Eq. (D.31).

b) Hermite polynomials

In this case also, we start by differentiating the generating function, given in Eq. (D.4), with respect to t:

$$\sum_{n=0}^{\infty} \frac{t^{n-1}}{(n-1)!} H_n(x) = 2(x-t)e^{2xt-t^2} = 2(x-t) \sum_{n=0}^{\infty} \frac{t^n}{n!} H_n(x). \tag{D.35}$$

This can be rewritten as

$$\sum_{n=0}^{\infty} \frac{t^n}{n!} H_{n+1}(x) = 2x \sum_{n=0}^{\infty} \frac{t^n}{n!} H_n(x) - 2 \sum_{n=0}^{\infty} \frac{t^n}{(n-1)!} H_{n-1}(x), \tag{D.36}$$

taking liberty of the fact that Hermite polynomials with negative indices are zero, by definition. Equating equal powers of t from both sides, we obtain

$$H_{n+1}(x) = 2xH_n(x) - 2nH_{n-1}(x), \tag{D.37}$$

which can be used to find all H_n's, starting from $H_0(x) = 1$.

c) Bessel functions

We start by differentiating Eq. (D.14) with respect to t:

$$\sum_{n=-\infty}^{\infty} nJ_n(x)t^{n-1} = \frac{x}{2}\left(1 + \frac{1}{t^2}\right)\exp\left(\frac{x}{2}\left(t - \frac{1}{t}\right)\right)$$

$$= \frac{x}{2} \sum_{n=-\infty}^{\infty} J_n(x)t^n + \frac{x}{2} \sum_{n=-\infty}^{\infty} J_n(x)t^{n-2}$$

$$= \frac{x}{2} \sum_{n=-\infty}^{\infty} (J_{n-1} + J_{n+1})\, t^{n-1}, \tag{D.38}$$

renaming the dummy index n in both sums in the last step. Since this should be valid for any value of t, we can equate the coefficients of equal powers of t to obtain

$$J_{n+1}(x) = \frac{2n}{x} J_n(x) - J_{n-1}(x), \tag{D.39}$$

which is a recurrence relation. It was used in §7.8.

D.3   Orthogonality of special functions

The special functions that we encountered in Appendix C are all solutions
of differential equations whose general form was given in Eq. (B.3). Suppose
that $c(x)$ depends on an integer n in some fashion, and for each value of n, we
obtain a polynomial solution. The form of $c(x)$ for a particular value of n will
be denoted by $c_n(x)$ and the corresponding solution by $y_n(x)$.

We multiply Eq. (B.3, p. 430) throughout by a function $w(x)$ which has the
property that

$$\frac{d}{dx}\Big(w(x)a(x)\Big) = w(x)b(x). \tag{D.40}$$

Then the solution y_n satisfies the differential equation

$$\frac{d}{dx}\left[w(x)a(x)\frac{dy_n}{dx}\right] + w(x)c_n(x)y_n = 0. \tag{D.41}$$

Multiplying Eq. (D.41) by $y_{n'}$ and integrating over some range of x, say from
x_1 to x_2, we obtain

$$\int_{x_1}^{x_2} dx\, w(x)c_n(x)y_n(x)y_{n'}(x) = -\int_{x_1}^{x_2} dx\, y_{n'}\frac{d}{dx}\left[w(x)a(x)\frac{dy_n}{dx}\right]. \tag{D.42}$$

Integrating by parts, we find that the right side is equal to

$$-w(x)a(x)y_{n'}\frac{dy_n}{dx}\bigg|_{x_1}^{x_2} + \int_{x_1}^{x_2} dx\, w(x)a(x)\frac{dy_n}{dx}\frac{dy_{n'}}{dx}. \tag{D.43}$$

If we choose the limits of integration, x_1 and x_2, to be two points where the
function $w(x)a(x)$ vanishes, we are left with

$$\int_{x_1}^{x_2} dx\, w(x)c_n(x)y_n(x)y_{n'}(x) = \int_{x_1}^{x_2} dx\, w(x)a(x)\frac{dy_n}{dx}\frac{dy_{n'}}{dx}. \tag{D.44}$$

We could have started with the differential equation appropriate for $y_{n'}$,
multiplied it by y_n, and integrated over x. In this case we would have obtained

$$\int_{x_1}^{x_2} dx\, w(x)c_{n'}(x)y_n(x)y_{n'}(x) = \int_{x_1}^{x_2} dx\, w(x)a(x)\frac{dy_n}{dx}\frac{dy_{n'}}{dx}. \tag{D.45}$$

Subtracting the two equations, we obtain

$$\int_{x_1}^{x_2} dx\, w(x)\Big[c_n(x) - c_{n'}(x)\Big]y_n(x)y_{n'}(x) = 0. \tag{D.46}$$

If further $c_n(x)$ is of the form

$$c_n(x) = C(x) + C_n, \qquad (\text{D.47})$$

where the first part is independent of n and the second part of x, we can take the factor $C_n - C_{n'}$ outside the integral and write

$$\int_{x_1}^{x_2} dx\ w(x) y_n(x) y_{n'}(x) = 0 \qquad \text{for } n \neq n'. \qquad (\text{D.48})$$

This is an orthogonality relation between the solutions of the differential equation. In what follows in this section, we will find the function $w(x)$ for specific differential equations using Eq. (D.40) and write down the orthogonality relations. We will also put the value of the integral for $n = n'$, i.e., the normalization relation, which is chosen only by convention. As we mentioned earlier, the normalization of the solutions of homogeneous differential equations is not determined by the equation.

a) Hermite polynomials

For the Hermite equation, Eq. (C.1), Eq. (D.40) is

$$\frac{dw(x)}{dx} = -2xw(x), \qquad (\text{D.49})$$

so that

$$w(x) = e^{-x^2}. \qquad (\text{D.50})$$

The solutions are valid for all x, from $-\infty$ to $+\infty$. The orthonormalization conditions are taken as

$$\int_{-\infty}^{+\infty} dx\ e^{-x^2} H_m(x) H_n(x) = \delta_{mn} 2^n n! \sqrt{\pi}. \qquad (\text{D.51})$$

For $m \neq n$, this is the orthogonality condition of Eq. (D.48). For $m = n$, this condition sets up the normalization with which we had written the Hermite polynomials in Eq. (C.7).

b) Legendre polynomials

For the Legendre differential equation, $w(x) = 1$. The orthonormalization condition of the polynomials, consistent with the normalization chosen in writing Eq. (C.18), is

$$\int_{-1}^{+1} dx\ P_m(x) P_n(x) = \frac{2}{2n+1}\ \delta_{mn}. \qquad (\text{D.52})$$

Note that the limits of the integration are from -1 to $+1$, the interval in which the Legendre polynomials are defined. Recall that $a(x) = 1 - x^2$ for this differential equation, which vanishes at both the limits.

c)   Associated Legendre polynomials

For these polynomials, we have to be somewhat innovative while using Eq. (D.48) since the concerned equation contains two parameters, l and m, as seen in Eq. (C.28). For a fixed value of m, we can pretend that m is a constant and therefore Eq. (D.48) will tell us that

$$\int_{-1}^{+1} dx\, P_l^m(x) P_{l'}^m(x) = 0 \qquad \text{for } l \neq l'. \tag{D.53}$$

For fixed l and different m, however, we cannot use Eq. (D.48) at all, since it is based on the special form of the derivative-independent term shown in Eq. (D.47), and the associated Legendre equation does not have the form shown there. So we need to go one step back, to Eq. (D.46), which does not depend on any special form of the terms in the differential equation. This form readily gives

$$\int_{-1}^{+1} dx\, \frac{P_l^m(x) P_l^{m'}(x)}{1 - x^2} = 0 \qquad \text{for } m \neq m'. \tag{D.54}$$

The two orthogonality conditions have some more formal similarity if we use $x = \cos\theta$:

$$\int_0^\pi d\theta\, \sin\theta\, P_l^m(\cos\theta) P_{l'}^m(\cos\theta) = 0 \qquad \text{for } l \neq l',$$

$$\int_0^\pi d\theta\, \frac{1}{\sin\theta}\, P_l^m(\cos\theta) P_l^{m'}(\cos\theta) = 0 \qquad \text{for } m \neq m'. \tag{D.55}$$

The normalizations for the associated Legendre functions are taken such that, for fixed m, we get

$$\int_{-1}^{1} dx\, P_k^m(x) P_l^m(x) = \frac{2(l+m)!}{(2l+1)(l-m)!}\, \delta_{kl}. \tag{D.56}$$

Note that if we put $m = 0$, this condition reduces to the normalization condition chosen for the Legendre polynomials, Eq. (D.52). On the other hand, for fixed values of l, the normalization condition is taken to be

$$\int_{-1}^{1} dx\, \frac{P_l^m(x) P_l^n(x)}{1 - x^2} = \begin{cases} 0 & \text{if } m \neq n, \\[2mm] \dfrac{(l+m)!}{m(l-m)!} & \text{if } m = n \neq 0. \end{cases} \tag{D.57}$$

For $m = n = 0$, this integral diverges. That is not an embarrassment since the normalization of the functions with $m = 0$ has already been fixed with Eq. (D.56).

☐ **Exercise D.4** *Show that the normalization prescribed in Eq. (D.56) follows from the definition of the functions given in Eq. (C.30).*

d) Laguerre polynomials

For the Laguerre equation, $a(x) = x$ and $b(x) = 1 - x$, so that we get

$$w(x) = e^{-x}. \tag{D.58}$$

The orthonormalization conditions are taken as

$$\int_0^\infty dx\, e^{-x} L_m(x) L_n(x) = \delta_{mn}. \tag{D.59}$$

In Eq. (C.89), we have written down the polynomials with this normalization in mind. Note that $w(x)a(x) = xe^{-x}$, which vanishes at both limits of integration.

e) Associated Laguerre polynomials

For the generalized Laguerre equation, $a(x) = x$ and $b(x) = l + 1 - x$. This gives

$$w(x) = x^l e^{-x}. \tag{D.60}$$

Therefore, the orthogonality conditions can be written in the form

$$\int_0^\infty dx\, x^l e^{-x} L_m^l(x) L_n^l(x) = \frac{(n + l)!}{n!}\delta_{mn}, \tag{D.61}$$

where the value of the integral for $m = n$ has been chosen to be compatible with the explicit forms given in Eq. (C.96). Concerning the limits of the integration, the comment made for Laguerre polynomials apply.

☐ **Exercise D.5** *Verify the following normalization conditions for Bessel functions:*

$$\int_{-\infty}^{+\infty} \frac{dx}{x}\, J_m(x) J_n(x) = 0 \qquad \text{for } m \neq n. \tag{D.62}$$

D.4 Asymptotic expansions of special functions

The series solutions described in Appendix C show how the functions behave for small values of their arguments. In the text, we often needed the behavior for large arguments, in order to discuss the wavefunctions at large distances from the origin. In this section, we give some examples of how the large-argument behavior can be obtained. We do not aim for a very general exposition of the techniques involved. Rather, we provide a quick recipe for obtaining some of the asymptotic expansions used in this book. A more general method will be described in Appendix E.

D.4.1 Spherical Bessel functions

The spherical Bessel differential equation was given in Eq. (C.57). Clearly, for very large x, we can neglect the quantity $n(n+1)$ that appears in that equation. For the resulting equation, let us try a solution

$$y \sim e^{ix} f(x). \tag{D.63}$$

Notice that in this section, we will use the symbol '$\sim$' to indicate that only the leading terms of two sides are equal for large values of the variable. Using Eq. (D.63), we obtain

$$\frac{dy}{dx} = \Big(f'(x) + if(x) \Big) e^{ix}, \tag{D.64a}$$

$$\frac{d^2y}{dx^2} = \Big(f''(x) + 2if'(x) - f(x) \Big) e^{ix}. \tag{D.64b}$$

Putting these into the spherical Bessel differential equation, we obtain

$$x\Big(f''(x) + 2if'(x) \Big) + 2\Big(f'(x) + if(x) \Big) \sim 0, \tag{D.65}$$

neglecting the $n(n+1)$ term as argued before. Assuming $f(x)$ is real for a real variable x, we can equate the real and the imaginary parts of this equation separately to zero. Both resulting equations have the solution $f(x) = 1/x$, implying that there is a solution of the spherical Bessel equation that behaves like

$$h^{(1)} \sim \frac{e^{ix}}{x} \tag{D.66}$$

at large x. The complex conjugate is also a solution, which can be checked in a similar way:

$$h^{(2)} \sim \frac{e^{-ix}}{x}. \tag{D.67}$$

These asymptotic forms have been extensively used, for example, in Chapter 13. There can, of course, be extra numerical factors multiplied with these shown solutions. They are decided upon by convention. They can affect the nature of the combinations $j_n = (h_n^{(1)} + h_n^{(2)})/2$ and $\eta_n = (h_n^{(1)} - h_n^{(2)})/2i$.

D.4.2 Bessel functions

For the Bessel equation, Eq. (C.41), we can likewise neglect the quantity α^2 and try a solution like that proposed in Eq. (D.63). Using Eq. (D.64) as before, we obtain

$$x\Big(f''(x) + 2if'(x)\Big) + \Big(f'(x) + if(x)\Big) \sim 0. \tag{D.68}$$

In this case, the solution will be $f(x) = 1/\sqrt{x}$, so that there are solutions of the Bessel equation that behave in the following manner for large values of the argument:

$$H^{(1)}(x) \sim \frac{e^{ix}}{\sqrt{x}}, \qquad H^{(2)}(x) \sim \frac{e^{-ix}}{\sqrt{x}}. \tag{D.69}$$

D.4.3 Airy function

For this function, we can try

$$y \sim e^{S(x)}. \tag{D.70}$$

Putting this into Eq. (C.33), we find that the exponent function $S(x)$ satisfies the differential equation

$$S'' + S'^2 - x = 0. \tag{D.71}$$

As a first approximation, if we assume that the second derivative is negligible, we find $S'^2 = x$. For $x \to +\infty$, this means $S' = \pm\sqrt{x}$ so that the leading term in the two solutions turn out to be

$$\exp(\pm\frac{2}{3}x^{3/2}). \tag{D.72}$$

The solution with the positive exponent is not of interest to us. The solution with the negative exponent is called Ai(x). We will derive a fuller expression for the asymptote in Appendix E.

D.4.4   Confluent hypergeometric functions

In this case, we try solutions of the form

$$y \sim e^x f(x). \tag{D.73}$$

If we put this into the confluent hypergeometric equation as given in Eq. (C.108), we find that the function $f(x)$ should satisfy the differential equation

$$b\frac{d^2 f}{dx^2} + (x - c + 2b)\frac{df}{dx} + (b - c)f = 0. \tag{D.74}$$

No approximation has been used so far. We now assume that $f(x) \sim x^\alpha$ for some α. In this case, $df/dx \sim x^{\alpha-1}$ and will be negligible for large x. So will be the second derivative. So, we obtain, for large x, the equation

$$x\frac{df}{dx} + (b - c)f = 0, \tag{D.75}$$

which gives $f(x) \sim x^{c-b}$. Thus, for large x, we find

$$y \sim e^x x^{c-b}. \tag{D.76}$$

This leading behavior can be identified in the expression for the confluent hypergeometric function written in Eq. (13.99, p. 347). The multiplying factors, as well as the subleading terms that appear in that equation, are not derived here.

D.4.5   General comments

In Appendix C, we have seen that many of the special functions can be seen as some special limit of the hypergeometric or confluent hypergeometric functions. Thus, once the asymptotic forms of these functions are known, we can infer from them the asymptotic forms of many of the other special functions that we have encountered.

For example, consider the Weber functions, or the parabolic cylinder functions. They were defined through the confluent hypergeometric functions in Eq. (C.113). So, their behavior for large values of the argument can also be understood from the behavior of the confluent hypergeometric functions in the same regime. The leading behavior is governed by the factor $e^{-x^2/2}$, which is what we see in Eq. (12.29).

These methods can be used even for finding some non-leading terms in the asymptotic forms. But the method involves some guesswork, or some prior knowledge, to begin with. There are other methods of really deriving the asymptotic form from scratch. We describe such a method in Appendix E.

Appendix E

Saddle-point method

The saddle-point method is widely used in many physics problems for approximately evaluating integrals in which the integrand has one or a few sharp peaks. The basic idea is that the region around the sharp peak gives the dominant contribution to the integral. So, if we can find the contribution around the peak, we have an estimate of the integral. The reason for the name of the method will be explained below.

E.1   Outline of the method

Suppose we have an integral of the form

$$I = \int_C dz \, g(z) e^{f(z)}, \tag{E.1}$$

where z is a complex variable, $f(z)$ and $g(z)$ are functions that are differentiable everywhere, and the integration is to be performed over a contour C in the complex plane. If we have a real integral, that can also be treated in this framework, with the contour C running along the real axis.

If $f(z)$ has a sharp maxima, the main contribution to the integral would come from there, as we said in the prelude to this appendix. There are now several issues to worry about. First, the original contour may not pass through the stationary points. This is not a big concern because we are considering integrals for which the integrand is analytic everywhere. Therefore, the result should be independent of the path, because the integral of a differentiable complex function around any closed path is zero. We need to deform the contour C to another contour C' that passes through the stationary points.

The second issue is that for differentiable complex functions, there is really no maxima. The Cauchy–Riemann conditions imply that both real and imaginary parts of a complex analytic function satisfy the 2D Laplace's equation. Therefore, if the second derivative is positive along one direction, it must be negative along the perpendicular direction. Thus, all stationary points are saddle points, and the method gets the name *saddle-point method* because of the association with these points.

So, passing the contour through the saddle point is not enough to ensure that we will obtain the dominant contribution to the integral. The deformed contour C' must go through the direction along which the function really attains a maxima and falls sharply at points away from the maxima. Our estimate of the integral will be best if we find the path where the maxima is the steepest, and for this reason the method is also called the *method of steepest descent*.

Near an extremum at $z = z_0$, we can write

$$f(z) = f(z_0) + \frac{1}{2}(z - z_0)^2 f''(z_0) + \cdots. \tag{E.2}$$

The dots denote terms with higher powers of $z - z_0$, which will be ignored. We will write

$$f''(z_0) = re^{i\phi}, \qquad z - z_0 = se^{i\zeta}. \tag{E.3}$$

The contour of integration can be approximated by a straight line in the complex plane near the extremum, and ζ is a constant for points along the contour. Then,

$$f(z) = f(z_0) + \frac{1}{2}rs^2 e^{i(\phi + 2\zeta)}. \tag{E.4}$$

The path along which $|f(z)|$ changes most rapidly is then given by the condition when the phase factor in the second derivative term is equal to -1, i.e., when

$$\zeta = -\phi/2 \pm \pi/2. \tag{E.5}$$

We need to choose a contour with this slope at the extremum in order to obtain the best result.

Once we choose the deformed contour with these restrictions, the rest is simple. The integration over the contour is now an integration over the variable s. Since the dominant contribution to the integral comes from the region $s = 0$, we do not care about the points far from that, and can take the range of integration from $-\infty$ to $+\infty$. So we can rewrite Eq. (E.1) as

$$I = g(z_0)e^{f(z_0)}e^{i\zeta} \int_{-\infty}^{+\infty} ds \, \exp(-\frac{1}{2}rs^2), \tag{E.6}$$

where the factor $e^{i\zeta}$ comes because Eqs. (E.3) and (E.5) give $dz = e^{i\zeta}ds$. We now encounter a Gaussian integral, which can be easily performed. Remembering that $r = |f''(z_0)|$, the result can be written as

$$I = g(z_0)e^{f(z_0)}e^{i\zeta} \times \sqrt{\frac{2\pi}{|f''(z_0)|}}. \tag{E.7}$$

This is the saddle-point evaluation of the integral in Eq. (E.1), performed with the assumptions described above. If there are more than one stationary points, the contributions from each of them ought to be added to obtain the estimate for the integral.

☐ **Exercise E.1** *The factorial function can be expressed as the integral*

$$n! = \int_0^\infty dt\ t^n e^{-t} = \int_0^\infty dt\ \exp(n\ln t - t). \tag{E.8}$$

Use the saddle-point method to show that for large n, one obtains the Stirling formula:

$$n! \sim \sqrt{2\pi n}\ n^n e^{-n}. \tag{E.9}$$

For the sake of completeness, we want to point out that the process described above does not work if $g(z)$, defined in Eq. (E.1), vanishes at the saddle point z_0 of $f(z)$. In this case, we need to go a little further and use the Taylor series expansion of $g(z)$ around z_0:

$$g(z) = (z - z_0)g'(z_0) + \frac{1}{2}(z - z_0)^2 g''(z_0) + \cdots. \tag{E.10}$$

Instead of writing Eq. (E.6), we should now include the factor of $g(z)$ inside the integral. Note that the term involving the first derivative of $g(z)$ will produce an odd integral, which would vanish. Thus, we will be left with

$$I = \frac{1}{2}e^{f(z_0)}e^{3i\zeta}g''(z_0)\int_{-\infty}^{+\infty} ds\ s^2 \exp(-\frac{1}{2}rs^2), \tag{E.11}$$

using the definitions of Eq. (E.3). The integral can be easily reduced to a gamma function, which gives

$$I = \frac{1}{2}e^{f(z_0)}e^{3i\zeta}g''(z_0) \times \sqrt{\frac{2\pi}{|f''(z_0)|^3}}. \tag{E.12}$$

In the rest of this appendix, we give some examples of the derivation of asymptotic forms of some functions that have been used in the text.

E.2   Airy function

In order to obtain the asymptotic form for the Airy function using the saddle-point method, we first need to express the Airy function as an integral. This can be easily done by using the Fourier transform, which is defined as

$$\text{Ai}(x) = \int_{-\infty}^{+\infty} \frac{dk}{2\pi} \, e^{ikx} \tilde{A}(k). \tag{E.13}$$

Then,

$$\frac{d^2}{dx^2}\text{Ai}(x) = \int_{-\infty}^{+\infty} \frac{dk}{2\pi} \, (-k^2)e^{ikx}\tilde{A}(k). \tag{E.14}$$

On the other hand, from the inverse relation

$$\tilde{A}(k) = \int_{-\infty}^{+\infty} dx \, e^{-ikx}\text{Ai}(x), \tag{E.15}$$

it is easily seen that

$$\frac{d}{dk}\tilde{A}(k) = \int_{-\infty}^{+\infty} dx \, (-ix)e^{-ikx}\text{Ai}(x), \tag{E.16}$$

so

$$x\text{Ai}(x) = i\int_{-\infty}^{+\infty} \frac{dk}{2\pi} \, e^{ikx}\frac{d\tilde{A}(k)}{dk}. \tag{E.17}$$

Putting Eqs. (E.14) and (E.17) into the Airy equation, Eq. (C.33), and using the orthogonality of the exponential functions, we obtain

$$-k^2\tilde{A}(k) - i\frac{d\tilde{A}(k)}{dk} = 0, \tag{E.18}$$

which gives

$$\tilde{A}(k) = e^{ik^3/3}. \tag{E.19}$$

Therefore, the integral representation of the Airy function of the first kind is given by

$$\text{Ai}(x) = \int_{-\infty}^{+\infty} \frac{dk}{2\pi} \, \exp\left(i(kx + k^3/3)\right). \tag{E.20}$$

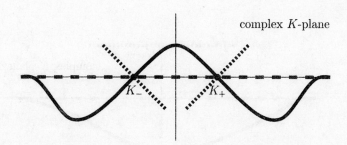

complex K-plane

Figure E.1 The dashed line is the original contour on which the integration was defined. The solid thick line is the deformed contour for finding the asymptotic form of the Airy function for large negative values of its argument. The dotted lines indicate ascending directions which must be avoided in applying the method.

We now try to find the asymptotic form for the integral for $x \to \pm\infty$. In order to use the powerful tools of complex integration, we promote k to a complex variable and call it K to avoid confusion. The integral really runs along the real line on the complex-K plane. However, since the integrand has no singularity, we can deform the contour in any way we want, as long as it starts from large negative values of $\mathrm{Re}(K)$ and ends up at large positive values.

The function $f(K)$, as defined in Eq. (E.1), is now

$$f(K) = i\left(Kx + \frac{1}{3}K^3\right). \tag{E.21}$$

We first try the case of large negative x. For this case, the saddle points are at

$$K_\pm = \pm\sqrt{|x|}. \tag{E.22}$$

The functional values at these two extrema are

$$f(K_\pm) = \pm\frac{2}{3}i|x|^{3/2}, \tag{E.23}$$

and the second derivatives are such that

$$r_\pm = 2\sqrt{|x|}, \qquad \phi_\pm = \mp\pi/2 \tag{E.24}$$

in the notation of Eq. (E.3). This means that we should choose a contour whose slope at the extrema are given by

$$2\zeta_+ - \pi/2 = \pm\pi \quad \Rightarrow \quad \zeta_+ = -\pi/4 \text{ or } 3\pi/4,$$
$$2\zeta_- + \pi/2 = \pm\pi \quad \Rightarrow \quad \zeta_- = \pi/4 \text{ or } -3\pi/4. \tag{E.25}$$

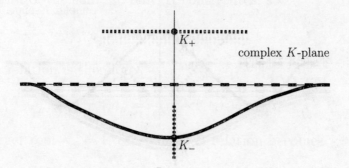

Figure E.2 The dashed line is the original contour on which the integration was defined. The solid thick line is the deformed contour for finding the asymptotic form of the Airy function for large positive values of its argument.

A contour with these slopes has been shown in Figure E.1.

In the notation of Eq. (E.1), we have $g(z) = 1/(2\pi)$ for the present case. Therefore, applying Eq. (E.7), we find the contributions to the integral from the two extrema. Adding them up, we obtain

$$\text{Ai}(x) \sim \frac{1}{2\pi} \sqrt{\frac{\pi}{\sqrt{|x|}}} \times \left(e^{\frac{2}{3} i |x|^{3/2}} e^{-i\pi/4} + e^{-\frac{2}{3} i |x|^{3/2}} e^{i\pi/4} \right)$$

$$\sim \frac{1}{2\pi^{1/2} |x|^{1/4}} \cos\left(\frac{2}{3} |x|^{3/2} - \frac{\pi}{4} \right). \tag{E.26}$$

It can obviously be written in terms of a sine function:

$$\text{Ai}(x) \sim \frac{1}{2\pi^{1/2} |x|^{1/4}} \sin\left(\frac{2}{3} |x|^{3/2} + \frac{\pi}{4} \right). \tag{E.27}$$

This form has been used in §11.4.

We now try the region for large positive x. The saddle points are at

$$K_\pm = \pm i\sqrt{x}, \tag{E.28}$$

with

$$f(K_\pm) = \mp \frac{2}{3} x^{3/2}. \tag{E.29}$$

The second derivative at the saddle points give

$$r_\pm = 2\sqrt{x}, \qquad \phi_+ = 0, \quad \phi_- = \pi. \tag{E.30}$$

This means that we need

$$\zeta_+ = \pi/2 \text{ or } -\pi/2, \qquad \zeta_- = 0 \text{ or } \pi. \tag{E.31}$$

But there is a problem if the contour goes through K_+. Suppose we start from $-\infty$ and go through K_+ with the slope $+\pi/2$. Then, it would be impossible to come back to the real axis without passing through the direction perpendicular to the direction of the steepest descent. As we pointed out earlier, this would invalidate the method. Therefore, we take a contour passing through only K_-, crossing the stationary point with zero slope. Then, putting in the appropriate values into Eq. (E.7), we obtain

$$\text{Ai}(x) \xrightarrow{x \to +\infty} \frac{1}{2\pi} \exp\left(-\frac{2}{3}x^{-3/2}\right) \times \frac{\sqrt{\pi}}{x^{1/4}}. \tag{E.32}$$

This asymptotic form was used in §11.4.

☐ **Exercise E.2** *We found the Fourier transform of Airy function by using the Airy equation. The Airy equation has two solutions. Why does the Fourier transform refer only to $\text{Ai}(x)$ and not $\text{Bi}(x)$?*

E.3 Bessel functions

In this case as well, our first task is to find an integral representation of the Bessel functions. We will restrict ourselves to Bessel functions of integer indices. Also, the power series solution given in Eq. (C.47) clearly shows that the values of the function for positive and negative values of the argument are related through an overall factor, so it is enough to find the asymptotic form for large positive x.

Using Eq. (D.14, p. 475), we easily see that

$$\oint_C \frac{dt}{t^{n+1}} \exp\left(\frac{x}{2}\left(t - \frac{1}{t}\right)\right) = 2\pi i J_n(x), \tag{E.33}$$

where the contour C encloses the origin. We can now put $t = e^{i\theta}$, transforming the integral to the form

$$J_n(x) = \frac{1}{2\pi} \int_{-\pi}^{+\pi} d\theta \, e^{i(x\sin\theta - n\theta)}. \tag{E.34}$$

Note that for real x, the imaginary part of the integrand integrates to zero, so the result is real.

Comparing with the notation introduced in Eq. (E.1), we have here

$$f(\theta) = i(x \sin\theta - n\theta). \tag{E.35}$$

For any x, the saddle points for $f(\theta)$ are located at

$$x \cos\theta = n. \tag{E.36}$$

For very large x, the saddle points are therefore at

$$\theta_{\pm} = \pm\frac{\pi}{2}. \tag{E.37}$$

At these points,

$$f(\theta_{\pm}) = \pm i(x - \frac{n\pi}{2}). \tag{E.38}$$

The second derivatives of the function $f(\theta)$ are given by

$$r_{\pm} = x, \qquad \phi_{\pm} = \mp\pi/2. \tag{E.39}$$

The calculation of the angle ζ is similar to what was done in Eq. (E.25), and accordingly the deformed contour also looks like that in Figure E.1 (p. 491). Adding the contributions from these two saddle points, we obtain

$$J_n(x) \overset{x\to+\infty}{\longrightarrow} \frac{1}{2\pi}\sqrt{\frac{2\pi}{x}} \times \left(e^{i(x-n\pi/2-\pi/4)} + e^{-i(x-n\pi/2-\pi/4)}\right)$$

$$= \sqrt{\frac{2}{\pi x}} \cos\left(x - \frac{n\pi}{2} - \frac{\pi}{4}\right). \tag{E.40}$$

This is the asymptotic form.

The asymptotic form for the spherical Bessel functions can now be easily obtained by using Eq. (C.60, p. 463). It gives

$$j_n(x) \overset{x\to+\infty}{\longrightarrow} \frac{1}{x}\cos\left(x - \frac{n\pi}{2} - \frac{\pi}{2}\right)$$

$$= \frac{1}{x}\sin\left(x - \frac{n\pi}{2}\right). \tag{E.41}$$

This is the form that was used in the partial wave analysis of scattering processes, in §13.4.

Index

⋆ Capitalization has been disregarded in the alphabetization of the index. Word boundaries have also been disregarded, except when there is a comma or a dash between two words. Thus, *Cauchy–Schwarz* comes before *Cauchy completeness*.

⋆ Some entries have been marked with a bullet sign (•), implying that we have not listed all occurrences of that word or phrase in the index. The index refers to pages where the word or phrase has been defined or discussed in detail, and also the page of its first appearance if that is different from the former.